Lecture Notes in Computer Science 2288

Edited by G. Goos, J. Hartmanis, and J. van Leeuwen

T0241812

Springer

Berlin
Heidelberg
New York
Barcelona
Hong Kong
London
Milan
Paris
Tokyo

Kwangjo Kim (Ed.)

Information Security and Cryptology – ICISC 2001

4th International Conference
Seoul, Korea, December 6-7, 2001
Proceedings

 Springer

Series Editors

Gerhard Goos, Karlsruhe University, Germany
Juris Hartmanis, Cornell University, NY, USA
Jan van Leeuwen, Utrecht University, The Netherlands

Volume Editor

Kwangjo Kim
International Research center for Information Security (IRIS)
Information and Communications University (ICU)
58-4, Hwaam-dong-Yusong-gu, Daejeon, 305-732, Korea
E-mail: kkj@icu.ac.kr

Cataloging-in-Publication Data applied for

Die Deutsche Bibliothek - CIP-Einheitsaufnahme

Information security and cryptology : 4th international conference ;
proceedings / ICISC 2001, Seoul, Korea, December 6 - 7, 2001. Kwangjo Kim
(ed.). - Berlin ; Heidelberg ; New York ; Barcelona ; Hong Kong ; London ;
Milan ; Paris ; Tokyo : Springer, 2002
    (Lecture notes in computer science ; Vol. 2288)
    ISBN 3-540-43319-8

CR Subject Classification (1998): E.3, G.2.1, D.4.6, K.6.5, F.2.1, C.2, J.1

ISSN 0302-9743
ISBN 3-540-43319-8 Springer-Verlag Berlin Heidelberg New York

Springer-Verlag Berlin Heidelberg New York
a member of BertelsmannSpringer Science+Business Media GmbH

http://www.springer.de

© Springer-Verlag Berlin Heidelberg 2002
Printed in Germany

Typesetting: Camera-ready by author, data conversion by Olgun Computergrafik
Printed on acid-free paper   SPIN 10846343   06/3142   5 4 3 2 1 0

Preface

Annually sponsored by the Korea Institute of Information Security and Cryptology (KIISC), the fourth International Conference on Information Security and Cryptology (ICISC 2001) was held at the 63 Building in Seoul, Korea, December 6–7, 2001. The 63 Building, consisting of 60 stories above the ground and 3 stories underground, stands soaring up into the sky on the island of Youido, the Manhattan of Korea, and ranks by far the tallest of all buildings in the country.

The program committee received 102 submissions from 17 countries and regions (Australia, Belgium, China, Denmark, France, Germany, India, Italy, Japan, Korea, The Netherlands, Spain, Taiwan, Thailand, Vietnam, UK, and USA), of which 32 were selected for presentation in 8 sessions. All submissions were anonymously reviewed by at least 3 experts in the relevant areas. There was one invited talk by David Pointcheval (ENS, France) on "Practical Security in Public-Key Cryptography".

We are very grateful to all the program committee members who devoted much effort and valuable time to reading and selecting the papers. These proceedings contain the final version of each paper revised after the conference. Since the revised versions were not checked by the program committee rigorously, the authors must bear full responsibility for the contents of their papers.

The program committee also requested the expert advice of their colleagues, including: Seigo Arita, Joonsang Baek, Aditya Bagcya, Colin Boyd, Denis Carabin, Donghyeon Cheon, Jung Hee Cheon, Joo Yeon Cho, Seung Bok Choi, Kilsoo Chun, Christophe Clavier, Jean-Sebastien Coron, James Edwards, Youichi Futa, Riccardo Focardi, Soichi Furuya, Steven Galbraith, Pierre Girard, Kishan Chand Gupta, Helena Handschuh, Matt Henricksen, Si-Hwan Hong, Hyo Sun Hwang, Kyubeom Hwang, Toshiya Itoh, Keiichi Iwamura, Marc Jóye, Sungwoo Kang, Chong Hee Kim, Hae Suk Kim, Jeeyeon Kim, Young-Baek Kim, Hyunjo Kwon, Dong Hoon Lee, Eonkyung Lee, Leonie Simpson, Mark Looi, Subhamoy Maitra, Wenbo Mao, Bill Millan, Shiho Moriai, Masahiko Motoyama, Yong Man No, Hanae Nozaki, Katsuyuki Okeya, Hai-Wen Ou, Pascal Paillier, Dong Jin Park, Haeryong Park, Ju Hwan Park, Ludovic Rousseau, Kazue Sako, Fumihiko Sano, Palash Sarkar, Kyungah Shim, Hideo Shimizu, K. Sikdar, Sang Gyoo Sim, Boyeon Song, Masakazu Sousya, Ron Steinfeld, Maenghee Sung, Mitsuru Tada, Tsuyoshi Takagi, Lawrence Teo, Toshio Tokita, Yasuyuki Tsukada, Christophe Tymen, Kapali Viswanathan, Ping Wang, Susanne Wetzel, Masato Yamamicya, Jun-hui Yang, Ding-feng Ye, and Dae Hyun Yum. We apologize for any omissions from this list.

Special thanks also goes to all members of IRIS (International Research center for Information Security, http://www.iris.re.kr) and C&IS (Cryptology and Information Security, http://caislab.icu.ac.kr) Lab. for their skillful and professional assistance in supporting the various tasks of the program chair. Byoungcheon Lee deserves special thanks for his help in publishing the proceedings.

We are also grateful to all the organizing committee members for their volunteer work.

Finally, we would like to thank all the authors who submitted their papers to ICISC 2001 (including those whose submissions were not successful), as well as the conference participants from around the world, for their support, which made this conference a big success.

December 2001   Kwangjo Kim

ICISC 2001

2001 International Conference on Information Security and Cryptology

63 Building, Seoul, Korea
December 6-7, 2001

Sponsored by

Korea Institute of Information Security and Cryptology (KIISC)
(www.kiisc.or.kr)

In cooperation with

Ministry of Information and Communication (MIC), Korea
Institute of Information Technology Assessment (IITA), Korea
and Korea Information Security Agency (KISA), Korea

Financially Supported by

BCQRE, SECUi.COM, and SOFTFORUM, Korea

General Chair

Sang Jae Moon   (Kyungpook National University, Korea)

Program Committee

Kwangjo Kim, Chair	(Information and Communications University, Korea)
Chae Hoon Lim, Vice Chair	(Future Systems, Korea)
Gail-Joon Ahn	(University of North Carolina at Charlotte, USA)
Zongduo Dai	(Academia Sinica, China)
Ed Dawson	(Queensland University of Technology, Australia)
Cunsheng Ding	(Hong Kong University of Science and Technology, China)
Markus Jakosson	(RSA Labs, USA)
Seungjoo Kim	(Korea Information Security Agency, Korea)
Xuejia Lai	(Secure Web & Intranet Solutions Group, Switzerland)
Chi Sung Laih	(National Cheng Kung University, Taiwan)
Kwok Yan Lam	(PrivyLink International, Singapore)
Pil Joong Lee	(POSTECH, Korea)
Jongin Lim	(Korea University, Korea)
Atsuko Miyaji	(JAIST, Japan)
David Naccache	(Gemplus Card International, France)
Tatsuaki Okamoto	(NTT, Japan)
Choonsik Park	(National Security Research Institute, Korea)
cangjoon Park	(BCQRE, Korea)
Bimal Roy	(Indian Statistical Institute, India)
Kouichi Sakurai	(Kyushu University, Japan)
Sung Won Sohn	(Electronics and Telecommunications Research Institute, Korea)
Nigel Smart	(University of Bristol, UK)
Moti Yung	(CertCo, USA)
Yuliang Zheng	(University of North Carolina at Charlotte, USA)

Organizing Committee

Youjin Song, Chair	(Dongguk University, Korea)
Kyo Il Chung	(Electronics and Telecommunications Research Institute, Korea)
Hyon-Cheol Chung	(Softforum, Korea)
Douglas Guen	(INICIS, Korea)
Jae Cheol Ha	(Korea Nazarene University, Korea)
Ki Yoong Hong	(SECUVE, Korea)
Souhwan Jung	(Soongsil University, Korea)
Moon-Soo Jang	(OULLIM Information Technology, Korea)
Ki Tae Kim	(The Korea Economic Daily, Korea)
Seok Woo Kim	(Hansei University, Korea)
Dong Hoon Lee	(Korea University, Korea)
Heon Lee	(Ministry of Information and Communication, Korea)
Hyung Woo Lee	(Cheonan University, Korea)
Im Yeong Lee	(Soonchunhyang University, Korea)
Jongin Lim	(Korea University, Korea)
Jong Sou Park	(Hankuk Aviation University, Korea)
Sung Jun Park	(BCQRE, Korea)
Gwangsoo Rhee	(Sookmyung Women's University, Korea)
Kyung Hyune Rhee	(Pukyong National University, Korea)
Dae Hyun Ryu	(Hansei University, Korea)
Jong Tae Shin	(ISR, Korea)
Jae Geol Yim	(Dongguk University, Korea)
E-Joon Yoon	(National Security Research Institute, Korea)

Table of Contents

Practical Security in Public-Key Cryptography

David Pointcheval

LIENS – CNRS, École Normale Supérieure, 45 rue d'Ulm, F-75230 Paris Cedex 05
David.Pointcheval@ens.fr
http://www.di.ens.fr/users/pointche

Abstract. Since the appearance of public-key cryptography in Diffie-Hellman seminal paper, many schemes have been proposed, but many have been broken. Indeed, for many people, the simple fact that a cryptographic algorithm withstands cryptanalytic attacks for several years is considered as a kind of validation. But some schemes took a long time before being widely studied, and maybe thereafter being broken.

A much more convincing line of research has tried to provide "provable" security for cryptographic protocols, in a complexity theory sense: if one can break the cryptographic protocol, one can "efficiently" solve the underlying problem. Unfortunately, very few practical schemes can be proven in this so-called "standard model" because such a security level rarely meets with efficiency. Moreover, for a long time the security proofs have only been performed in an asymptotic framework, which provides some confidence in the scheme but for very huge parameters only, and thus for unpractical schemes.

A recent trend consists in providing very efficient reductions, with a practical meaning: with usual parameters (such as 1024-bit RSA moduli) the computational cost of any attack is actually 2^{72}, given the state of the art about classical problems (*e.g.* integer factoring).

In this paper, we focus on practical schemes together with their "reductionist" security proofs. We cover the two main goals that public-key cryptography is devoted to solve: authentication with digital signatures and confidentiality with public-key encryption schemes.

1 Introduction

1.1 Motivation

Since the beginning of public-key cryptography, with the seminal Diffie-Hellman paper [14], many suitable algorithmic problems for cryptography have been proposed (*e.g.* one-way –possibly trapdoor– functions). Then, many cryptographic schemes have been designed, together with more or less heuristic proofs of their security relative to the intractability of these problems. However, most of those schemes have thereafter been broken.

The simple fact that a cryptographic algorithm withstands cryptanalytic attacks for several years is often considered as a kind of validation procedure, but some schemes take a long time before being broken. The Chor-Rivest cryptosystem [10] illustrates this fact quite well. This scheme based on the knapsack

K. Kim (Ed.): ICICS 2001, LNCS 2288, pp. 1–17, 2002.

problem took more than 10 years to be totally broken [42] whereas before the effective attack it was believed to be very hard since all the classical techniques against the knapsack problems, such as LLL [26], had failed because of the high density of the involved instances. Therefore, the lack of attacks at some time should never be considered as a security validation of any proposal.

1.2   Provable Security and Practical Security

A completely different paradigm is provided by the concept of "provable" security. A significant line of research has tried to provide proofs in the framework of complexity theory (*a.k.a.* "reductionist" security proofs [3]): the proofs provide reductions from a well-studied problem to an attack against a cryptographic protocol. At the beginning, people just tried to define the security notions required by actual cryptographic schemes, and then to design protocols which achieve these notions. The techniques were directly derived from the complexity theory, providing polynomial reductions. However, their aim was essentially theoretical, and thus they were trying to minimize the required assumptions on the primitives (one-way functions or permutations, possibly trapdoor, etc). Therefore, they just needed to exhibit polynomial reductions from the basic assumption on the primitive into an attack of the security notion, in an asymptotic way.

However, such a result has no practical impact on actual security of proposed schemes. Indeed, even with a polynomial reduction, one may be able to break the cryptographic protocol within few hours, whereas the reduction just leads to an algorithm against the underlying problem which requires many years. Therefore, those reductions only prove the security when very huge (and thus maybe unpractical) parameters are used, under the assumption that no polynomial time algorithm exists to solve the underlying problem.

For a few years, more efficient reductions have been expected, under the denominations of either "exact security" [7] or "concrete security" [31], which provide more practical security results. The perfect situation is reached when one manages to prove that, from an attack, one can describe an algorithm against the underlying problem, with almost the same success probability within almost the same amount of time. We have then achieved "practical security". Unfortunately, in many cases, provable security is at the cost of an important loss in terms of efficiency for the cryptographic protocol, or relies on a weaker computational problem. Therefore, a classical way to give some convincing evidences about the security of an efficient scheme relative to a strong computational problem is to make some hypotheses on the adversary's behavior: the attack is generic, independent of the actual implementation of some objects:

- of the hash function, in the "random oracle model" [17,5];
- of the group, in the "generic (group) model" [28,39].

1.3   Organization of the Paper

In the next section, we describe more formally what a signature scheme and an encryption scheme are. Moreover, we make precise the security notions one wants the schemes to achieve. Such a formalism is the first step towards provable security. In section 3, we present some classical assumptions on which the security may rely. In sections 4 and 5, we describe several signature and encryption schemes with their formal security results and some detailed security proofs.

2   A First Formalism

2.1   Digital Signature Schemes

Definitions. A signature scheme is defined by the three following algorithms:

- The *key generation algorithm* K. On input 1^k, the algorithm K produces a pair (k_p, k_s) of matching public and private keys. Algorithm K is probabilistic. The input k is called the security parameter.
- The *signing algorithm* Σ. Given a message m and a pair of matching public and private keys (k_p, k_s), Σ produces a signature σ. The signing algorithm might be probabilistic.
- The *verification algorithm* V. Given a signature σ, a message m and a public key k_p, V tests whether σ is a valid signature of m with respect to k_p.

Forgeries and Attacks. In this subsection, we formalize some security notions which capture the main practical situations. On the one hand, the goals of the adversary may be various [24]:

- Disclosing the private key of the signer. It is the most serious attack. This attack is termed *total break*.
- Constructing an efficient algorithm which is able to sign messages with good probability of success. This is called *universal forgery*.
- Providing a new message-signature pair. This is called *existential forgery*.

On the other hand, various means can be made available to the adversary, helping her into the forgery. We focus on two specific kinds of attacks against signature schemes: the *no-message attacks* and the *known-message attacks*. In the first scenario, the attacker only knows the public key of the signer. In the second one, the attacker has access to a list of valid message-signature pairs. According to the way this list was created, we usually distinguish many subclasses, but the strongest is the *adaptive chosen-message attack*, where the attacker can ask the signer to sign any message of her choice. She can therefore adapt her queries according to previous answers.

　　When one designs a signature scheme, one wants to computationally rule out existential forgeries even under adaptive chosen-message attacks. More formally, one wants that the success probability of any adversary $\mathbf{A}$ with a reasonable amount of time is small, where

$$\mathsf{Succ}^{\mathsf{cma}}(\mathbf{A}) = \Pr\left[(k_p, k_s) \leftarrow K(1^k), (m, \sigma) \leftarrow \mathbf{A}^{\Sigma_{k_s}}(k_p) : V(k_p, m, \sigma) = 1\right].$$

　　We remark that since the adversary is allowed to play an adaptive chosen-message attack, the signing algorithm is made available, without any restriction, hence the oracle notation $\mathbf{A}^{\Sigma_{k_s}}$. Of course, in its answer, there is the natural restriction that the returned signature has not been obtained from the signing oracle Σ_{k_s} itself.

2.2   Public-Key Encryption

The aim of a public-key encryption scheme is to allow anybody who knows the public key of Alice to send her a message that she will be the only one able to recover, granted her private key.

Definitions. A public-key encryption scheme is defined by the three following algorithms:

- The *key generation algorithm K*. On input 1^k, the algorithm K produces a pair (k_p, k_s) of matching public and private keys. Algorithm K is probabilistic.
- The *encryption algorithm E*. Given a message m and a public key k_p, E produces a ciphertext c of m. This algorithm may be probabilistic. In this latter case, we can write $E(k_p, m; r)$ where r is the random tape.
- The *decryption algorithm D*. Given a ciphertext c and the private key k_s, D gives back the plaintext m. This algorithm is necessarily deterministic.

Security Notions. As for signature schemes, the goals of the adversary may be various. The first common security notion that one would like for an encryption scheme is *one-wayness* (OW): with just public data, an attacker cannot get back the whole plaintext of a given ciphertext. More formally, this means that for any adversary $\mathbf{A}$, her success in inverting E without the private key should be negligible over the message space $\mathbf{M}$ and the internal random coins of the adversary and the encryption algorithm:

$$\mathsf{Succ}^{\mathsf{ow}}(\mathbf{A}) = \Pr_m[(k_p, k_s) \leftarrow K(1^k) : \mathbf{A}(k_p, E(k_p, m)) = m].$$

However, many applications require more, namely the *semantic security* (IND), *a.k.a. polynomial security/indistinguishability of encryptions* [22]. This security notion means computational impossibility to distinguish between two messages, chosen by the adversary, one of which has been encrypted, with a probability significantly better than one half: her advantage $\mathsf{Adv}^{\mathsf{ind}}(\mathbf{A})$, formally defined as

$$2 \times \Pr_b \left[\begin{array}{l} (k_p, k_s) \leftarrow K(1^k), (m_0, m_1, s) \leftarrow \mathbf{A}_1(k_p), \\ c = E(k_p, m_b) : \mathbf{A}_2(m_0, m_1, s, c) = b \end{array} \right] - 1,$$

where the adversary $\mathbf{A}$ is seen as a 2-stage attacker $(\mathbf{A}_1, \mathbf{A}_2)$, should be negligible.

A later notion is *non-malleability* (NM) [15]. To break it, given a ciphertext, the adversary tries to produce a new ciphertext such that the plaintexts are meaningfully related. This notion is stronger than the above semantic security, but it is equivalent to the latter in the most interesting scenario [4] (the CCA attacks, see below). Therefore, we will just focus on one-wayness and semantic security.

On the other hand, an attacker can play many kinds of attacks, according to the available information: since we are considering asymmetric encryption, the adversary can encrypt any plaintext of her choice, granted the public key, hence the *chosen-plaintext attack* (CPA). She may furthermore have access to more information, modeled by partial or full access to some oracles: a plaintext-checking oracle which, on input a pair (m, c), answers whether c encrypts the message m. This attack has been named the *Plaintext-Checking Attack* (PCA) [32]; a validity-checking oracle which, on input a ciphertext c, just answers whether it is a valid ciphertext or not (the so-called *reaction attacks* [21,8]); or the decryption oracle itself, which on any ciphertext, except the challenge ciphertext, answers the corresponding plaintext (*non-adaptive [27]/adaptive [36] chosen-ciphertext attacks*). This latter scenario which allows adaptively chosen ciphertexts as queries to the decryption oracle is the strongest attack, and is named the *chosen-ciphertext attack* (CCA).

A general study of these security notions and attacks was conducted in [4]. We refer the reader to this paper for more details.

3 The Basic Assumptions

3.1 Computational Assumptions

For asymmetric cryptography, no security can be unconditionally guaranteed. Therefore, for any cryptographic protocol, security relies on a computational assumption: the existence of one-way functions, or permutations, possibly trapdoor.

Integer Factoring. The most famous intractable problem is factorization of integers: while it is easy to multiply two prime integers p and q to get the product $N = p \cdot q$, it is not simple to decompose N into its prime factors p and q. Unfortunately, it just provides a one-way function, without any possibility to invert the process. In 1978, Rivest, Shamir and Adleman [37] defined the so-called *RSA problem*: Let $N = pq$ be the product of two large primes of similar sizes and e an integer relatively prime to $\varphi(N)$. For a given $y \in \mathbb{Z}_N^\star$, find $x \in \mathbb{Z}_N^\star$ such that $x^e = y \bmod N$. The RSA assumption then says that this problem is intractable for any modulus $N = pq$, large enough (presumably as hard as factoring the modulus): the success probability $\mathsf{Succ}^{\mathsf{rsa}}(\mathbf{A})$ of any adversary $\mathbf{A}$ within a reasonable running time is small.

Discrete Logarithm. Some other classical problems are related to the discrete logarithm. The setting is quite general: one is given a finite cyclic group $\mathcal{G}$ of prime order q (such as a subgroup of $(\mathbb{Z}_p^\star, \times)$ for $q \mid p-1$, or an elliptic curve, etc) and a generator $\mathbf{g}$ (*i.e.* $\mathcal{G} = \langle \mathbf{g} \rangle$). In such a group, one considers the following problems (using the additive notation):

- the **Discrete Logarithm** problem (**DL**): given $\mathbf{y} \in \mathcal{G}$, compute $x \in \mathbb{Z}_q$ such that $\mathbf{y} = x \cdot \mathbf{g} = \mathbf{g} + \ldots + \mathbf{g}$ (x times), then one writes $x = \log_{\mathbf{g}} \mathbf{y}$.
- the **Computational Diffie-Hellman** problem (**CDH**): given two elements in the group $\mathcal{G}$, $\mathbf{a} = a \cdot \mathbf{g}$ and $\mathbf{b} = b \cdot \mathbf{g}$, compute $\mathbf{c} = ab \cdot \mathbf{g}$. Then one writes $\mathbf{c} = \mathbf{DH}(\mathbf{a}, \mathbf{b})$.
- the **Decisional Diffie-Hellman** problem (**DDH**): given three elements in the group $\mathcal{G}$, $\mathbf{a} = a \cdot \mathbf{g}$, $\mathbf{b} = b \cdot \mathbf{g}$ and $\mathbf{c} = c \cdot \mathbf{g}$, decide whether $\mathbf{c} = \mathbf{DH}(\mathbf{a}, \mathbf{b})$.

It is clear that they are sorted from the strongest problem to the weakest one. Very recently, Okamoto and the author [33] defined a new variant of the Diffie-Hellman problem, which we called the *Gap Diffie-Hellman problem* (**GDH**), where one wants to solve the **CDH** problem with an access to a **DDH** oracle.

3.2 Ideal Objects

As already remarked, one often has to make some assumptions about the adversary's behavior. Let us present two classical models.

The Generic Model. Generic algorithms [28,39], as introduced by Nechaev and Shoup, encompass group algorithms that do not exploit any special property of the encodings of group elements other than the property that each group element is encoded by a unique string. Remark that such algorithms are the only known for well-chosen elliptic curves. However, it is a strong and non-realistic restriction when one works in a subgroup of $\mathbb{Z}_p^\star$.

A *generic* algorithm **A** over a group $\mathcal{G}$ is a probabilistic algorithm that takes as input an *encoding list* $\{\sigma(x_1), \cdots, \sigma(x_k)\}$, where each x_i is in $\mathcal{G}$. An encoding of a standard group $\mathcal{G}$ is an injective map from $\mathcal{G}$ into a set of bit-strings S. While it executes, the algorithm may consult an oracle for further encodings. Oracle calls consist of triples $\{i, j, \epsilon\}$, where i and j are indices of the encoding list and ϵ is $\pm$. The oracle returns the string $\sigma(x_i \pm x_j)$, according to the value of ϵ and this bit-string is appended to the list, unless it was already present.

An interesting result in this model is the complexity lower-bound for breaking the **DL** problem. Similar results have been proven for all the above Diffie-Hellman problems [39,1]. The consequence is that all these problems (**DL**, **CDH**, **DDH** and **GDH**) require an expected time in the square root of the order q to be solved by any generic algorithm.

Theorem 1. *Let $\mathcal{G}$ be a standard cyclic group of prime order q. Let **A** be a generic algorithm over $\mathcal{G}$ that makes at most n queries to the group-oracle. If $x \in \mathcal{G}$ and an encoding σ are chosen at random, then the probability that **A** returns x on input $\{\sigma(1), \sigma(x)\}$ is less than $1/q + (n+2)^2/q$.*

Proof. The idea of the proof is to identify the probabilistic space consisting of σ and x with the space $S^{n+2} \times \mathcal{G}$, where S is the set of bit-string encodings. Given a tuple $\{z_1, \cdots, z_{n+2}, x\}$ in this space, z_1 and z_2 are used as $\sigma(1)$ and $\sigma(x)$, the successive z_i are used in sequence to answer the oracle queries, modeled by formal linear relations of 1 and x, *i.e.*, linear polynomials $P_i = a_i + b_i X$. This interpretation may yield inconsistencies as it does not take care of possible collisions between oracle queries when evaluating the polynomials P_i in x, but with probability less than $(n + 2)^2/q$. Eventually, let us note that the output of a computation corresponding to a good sequence $\{z_1, \cdots, z_{n+2}, x\}$ (which does not make two polynomials to collude in x) does not depend on x. $\square$

The Random Oracle Model. The "random oracle model" was the first to be introduced in the cryptographic community [17,5]: the hash function is formalized by an oracle which produces a truly random value for each new query. Of course, if the same query is asked twice, identical answers are obtained.

This model has been strongly accepted by the community, and is considered as a good one, in which proofs of security give a good taste of the actual security level. Even if it does not provide a formal proof of security (as in the standard model, without any ideal assumption) it is argued that proofs in this model ensure security of the overall design of the scheme provided that the hash function has no weakness.

More formally, this model can also be seen as a restriction on the adversary's capabilities. Indeed, it simply means that the attack is generic without considering any particular instantiation of the hash functions.

4   Provably Secure Digital Signature Schemes

4.1   Basic Signature Schemes

The Plain-RSA Signature. Two years after the Diffie-Hellman paper [14], Rivest, Shamir and Adleman [37] proposed the first signature scheme based on the "trapdoor one-way permutation paradigm", using the RSA function: the key generation algorithm produces a large composite number $N = pq$, a public key e, and a private key d such that $e \cdot d = 1 \bmod \varphi(N)$. The signature of a message m, encoded as an element in $\mathbb{Z}_N$, is its eth root, $\sigma = m^{1/e} = m^d \bmod N$. The verification algorithm simply checks whether $m = \sigma^e \bmod N$.

However, the RSA scheme is not secure by itself since it is subject to existential forgery: it is easy to create a valid message-signature pair, without any help of the signer, first randomly choosing a certificate σ and getting the signed message m from the public verification relation, $m = \sigma^e \bmod N$.

The El Gamal Signature Scheme. In 1985, El Gamal proposed the first digital signature scheme based on the **DL** problem [16], but with no formal security analysis: the key generation algorithm produces a large prime p, as well as an element g in $\mathbb{Z}_p^\star$ of large order. It also creates a pair of keys, the private key $x \in \mathbb{Z}_{p-1}^\star$ and the public key $y = g^x \bmod p$. The signature of a message m is a pair (r, s), where $r = g^k \bmod p$, with a random $k \in \mathbb{Z}_{p-1}^\star$, and $s = (m - xr)/k \bmod p - 1$. This pair satisfies $g^m = y^r r^s \bmod p$, which is checked by the verification algorithm. Unfortunately, as above, existential forgeries are easy.

4.2   DL-Based Signatures

In 1986 a new paradigm for signature schemes was introduced. It is derived from fair zero-knowledge identification protocols involving a prover and a verifier [23], and uses hash functions in order to create a kind of virtual verifier. The first application was derived from the Fiat–Shamir identification scheme [17]. This paradigm has also been applied by Schnorr [38], and provided the most efficient El Gamal-like scheme, with no easy existential forgery.

The security results for that paradigm have been considered as folklore for a long time but without any formal validation. However, Stern and the author [35] formally proved the above paradigm when H is assumed to behave like a random oracle. The proof is based on the by now classical *oracle replay technique* [35]. However, for the Schnorr's signature scheme, one can just formally prove that if an adversary manages to perform an existential forgery under an adaptive chosen-message attack within an expected time T, after q_h queries to the random oracle and q_s queries to the signing oracle, then the discrete logarithm problem can be solved within an expected time less than $207q_hT$. Actually, this security result is not practical, since q_h may be huge.

This technique has been applied on several other variants of the El Gamal [16] signature scheme, such as the Korean Standard KCDSA [25]. However, the American Standard DSA [29] does not fit with any of these designs. Therefore, this widely used scheme never got any formal security proof (even with a costly reduction). Recently, Brown [9] considered this standard in the generic model, which provides a practical result, under the assumption of generic adversaries. However, this makes this result possibly suitable for ECDSA only [2].

Description of ECDSA. The key generation algorithm defines an elliptic curve, and a point $\mathbf{g}$ of large prime order q. It also creates a pair of keys, the private key $x \in \mathbb{Z}_q$ and the public key $\mathbf{y} = x \cdot \mathbf{g}$. The signature of a message m is a pair (r, s): $\mathbf{r} = k \cdot \mathbf{g}$, with a random $k \in \mathbb{Z}_q^{\star}$, $r = x_{\mathbf{r}} \bmod q$, where $x_{\mathbf{r}}$ is the x-coordinate of $\mathbf{r}$, $e = H(m)$ and $s = k^{-1}(e + xr) \bmod q$. This pair satisfies $r = x_{\mathbf{r}'} \bmod q$ where $\mathbf{r}' = es^{-1} \cdot \mathbf{g} + rs^{-1} \cdot \mathbf{y}$, with $e = H(m)$, which is checked by the verification algorithm. It involves a hash function H which outputs h-bit long digests.

Theorem 2. *Let $\mathcal{G}$ be a standard cyclic group of prime order q. Let S be a set of bit-string encodings. Let $\mathbf{A}$ be a generic algorithm over $\mathcal{G}$ that makes at most q_s queries to the signing oracle and n queries to the group-oracle, with a running time bounded by t.*

If $\mathbf{A}$ can perform an existential forgery with probability greater than ε, for random x and random encoding σ, on input $\{\mathbf{g} = \sigma(1), \mathbf{y} = \sigma(x)\}$, then one can extract a collision for H with probability $\varepsilon' \geq \varepsilon - (n + q_s + 2)(n + 2)/q$, within almost the same time.

Proof. The proof uses the same technique as for the theorem 1. Let $\mathbf{A}$ be a generic attacker able to forge some message M with a signature (r, s). We describe several games, which differ just a little bit between each other [40].

Game$_0$: This is the game the generic adversary plays, with a random encoding σ, and a random pair of keys. The adversary eventually outputs a message M and a signature (r, s). We denote by S_0 the event $V(k_p, M, (r, s)) = 1$ (as well as S_i in any Game$_i$ below.) By definition, we have $\Pr[S_0] = \varepsilon$.

Game$_1$: In this game, we simulate the encoding and the group-oracle using a random sequence $\{z_1, \cdots, z_{n+2}, x\}$, modeling oracle queries by linear polynomials $P_i = a_i + b_i X$: $| \Pr[S_1] - \Pr[S_0] | \leq (n + 2)^2 / q$.

Game$_2$: We modify the random choice of the encoding oracle answers z_i. For all the queries $\sigma(b_\ell X + a_\ell)$, one gets a random $e_\ell \in_R \mathbb{Z}_q$, as well as a random $z_\ell \in S$ such that the x-coordinate of the corresponding point is equal to $b_\ell a_\ell^{-1} e_\ell \bmod q$. For convenient compact encoding sets, this simulation can be perfect: $\Pr[S_2] = \Pr[S_1]$.

Game$_3$: We modify the random choice of the e_ℓ when this latter is smaller than 2^h, where h is the bit-length of H-output: one gets a random message M_ℓ and computes $e_\ell = H(M_\ell) \bmod q$. Under the assumption of the uniformity of the output of H, which is related to the collision-resistance, $\Pr[S_3] = \Pr[S_2]$.

Game$_4$: In this game, we simulate the signing oracle, which can be perfectly performed by defining some values of the encoding, unless they have already been defined before: $| \Pr[S_4] - \Pr[S_3] | \leq n q_s / q$.

In this latter game, one can easily see that an existential forgery $(M, (r, s))$ leads to a collision for H, between M and some M_ℓ, if the bit-size of q is larger than h: $r = \sigma(H(M)s^{-1} + xrs^{-1}) = (rs^{-1}) / (H(M)s^{-1}) \times H(M_\ell) \bmod q$. □

Therefore, under the collision-resistance of H, implemented by SHA-1 [30], and $q > 2^{160}$, one gets a very tight security result against generic adversaries. However, this strong generic model is not as convincing as the random oracle model. Studies on elliptic curves may reveal non-generic attacks.

4.3   RSA-Based Signatures

In 1996, Bellare and Rogaway [7] proposed some signature schemes, based on the RSA assumption, provably secure in the random oracle model. The first scheme is the by now classical hash-and-decrypt paradigm (*a.k.a.* the Full-Domain Hash paradigm): instead of directly signing m using the RSA function, one first hashes it using a full-domain hash function $H : \{0,1\}^\star \to \mathbb{Z}_N$, and computes the eth root, $\sigma = H(m)^d \bmod N$. Everything else is straightforward. For this scheme, named FDH-RSA, one can prove in the random oracle model [7,11,5]: for any adversary, her probability for an existential forgery under a chosen-message attack within a time t, after q_h and q_s queries to the hash function and the signing oracle respectively, is upper-bounded by $3q_s \mathsf{Succ}^{rsa}(t + (q_s + q_h)T_{exp})$, where T_{exp} is the time for an exponentiation to the power e, modulo N. This is quite bad because of the factor q_s. This factor is better than the factor q_h, as it was in the original proof [7], and for the **DL**-based signature schemes, but it is still too bad for practical security. Therefore, Bellare and Rogaway proposed a better candidate, the Probabilistic Signature Scheme (PSS): the key generation is still the same, but the signature process involves three hash functions

$$F : \{0,1\}^{k_2} \to \{0,1\}^{k_0}, G : \{0,1\}^{k_2} \to \{0,1\}^{k_1} \text{ and } H : \{0,1\}^{\star} \to \{0,1\}^{k_2},$$

where $k = k_0 + k_1 + k_2 + 1$ is the bit-length of the modulus N. For each message m to be signed, one chooses a random string $r \in \{0,1\}^{k_1}$. One first computes $w = H(m,r)$, $s = G(w) \oplus r$ and $t = F(w)$. Then one concatenates $y = 0\|w\|s\|t$, where $a\|b$ denotes the concatenation of the bit strings a and b. Finally, one computes the eth root, $\sigma = y^d \bmod N$.

The verification algorithm first computes $y = \sigma^e \bmod N$, and parses it as $y = b\|w\|s\|t$. Then, one can get $r = s \oplus G(w)$, and checks whether $b = 0$, $w = H(m,r)$ and $t = F(w)$.

About RSA–PSS, Bellare and Rogaway proved the security in the random oracle model.

Theorem 3. *Let $\mathbf{A}$ be a CMA-adversary against RSA–PSS. Let us consider any adversary $\mathbf{A}$ which produces an existential forgery within a time t, after q_F, q_G, q_H and q_s queries to the hash functions F, G and H and the signing oracle respectively. Then her success probability is upper-bounded by*

$$\mathsf{Succ}^{\mathsf{rsa}}(t + (q_s + q_H)k_2 \cdot T_{exp}(k)) + \frac{1}{2^{k_2}} + (q_s + q_H) \cdot \left(\frac{q_s}{2^{k_1}} + \frac{q_F + q_G + q_H + q_s + 1}{2^{k_2}} \right),$$

with $T_{exp}(k)$ the time for an exponentiation modulo a k-bit integer.

Proof. First, we assume the existence of an adversary $\mathbf{A}$ that produces an existential forgery with probability ε within time t, after q_F, q_G and q_H queries to the random oracles F, G and H and q_s queries to the signing oracle. This is the game of the real-world attack (denoted below Game_0). In any Game_i, we denote by S_i the event $V(\mathsf{k_p}, m, \sigma) = 1$.

Game_1: In this game, we replace the random oracles F and G by random answers for any new query. This game is clearly identical to the previous one: $\Pr[S_1] = \Pr[S_0]$.

Game_2: Then, we replace the random oracle H by the following simulation. For any new query (m,r), one chooses a random $u \in \mathbb{Z}_N$ and computes $z = u^e \bmod N$, until the most significant bit of z is 0, but at most k_2 times (otherwise one aborts). Thereafter, z is parsed into $0 \| w \| s \| t$, and one defines $F(w) \leftarrow t$, $G(w) \leftarrow s \oplus r$ and $H(m,r) \leftarrow w$. Finally, one returns w. Let us remark that the number of calls to H is upper-bounded by $q_s + q_H$. This game may only differ from the previous one if during some H-simulations,
  - z is still in the bad range, even after the k_2 attempts;
  - $F(w)$ or $G(w)$ have already been defined.

$$| \Pr[S_2] - \Pr[S_1] | \leq (q_H + q_s) \times \left(\frac{1}{2^{k_2}} + \frac{q_F + q_G + q_H + q_s}{2^{k_2}} \right).$$

Game_3: Now, we simply abort if the signing oracle makes a $H(m,r)$-query for some (m,r) that has already been asked to H. Furthermore, for any new query (m,r) directly asked by the adversary, one computes $z = yu^e \bmod N$, instead of $z = u^e \bmod N$. The distribution of the z is exactly the same as before. Thus the above abortion makes the only difference, which gives $| \Pr[S_3] - \Pr[S_2] | \leq q_s(q_H + q_s)/2^{k_1}$.

Game$_4$: In the last game, we replace the signing oracle by an easy simulation, returning the value u involved in the answer $H(m, r) = z = u^e \bmod N$. The simulation is perfect, then $\Pr[S_4] = \Pr[S_3]$.

The event S_4 means that, at the end of **Game$_4$**, the adversary outputs a valid message/signature (m, σ). But in this game, it is only possible either by chance, or by inverting RSA: $\Pr[S_4] \leq \mathsf{Succ}^{\mathsf{rsa}}(t', k) + 2^{-k_2}$, where t' is the running time of the adversary and the simulations: $t' \leq t + (q_s + q_H)k_2 \cdot T_{exp}(k)$.   □

The important point in this security result is the very tight link between success probabilities, but also the almost linear time of the reduction. Thanks to this exact and efficient security result, RSA–PSS has become the new PKCS #1 v2.0 standard for signature.

Recently, Cramer and Shoup [13] proposed the first efficient signature scheme with a security proof in the standard model, and thus no ideal assumption. However, the security relies on a stronger computational assumption, the intractability of the so-called *flexible RSA problem*: Let $N = pq$ be the product of two large primes of similar sizes. For a given $y \in \mathbb{Z}_N^\star$, find a prime exponent e and $x \in \mathbb{Z}_N^\star$ such that $x^e = y \bmod N$.

The key generation algorithm produces a large composite number $N = pq$, where p and q are strong primes ($p = 2p'+1$ and $q = 2q'+1$, with p' and q' some prime integers). It also generates two non-quadratic residues $h, x \in \mathbb{Z}_N^\star$, and an $(\ell+1)$-bit prime integer e'. For signing a message m, one chooses a random non-quadratic residue $y \in \mathbb{Z}_N^\star$ and an $(\ell+1)$-bit prime integer e. Then one computes $x' = (y')^{e'}h^{-H(m)} \bmod N$, and solves the equation $y^e = xh^{H(x')} \bmod N$ for the unknown y. The signature is the triple (e, y, y'), with e an odd $(\ell+1)$-bit integer, which satisfies $(y')^{e'} = x'h^{H(m)} \bmod N$ and $y^e = xh^{H(x')} \bmod N$. The verification algorithm simply checks the above properties for the triple.

Even if the security holds in the standard model, the reduction is quite expensive (at least quadratic in the number q_s of queries asked to the signing oracle) and furthermore it is not tight, since once again, a factor q_s appears between the success probability for solving the *flexible RSA problem* and for breaking the signature scheme. Therefore, this security does not mean anything for practical parameters.

5   Provably Secure Public-Key Encryption Schemes

5.1   Basic Encryption Schemes

The Plain-RSA Encryption. The RSA primitive [37] can also be used for encryption: the key generation algorithm produces a large composite number $N = pq$, a public key e, and a private key d such that $e \cdot d = 1 \bmod \varphi(N)$. The encryption of a message m, encoded as an element in $\mathbb{Z}_N$, is $c = m^e \bmod N$. This ciphertext can be decrypted thanks to the knowledge of d, $m = c^d \bmod N$. Clearly, this encryption is OW-CPA, relative to the RSA problem. The determinism makes a plaintext-checking oracle useless. Indeed, the encryption of a message m, under a public key k_p is always the same, and thus it is easy to check whether a ciphertext c really encrypts m, by re-encrypting this latter. Therefore the RSA-encryption scheme is OW-PCA relative to the RSA problem as well. Because of this determinism, it cannot be semantically secure: given the encryption c of either m_0 or m_1, the adversary simply computes $c' = m_0^e \bmod N$ and checks whether $c' = c$ or not to make the decision.

The El Gamal Encryption Scheme. In 1985, El Gamal [16] also designed a **DL**-based public-key encryption scheme, inspired by the Diffie-Hellman key exchange protocol [14]: given a cyclic group $\mathcal{G}$ of prime order q and a generator **g**, the generation algorithm produces a random element $x \in \mathbb{Z}_q^*$ as private key, and a public key $\mathbf{y} = x \cdot \mathbf{g}$. The encryption of a message m, encoded as an element **m** in $\mathcal{G}$, is a pair $(\mathbf{c} = a \cdot \mathbf{g}, \mathbf{d} = a \cdot \mathbf{y} + \mathbf{m})$. This ciphertext can be easily decrypted thanks to the knowledge of x, since $a \cdot \mathbf{y} = ax \cdot \mathbf{g} = x \cdot \mathbf{c}$, and thus $\mathbf{m} = \mathbf{d} - x \cdot \mathbf{c}$. This encryption scheme is well-known to be OW-CPA relative to the **CDH** problem. It is also IND-CPA relative to the **DDH** problem [41]. About OW-PCA, it relies on the new **GDH** problem [33]. However, it does not prevent adaptive chosen-ciphertext attacks because of the homomorphic property.

As we have seen above, the expected security level is IND-CCA. We wonder if we can achieve this strong security with practical encryption schemes.

5.2   The Optimal Asymmetric Encryption Padding

In 1994, Bellare and Rogaway proposed a generic conversion [6], in the random oracle model, the "Optimal Asymmetric Encryption Padding" (OAEP), which was claimed to apply to any family of trapdoor one-way permutations, such as RSA. The key generation produces a one-way permutation $f : \{0,1\}^k \to \{0,1\}^k$, the public key. The private key is the inverse permutation g, which requires a trapdoor to be actually computed. The scheme involves two hash functions

$$G : \{0,1\}^{k_0} \to \{0,1\}^{n+k_1} \quad \text{and} \quad H : \{0,1\}^{n+k_1} \to \{0,1\}^{k_0},$$

where $k = k_0 + k_1 + n + 1$. For any message $m \in \{0,1\}^n$ to be encrypted, instead of computing $f(m)$, as done with the above plain-RSA encryption, one first modifies m. For that, one chooses a random string $r \in \{0,1\}^{k_0}$; one computes $s = (m\|0^{k_1}) \oplus G(r)$ and $t = r \oplus H(s)$; finally, one computes $c = f(s\|t)$.

The decryption algorithm first computes $P = g(c)$, granted the private key, the trapdoor to compute g, and parses it as $P = s\|t$. Then, one can get $r = t \oplus H(s)$, and $M = s \oplus G(r)$, which is finally parsed into $M = m\|0^{k_1}$, if the k_1 least significant bits are all 0.

For a long time, the OAEP conversion has been widely believed to provide an IND-CCA encryption scheme from any trapdoor one-way permutation. However, the sole proven result was the semantic security against non-adaptive chosen-ciphertext attacks (*a.k.a.* lunchtime attacks [27]). Recently, Shoup [40] showed that it was very unlikely that a stronger security result could be proven.

However, because of the wide belief of a strong security level, RSA–OAEP became the new PKCS #1 v2.0 for encryption after an effective attack against the PKCS #1 v1.5 [8].

Fortunately, Fujisaki, Okamoto, Stern and the author [20] provided a complete security proof of IND-CCA-security for OAEP in general, but also for RSA–OAEP in particular under the RSA assumption.

The proof is a bit intricate, so we refer the reader to [20] for more information. However, our reduction is worse than the incomplete one originally proposed by Bellare and Rogaway [6]: an attacker in time t with advantage ε against RSA–OAEP can be used to break RSA with probability almost ε^2, but within a time bound $t + q_h^2 \times \mathcal{O}(k^3)$, where q_h is the total number of queries asked to the hash functions. Because of the quadratic term q_h^2, this reduction is meaningful for huge moduli only, more than 4096-bit long!

5.3 A Rapid Enhanced-Security Asymmetric Cryptosystem Transform

Anyway, there is no hope to use OAEP with any **DL**-based primitive, even with huge parameters, because of the "permutation" requirement which limits the application of OAEP to RSA only. More general conversions have recently been proposed, first by Fujisaki and Okamoto [18,19], then by the author [34], that apply to any **OW-CPA** scheme to make it into an **IND-CCA** one, still in the random oracle model. But the last one proposed by Okamoto and the author [32] is the most efficient: REACT (see figure 1). It applies to any encryption scheme $\mathbf{S} = (K, E, D)$

$$E : \mathbf{PK} \times \mathbf{M} \times \mathbf{R} \to \mathbf{C} \qquad D : \mathbf{SK} \times \mathbf{C} \to \mathbf{M},$$

where **PK** and **SK** are the sets of the public and private keys, **M** is the message space, **C** is the ciphertext space and **R** is the random coin space. We also need two hash functions G and H,

$$G : \mathbf{M} \to \{0,1\}^\ell, H : \mathbf{M} \times \{0,1\}^\ell \times \mathbf{C} \times \{0,1\}^\ell \to \{0,1\}^\kappa,$$

where κ is the security parameter, while ℓ denotes the size of the messages to encrypt. About the converted scheme $\mathbf{S}' = (K', E', D')$, one can claim the following security result:

Theorem 4. *Let* **A** *be a CCA-adversary against the semantic security of* **S**'. *If* **A** *can get an advantage* ε *after* q_D, q_G *and* q_H *queries to the decryption oracle and to the random oracles* G *and* H *respectively, within a time* t, *then one can invert* E *after less than* $q_G + q_H$ *queries to the Plaintext-Checking Oracle with probability greater than* $\varepsilon/2 - q_D/2^\kappa$, *within a time* $t + (q_G + q_H)T_{PCA}$, *where* T_{PCA} *denotes the time required by the PCA oracle to answer any query.*

Proof. We consider a sequence of games in which the adversary $\mathbf{A} = (\mathbf{A}_1, \mathbf{A}_2)$ is involved. In each game, we use a random bit β and we are given $\alpha = E(\mathsf{k_p}, \rho; w)$,

K': **Key Generation** $\rightarrow (k_p, k_s)$
$(k_p, k_s) \leftarrow K(1^k)$
E': **Encryption of** $m \in \mathbf{M'} = \{0,1\}^\ell \rightarrow (a,b,c)$
$R \in \mathbf{M}$ and $r \in \mathbf{R}$ are randomly chosen
$a = E(k_p, R; r) \qquad b = m \oplus G(R) \qquad c = H(R,m,a,b)$
D': **Decryption of** $(a,b,c) \rightarrow m$
Given $a \in \mathbf{C}$, $b \in \{0,1\}^\ell$ and $c \in \{0,1\}^\kappa$
$R = D(k_s, a) \qquad m = b \oplus G(R)$
if $c = H(R,m,a,b)$ and $R \in \mathbf{M} \rightarrow m$ is the plaintext

Fig. 1. Rapid Enhanced-security Asymmetric Cryptosystem Transform $\mathbf{S'}$

for random ρ, w. The adversary runs in two stages: given k_p, $\mathbf{A}_1$ outputs a pair of messages (m_0, m_1), one encrypts $C' = (a', b', c') = E'(k_p, m_\beta)$; on input C', $\mathbf{A}_2$ outputs a bit β'. In both stages, the adversary has access to the random oracles G and H, but also to the decryption oracle. In each game, we denote by S_i the event $\beta' = \beta$.

Game$_0$: This is the above real-world game: $\Pr[S_0] = (1 + \varepsilon)/2$.

Game$_1$: In this game, we simulate the oracles G and H in a classical way, returning new random values for any new query: $\Pr[S_1] = \Pr[S_0]$.

Game$_2$: Then, we replace the decryption oracle by the following simulation. For each query (a,b,c), one looks at all the pairs (R,m) such that $H(R,m,a,b)$ has been asked. For any such R, one asks the Plaintext-Checking Oracle whether a is a ciphertext of R. Then it computes $K = G(R)$. If $b = K \oplus m$ then one outputs m as the plaintext of the triple (a,b,c). Otherwise, one rejects the ciphertext. One may remark that the probability of a wrong simulation is less than $1/2^\kappa$ (if the adversary guessed the H-value), therefore $|\Pr[S_2] - \Pr[S_1]| \le q_D/2^\kappa$.

Game$_3$: Now, we modify the computation of C', given (m_0, m_1). Indeed, we set $a' \leftarrow \alpha$, and $b \in_R \{0,1\}^\ell$, $c \in_R \{0,1\}^\kappa$. Without asking $G(\rho)$ nor $H(\rho, m_i, a', b')$, the adversary cannot see the difference. In such a case we simply stop the game. Anyway, ρ would be in the list of queries asked to G or to H. It can be found after $q_G + q_H$ queries to the Plaintext-Checking Oracle: $|\Pr[S_3] - \Pr[S_2]| \le \mathsf{Succ}^{\mathsf{ow}}(\mathbf{A'})$, where $\mathbf{A'}$ is a PCA-adversary.

In this latter game, one can easily see that without having asked $G(\rho)$ or $H(\rho, m_i, a', b')$ to get any information about the encrypted message m, the advantage of the adversary is 0. This concludes the proof. $\qquad\square$

Hybrid Cryptosystems. In this REACT conversion, one can improve efficiency, replacing the one-time pad by any symmetric encryption scheme, using $K = G(R)$ as a session key. Moreover, the symmetric encryption scheme is just required to be semantically secure under passive attacks, a very weak requirement. With RSA, but also any other deterministic primitive, the construction can be further improved, with just $c = H(R,m)$, or equivalently $c = H(R,b)$.

5.4   Practical Security

As for PSS only, but which was very specific to RSA, the security proof of REACT is both tight with a very efficient reduction in the widely admitted random oracle model: the cost of the reduction is linear in the number of oracle queries. Furthermore, the success probabilities are tightly related. Therefore, this scheme is perfectly equivalent to the difficulty of the underlying problem, without having to use larger parameters. For example, RSA–REACT with a 1024-bit modulus actually provides a provable security level in 2^{72}, whereas 1024-bit RSA–OAEP would provide a security level in 2^{36} only!

We cannot deal with provably secure encryption schemes without referring to the first efficient scheme proven in the standard model, proposed three years ago by Cramer and Shoup [12]. Actually, this encryption scheme achieves IND-CCA, with a both tight and very efficient reduction, but to the **DDH** problem. Furthermore, the encryption and decryption processes are rather expensive (more than twice as much as other constructions in the random oracle model.)

6   Conclusion

In this paper, we reviewed several encryption and signature schemes with their security proofs. The security results are various, and have to be carefully considered since some of them are meaningless for usual sizes. Fortunately, several schemes have a practical significance. However, when one needs such a cryptographic scheme, one first has to decide between (unrelated) assumptions: a computational problem (*e.g.*, RSA, **CDH**, **GDH**) in the random oracle model; a decisional problem (*e.g.*, **DDH**) in the standard model; or the generic model. Second, the efficiency may also be a major criterion. Therefore security is still a matter of subtle trade-offs, until one finds a very efficient and secure scheme, relative to a strong problem in the standard model.

References

1. M. Abdalla, M. Bellare, and P. Rogaway. The Oracle Diffie-Hellman Assumptions and an Analysis of DHIES. In *CT – RSA '01*, LNCS 2020, pages 143–158. Springer-Verlag, Berlin, 2001.
2. American National Standards Institute. Public Key Cryptography for the Financial Services Industry: The Elliptic Curve Digital Signature Algorithm. ANSI X9.62-1998. January 1999.
3. M. Bellare. Practice-Oriented Provable Security. In *ISW '97*, LNCS 1396. Springer-Verlag, Berlin, 1997.
4. M. Bellare, A. Desai, D. Pointcheval, and P. Rogaway. Relations among Notions of Security for Public-Key Encryption Schemes. In *Crypto '98*, LNCS 1462, pages 26–45. Springer-Verlag, Berlin, 1998.
5. M. Bellare and P. Rogaway. Random Oracles Are Practical: a Paradigm for Designing Efficient Protocols. In *Proc. of the 1st CCS*, pages 62–73. ACM Press, New York, 1993.

6. M. Bellare and P. Rogaway. Optimal Asymmetric Encryption – How to Encrypt with RSA. In *Eurocrypt '94*, LNCS 950, pages 92–111. Springer-Verlag, Berlin, 1995.

7. M. Bellare and P. Rogaway. The Exact Security of Digital Signatures – How to Sign with RSA and Rabin. In *Eurocrypt '96*, LNCS 1070, pages 399–416. Springer-Verlag, Berlin, 1996.

8. D. Bleichenbacher. A Chosen Ciphertext Attack against Protocols based on the RSA Encryption Standard PKCS #1. In *Crypto '98*, LNCS 1462, pages 1–12. Springer-Verlag, Berlin, 1998.

9. D. R. L. Brown. The Exact Security of ECDSA. January 2001. Available from http://grouper.ieee.org/groups/1363/.

10. B. Chor and R. L. Rivest. A Knapsack Type Public Key Cryptosystem based on Arithmetic in Finite Fields. In *Crypto '84*, LNCS 196, pages 54–65. Springer-Verlag, Berlin, 1985.

11. J.-S. Coron. On the Exact Security of Full-Domain-Hash. In *Crypto '00*, LNCS 1880, pages 229–235. Springer-Verlag, Berlin, 2000.

12. R. Cramer and V. Shoup. A Practical Public Key Cryptosystem Provably Secure against Adaptive Chosen Ciphertext Attack. In *Crypto '98*, LNCS 1462, pages 13–25. Springer-Verlag, Berlin, 1998.

13. R. Cramer and V. Shoup. Signature Scheme based on the Strong RSA Assumption. In *Proc. of the 6th CCS*, pages 46–51. ACM Press, New York, 1999.

14. W. Diffie and M. E. Hellman. New Directions in Cryptography. *IEEE Transactions on Information Theory*, IT–22(6):644–654, November 1976.

15. D. Dolev, C. Dwork, and M. Naor. Non-Malleable Cryptography. *SIAM Journal on Computing*, 30(2):391–437, 2000.

16. T. El Gamal. A Public Key Cryptosystem and a Signature Scheme Based on Discrete Logarithms. *IEEE Transactions on Information Theory*, IT–31(4):469–472, July 1985.

17. A. Fiat and A. Shamir. How to Prove Yourself: Practical Solutions of Identification and Signature Problems. In *Crypto '86*, LNCS 263, pages 186–194. Springer-Verlag, Berlin, 1987.

18. E. Fujisaki and T. Okamoto. How to Enhance the Security of Public-Key Encryption at Minimum Cost. In *PKC '99*, LNCS 1560, pages 53–68. Springer-Verlag, Berlin, 1999.

19. E. Fujisaki and T. Okamoto. Secure Integration of Asymmetric and Symmetric Encryption Schemes. In *Crypto '99*, LNCS 1666, pages 537–554. Springer-Verlag, Berlin, 1999.

20. E. Fujisaki, T. Okamoto, D. Pointcheval, and J. Stern. RSA–OAEP is Secure under the RSA Assumption. In *Crypto '01*, LNCS 2139, pages 260–274. Springer-Verlag, Berlin, 2001.

21. C. Hall, I. Goldberg, and B. Schneier. Reaction Attacks Against Several Public-Key Cryptosystems. In *Proc. of ICICS'99*, LNCS, pages 2–12. Springer-Verlag, 1999.

22. S. Goldwasser and S. Micali. Probabilistic Encryption. *Journal of Computer and System Sciences*, 28:270–299, 1984.

23. S. Goldwasser, S. Micali, and C. Rackoff. The Knowledge Complexity of Interactive Proof Systems. In *Proc. of the 17th STOC*, pages 291–304. ACM Press, New York, 1985.

24. S. Goldwasser, S. Micali, and R. Rivest. A Digital Signature Scheme Secure Against Adaptative Chosen-Message Attacks. *SIAM Journal of Computing*, 17(2):281–308, April 1988.

25. KCDSA Task Force Team. The Korean Certificate-based Digital Signature Algorithm. August 1998. Available from http://grouper.ieee.org/groups/1363/.
26. A. K. Lenstra, H. W. Lenstra, and L. Lovász. Factoring Polynomials with Rational Coefficients. *Mathematische Annalen*, 261(4):515–534, 1982.
27. M. Naor and M. Yung. Public-Key Cryptosystems Provably Secure against Chosen Ciphertext Attacks. In *Proc. of the 22nd STOC*, pages 427–437. ACM Press, New York, 1990.
28. V. I. Nechaev. Complexity of a Determinate Algorithm for the Discrete Logarithm. *Mathematical Notes*, 55(2):165–172, 1994.
29. NIST. Digital Signature Standard (DSS). Federal Information Processing Standards PUBlication 186, November 1994.
30. NIST. Secure Hash Standard (SHS). Federal Information Processing Standards PUBlication 180–1, April 1995.
31. K. Ohta and T. Okamoto. On Concrete Security Treatment of Signatures Derived from Identification. In *Crypto '98*, LNCS 1462, pages 354–369. Springer-Verlag, Berlin, 1998.
32. T. Okamoto and D. Pointcheval. REACT: Rapid Enhanced-security Asymmetric Cryptosystem Transform. In *CT – RSA '01*, LNCS 2020, pages 159–175. Springer-Verlag, Berlin, 2001.
33. T. Okamoto and D. Pointcheval. The Gap-Problems: a New Class of Problems for the Security of Cryptographic Schemes. In *PKC '01*, LNCS 1992. Springer-Verlag, Berlin, 2001.
34. D. Pointcheval. Chosen-Ciphertext Security for any One-Way Cryptosystem. In *PKC '00*, LNCS 1751, pages 129–146. Springer-Verlag, Berlin, 2000.
35. D. Pointcheval and J. Stern. Security Arguments for Digital Signatures and Blind Signatures. *Journal of Cryptology*, 13(3):361–396, 2000.
36. C. Rackoff and D. R. Simon. Non-Interactive Zero-Knowledge Proof of Knowledge and Chosen Ciphertext Attack. In *Crypto '91*, LNCS 576, pages 433–444. Springer-Verlag, Berlin, 1992.
37. R. Rivest, A. Shamir, and L. Adleman. A Method for Obtaining Digital Signatures and Public Key Cryptosystems. *Communications of the ACM*, 21(2):120–126, February 1978.
38. C. P. Schnorr. Efficient Signature Generation by Smart Cards. *Journal of Cryptology*, 4(3):161–174, 1991.
39. V. Shoup. Lower Bounds for Discrete Logarithms and Related Problems. In *Eurocrypt '97*, LNCS 1233, pages 256–266. Springer-Verlag, Berlin, 1997.
40. V. Shoup. OAEP Reconsidered. In *Crypto '01*, LNCS 2139, pages 239–259. Springer-Verlag, Berlin, 2001.
41. Y. Tsiounis and M. Yung. On the Security of El Gamal based Encryption. In *PKC '98*, LNCS. Springer-Verlag, Berlin, 1998.
42. S. Vaudenay. Cryptanalysis of the Chor-Rivest Scheme. In *Crypto '98*, LNCS 1462, pages 243–256. Springer-Verlag, Berlin, 1998.

A New Cryptanalytic Method
Using the Distribution Characteristics
of Substitution Distances

Beomsik Song, Huaxiong Wang, and Jennifer Seberry

Centre for Computer Security Research
School of Information Technology and Computer Science
University of Wollongong
Wollongong 2522, Australia
{bs81,huaxiong,jennifer_seberry}@uow.edu.au

Abstract. In this paper, we suggest a new method for cryptanalysis of the basic structures of the block ciphers having SP network structure. The concept of the substitution difference is introduced and the distribution characteristics of substitution distances in an S-box is developed. This gives clues for cryptanalysis of the cipher. We then examine if this method is applicable to cryptanalysis of *Rijndael*. We present the method for cryptanalysis of the first round of *Rijndael* including the initial Round-Key addition part in order to illustrate our new method.

Keywords: Cryptanalysis, Substitution-Permutation(SP) Networks, Subtitution Distance, Rijndael.

1 Introduction

Cryptanalytic research has been used to promote the development of more secure ciphers. Some powerful cryptanalytic tools including DC (*Differential Cryptanalysis*) [3,4] and LC (*Linear Cryptanalysis*) [7] can be said to have played an important role in strengthening the design criteria for block ciphers. For this reason, we assert that these cryptanalytic methods have brought about the design of new block ciphers believed to be more secure than DES (*Data Encryption Standard*) [2]. However, it does not be necessarily follow that because these newer ciphers are designed to resist the existing cryptanalytic methods, they are permanently secure. This is because cryptanalytic methods are developing as well.

The purpose of this paper is to suggest a new method for cryptanalysis of the basic structures of the block ciphers having SP network structure. To do this, some basic concepts are defined to establish the main ideas, and the main concept of the Substitution Distance is established on the basis of these basic concepts. And after that, the method of applying this concept to cryptanalysis is introduced. This method consists of two parts, which are finding possible keys with a known plaintext, and the determination of the keys actually used from those possible keys with some additional chosen (or known) plaintexts.

K. Kim (Ed.): ICICS 2001, LNCS 2288, pp. 18–31, 2002.

Especially, this method is focused on how to choose plaintexts which are the most likely to select the keys actually used from possible keys.

We find out that suitable input differences between the known plaintext used for finding possible keys and additional chosen (or known) plaintexts used for the determination of the keys actually used can be obtained from the distribution characteristics of the values of substitution distances in the S-box of the cipher algorithm. This means that if we know the substitution distances in the S-box of a given cipher algorithm, in cryptanalysis of the cipher algorithm we can choose plaintexts having suitable input differences from the plaintext used for finding possible keys. It does not depend on the keys.

In the last part of this paper, we introduce the method for cryptanalysis of the first round of the AES algorithm ($Rijndael$), including the initial Round-Key addition part, in order to briefly illustrate this cryptanalytic method. The new method we establish is unlike previous cryptanalytic methods on $Rijndael$ [5,6].

2 Preliminaries

2.1 Definitions

Definition 1 (Sequence Difference, Input Difference, Input-Output Difference). *The sequence difference is defined as the bitwise XOR of two binary sequences X and X'. If X and X' are input sequences of an algorithm, we say $X \oplus X'$ is the input difference denoted by ΔX. If X is an input sequence of an algorithm, and Y is the corresponding output sequence, then we say that $X \oplus Y$ is the input-output difference denoted by $IOD(X, Y)$.*

Definition 2 (Substitution Difference). *Let a sequence X be an input sequence of a substitution table (S-box) or function, and the sequence $S(X)$ be the corresponding output sequence of the S-box. Then we say that the XOR of X and $S(X)$ is the substitution difference between X and $S(X)$ denoted by $SD(X, S(X))$. This means that*

$$SD(X, S(X)) = X \oplus S(X) = (x_0, x_1, x_2, \cdots, x_n) \oplus (s_0, s_1, s_2, \cdots, s_{n-1}).$$

Example 1. Suppose that substitution of each 3 bits is given by $(0\ 0\ 0 \rightarrow 1\ 1\ 0), (0\ 0\ 1 \rightarrow 0\ 1\ 0), (0\ 1\ 0 \rightarrow 1\ 1\ 1), (0\ 1\ 1 \rightarrow 0\ 0\ 0), (1\ 0\ 0 \rightarrow 1\ 0\ 1), (1\ 0\ 1 \rightarrow 0\ 0\ 1), (1\ 1\ 0 \rightarrow 1\ 0\ 0), (1\ 1\ 1 \rightarrow 0\ 1\ 1)$. Then we say that the substitution difference $SD((1\ 1\ 0), (1\ 0\ 0))$ between the input sequence $X = (1\ 1\ 0)$ and the corresponding output sequence $S(X) = (1\ 0\ 0)$ is

$$(1\ 1\ 0) \oplus (1\ 0\ 0) = (0\ 1\ 0).$$

The substitution differences are

$$(1\ 1\ 0), (0\ 1\ 1), (1\ 0\ 1), (0\ 1\ 1), (0\ 0\ 1), (1\ 0\ 0), (0\ 1\ 0), \text{and } (1\ 0\ 0) \tag{1}$$

in turn.

Definition 3 (Substitution Distance). *Let*

$$SD(X, S(X)) = X \oplus S(X) = (d_0, d_1, d_2, \cdots, d_{n-1})$$

be a substitution difference between the input sequence X and the corresponding output sequence $S(X)$. And let

$$SD(X', S(X')) = X' \oplus S(X') = (d'_0, d'_1, d'_2, \cdots, d'_{n-1})$$

be another substitution difference between the input sequence X' and the corresponding output sequence $S(X')$. Then the substitution distance between $SD(X, S(X))$ and $SD(X', S(X'))$ is defined as

$$SD(X, S(X)) \oplus SD(X', S(X'))$$

and denoted by

$$SDT[SD(X, S(X)), \ SD(X', S(X'))].$$

This means that

$$SDT[SD(X, S(X)), SD(X', S(X'))] = X \oplus S(X) \oplus X' \oplus S(X')$$
$$= (d_0, d_1, d_2, \cdots, d_{n-1}) \oplus (d'_0, d'_1, d'_2, \cdots, d'_{n-1}).$$

Example 2. Suppose that substitution differences are as described in (1). Then we say that the substitution distance $SDT[(1\ 1\ 0), (0\ 1\ 1)]$ between the first substitution difference $(1\ 1\ 0)$ and the second substitution difference $(0\ 1\ 1)$ is $(1\ 0\ 1)$. The substitution distances between the first substitution difference and other substitution differences are

$$(1\ 0\ 1), (0\ 1\ 1), (1\ 0\ 1), (1\ 1\ 1), (0\ 1\ 0), (1\ 0\ 0), \text{and } (0\ 1\ 0)$$

in turn.

2.2   Notation

Notations are defined as follows:

$$
\begin{array}{ll}
X = (x_0, x_1, x_2, \cdots, x_{n-1}) & : \text{a binary sequence.} \\
X' = (x'_0, x'_1, x'_2, \cdots, x'_{n-1}) & : \text{another binary sequence.} \\
\Delta X = X \oplus X' & : \text{the input difference between } X \text{ and } X'. \\
S(X) = (s_0, s_1, s_2, \cdots, s_{n-1}) & : \text{the output sequence, of the } S\text{-box, corresponding to } X. \\
S(X') = (s'_0, s'_1, s'_2, \cdots, s'_{n-1}) & : \text{the output sequence, of the } S\text{-box, corresponding to } X'. \\
SD(X, S(X)) = X \oplus S(X) & : \text{the substitution difference between } X \text{ and } S(X). \\
SD(X', S(X')) = X' \oplus S(X') & : \text{the substitution difference between } X' \text{ and } S(X'). \\
SDT[SD(X, S(X)), SD(X', S(X'))] & : \text{the substitution distance between } SD(X, S(X)) \text{ and } SD(X', S(X')). \\
Y = (y_0, y_1, y_2, \cdots, y_{n-1}) & : \text{the output sequence, of the algorithm, corresponding to } X. \\
Y' = (y'_0, y'_1, y'_2, \cdots, y'_{n-1}) & : \text{the output sequence, of the algorithm, corresponding to } X'. \\
IOD(X, Y) & : \text{the input-output difference between } X \text{ and } Y.
\end{array}
$$

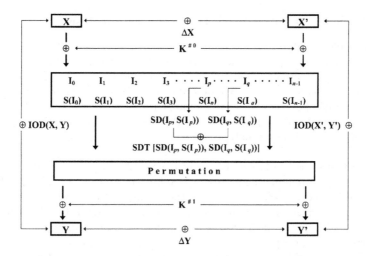

Fig. 1. The factors influencing the input-output difference.

2.3   Relationships between the Basic Concepts

In order to describe the *Substitution-Distance Cryptanalysis* clearly, the relationships between the basic concepts described above are shown in Figure 1 on the basic structure of the *SP*-network-structure block cipher algorithm. This structure will be used throughout this paper.

3   Principles of Substitution-Distance Cryptanalysis

3.1   Requirements for Substitution-Distance Cryptanalysis

In this paper, we call the structures like Figure 1 the basic structures of the *SP*-network-structure block cipher algorithms. The purpose of *Substitution-Distance Cryptanalysis* is to analyse the basic structures of *SP*-network-structure block cipher algorithms. In other words, it is the purpose of *Substitution-Distance Cryptanalysis* to find the keys used in the basic structures of *SP*-network block cipher algorithms consisting of the substitution, the permutation and the addition of round keys. In order to do this, a known pair of a plaintext and the corresponding ciphertext are used for finding possible keys, and some other chosen (or known) pairs of plaintext and corresponding ciphertext are used for the determination of the keys actually used from those possible keys. The number of plaintext-ciphertext pairs for finding keys actually used depends on the distribution characteristics of the values of the substitution distances in the *S*-box of the given cipher. Of course, carefully chosen pairs of a plaintext and the corresponding ciphertext will be more helpful than known pairs of a plaintext and the corresponding ciphertext to lighten the difficulty of cryptanalysis.

3.2   Principle of Substitution-Distance Cryptanalysis

The principle of *Substitution-Distance Cryptanalysis* consists of two parts, which are the finding of all possible keys with a known plaintext, and the determination of the keys actually used from those possible keys with some additional known (or chosen) plaintexts. We describe how to choose plaintexts having the suitable input differences with the plaintext used to find the possible keys. For the purpose of this principle, we ignore the step of the permutation because it has no effect on this cryptanalytic method.

Finding the Possible Keys. For a given algorithm like Figure 1, if a plaintext X and the corresponding ciphertext Y are known, we can find all possible key pairs of $K_i^{\#0}$ and $K_i^{\#1}$ ($0 \leq i < n$, n is the number of the substitution paths in the S-box) to transform X into Y, by the following method. Let the number of the substitution paths in the S-box of the algorithm be n, an input be X, and the corresponding output be Y. Then, if we suppose that the substitution path $(I_e, S(I_e))$ in the S-box is used for the encryption, we can make the formulas

$$I_e = X \oplus K_e^{\#0} \tag{2}$$

and

$$Y = S(I_e) \oplus K_e^{\#1}. \tag{3}$$

Also, because we know the values of X, Y, I_e, and $S(I_e)$, we can obtain $K_e^{\#0}$ and $K_e^{\#1}$ from (2) and (3). Now, if we consider all possible substitution paths from $(I_0, S(I_0))$ to $(I_{n-1}, S(I_{n-1}))$, we can obtain all possible key pairs of $K_i^{\#0}$ and $K_i^{\#1}$ ($0 \leq i < n$) to transform X into Y. The number of all possible key pairs of $K_i^{\#0}$ and $K_i^{\#1}$ will be equal to the number of substitution paths in the S-box if the S-box uses a one-to-one correspondence. For this reason, we can describe the set of all possible key pairs

$$K_{set} = \{(K_i^{\#0}, K_i^{\#1}) | I_i = X \oplus K_i^{\#0}, Y = S(I_i) \oplus K_i^{\#1}\},$$

where $S_{set} = \{(I_i, S(I_i)) | 0 \leq i < n\}$. The cardinality of the K_{set} is equal to that of S_{set}, that is

$$|K_{set}| = |S_{set}|.$$

Determination of the Keys Actually Used. Now, we have to select the key pair actually used of $K_r^{\#0}$ and $K_r^{\#1}$ from K_{set}, the set of the possible key pairs. In order to do this we need more clues. One method may be to use another known pair of a plaintext and the corresponding ciphertext. This is because if we know another pair of a plaintext X' and the corresponding ciphertext Y', we can obtain not only another set of possible key pairs

$$K'_{set} = \{(K_i'^{\#0}, K_i'^{\#1}) | I_i = X' \oplus K_i'^{\#0}, Y' = S(I_i) \oplus K_i'^{\#1}\}$$

but also the intersection

$$PK_{set} = \{(K_j^{\#0}, K_j^{\#1})|(K_j^{\#0}, K_j^{\#1}) \in K_{set} \cap K'_{set}\}$$

of sets K_{set} and K'_{set}. It is obvious that $|PK_{set}| \leq |K_{set}|$ and $|PK_{set}| \leq |K'_{set}|$. This means that the number of possible key pairs can be reduced by the consideration of another known pair of a plaintext and the corresponding ciphertext. However, because the number of elements of PK_{set} must be at least 2 by Theorem 1, that is $|PK_{set}| \geq 2$, we can not perfectly determine the key pair actually used even now. This is because we can only select the key pair actually used when the cardinality of PK_{set} is 1.

Theorem 1. *For the basic structure of the SP-network-structure block cipher algorithm, let X and X' be encrypted with the key pair $K_r^{\#0}$ and $K_r^{\#1}$, giving outputs Y and Y' respectively. Then there must be another key pair $K_r'^{\#0}$ and $K_r'^{\#1}$ which gives the same output.*

Proof. Let the set consisting of all n substitution paths in the S-box be S_{set}, that is,

$$S_{set} = \{(I_i, S(I_i))|0 \leq i < n\}.$$

Then the set consisting of all possible key pairs to transform X into Y can be described by

$$K_{set} = \{(K_i^{\#0}, \ K_i^{\#1})|I_i = X \oplus K_i^{\#0}, \ Y = S(I_i) \oplus K_i^{\#1}\},$$

and another set consisting of all possible key pairs to transform X' into Y' can be described by

$$K'_{set} = \{(K_i'^{\#0}, \ K_i'^{\#1})|I_i = X' \oplus K_i'^{\#0}, \ Y' = S(I_i) \oplus K_i'^{\#1}\}.$$

Now, let the key pair $(K_r^{\#0}, K_r^{\#1})$ from K_{set} be the key pair actually used to transform X into Y and X' into Y'. Then the path $(I_r, S(I_r))$ and the path $(I_r \oplus \Delta X, S(I_r \oplus \Delta X))$ in the S-box have to be separately used (ΔX is an input difference between X and X' : $\Delta X = X \oplus X'$). On the other hand, it is obvious that among the elements of K'_{set} there is already the key pair $(K_r'^{\#0}, K_r'^{\#1})$ to transform X' into Y' with the substitution path $(I_r, S(I_r))$. This key pair can also transform the input X into the output Y with the substitution path $(I_r \oplus \Delta X, S(I_r \oplus \Delta X))$. After all, both of these key pairs $(K_r^{\#0}, K_r^{\#1})$ and $(K_r'^{\#0}, K_r'^{\#1})$ can transform not only X into Y, but also X' into Y'.

For this reason, it is necessary to use at least two known plaintexts to determine the key pair actually used from the possible key pairs to transform X into Y. In fact, in many cases, we may need many known plaintexts for the selection of the key pair actually used because we have to obtain the sets of $K'_{set}, K''_{set} \cdots$ to make the cardinality of the intersection $PK_{set}(K_{set} \cap K'_{set} \cap K''_{set} \cdots) = 1$.

However, it is not easy even to collect enough pairs of a plaintext and the corresponding ciphertext. Because of this, in order to minimize the effort in the determination of the key pair actually used we need to carefully choose the plaintexts, so that the cardinality of PK_{set} becomes 1 easily. Theorem 2 shows the relation between X', $|K_{set}|$, $|K'_{set}|$ and $|PK_{set}| = |K_{set} \cap K'_{set}|$.

Theorem 2. *For the basic structure of the SP-network-structure block cipher algorithm, let X be a plaintext, Y be the corresponding ciphertext,*

$$S_{set} = \{(I_i, S(I_i)) | 0 \leq i < n\}$$

be a set consisting of all n substitution paths in the S-box, and

$$K_{set} = \{(K_i^{\#0}, K_i^{\#1}) | I_i = X \oplus K_i^{\#0}, \ Y = S(I_i) \oplus K_i^{\#1}\}$$

be the set consisting of all possible key pairs to transform X into Y. And let another plaintext be X' and the corresponding ciphertext by each key pair in K_{set} be $Y_0', Y_1', Y_2', \cdots, Y_{n-1}'$ respectively. Then $|PK_{set}| = |K_{set} \cap K_{setp}'|$ is equal to the number of key pairs in K_{set} to transform X' into Y_p', where

$$K_{setp}' = \{(K_i'^{\#0}, K_i'^{\#1}) | I_i = X' \oplus K_i'^{\#0}, \ Y_p' = S(I_i) \oplus K_i'^{\#1}, \ and \ 0 \leq i < n\}.$$

Proof. Trivial by the definition of intersection.

As shown in Theorem 2, in order to make $|PK_{set}| = |K_{set} \cap K_{set}'|$ as small as possible, we need to use plaintexts which can be transformed into diverse ciphertexts by key pairs in K_{set} when we select the key pair actually used from the possible key pairs. In other words, we need to use plaintexts that may bring well distributed ciphertexts by key pairs in K_{set} (ciphertexts which are not concentrated on a certain value). These plaintexts can generate $K_{set}', K_{set}'', K_{set}''', \cdots$ to make $|PK_{set}|$ as small as possible. The next step is to find how to obtain these plaintexts which can be transformed into well distributed ciphertexts by the key pairs in K_{set}. Even if these plaintexts can be obtained from the direct computation for all plaintexts with the key pairs in K_{set}', we need to establish a more effective method of finding these plaintexts. Theorem 6 shows which plaintexts can be transformed into well distributed ciphertexts. This idea is based on the relation between the input differences of the S-box and the distribution characteristics of the substitution distances in the S-box of a given cipher algorithm. Theorems 3, 4, and 5 enable Theorem 6 to be established.

Theorem 3. *For the basic structure of the SP-network-structure block cipher algorithm, let X be a plaintext, Y be the corresponding ciphertext,*

$$S_{set} = \{(I_i, S(I_i)) | 0 \leq i < n\}$$

be the set consisting of all n substitution paths in the S-box, and

$$K_{set} = \{(K_i^{\#0}, K_i^{\#1}) | I_i = X \oplus K_i^{\#0}, \ Y = S(I_i) \oplus K_i^{\#1}\}$$

be the set consisting of all possible key pairs to transform X into Y. And let another plaintext, having the input difference ΔX with X, be X' and the corresponding ciphertexts for each key pair in K_{set} be $Y_0', Y_1', Y_2', \cdots, Y_{n-1}'$ respectively. Then

$$Y_p' = X' \oplus SD(I_p \oplus \Delta X, S(I_p \oplus \Delta X)) \oplus (K_p^{\#0}, K_p^{\#1}).$$

Proof. Because Input-Output Difference $IOD(X, Y) =$ Substitution Difference $SD(I, S(I)) \oplus$ Key Value $(K^{\#0}, K^{\#1})$ in this algorithm, we can describe

$$Y = X \oplus SD(I_p, S(I_p)) \oplus (K_p^{\#0}, K_p^{\#1}).$$

If there is another plaintext X' having the input difference ΔX with X, then $X' = X \oplus \Delta X$. Here, because X passes through the substitution path $(I_p, S(I_p))$ and is transformed into Y by XOR with a key pair $(K_p^{\#0}, K_p^{\#1})$, $X'(= X \oplus \Delta X)$ has to pass through the substitution path $(I_p \oplus \Delta X, S(I_p \oplus \Delta X))$ and Y_p' is equal to $X' \oplus SD(I_p \oplus \Delta X, S(I_p \oplus \Delta X)) \oplus (K_p^{\#0}, K_p^{\#1})$ by XOR with the same key pair $(K_p^{\#0}, K_p^{\#1})$.

Theorem 4. *In Theorem 3,*

$$\begin{aligned} Y_p' &= X' \oplus SD(I_p, S(I_p)) \oplus SDT[SD(I_p, S(I_p)), SD(I_p \oplus \Delta X, S(I_p \oplus \Delta X))] \\ &\oplus (K_p^{\#0}, K_p^{\#1}). \end{aligned}$$

Proof. As $SDT[SD(I_p, S(I_p)), SD(I_p \oplus \Delta X, S(I_p \oplus \Delta X))] = SD(I_p, S(I_p)) \oplus SD(I_p \oplus \Delta X, S(I_p \oplus \Delta X))$ by Definition 3, it is trivial.

Theorem 5. *For the basic structure of the SP-network-structure block cipher algorithm, let X be a plaintext, Y be the corresponding ciphertext,*

$$S_{set} = \{(I_i, S(I_i)) | 0 \leq i < n\}$$

be the set consisting of all n substitution paths in the S-box, and

$$K_{set} = \{(K_i^{\#0}, K_i^{\#1}) | I_i = X \oplus K_i^{\#0}, Y = S(I_i) \oplus K_i^{\#1}\}$$

be the set consisting of all possible key pairs to transform X into Y. And let another plaintext having the input difference ΔX with X be X' and the corresponding ciphertexts by each key pair in K_{set} be $Y_0', Y_1', Y_2', \cdots, Y_{n-1}'$ respectively. Then, if

$$\begin{aligned} SDT[SD(I_p, S(I_p)), SD(I_p \oplus \Delta X, S(I_p \oplus \Delta X))] \\ = SDT[SD(I_q, S(I_q)), SD(I_q \oplus \Delta X, S(I_q \oplus \Delta X))], \end{aligned}$$

$$Y_p' = Y_q'.$$

Proof. By Theorem 4

$$\begin{aligned} Y_p' &= X' \oplus SD(I_p, S(I_p)) \oplus SDT[SD(I_p, S(I_p)), SD(I_p \oplus \Delta X, S(I_p \oplus \Delta X))] \\ &\oplus (K_p^{\#0}, K_p^{\#1}) \end{aligned}$$

and

$$\begin{aligned} Y_q' &= X' \oplus SD(I_q, S(I_q)) \oplus SDT[SD(I_q, S(I_q)), SD(I_q \oplus \Delta X, S(I_q \oplus \Delta X))] \\ &\oplus (K_q^{\#0}, K_q^{\#1}). \end{aligned}$$

Also, $SD(I_p, S(I_p)) \oplus (K_p^{\#0}, K_p^{\#1}) = SD(I_q, S(I_q)) \oplus (K_q^{\#0}, K_q^{\#1})$ in the above two formulas because both parts are the input-output difference between X and Y. Therefore, if

$$SDT[SD(I_p, S(I_p)), SD(I_p \oplus \Delta X, S(I_p \oplus \Delta X))]$$
$$= SDT[SD(I_q, S(I_q)), SD(I_q \oplus \Delta X, S(I_q \oplus \Delta X))],$$

$$Y_p' = Y_q'.$$

Theorem 6. *For the basic structure of the SP-network-structure block cipher algorithm, let X be a plaintext, Y be the corresponding ciphertext,*

$$S_{set} = \{(I_i, S(I_i))|0 \le i < n\}$$

be the set consisting of all n substitution paths in the S-box, and

$$K_{set} = \{(K_i^{\#0}, K_i^{\#1})|I_i = X \oplus K_i^{\#0}, Y = S(I_i) \oplus K_i^{\#1}\}$$

be the set consisting of all possible key pairs to transform X into Y. And let another plaintext, having the input difference ΔX with X, be X' and the corresponding ciphertexts by each key pair in K_{set} be $Y_0', Y_1', Y_2', \cdots, Y_{n-1}'$ respectively. Then, the distribution characteristics of the ciphertexts $Y_0', Y_1', Y_2', \cdots, Y_{n-1}'$ is equivalent to that of

$$SDT[SD(I_i, S(I_i)), \ SD(I_i \oplus \Delta X, S(I_i \oplus \Delta X))], where \ 0 \le i < n.$$

Proof. Trivial by Theorem 5.

As shown in Theorems 3, 4, 5 and 6, if we know ΔXs which make the substitution distances $SDT[SD(I_i, S(I_i)), SD(I_i \oplus \Delta X, S(I_i \oplus \Delta X)]$ in a S-box as well distributed as possible, where $0 \le i < n$, we also obtain the plaintexts which can be transformed into well distributed ciphertexts for all key pairs in K_{set}. Because it does not depend on keys to choose ΔXs, we can obtain these off-line without any restriction of time. Hence, we can know which plaintexts have to be additionally chosen in the determination of the key pair really used, when we have a pair of a plaintext and the corresponding ciphertext used for finding the possible key pairs.

This means that we can choose the plaintexts to make $|PK_{set}| = |K_{set} \cap K'_{set}|$ smallest by the distribution characteristics of substitution distances, and reduce the number of additional plaintexts that we need for the selection of the key pair actually used. Figure 2 is a pseudo code showing the above idea of selecting ΔXs in the S-box substituting 8-bit inputs into 8-bit outputs.

4   Applying to *Rijndael* in Standard Case

4.1   Outline of *Rijndael* in Standard Case

As figure 3 shows, *Rijndael* has SP network structure and processes data blocks of 128bits in the standard case [1,5]. The bytewise substitution by the ByteSub

```
for(ΔX=0x01; ΔX≤0xff; ΔX⁺⁺)  {
   for (t=0; t<256 ;t⁺⁺ ) count[t]=0x00;
   for (i=0; i<256 ;i⁺⁺ ){
      t=SDT [SD(Iᵢ,S(Iᵢ)), SD((Iᵢ⊕ΔX), S(Iᵢ⊕ΔX)];
      count[t]= count[t]+1;
                        }
   value[ΔX]=the greatest count[t];
                        }

arrange ΔXs in increasing order of value[ ];

select the ΔX to make the number of  value[ΔX]
smallest;
```

Fig. 2. Finding the most suitable input differences ΔXs.

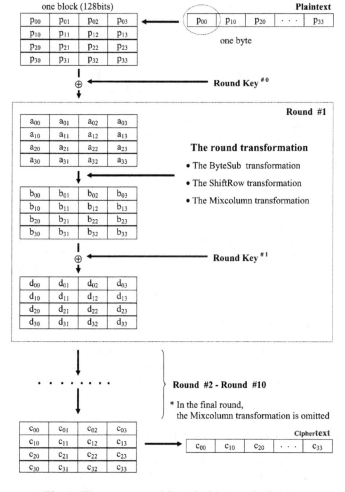

Fig. 3. The structure of *Rijndael* in standard case.

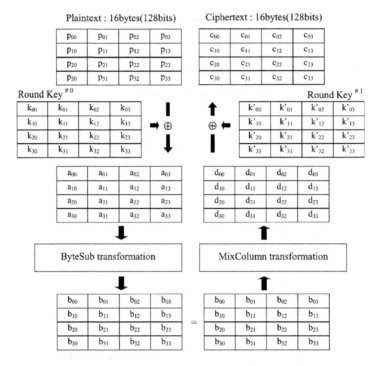

Fig. 4. The first round of *Rijndael* including the initial Round-Key addition.

transformation, the cyclic shift of the four bytes in each row by the ShiftRow transformation, and the mix of the four bytes in each column by the MixColumn transformation are performed every round. After these operations, a 128-bit round key extended from an user key is XORed in the last part of every round. The MixColumn transformation is omitted in the last round, but before the first round a 128-bit round key is XORed.

4.2   Structure of the First Round of *Rijndael*

Figure 4 shows the first round of *Rijndael* including the initial Round-Key addition part. Because the ShiftRow transformation in *Rijndael* has no effect on the cryptanalytic method suggested in this paper, that operation is omitted in Figure 4.

4.3   Finding *Round Key*$^{\#0}$ and *Round Key*$^{\#1}$

Preparations. To find *Round Key*$^{\#0}$ and *Round Key*$^{\#1}$ start with finding the substitution differences in the *S*-box. In the case of *Rijndael*, the *S*-box has a non-linear byte substitution. The *S*-box consists of 256 substitution paths,

$$
\begin{pmatrix} d_{0c} \\ d_{1c} \\ d_{2c} \\ d_{3c} \end{pmatrix} = \begin{pmatrix} 02 & 03 & 01 & 01 \\ 01 & 02 & 03 & 01 \\ 01 & 01 & 02 & 03 \\ 03 & 01 & 01 & 02 \end{pmatrix} \begin{pmatrix} b_{0c} \\ b_{1c} \\ b_{2c} \\ b_{3c} \end{pmatrix}
$$

Fig. 5. The MixColumn transformation of *Rijndael*.

and this S-box substitutes the value of each input byte into a new value (*eg* : $0x46 \rightarrow 0x5a$) every round through the ByteSub transformation. However, after the substitution, because all four bytes in each column are mixed affecting each other by the matrix formula in Figure 5 through the MixColumn transformation, all four bytes in each column need to be considered together in finding the substitution characteristics of the S-box.

In other words, we need to modify the round transformation properly to obtain the four byte substitution differences and the substitution distances between these substitution differences. Because we know both the subtitution differences in the S-box used in the ByteSub transformation and the matrix formula used in the MixColumn transformation, we can obtain the outputs of these operations for all sorts of four byte inputs from $(0x00, 0x00, 0x00, 0x00)$ to $(0xff, 0xff, 0xff, 0xff)$. Now, the new S-box having all $(256)^4$ substitution paths can be obtained. This new S-box can replace the above two operations because the substitution effect of this new S-box and the substitution effect of the above two operations are the same for all four byte inputs. As shown above, it will be helpful in the application of *Substitution-Distance Cryptanalysis* to change complicated round transformations in a given algorithm into a simple S-box. The next step is to find the ΔXs to obtain the well distributed values of the substitution distances in the newly generated S-box. This is because if we know these ΔXs, by Theorem 6 we can choose the plaintexts to be used for the determination of the key pair actually used when we obtain the possible key pairs with a known plaintext.

Figure 6 is a pseudo code to describe this step briefly.

Finding Possible Key Pairs of *Round Key*$^{\#0}$ and *Round Key*$^{\#1}$. In order to simplify the discussion, the method to find the first column (4 bytes) in *Round Key*$^{\#0}$ and the first column (4 bytes) in *Round Key*$^{\#1}$ is described first. In other words, the method of finding $(k_{00}, k_{10}, k_{20}, k_{30})$ and $(k'_{00}, k'_{10}, k'_{20}, k'_{30})$ is explained. Let a pair comprising $X = (p_{00}, p_{10}, p_{20}, p_{30})$ and the corresponding ciphertext $Y = (c_{00}, c_{10}, c_{20}, c_{30})$ be known as in Figure 4. Then, by the method referred in 3.2 we can obtain all $(256)^4$ possible key pairs of $(k_{00}, k_{10}, k_{20}, k_{30})$ and $(k'_{00}, k'_{10}, k'_{20}, k'_{30})$ by considering all $(256)^4$ possible substitution paths from I_i to $S(I_i)$ (I_i is $0x00000000$ to $0xffffffff$). The result is the set

$$
K_{set} = \{(K_i^{\#0}, K_i^{\#1}) | I_i = X \oplus K_i^{\#0}, \ Y = S(I_i) \oplus K_i^{\#1} \text{and } 0 \leq i < 256^4\}.
$$

for(ΔX=0x00000001; ΔX$\leq$0xffffffff; ΔX^{++}){

  for (t=0; t<256^4 ;t^{++}) count[t]=0x00000000;
  for (i=0; i<256^4 ;i^{++}){
    t=SDT [SD(I$_i$,S(I$_i$)), SD((I$_i\oplus\Delta$X), S(I$_i\oplus\Delta$X)];
    count[t]= count[t]+1;
            }
  value[ΔX]=the greatest count[t];

                  }
  arrange ΔXs in increasing order of value[ ];
  select the ΔX to make the number of value[ΔX]
  smallest;

Fig. 6. Finding the most suitable input differences ΔXs in *Rijndael*.

Determining the Key Pairs Actually Used of $Round\,Key^{\#0}$ and $Round\,Key^{\#1}$.

As mentioned above, we can find the set K_{set} consisting of all possible key pairs of the first column in *Round Key$^{\#0}$*and the first column in *Round Key$^{\#1}$*. Now, we have to determine the key pair actually used from those 256^4 key pairs. In order to do this, we need to use some more known (or chosen) plaintexts. This is because if we know these plaintexts we can select the key pair actually used from the intersection of $K_{set}, K'_{set}, K''_{set} \cdots$ (these sets can be made by the step of finding all possible keys). If we can obtain known plaintexts having the input differences ΔXs which yield the well distributed values of

$$SDT[SD(I_i, S(I_i)), SD(I_i \oplus \Delta X, S(I_i \oplus \Delta X))], \text{where } 0 \leq i < 256^4,$$

we can determine the key pair actually used with fewer plaintexts by Theorems 2 and 6. In this case, if we have plaintexts, which ensure $|K_{set} \cap K'_{set} \cap K''_{set} \cdots|=1$, we may determine the key pair actually used. On the other hand, if we do not have known plaintexts satisfing the above condition, we need to obtain chosen plaintexts, which have the input differences ΔXs with the known plaintext used for finding possible key pairs. These plaintexts are just the plaintexts to yield the well distributed values of

$$SDT[SD(I_i, S(I_i)), SD(I_i \oplus \Delta X, S(I_i \oplus \Delta X))], \text{where } 0 \leq i < 256^4.$$

If we know the most suitable ΔXs from the step of the Preparation, we can choose the plaintexts. And then, we can finally find the key pair actually used by the intersection of $K_{set} \cap K'_{set} \cap K''_{set} \cdots$ when $|K_{set} \cap K'_{set} \cap K''_{set} \cdots| = 1$. Remember that the plaintexts having the suitable input differences ΔXs with the plaintext X used for finding possible key pairs have to be chosen. Otherwise, we may need many more chosen plaintexts than minium. At the moment, because we have not tried to look over the substitution distances and to inspect the suitable input differences ΔXs, we can not know exactly the number of plaintexts we need for the selection of the key pair actually used. However, if we find out these characteristics, we may know the number of plaintexts actually needed

Applying to All Columns. As shown previously, we can select the key pairs actually used in the first columns in $Round\ Key^{\#0}$ and $Round\ Key^{\#1}$. By the application of the same method to the second column, the third column and the last column, we can find all the columns in $Round\ Key^{\#0}$ and $Round\ Key^{\#1}$ (256 bits). This means that we can find the $Round\ Key^{\#0}$ and $Round\ Key^{\#1}$ by considering 4×256^4 possible key pairs of 4 bytes.

5   Conclusions

We have shown that the Substitution Distances between the substitution paths in the S-box can be used as clues for cryptanalysis of the block ciphers having SP network structure. Specifically, we show how to find all possible keys with a known plaintext on the basic structure of the SP network block cipher algorithms. We explain how to select the key actually used from these possible keys by the use of the distribution characteristics of the values of the substitution distances in the S-box of the algorithm. In other words, we show which input differences ΔXs are the most suitable in the selection of the plaintexts to be used for the determination of the keys actually used. Along with this, we introduce the possibility of applying these principles to cryptananalysis of the first round of *Rijndael* in order to briefly illustrate this cryptanalytic method. As a result of our efforts, we find out that *Substitution-Distance Cryptanalysis* can be applied for cryptanalysis of the basic structure of SP-network-structure block cipher algorithms. Remember that if we know the most suitable input differences (ΔXs : they do not depend on keys), we may find the keys used for the encryption on the basic round of the SP-network-structure block cipher algorithms. For this reason, it is useful for further study to find suitable ΔXs for the SP-network-structure block cipher algorithms including *Rijndael*.

References

1. "Advanced Encryption Standard(AES)", FIPS-Pub. ZZZ, NIST, http://csrc.nist.gov/publications/drafts/dfips-AES.pdf, 2001.
2. "Data Encryption Standard", FIPS 46-2, NIST, 1993.
3. E. Biham and A. Shamir, "Differential cryptanalysis of DES-like cryptosystems", *J. Cryptology*, Vol.4, pp.3-72, 1991.
4. E. Biham and A. Shamir, "Differential cryptanalysis of the full 16-round DES", *Advances in Cryptology-Crypto'92*, Lecture Notes in Computer Science, Springer-Verlag, pp.487-496, 1992.
5. J. Daemen and V. Rijmen, "*AES* Proposal: Rijndael", http:// csrc.nist.gov/encryption/aes/rijndael/Rijndael.pdf, 1999.
6. N. Ferguson, et al, "Improved Cryptanalysis of Rijndael", *the Fast Software Encryption Workshop '2000*, Preproceeding, 2000.
7. M. Matsui, "Linear cryptanalysis method for DES cipher", *Advances in Cryptology -Eurocrypt'93*, Lecture Notes in Computer Science, Springer-Verlag, pp.386-397, 1993.

Truncated Differential Cryptanalysis of Camellia

Seonhee Lee[1], Seokhie Hong[1], Sangjin Lee[1], Jongin Lim[1], and Seonhee Yoon[2]

[1] Center for Information Security Technologies(CIST),
Korea University, Anam Dong, Sungbuk Gu,
Seoul, Korea
{smilesun,hsh,sangjin,jilim}@cist.korea.ac.kr
[2] Korea Information Security Agency(KISA),
5th FL., Dong-A Tower, 1321-6, Seocho-Dong, Seocho-Gu,
Seoul 137-070, Korea
shyoon@kisa.or.kr

Abstract. Camellia is a block cipher cooperatively designed by NTT and Mitsubshi Electric Corporation and submitted to NESSIE. In this paper, we present truncated differential cryptanalysis of modified Camellia reduced to 7 and 8 rounds. For modified Camellia with 7 rounds we can find 8-bit key with $3 \cdot 2^{81}$ plaintexts and for modified Camellia with 8 rounds we can find 16-bit key with $3 \cdot 2^{82}$ plaintexts.

1 Introduction

Camellia is a 128-bit Block Cipher designed by NTT and Mitsubshi Electric Corporation, which can be seen as a modified version of E2 [2]. E2 was analyzed by truncated differential cryptanalysis by Matsui et al.[7]. In [7] they used "cyclic byte characteristic" whose input pattern is the same as output pattern, which came from the flaw of byte permutation of p-function. The SPS-structure in round function of E2 is not more secure than SP-structure in the viewpoint of truncated differential cryptanalysis and that it has bad influence on speeding up. To supplement these flaws Aoki et al. changed SPS-structure into SP-structure, amended byte permutation of p-function and added FL/FL^{-1}-function every six round except the last round so that Camellia has no "cyclic byte characteristic" with high probability. In [5] Knudsen showed that the first seven rounds of Camellia without the functions FL/FL^{-1} can be distinguished from a random permutation using about 2^{68} chosen plaintexts using cyclic byte characteristic with low probability. Since p-function of Camellia has a branch number 5, we focused on finding byte characteristic of round function, which can be made to be iterated byte characteristic. Four "byte characteristics" was found with nonzero three bytes of input difference and nonzero two bytes of output difference. Using a characteristic out of them we can attack on 7 rounds and 8 rounds Camellia without FL/FL^{-1}-function.

This paper is organized as follows. In section 2, we will briefly explain the structure of Camellia. Truncated differential cryptanalysis of 7 rounds Camellia without FL/FL^{-1}-function and 8 rounds Camellia without FL/FL^{-1}-function will be shown in section 3 and 4, respectively.

K. Kim (Ed.): ICICS 2001, LNCS 2288, pp. 32–38, 2002.
© Springer-Verlag Berlin Heidelberg 2002

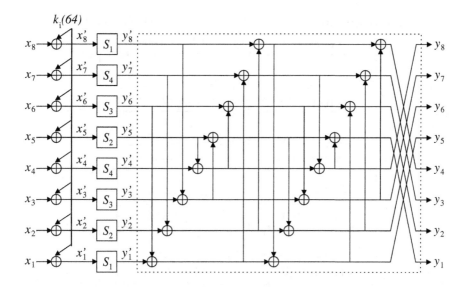

Fig. 1. The round function of Camellia

2   Preliminaries and The F-Function of Camellia

In this section, we briefly describe Camellia and define some notations. For more details, see [1,8]. Camellia is a 128-bit block cipher with 128-bit, 192-bit and 256-bit key. Considering Camellia with only 128-bit key, it has $18(=3 \times 6)$ round Feistel structure with FL/FL^{-1}-function every six round except the last round and it has prewhitening and postwhitening process in which plaintext and ciphertext are XORed with each 128-bit key. But we will consider Camellia without FL/FL^{-1}-function and call it modified Camellia. Figure 1 shows the round function of Camellia. In the round function F, four s-boxes, namely s_1-box, s_2-box, s_3-box, s_4-box, are used. Since Camellia is byte-oriented block cipher, we will denote non-zero one byte difference by "1", zero one byte difference by "0" and fixed non-zero one byte difference by a, b, c and d and non-fixed non-zero one byte difference by α and β.

3   Attack on Camellia Reduced to 7 Rounds

In this section we describe truncated differential cryptanalysis of modified Camellia reduced to 7 rounds. In Figure 1, p-function can be represented by 8×8 binary matrix with branch number 5, so we focused on finding byte characteristic of round function with nonzero three bytes of input difference and nonzero two bytes of output difference, which can be made to be iterated byte characteristic. We found four "byte characteristics" as followings;

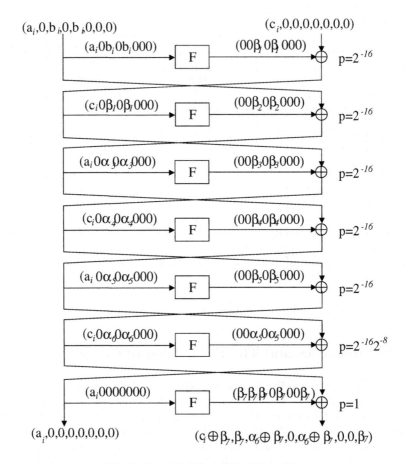

Fig. 2. 7-round differential of Camellia

$$(1\ 0\ 1\ 0\ 1\ 0\ 0\ 0) \rightarrow (0\ 0\ 1\ 0\ 1\ 0\ 0\ 0), \quad p = 2^{-16} \tag{1}$$

$$(1\ 0\ 1\ 0\ 0\ 0\ 1\ 0) \rightarrow (1\ 0\ 0\ 0\ 0\ 0\ 1\ 0), \quad p = 2^{-16} \tag{2}$$

$$(0\ 1\ 0\ 1\ 0\ 1\ 0\ 0) \rightarrow (0\ 0\ 0\ 1\ 0\ 1\ 0\ 0), \quad p = 2^{-16} \tag{3}$$

$$(0\ 1\ 0\ 1\ 0\ 0\ 0\ 1) \rightarrow (0\ 1\ 0\ 0\ 0\ 0\ 0\ 1), \quad p = 2^{-16} \tag{4}$$

A "byte characteristic" (1) will take place only when $\triangle y'_1 = \triangle y'_3 = \triangle y'_5$. Here $\triangle y'_i$ means output difference of s-box. Thus characteristic (1) happens with probability $(\frac{1}{255})^2$. For simplicity, we will regard this probability as 2^{-16}. We use a round characteristic (1) to make a 7-round characteristic.

Consider a characteristic as like Figure 2. In Figure 2, α_i and β_i are non-fixed values. To occur a cancellation after 6th round, the probability of 6th round characteristic goes up by 2^{-8} comparing other round probability so becomes 2^{-24}. Consequently, seven rounds characteristic hold with approximate probability $(2^{-16})^5 \cdot 2^{-24} = 2^{-104}$. We will find the first byte of 7th round key using this byte characteristic.

To begin with let's compute signal to noise (S/N) of this characteristic. If the round function is random permutation, the same output difference is expected to appear with probability $(2^{-8})^{14} = 2^{-112}$. Thus S/N becomes

$$\frac{2^8 \cdot 2^{-104}}{1 \cdot (2^{-104} + 2^{-112})} \approx 2^8 > 1.$$

Let $\mathbb{P}_i = (P_i, P_i')$ be a plaintext pair with plaintext difference $P_i \oplus P_i' = (a_i, 0, b_i, 0, b_i, 0, 0, 0, c_i, 0, 0, 0, 0, 0, 0, 0)$ for some a_i, b_i and $c_i \in Z_2^8$, $\widetilde{\mathbb{P}}_i = (P_i', P_i)$ and C_i and C_i' be the ciphertext of P_i and P_i', respectively. Let $n = 3 \times 2^{104}$ and define

$$S = \{\mathbb{P}_i(i = 1, \cdots, n) | \mathbb{P}_i \neq \mathbb{P}_j \text{ and } \mathbb{P}_i \neq \widetilde{\mathbb{P}}_j \text{ for all } i \neq j\}.$$

From definition of S we can expect to obtain three plaintext pairs, namely, $\mathbb{P}_i, (i = 1, 2, 3)$, whose output differences are of the form $(a_i, 0, 0, 0, 0, 0, 0, 0, c_i \oplus \beta_7, \beta_7, \alpha_6 \oplus \beta_7, 0, \alpha_6 \oplus \beta_7, 0, 0, \beta_7), (i = 1, 2, 3)$, respectively, where α_6 and β_7 are non-fixed value. From each pairs we can guess the first byte of 7th round key by checking whether the first byte of output difference of 7th round equal to c_i exclusive-or the 9th byte of ciphertext difference. Let's consider the following algorithm.

- Input : A set S defined as above.
- Output : The first byte of 7th round key.
- Algorithm
  1. Set $T[i] = 0$ for $i = 0, \cdots, 255$.
  2. Find all pairs (P_i, P_i') such that $C_i \oplus C_i' = (a_i, 0, 0, 0, 0, 0, 0, 0, c_i \oplus \beta_7, \beta_7, \alpha_6 \oplus \beta_7, 0, \alpha_6 \oplus \beta_7, 0, 0, \beta_7)$.
  3. For each possible first byte of 7th round key, $k = 0, \cdots, 255$, do the followings;
     (a) Compute output difference of 7th round and check whether the first byte output difference of 7th round equal to c_i exclusive-or the 9th byte of ciphertext difference.
     (b) If so, increment $T[i]$ by 1.
  4. If $T[j] \geq 3$ for some j, then output j

In step 4 the right key must be counted at least three times and the probability that a wrong key is counted once is 2^{-8}, so we can say with high probability that a wrong key cannot be counted more two times. How many plaintexts are need to collect 3×2^{104} plaintext pairs? To answer this question consider a set T, T' where

$$T = \{(x_1, x_2, x_3, x_4) | x_i \in \{0, \cdots, 255\}\}$$

and

$$T' = \{(t_1, t_2) \in T \times T | t_1 \oplus t_2 = (a_i, b_i, b_i, c_i) \text{ for given } a_i, b_i \text{ and } c_i\}.$$

Then

$$|T| = 2^{32}, \ |T'| = 2^8 \cdot 2^8 \cdot 2^8 \cdot 2^8 = 2^{32}$$

for given a_i, b_i and c_i. But the possible number of a_i, b_i and c_i are $2^8 - 1 \approx 2^8$, respectively. Hence we can obtain $\frac{1}{2} \cdot 2^{24} \cdot 2^{32}$ different input pairs from 2^{32} inputs. So we need $3 \cdot 2^{81}$ plaintexts to collect 3×2^{104} input pairs. In order to compute time complexity, let's define one round encryption as one encryption. Since we need only one s-box encryption for each one byte key in step 3 (a), we have only $\frac{1}{8} \cdot (2 \cdot 3 \cdot 2^8) = 3 \cdot 2^6$ encryptions to compute step 3 (a) in previous algorithm. Consequently, we can get one byte of 7th round key with $3 \cdot 2^{81}$ plaintexts and $3 \cdot 2^6 = 192$ encryptions.

4   Attack on Camellia Reduced to 8 Rounds

In this section we will attack modified Camellia reduced to eight rounds by prepending one round structure to a 7-round characteristic. Consider a characteristic as like Figure 3 made by prepending one round structure to a 7-round characteristic. By using this byte characteristic we will find the first byte of initial round key and the first byte of 8th round key. Let $P_L{}^i$ and $P'_L{}^i$ be two 64-bit values so that $P_L^i \oplus P'_L{}^i = (a_i, 0, 0, 0, 0, 0, 0, 0)$ for some a_i, where $1 \le i \le 3 \cdot 2^{48}$. For each pair $(P_L^i, P'_L{}^i)$ and each possible first byte value of initial round key choose right pair $(P_{R,j}^i, P'{}_{R,j}^i)$ $(1 \le j \le 2^{56})$ so that output difference after one round is of form $(b_i, 0, c_j^i, 0, c_j^i, 0, 0, 0, a_i, 0, 0, 0, 0, 0, 0, 0)$. Thus we can collect 3×2^{104} plaintext pairs whose output difference after one round are of the form $(b_i, 0, c_j^i, 0, c_j^i, 0, 0, 0, a_i, 0, 0, 0, 0, 0, 0, 0)$ for each possible one byte key. So the previous algorithm in section 3 can be applied to find the first byte of 8th round key. If 3×2^{104} plaintext pairs are given, we can find the first byte key of 8th round for each one byte candidate key of initial round. Hence If $2 \cdot 3 \times 2^{104}$ plaintext pairs are given, we can find 16 bit (2 byte) key. By similar argument in section 3 we can obtain 2^{71} plaintext pairs from 2^{48} plaintexts. So we only need $3 \cdot 2^{81}$ plaintexts to collect 3×2^{104} plaintext pairs. In order to get 2 byte key, we need only $\frac{1}{8} \cdot (2 \cdot 3 \cdot 2^{48} \cdot 2^8) + \frac{1}{8} \cdot (2 \cdot 6 \cdot 2^8) = 3 \cdot (2^{54} + 2^7) \approx 3 \cdot 2^{54}$ encryptions. Therefore we can get two byte key - one byte of first round and one byte of 8th round - with $2 \cdot 3 \cdot 2^{81}$ plaintexts and about $3 \cdot 2^{54}$ encryptions.

5   Conclusion

In this paper, we studied truncated differential cryptanalysis for modified Camellia reduced to seven and eight rounds. For modified Camellia with 7 rounds we can find one byte key with $3 \cdot 2^{81}$ plaintexts and 192 encryptions and for modified Camellia with 8 rounds we can find two bytes key with $3 \cdot 2^{82}$ plaintexts and about $3 \cdot 2^{54}$ encryptions.

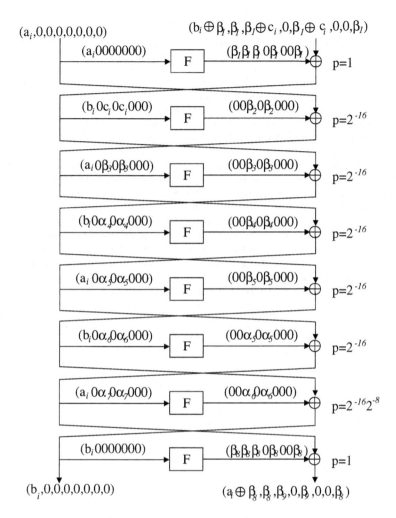

Fig. 3. 8 round differential of Camellia

Table 1. Complexity of reduced version of Camellia

	♯{*needed plaintexts*}	time complexity	♯{*required memory*}
7 round	$3 \cdot 2^{81}$	192	2^8
8 round	$2 \cdot 3 \cdot 2^{81}$	$3 \cdot 2^{54}$	2^{16}

References

1. K. Aoki and M. Matsui S. Moriai J. Nakajima T. Tokita. T. Ichikawa, M. Kanda. *Camellia – a 128-bit block cipher suitable for multiple platforms*(extended abstract), pages 39–56, 2000.
2. Aoki Ueda Takashima Ohta Matsumoto Kanda, Moriai. *E2 — a new 128-bit block cipher.* In IEICE Transactions Fundamentals of Electronics, Communications and Computer Sciences, volume E83-A, No. 1, pages 48–59, 2000.
3. M. Kanda and T. Matsumoto. *Security of camellia against truncated differential cryptanalysis.* In Fast Software Encryption - Eighth International Workshop, pre-proceedings FSE'2001.
4. L. R. Knudsen. *Truncated and higher order differential.* In B. Preneel, editor, Fast Software Encryption - Second International Workshop, volume 1008 of Lecture Notes in Computer Science, pages 196–211. Springer-Verlag, 1995.
5. L. R. Knudsen. *Analysis of camellia.* 2000.
6. L. R. Knudsen and T. A. Berson. *Truncated differentials of SAFER.* In D. Gollmann, editor, Fast Software Encryption – Third International Workshop, volume 1039 of Lecture Notes in Computer Science, pages 15–25. Springer-Verlag, 1996.
7. M. Matsui and T. Tokita. *Cryptanalysis of a reduced version of the block cipher e2.* In Fast Software Encryption – 6-th international workshop, FSE'99, Lecture Notes in Computer Science, pages 71–80. Springer-Verlag, 1999.
8. Mitsubishi. NTT. *Camellia - 128-bit block cipher,* 2000.

Improved Impossible Differential Cryptanalysis of Rijndael and Crypton

Jung Hee Cheon[1], MunJu Kim[2], Kwangjo Kim[1],
Jung-Yeun Lee[1], and SungWoo Kang[3]

[1] IRIS, Information and Communications University, Korea
{jhcheon,kkj,bushman}@icu.ac.kr
[2] Mathematics Department, Brown University, USA
mjkim@math.brown.edu
[3] Korea Information Security Agency, Korea
swkang@kisa.or.kr

Abstract. Impossible differential attacks against Rijndael and Crypton have been proposed up to 5-round. In this paper we expand the impossible differential attacks to 6-round. Although we use the same 4-round impossible differential as in five round attacks, we put this impossible differential in the middle of 6-round. That is, we will consider one round before the impossible differential and one more round after. The complexity of the proposed attack is bigger than that of the Square attack, but still less than that of the exhaustive search.

1 Introduction

The ciphers Rijndael [6] and Crypton [7] were submitted to the AES (Advanced Encryption Standard) candidates and Rijndael was later selected as the AES [1]. Both of them are based on the Square cipher [5] and so have SPN (Substitution-Permutation Network) structure. The original design of Square cipher concentrates on the resistance against differential and linear cryptanalysis. So it's known that two ciphers have the resistance against those attacks. Although these ciphers have those merits, they have a weakness which results from the characteristic of the optimal linear layer. The known attacks against each cipher, using this weakness, are impossible differential attack [3,9] and Square attack [3,4], which were described by the designers of the Square cipher. These attacks are chosen plaintext attacks and are independent of the specific choice of Sbox, the multiplication polynomial of MixColumn, and the key schedule. They are only related to the characteristic of the linear layer. That is the branch number which is introduced by the designers to explain the diffusion power.

Definition 1 (branch number).
Let $W(\cdot)$ be the byte weight function. The branch number of a linear transformation $F : \mathbb{Z}_2{}^8 \to \mathbb{Z}_2{}^8$ is

$$min_{a\neq 0, a\in\mathbb{Z}_2{}^8}(W(a) + W(F(a))).$$

K. Kim (Ed.): ICICS 2001, LNCS 2288, pp. 39–49, 2002.
© Springer-Verlag Berlin Heidelberg 2002

This branch number makes the following property for each cipher.

- Rijndael: Rijndael has a branch number 5, that is, if a state is applied with a single nonzero byte, the output has 4 nonzero bytes.
- Crypton: Crypton has a branch number 4 (designer refers this as diffusion order), namely, if a state is applied with a single nonzero byte, the output has 3 nonzero bytes

Next, each attack is described as the following:

1. Square attack
   Using the above property of branch number, we can deduce the characteristic for the relation of input and output of each cipher reduced to 4 rounds: if two plaintext differ by one byte then before the third MixColumn the data differ by all 16 bytes. It leads to the following interesting properties: Consider a set 256 plaintexts which are equal in all bytes except for one and in this one assume all the possible values. Because of the property the inputs of the third MixColumn assume all 256 possible values in each byte. So the XOR of them in each byte is 0. Since the MixColumn is a linear transformation, this property holds after the MixColumn too. This is the property on which Square attack is based. Here we guess one byte in the round key of the fourth round and decrypt the forth round in the corresponding byte in all of the 256 ciphertexts. We get the 256 inputs of the ByteSubstitution of the fourth round. If the key is right, then XOR of them is equal to 0. Through this way we can derive the forth round key.
   This attack was extended to Rijndael reduced to 5 and 6 rounds by adding one round in the beginning or in the end or both of them. At recent, this attack succeed to the 7 round Rijndael assuming the whole seventh round key.
2. Impossible differential attack
   Consider two plaintexts which differ by only one byte. Then, the corresponding ciphertexts of the 4-round variant should differ in the special combinations of bytes. We call those combinations impossible differential. This is the property on which impossible differential attack is based. The 5-round impossible differential attack is briefly introduced as the followings: Let's add one round at the end of the 4-round impossible differential and decrypt one round with assuming the fifth round key. Then, if there appears the impossible differential among them, that is a wrong key. Continuing this process we can find the fifth round key.

In this paper, we propose impossible differential attacks against each cipher reduced to six rounds. Our attacks are based on the four round impossible differentials each of which was used in the impossible differential attack against each cipher reduced to five rounds [3,9]. While the previous attacks have one additional round with the four round impossible differential, the proposed method has additional two rounds. In this method, we assume two round keys (the first round key and the last round key) and get rid of all wrong key pairs using the impossible differentials. The complexity of the proposed attack is larger than that of the Square attack against each cipher reduced to six rounds, but still less than that of the exhaustive attack. We expect that this method may be applicable to other ciphers with SPN structure.

The rest of the paper is organized as follows: The description and 5-round impossible differential attack of Rijndael and Crypton is given in section 2 and 3, respectively. In section 4 we conclude by summarizing the efficiency of our attack together with those of previous works.

2 Rijndael

2.1 Description of Rijndael

Rijndael is a block cipher. The length of the block and the length of the key can be specified to be 128, 192 or 256 bits, independently of each other. In this paper we discuss the variant with 128-bit blocks and 128-bit keys. In this variant, the cipher consists of 10 rounds. We represent 128-bit data in 4×4 matrix as in Fig. 1.

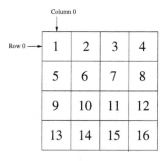

Fig. 1. Byte Coordinate of 128-bit

Every round except for the last consists of 4 transformation:

- *ByteSubstitution* is applied to each byte separately and is a nonlinear byte-wise substitution to use the Sbox.
- *ShiftRow* is a cyclic shift of the bytes of each row by 0, 1, 2, or 3, respectively.
- *MixColumn* is a linear transformation applied to columns of the matrix. The branch number of this layer is 5.
- *AddRoundKey* is a key XOR.

Before the first round AddRoundKey is performed using the key as the round key. In the last round the MixColumn is omitted.

Observe that MixColumn is a linear transformation over four bytes of input differences, since it is a linear transformation over four input bytes. If three bytes of output difference are zero, the choice of input differences is $2^8 - 1$ since MC is invertible. Hence the probability that output difference is zero at three bytes is $4 \times (2^8 - 1)/2^{32}$ since there are four choices on nonzero byte of output difference and the output difference is not zero for nonzero input difference.

Lemma 1. *The output of MC (or MC^{-1} transformation) has zero difference in three bytes with probability about 2^{-22} over all possible input pairs.*

This lemma holds even if some values of input differences for fixed bytes are restricted. This lemma will be used when analyzing the complexity of the proposed impossible differential attack.

2.2   Impossible Differential

We use the same impossible differential described in [3]. See Fig. 2.

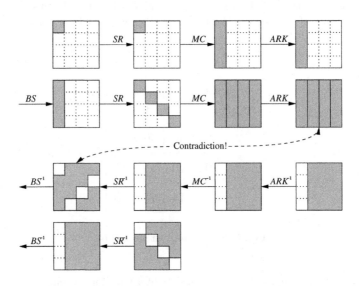

Fig. 2. Four Rounds Impossible Differential of Rijndael

Property 1 (Impossible Differential of Rijndael). Given plaintext pair which are equal at all bytes but one, the ciphertexts after 4-round cannot be equal in any of the following *prohibited* combinations of bytes: (1,6,11,16), (2,7,12,13), (3,8,9,14), nor (4,5,10,15).

This property follows from the property of MixColumn transformation: if two inputs of this transformation differ by one byte then the corresponding outputs differ by all the four bytes.

2.3   Rijndael Reduced to Six Rounds

In this subsection, we describe an impossible differential cryptanalysis of Rijndael reduced to six rounds. The attack is based on the four round impossible differential with additional one round at each of the beginning and the end as in Fig. 3. Note that the last round of Rijndael does not have MixColumn transformation before KeyAddition.

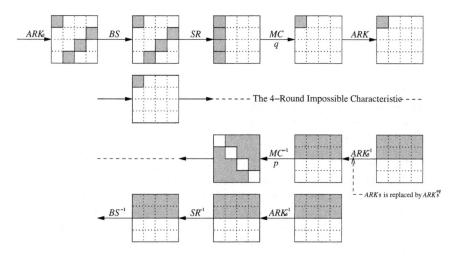

Fig. 3. Impossible Attack against Rijndael Reduced to Six Rounds

The procedure is as follow:

1. A structure is defined as a set of plaintexts which have certain fixed values in all but the four bytes (1,8,11,14). One structure consists of 2^{32} plaintexts and proposes $2^{32} \times 2^{32} \times \frac{1}{2} = 2^{63}$ pairs of plaintexts.
2. Take $2^{59.5}$ structures ($2^{91.5}$ plaintexts, $2^{122.5}$ plaintext pairs). Choose pairs whose ciphertext pairs have zero difference at the row 2 and 3. The expected number of such pairs is $2^{122.5} \times 2^{-64} = 2^{58.5}$.
3. Assume a 64-bit value at the row 0 and 1 of the last round key K_6.
4. For each ciphertext pair (C, C^*), compute $C_5 = BS^{-1} \circ SR^{-1}(C \oplus K_6)$ and $C_5^* = BS^{-1} \circ SR^{-1}(C^* \oplus K_6)$ and choose pairs whose difference $MC^{-1}(C_5 \oplus C_5^*)$ is zero at the prohibited four bytes (1,6,11,16), (2,7,12,13), (3,8,9,14) or (4,5,10,15) after the inverse of MC transformation. Since the probability is about $p = 2^{-32} \times 4 = 2^{-30}$, the expected number of the remaining pairs is $2^{58.5} \times 2^{-30} = 2^{28.5}$.
5. For a pair (P, P^*) with such ciphertext pairs and 32-bit value at the four bytes (1,8,11,14) of the initial key K_0, calculate

$$MC \circ SR(BS(P \oplus K_0) \oplus BS(P^* \oplus K_0))$$

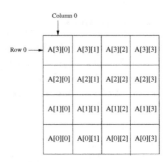

Fig. 4. Byte Coordinate of 128-bit

and choose pairs whose difference is zero except only one byte after MC transformation. The probability is about $q = 2^{-24} \times 4 = 2^{-22}$ since MC is linear for each byte of input values.

6. Since such a difference is impossible, every key that proposes such a difference is a wrong key. After analyzing $2^{28.5}$ ciphertext pairs, there remain only about $2^{32}(1 - 2^{-22})^{2^{28.5}} \approx 2^{32}e^{-2^{6.5}} \approx 2^{-98.5}$ wrong values of the four bytes of K_0.

7. Unless the initial assumption on the final round key K_6 is correct, it is expected that we can get rid of the whole 32-bit value of K_0 for each 64-bit value of K_6 since the wrong value (K_0, K_6) remains with the probability $2^{-34.5}$. Hence if there remains a value of K_0, we can assume the key K_6 is a right key. So if we repeat Step 2 through Step 5 after changing the row 2 and 3 into the row 0 and 1, we can get the whole value of K_6.

8. Step 4 requires about $2^{123.5} (= 2 \times 2^{64} \times 2^{58.5})$ one round operations. Step 5 requires about 2^{119} one round operations since

$$2^{64} \times 2 \times 2^{32}\{1 + (1 - 2^{-22}) + (1 - 2^{-22})^2 + \cdots + (1 - 2^{-22})^{2^{28.5}}\} \approx 2^{119}.$$

Consequently, since we repeat this procedure two times, this attack requires about $2^{91.5}$ chosen plaintexts and 2^{122} encryptions of Rijndael reduced to 6 round.

3   Crypton

3.1   Description of Crypton

Crypton is a 128-bit block cipher. We represent 128-bit data in 4×4 matrix as in Fig. 4. The component functions, σ, τ, π, and γ, are as follows.

- γ is a nonlinear byte-wise substitution. There are two versions of γ: γ_o is for odd rounds and γ_e is for even rounds.
- π is a linear bit permutation. It bit-wisely mixes each column (4 bytes). In fact, there are two versions of π: π_o in odd rounds and π_e in even rounds. One important fact is both versions have the branch number 4 as maps from 4-byte input to 4-byte output [7].

- τ is a linear transposition. It simply moves the byte at $A[i][j]$ to $A[j][i]$.
- σ is a key XOR. We will use notation σ_K when the given key is K.

The $2n$-round encryption of Crypton can be described as

$$\phi_e \circ \rho_{e_{K_{2n}}} \circ \rho_{o_{K_{2n-1}}} \cdots \circ \rho_{e_{K_2}} \circ \rho_{e_{K_1}} \circ \sigma_{K_0},$$

where $\rho_{o_{K_i}} = \sigma_{K_i} \circ \tau \circ \pi_o \circ \gamma_o$ for odd rounds and $\rho_{e_{K_i}} = \sigma_{K_i} \circ \tau \circ \pi_e \circ \gamma_e$ for even rounds, and the linear output transformation $\phi_e = \tau \circ \pi_e \circ \tau$ is used at the last round.

3.2   Impossible Differential of Crypton

We introduce a four round impossible differential of Crypton. Fig. 5 describes one pattern of impossible differentials. The impossible differentials comes from the following observation.

1. If an input pair has zero difference at a byte, then the output pair after σ, γ, σ^{-1}, or γ^{-1} also has zero difference at the byte.
2. If an input pair has zero difference at byte$[i][j]$, then the output pair after τ or τ^{-1} has zero difference at byte$[j][i]$.
3. π, the word transformation has the branch number 4 as a map from 4-byte input to 4-byte output. That is, if input pair has only one nonzero difference out of four bytes, then the output difference has at least three nonzero difference.

Property 2 (Impossible Differential). Given input pair to τ whose difference is zero at all bytes but one, the output difference after the four round starting with τ and ending with γ cannot be zero at all but two rows in the left two columns.

3.3   An Attack against Crypton Reduced to Six Rounds

In this subsection, we describe an impossible differential cryptanalysis on Crypton reduced to 6 rounds. The attack is based on the four round impossible differential. We compose the 6 rounds as in Fig. 6. One thing we need to notice is that we replace the 5th and 6th round key addition σ_{K_5} and σ_{K_6} by σ_{null} which is a key addition with zero key. We can compose the same encryption system by putting $\sigma_{K_5^{eq}}$ and $\sigma_{K_6^{eq}}$ between γ and π of their rounds. Here K^{eq} means the equivalent key *i.e.* $\pi^{-1} \circ \tau^{-1}(K)$

The procedure is as follow:

1. A structure is defined as a set of plaintexts which have certain fixed values in the column 1, 2, and 3. One structure consists of 2^{32} plaintexts and proposes $2^{32} \times 2^{32} \times \frac{1}{2} = 2^{63}$ pairs of plaintexts.
2. Take 2^{59} structures (2^{91} plaintexts, 2^{122} plaintext pairs). Choose plaintext pairs (P, P^*) such that the pairs (C_6, C_6^*) has zero difference at the row 0 and 1, where $C_6 = \pi_e^{-1} \circ \tau^{-1} \circ \sigma_{null} \circ \tau^{-1} \circ \pi_e^{-1} \circ \tau^{-1}(C)$ and C is a ciphertext of P. The expected number of such pairs is $2^{122} \times 2^{-64} = 2^{58}$.

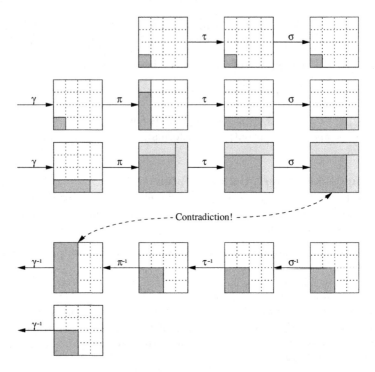

Fig. 5. Four Round Impossible Differential of Crypton

3. Assume a 64-bit value of the row 2 and 3 of the last round key K_6^{eq}.
4. For each pair (C_6, C_6^*) satisfying Step 2, compute $C_5 = \gamma_e^{-1}(C_6 \oplus K_6^{eq})$ and $C_5^* = \gamma_e^{-1}(C_6^* \oplus K_6^{eq})$ and choose pairs whose difference $\pi_o^{-1} \circ \tau^{-1}(C_5 \oplus C_5^*)$ is zero at any two rows. Since the probability is about $p = 2^{-32} \times 6 \approx 2^{-29.5}$, the expected number of the remaining pairs is $2^{58} \times 2^{-29.5} = 2^{28.5}$.
5. For a pair (P, P^*) satisfying Step 4, consider 32-bit values of the first column of K_0 such that $\pi(\gamma_o(P \oplus K_0) \oplus \gamma_o(P^* \oplus K_0))$ is zero at all but one byte of the first column. The probability is $q = 2^{-24} \times 4 = 2^{-22}$
6. Since such a difference is impossible, every key that proposes such a difference is a impossible key with the chosen key K_6^{eq} in Step 3. After analyzing $2^{28.5}$ plaintext pairs, there remain only about $2^{32}(1 - 2^{-22})^{2^{28.5}} \approx 2^{32}e^{-2^{6.5}} \approx 2^{-98.5}$ possible values for the four bytes of the first column of K_0, which means no possibility.
7. Unless the initial assumption on the final round key K_6^{eq} is correct, it is expected that we can get rid of the whole 32-bit values of K_0 for each 64-bit value of K_6^{eq} since the wrong value (K_0, K_6^{eq}) remains with the probability $2^{64} \times 2^{-98.5} = 2^{-34.5}$. Hence if there remains a value of K_0, we can assume the key K_6^{eq} is a right key. So if we repeat Step 2 through Step 6 after changing the row 0 and 1 with the row 2 and 3, we can get the whole value of K_6^{eq}.

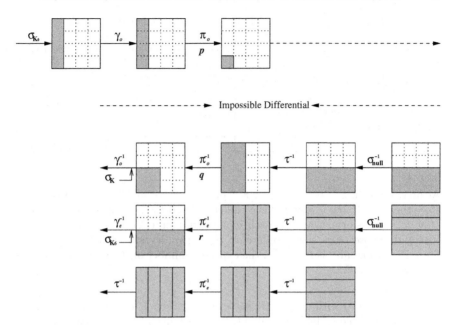

Fig. 6. Impossible Attack against Crypton Reduced to Six Rounds

8. Step 2 requires about $2^{123} (= 2 \times 2^{122})$ of $\pi_e^{-1} \circ \pi_e^{-1} \circ \tau^{-1} \circ 6round$. Step 4 requires about $2^{125.5} (= 2^{64} \times 2 \times 2^{60.5})$ of $\pi_o^{-1} \circ \tau^{-1} \circ \sigma^{-1}$. Step 6 requires about 2^{119} of $\pi_o \circ \gamma_o$ operations since

$$2^{64} \times 2 \times = 2^{32} \{1 + (1 - 2^{-22}) + (1 - 2^{-22})^2 + \cdots + (1 - 2^{-22})^{2^{28.5}}\} \approx 2^{119}.$$

Consequently, since we repeat this procedure twice, this attack requires about 2^{91} chosen plaintexts and 2^{124} encryptions of Crypton reduced to 6 rounds.

3.4   A Variant of the Attack Using Memory

In this subsection, we describe a variant of the impossible differential cryptanalysis of the former subsection using memory.

Precomputation Stage

Take all $2^{32} \times 2^8 \times 4 = 2^{42}$ pairs of four bytes in the first column which differ only in one byte(this is the data after one round encryption except the first round key addition). For these pairs, we undo the encryption of the first round, i.e., perform τ^{-1}, π^{-1} and γ^{-1}, and create a hash table containing one of the inputs of γ transformation and the XOR of two inputs $x \oplus y$, indexed by $x \oplus y$, where x, y are the inputs of the γ transformation.

Table 1. Complexity of 6-Round Impossible Differential Attack

Cipher	Attack	Round	Chosen Ciphertexts	Complexity
Rijndael	Square attack	6 round	2^{32}	2^{72}
	Impossible differential attack	5 round	$2^{29.5}$	2^{31}
		6 round	$2^{91.5}$	2^{122}
Crypton	Square attack	6 round	$2^{32}(2^{32}\,Mem.)$	2^{56}
	Impossible differential attack	5 round	$2^{83.4}$	2^{43}
		6 round	2^{91}	2^{124}

Step 4'

At first, perform $\pi^{-1} \circ \gamma^{-1} \circ \pi^{-1} \circ \tau^{-1}$ operations for all $2^{93.5}$ ciphertexts, and store the pairs (C, U_C) where C is a ciphertext and U_C is a corresponding result after π^{-1} transformation. For each ciphertext pair (C, C^*), compare U_C and U_{C^*} and choose pairs whose difference $U_C \oplus U_{C^*}$ is zero at the row 0 and 1. Since the probability is $p = 2^{-32}$, the expected number of the remaining pairs is $2^{60.5} \times 2^{-32} = 2^{28.5}$.

Step 5'

For a pair (P, P^*) with ciphertext pairs passing Step 4, we compute $x \oplus y$ and use the hash table to fetch the about 2^{10} possibility of x which correspond to the computed $x \oplus y$. This process identities about 2^{10} wrong key by XORing the plaintexts and the x's.

If we replace Step 4 and Step 5 by Precomputation stage, Step 4' and Step 5', the complexity is as follows:

- Step 4 requires $2^{93.5}$ $\pi^{-1} \circ \gamma^{-1} \circ \pi^{-1} \circ \tau^{-1}$ operations and $2^{-124.5}$ operations of two times memory access and comparison which is about 2^{14} times faster than the 6-round Crypton . Hence Step 4 is equivalent to about $2^{110.5}$ encryptions. In addition, Step 4 requires $2^{93.5} \times 128 \times 128 = 2^{104.5}$ memory.
- In Step 5', for $2^{28.5}$ remaining plaintext pairs we get rid of 2^{10} impossible keys by XORing the plaintext and x's which requires about 2^{32} encryptions for each assumed key K_6^{eq}.
- To sum up, this attack requires $2^{93.5}$ plaintexts, $2^{110.5}$ encryptions and $2^{104.5}$ bytes of memory.

4   Conclusion

In this paper, we described an impossible differential attack against Rijndael and Crypton reduced to 6 rounds. The attack against Rijndael reduced to 6 rounds requires about $2^{91.5}$ chosen plaintexts and 2^{122} encryptions. The attack against Crypton reduced to 6 rounds requires about 2^{91} chosen plaintexts and 2^{124} encryptions, or $2^{93.5}$ plaintexts, $2^{110.5}$ encryptions and $2^{104.5}$ bytes of memory. We summarize the complexities of our attacks together with those of previous works in Table 1. We expect that this method can be applied to other block ciphers with SPN Structure.

References

1. The Advanced Encryption Standard, *http://www.nist.gov/aes*
2. E. Biham and A. Shamir, "Differential Cryptanalysis of DES-like Cryptosystems," J. of Cryptology, Vol. 3, pp.27-41, 1990.
3. E. Biham and N. Keller, "Cryptanalysis of Reduced Variants of Rijndael," *http://csrc.nist.gov/encryption /aes/round2/conf3/aes3papers.html*
4. C. D'Halluin, G. Bijnens, V. Rijmen, and B. Preneel, "Attack on Six Rounds of Crypton," Proc. of Fast Software Encryption'99, Lecture Notes in Computer Science Vol. 1636, pp. 46 – 59, Springer-Verlag, 1999.
5. J. Daemen, L. Knudsen, and V. Rijmen, "The Block Cipher Square," Proc. of Fast Software Encryption'97, Lecture Notes in Computer Science Vol. 1267, pp. 149–165, 1997.
6. J. Daemen and V. Rijmen, "AES Proposal: Rijndael," *http://csrc.nist.gov /encryption/aes/rijndael/*
7. C. Lim, "A Revised Version of Crypton - Crypton 1.0," Proc. of Fast Software Encryption'99, Lecture Notes in Computer Science Vol. 1636, pp. 31 – 45, Springer-Verlag, 1999.
8. M. Matsui, "Linear Cryptanalysis Method for DES cipher," Proc. of Eurocrypt'93, Lecture Notes in Computer Science Vol. 765, pp.386 – 397, Springer-Verlag, 1993.
9. H. Seki and T. Kaneko, "Cryptanalysis of Five Rounds of CRYPTON Using Impossible Differentials," Proc. of Asiacrypt'99, Lecture Notes in Computer Science Vol. 1716, pp.43-51,1999.
10. Stefan Lucks, "Attacking Seven Rounds of Rijndael under 192-bit and 256-bit Keys," Proc. of Third AES Candidate Conference, AES3, 2000.

Cryptanalysis of Nonlinear Filter Generators with $\{0,1\}$-Metric Viterbi Decoding[*]

Sabine Leveiller[1,2], Joseph Boutros[2], Philippe Guillot[3], and Gilles Zémor[2]

[1] Thales Communication
66, rue du Fossé Blanc, 92231 Genevilliers, France
[2] Ecole Nationale Supérieure des Télécommunications
46, rue Barrault, 75013 Paris, France
{leveille,boutros,zemor}@enst.fr
[3] Canal-Plus Technologies
34, place Raoul Dautry, 75015 Paris, France
pguillot@canal-plus.fr

Abstract. This paper presents a new deterministic attack against stream ciphers based on a nonlinear filter key-stream generator. By "deterministic" we mean that it avoids replacing the non-linear Boolean function by a probabilistic channel. The algorithm we present is based on a trellis, and essentially amounts to a Viterbi algorithm with a $\{0,1\}$-metric. The trellis is derived from the Boolean function and the received key-stream. The efficiency of the algorithm is comparable to Golic et al.'s recent "generalized inversion attack" but uses an altogether different approach: it brings in a novel cryptanalytic tool by calling upon trellis decoding.

Keywords: Boolean functions, stream ciphers, filter generator, Viterbi algorithm, Fourier transform.

---

[*] A full version of this paper was published in: B. Honary (Ed.) Cryptography and Coding. Proceedings 8th IMA International Conference, Cirencester, UK, December 2001. LNCS 2260, Springer 2001.

K. Kim (Ed.): ICICS 2001, LNCS 2288, p. 50, 2002.
© Springer-Verlag Berlin Heidelberg 2002

An IND-CCA2 Public-Key Cryptosystem with Fast Decryption

Johannes Buchmann[1], Kouichi Sakurai[2], and Tsuyoshi Takagi[1]

[1] Technische Universität Darmstadt,
Fachbereich Informatik,
Alexanderstr.10, D-64283 Darmstadt, Germany
{buchmann,ttakagi}@cdc.informatik.tu-darmstadt.de
[2] Kyushu University
Department of Computer Science and Communication Engineering
Hakozaki, Fukuoka 812-81, Japan
sakurai@csce.kyushu-u.ac.jp

Abstract. We propose an IND-CCA2 public-key cryptosystem with fast decryption, called the NICE-X cryptosystem. Its decryption time is the polynomial time of degree 2 by the bit-length of a public-key D, i.e., $\mathcal{O}((\log|D|)^2)$, and the cost of two hash functions. The NICE-X is an enhancement of the NICE cryptosystem, which is constructed over the quadratic class group $Cl(D)$. We first show that the one-wayness of the encryption of the NICE cryptosystem is as intractable as the Smallest Kernel Equivalent Problem (SKEP). We also prove that the NICE cryptosystem is IND-CPA under the Decisional Kernel Problem (DKP). Then we prove that the NICE-X cryptosystem is IND-CCA2 under the SKEP in the random oracle model. Indeed, the overhead of the decryption of the NICE-X from the NICE is only the cost of one ideal multiplication and two hash functions. Our conversion technique from the NICE to the NICE-X is based on the REACT. However we modify it to be suitable for the NICE. A message of the NICE-X is encrypted with the random mask of the encryption function of the NICE, instead of the encrypted key. Then the reduced security problem of the NICE-X is enhanced from the Gap-SKEP to the SKEP.

Keywords: Public-key cryptosystem, Chosen ciphertext attack, NICE cryptosystem, factoring algorithm, fast decryption

1 Introduction

The simplest requirement for the security of a public-key cryptosystem is the one-wayness of the encryption function. However, after the invention of a chosen ciphertext attack against the PKCS # 1, version 1.5 by Bleichenbacher [Ble98], a chosen ciphertext security becomes a standard requirement for a public-key cryptosystem. In this paper we consider the indistinguishability against chosen ciphertext (or plaintext) attack (IND-CCA2 or IND-CPA) [RS91] [BDPR98]. We also use the random oracle model [BR93,BR94]. Although the security of the

K. Kim (Ed.): ICICS 2001, LNCS 2288, pp. 51–71, 2002.
© Springer-Verlag Berlin Heidelberg 2002

random oracle model does not imply the security of the real world (see [CGH98]), the random oracle model is a useful technique to design a cryptosystem.

Several IND-CCA2 public-key cryptosystems have been proposed, for example [TY98] [SG98] [CS98] [Poi99a] [PP99] [FOPS01] (see appendix D). The IND-CCA2 security of these schemes is proved under specific number-theoretic problems. Then general conversion techniques, which enhance a public-key cryptosystem to be an IND-CCA2 scheme, have been also proposed. Fujisaki and Okamoto [FO99b] and Pointcheval [Poi00] proposed a generic technique to convert a one-way public-key cryptosystem into an IND-CCA2 scheme in the random oracle model. Baek et al proposed a conversion whose ciphertext is smaller than that of the Pointcheval scheme [BLK00]. Fujisaki and Okamoto proposed a generic conversion from an IND-CPA public-key cryptosystem to an IND-CCA2 scheme in the random oracle model [FO99a]. Recently Okamoto and Pointcheval presented a conversion 'REACT' which allows to convert a one-way public-key cryptosystem into an IND-CCA2 scheme in the random oracle model [OP01b]. The REACT is based on the paper by Bellare and Rogaway [BR93]. The REACT requires only two extra hash functions for its encryption and decryption and its security reduction is very tight. The converted scheme is IND-CCA2 under the Gap-problem [OP01a].

In order to construct a public-key cryptosystem, we also have to consider the efficiency of the scheme. A public-key cryptosystem with fast decryption, called the NICE cryptosystem, is proposed [PT00]. Its decryption time is the polynomial time of degree 2 by the bit-length of a public-key D. Even if the key length of the NICE cryptosystem becomes larger in the future, the running time of the decryption process remains fast. The NICE cryptosystem is constructed over quadratic class groups $Cl(D)$ of a discriminant $D = \Delta q^2$, where Δ is the fundamental discriminant of D and q is a prime integer. The security of the NICE cryptosystem is related to the problem of factoring the discriminant D. However there was no exact analysis of its security. A chosen ciphertext attack against the NICE cryptosystem is proposed by Jaulmes and Joux [JJ00]. In this paper we define several number-theoretic problems over the class group $Cl(D)$, which are related to the NICE cryptosystem, and discuss their relationships. We then investigate the security of the NICE by reducing these number-theoretic problems. Let Φ be a homomorphism $\Phi : Cl(D) \to Cl(\Delta)$. We say that two elements $\mathcal{A}$ and $\mathcal{B}$ of $Cl(D)$ are kernel-equivalent if $\Phi(\mathcal{A}) = \Phi(\mathcal{B})$ holds. We introduce two intractable problems. The first one is the Smallest Kernel Equivalent Problem SKEP, which finds the smallest kernel-equivalent element for a given element of $Cl(D)$. The other one is the Decisional Kernel Problem DKP, which distinguishes whether an element of $Cl(D)$ is in the kernel $Ker(\Phi)$. We prove that an algorithm, which solves the SKEP, can also solve the DKP.

First, we prove two results about the security of the NICE cryptosystem: (1)the one-wayness of the NICE cryptosystem is as intractable as the SKEP, and (2)the NICE cryptosystem is IND-CPA under the DKP. Then, the conversion techniques by Fujisaki-Okamoto [FO99b] or Pointcheval [Poi00] make the NICE cryptosystem IND-CCA under SKEP in the random oracle model. The conversion technique by Fujisaki and Okamoto [FO99a] enhances the NICE cryptosystem to IND-CCA2 under the DKP in the random oracle model. However, in the decryption process, the two conversions check whether the whole encryption computation is correctly generated. The encryption time of the NICE cryptosystem is estimated by cubic complexity $\mathcal{O}(\log |D|)^3$. Therefore, the decryption time of the converted scheme is not any more of quadratic order $\mathcal{O}(\log |D|)^2$ and thus we lose the advantage of the NICE cryptosystem. When we apply the REACT to the NICE cryptosystem, the decryption time of the REACT-NICE cryptosystem remains that of the NICE cryptosystem $\mathcal{O}(\log |D|)^2$ and the cost of two hash functions. However the REACT-NICE cryptosystem is IND-CCA2, relative to the Gap-SKEP, which consists in solving the SKEP with access to the DKP. The Gap-SKEP is stronger assumption than the SKEP.

We modify the REACT to be suitable for the NICE cryptosystem and we propose the NICE-X cryptosystem. The NICE-X cryptosystem is IND-CCA2 under only the SKEP and its security reduction is very tight. The decryption time of the NICE-X is of quadratic complexity $\mathcal{O}(\log |D|)^2$ and the cost of two hash functions. Let $\mathcal{M}$ be a message class of the NICE. Then the cipher class $\mathcal{C}$ is computed by $\mathcal{C} = \mathcal{M}\mathcal{P}^r$, where r is a random integer and $\mathcal{P}$ is the public key. Thus the encryption of the NICE cryptosystem $\mathcal{M} \to \mathcal{C}$ is probabilistic. Let $c = \mathcal{E}(m, r)$ be a probabilistic encryption function of a message m with a random mask r. The REACT encrypts a message k using the encryption scheme $\mathcal{E}$ as follows: $A = \mathcal{E}(m, r), B = k \oplus g(m), H = h(k, m, A, B)$, where h, g are some hash functions. On the contrary, the NICE-X encrypts a message k with the random mask r instead of m, namely $B = k \oplus g(\mathcal{P}^r)$ instead of $B = k \oplus g(\mathcal{M})$. After this modification, we can prove that the IND-CCA2 security of the NICE-X is based on the SKEP. For the proof of the security, we use the following special property of the NICE cryptosystem. There is a distinguisher that checks whether $\mathcal{P}^{r'} \in Ker(\Phi)$ is equivalent to the random mask $\mathcal{P}^r$ of the cipher class $\mathcal{C}$. If yes, we have $r = r'$. Moreover if we find r, then the ideal $\mathcal{M}$ can be computed by $\mathcal{C}(\mathcal{P}^r)^{-1}$ and the SKEP can be solved. This property is not true for the ElGamal encryption scheme. Indeed, for a ciphertext $c = (g^r, my^r)$, the random mask of the ciphertext c is y^r, where $y = g^s$ is a public key. To check whether $t \in\ <g>$ is the random mask of a ciphertext $c = (g^r, my^r)$ is equivalent to check whether (g^r, g^r, t) is a Diffie-Hellman triple. Even if we modify the ElGamal-REACT with $B = k \oplus g(y^r)$, its security is still based on the Gap-Diffie-Hellman problem. The NICE-X is the first example to utilize the random mask of an encryption function for proving the IND-CCA2 security of the converted scheme.

The paper is organized as follows. In section 2 we review the algorithm of the NICE cryptosystem. In section 3 we define several security notations for the NICE cryptosystem. In section 4 we formalize four number theoretic problems and investigate their relationships. In section 5 the security of the original NICE cryptosystem is discussed. In section 6 we propose the NICE-X cryptosystem and prove that it is secure against the chosen ciphertext attack. In section 7 we compare the security and efficiency of the NICE/NICE-X cryptosystem with those of other cryptosystems.

Key Generation
k, a security parameter
$D = \Delta q^2$, non-fundamental discriminant
(p, q: primes with $2^{k-1} < p < q < 2^k$, $\Delta = -p$)
$\mathcal{P}$, a generator of $Ker(\Phi)$
Public-key: $(D, \mathcal{P})$, Secret key: (Δ, q)

Encryption of $\mathcal{M}$
$\mathcal{M} \in \mathsf{SI}(D)$, a message class
$r \in_R \{0, 1, .., 2^k\}$
$\mathcal{C} = \mathcal{E}_{NICE} = \mathcal{M}\mathcal{P}^r \in Cl(D)$, a cipher class

Decryption of $\mathcal{C}$
$\mathcal{M} = \mathcal{D}_{NICE}(\mathcal{C}) = \Psi(\Phi(\mathcal{C}))$

Fig. 1. The NICE Cryptosystem

2   The Original NICE Cryptosystem

In this section we review the NICE cryptosystem [PT00] [HPT99]. The NICE cryptosystem has two interesting properties. The first property is that the trap-door mechanism is different from the previously reported public-key cryptosystems, including the RSA cryptosystem and the ElGamal based cryptosystems. The second property is that the decryption process of the NICE cryptosystem is very fast. It is of quadratic bit complexity in the length of the public key.

We give a description of the NICE cryptosystem in the following (see also figure 1). Let Δ be a negative fundamental discriminant such that $\Delta = -p$, where p is a prime number and $p \equiv 3 \bmod 4$. Let q be a prime number with $2^{k-1} < p < q < 2^k$ for an integer k, where k is a security parameter of the NICE cryptosystem. Let $\mathcal{O}_\Delta = \mathbb{Z} + \frac{\Delta + \sqrt{\Delta}}{2}\mathbb{Z}$ be the quadratic order. The NICE cryptosystem is constructed using the class group $Cl(D)$ with non-fundamental discriminant $D = q^2\Delta$. We use the following maps:

$$\Phi : \quad Cl(D) \longrightarrow Cl(\Delta)$$
$$[\mathfrak{a}] \longmapsto [\mathfrak{a}\mathcal{O}_\Delta].$$

$$\Psi : \quad Cl(\Delta) \longrightarrow Cl(D)$$
$$[\mathfrak{A}] \longmapsto [Red(\mathfrak{A}) \cap \mathcal{O}_D].$$

Φ is a surjective homomorphism of groups. In general Ψ is not the inverse of Φ.

The keys used in the NICE cryptosystem are as follows: Let $Ker(\Phi)$ be the kernel of Φ and let $\mathcal{P}$ be the generator of $Ker(\Phi)$. Then the public key of the NICE cryptosystem is $(D, \mathcal{P})$ and the secret key is (Δ, q). The encryption process of the NICE cryptosystem is as follows: Denote by $\mathsf{SI}(D)$ the set of all ideal classes $\mathcal{A} \in Cl(D)$ with $N(\mathcal{A}) < |D|^{1/6}\sqrt{1/8} < \sqrt{|\Delta|/4}$, where the norm of an ideal class is the norm of the reduced ideal contained in its ideal class. A message class $\mathcal{M}$ is chosen from $\mathsf{SI}(D)$. There is an algorithm, which embeds an integer to the message class [PT00]. The message class $\mathcal{M}$ is encrypted as follows: Choose a random integer r in $\{0, 1, ..., 2^k\}$ and pick up the public key $(D, \mathcal{P})$ of the receiver, then the cipher class $\mathcal{C}$ is computed by

$$\mathcal{C} = \mathcal{E}_{NICE}(\mathcal{M}) = \mathcal{M}\mathcal{P}^r \in Cl(D). \tag{1}$$

Here we can precompute the ideal class $\mathcal{P}^r$, then we need only an ideal multiplication and a reduction to encrypt a message class $\mathcal{M}$. The cipher class $\mathcal{C}$ is decrypted as follows: $\mathcal{M} = \mathcal{D}_{NICE}(\mathcal{C}) = \Psi(\Phi(\mathcal{C}))$. Indeed, the person who knows the secret key (Δ, q) can compute the maps Φ, Ψ. He/she can decrypt $\mathcal{C}$ by computing $\mathcal{M} = \Psi(\Phi(\mathcal{C}))$. This can be ensured by the following Theorem 1 (see [PT00]).

Theorem 1. *Let $(D, \mathcal{P})$ be the public key and (Δ, q) be the secret key of the NICE cryptosystem. Let $\mathcal{M} \in SI(D)$. Let $\mathcal{C} = \mathcal{M}\mathcal{P}^r$ be the cipher class of $\mathcal{M}$. Then we have the relationship $\mathcal{M} = \Psi(\Phi(\mathcal{C}))$.*

The decryption process of the NICE cryptosystem consists of the maps Φ and Ψ. The running time of these maps is of quadratic bit complexity $O((\log|D|)^2)$. Therefore the decryption process of the NICE cryptosystem is of quadratic bit complexity $O((\log|D|)^2)$.

3   Security Notations

In this section we define several security notations concerning the NICE cryptosystem. We will define the one-wayness of NICE, and the indistinguishability of a chosen ciphertext (plaintext) attack against the NICE. More comprehensive definition of the security can be found in the reference [BDPR98].

Let $A(x_1, x_2, ..., x_n)$ be a probabilistic algorithm on input $(x_1, x_2, ..., x_n)$. Denote by $Pr[x \leftarrow S; y \leftarrow T; ... : p(x, y, ...)]$ the probability which the predicate $p(x, y, ...)$ is true for experiment $x \leftarrow S, y \leftarrow T, ...$, where $S, T, ..$ are probabilistic spaces. Let $x \leftarrow_R S$ be an operation of picking x at randomly and uniformly from finite set S. A fucntion $\epsilon : \mathbb{N} \rightarrow \mathbb{R}$ is negligible, if for every constant $c \geq 0$, there exists an integer k_c such that $\epsilon \leq k^{-c}$ for all $k \geq k_c$.

The t-one-wayness adversary A of the NICE cryptosystem computes, on input any cipher class $\mathcal{C}$, the original message class $\mathcal{M}$, in time t, where $\mathcal{C} = \mathcal{E}_{NICE}(\mathcal{M})$. Let k be the security parameter of the NICE cryptosystem. The success probability of the one-wayness adversary A is defined by

$$\mathsf{Succ}^{\mathsf{ow}}_{A, NICE}(k) = Pr_{\mathcal{M} \in SI(D)}[\mathcal{C} \leftarrow \mathcal{E}_{NICE}(\mathcal{M}) : A(\mathcal{C}) = \mathcal{M}]. \tag{2}$$

Definition 1. *The NICE cryptosystem is called (t, ε)-one-wayness if the success probability $\mathsf{Succ}^{\mathsf{ow}}_{A, NICE}(k)$ of t-one-wayness adversary A is upper-bounded by negligible ϵ.*

Next we explain about the chosen ciphertext (plaintext) attack against the NICE cryptosystem. A chosen ciphertext attack A consists of two algorithms A_1

and A_2: the first algorithm A_1 computes, on input the public-key $(D, \mathcal{P})$, a triple $(\mathcal{M}_0, \mathcal{M}_1, st)$, where $\mathcal{M}_0, \mathcal{M}_1$ are message classes and st is state information. Let $\mathcal{C}$ be a challenge which is an encryption of $\mathcal{M}_b$, where b is randomly chosen from $\{0, 1\}$. Then the second algorithm A_2 determines the bit b' on input the state information st and the challenge $\mathcal{C}$. During the algorithm, the attacker A is allowed to ask at most q_D queries to the decryption oracle. We write the attacker $A^{\mathcal{D}_{NICE}}$. If the attacker is not allowed to ask the decryption oracle, then the chosen ciphertext attack is called a chosen plaintext attack. If the algorithm A runs in time t, then we say that A is (t, q_D)-chosen ciphertext (or plaintext) attack algorithm.

Moreover, we consider the random oracle model version of the chosen ciphertext attack A. Let Has be the set of all hash functions, which map some appropriate domain to appropriate range. During the attack the attacker is allowed to ask at most q_H queries to the random oracle $H \in$ Has. Moreover the encryption function $\mathcal{E}_{NICE}$ and the decryption function $\mathcal{D}_{NICE}$ may depend on the random oracle H, which we write as $\mathcal{E}^H_{NICE}$ and $\mathcal{D}^H_{NICE}$, respectively. We write the attack as A^H. If the attack A^H runs in time t at most q_H queries to the random oracle H, then we say that A^H is (t, q_H, q_D)-chosen ciphertext (or plaintext) attack algorithm.

The advantage of the chosen ciphertext (or plaintext) attack adversary A is defined as follows:

$$Adv^{\mathsf{IND-CPA}}_{A,NICE}(k) = 2Pr[G, H \leftarrow \mathsf{Hash}; (\mathcal{M}_0, \mathcal{M}_1, st) \leftarrow A^{G,H}_1(D, \mathcal{P}); b \leftarrow_R \{0, 1\};$$
$$\mathcal{C} \leftarrow \mathcal{E}_{NICE}(\mathcal{M}_b) : A^{G,H}_2(st, \mathcal{M}_0, \mathcal{M}_1, \mathcal{C}) = b] - 1.$$
$$Adv^{\mathsf{IND-CCA}}_{A,NICE}(k) = 2Pr[G, H \leftarrow \mathsf{Hash}; (\mathcal{M}_0, \mathcal{M}_1, st) \leftarrow A^{G,H,\mathcal{D}_{NICE}}_1(D, \mathcal{P});$$
$$b \leftarrow_R \{0, 1\}; \mathcal{C} \leftarrow \mathcal{E}_{NICE}(\mathcal{M}_b) : A^{G,H,\mathcal{D}_{NICE}}_2(st, \mathcal{M}_0, \mathcal{M}_1, \mathcal{C}) = b] - 1.$$

Definition 2. *We say that the NICE cryptosystem is $(t, \varepsilon, q_H, q_G, q_D)$-secure against the chosen ciphertext (or plaintext) attack in the random oracle model, if the advantage $Adv^{\mathsf{IND-CCA}}_{A,NICE}(k)$ (or $Adv^{\mathsf{IND-CPA}}_{A,NICE}(k)$) of (t, q_H, q_G, q_D)-chosen ciphertext (or plaintext) attack algorithm, which asks at most q_H queries to random oracle $G(\cdot)$ and at most q_G queries to random oracle $H(\cdot)$, is upper-bounded by negligible ε.*

4   Number Theoretic Problems

In this section we investigate several number theoretic problems in $Cl(D)$, which are related to the NICE cryptosystem. The security of the NICE cryptosystem is based on the factoring of the non-fundamental discriminant $D = \Delta q^2$ and on solving the discrete logarithm problem over class group $Cl(D)$. If the discriminant $D = \Delta q^2$ is factored, then the NICE cryptosystem is totally broken.

(1) Factoring Problem (FP): Given an non-fundamental discriminant $D = \Delta q^2$, factor the discriminant D.

The number field sieve [LL91] and the elliptic curve method [Len87] are the different types of factoring algorithms, which have to be taken care of. The number field sieve is the fastest factoring algorithm, and its running time depends on the total bit length of the composite number $|D|$; it is of the order of $L_{|D|}[1/3, (64/9)^{1/3}]$. The elliptic curve method depends on the size of the prime factor of D and the expected running time is $L_r[1/2, 2^{1/2}]$, where r is the divisor of D. For example, if the discriminant D is larger than 1024 bits with same bit-length divisors Δ, q, to factor D by the two algorithms is not feasible.

(2) Quadratic Order Discrete Logarithm Problem (QODLP): Given a discriminant $D = \Delta q^2$ and two ideal classes $\mathcal{A}, \mathcal{B} \in Cl(D)$, compute the discrete logarithm r of $\mathcal{A} = \mathcal{B}^r \in Cl(D)$.

The fastest algorithm to solve the QODLP is the Hafner-McCurley algorithm [HM89]. Its running time is $L_{|D|}[1/2, 2^{1/2}]$, which is much slower than factoring D. It is known that we can solve the FP using a algorithm to solve QODLP in polynomial time of $\log D$ [BW88].

Next, we would like to define two more number-theoretic problems. At first, we observe a unique representation of elements of $Cl(D)$ using a kernel element in $Ker(\Phi)$. When we know the prime divisors Δ, q, we can represent it with a simple form. If two classes $\mathcal{A}, \mathcal{B}$ of $Cl(D)$ are equivalent in the subgroup $Cl(\Delta)$ such that $\Phi(\mathcal{A}) = \Phi(\mathcal{B})$, then we call them *kernel equivalent*. Because the group $Ker(\Phi)$ is a cyclic subgroup of $Cl(D)$, two kernel-equivalent classes $\mathcal{A}$ and $\mathcal{B}$ are related by $\mathcal{A} = \mathcal{B}\mathcal{P}^r$, where $\mathcal{P}$ is a generator of $Ker(\Phi)$ and $0 \le r < q - (\Delta/q)$. Here we have the following lemma.

Lemma 1. *For any ideal class $\mathcal{A} \in Cl(\Delta)$, the norm of the ideal class $\Psi(\mathcal{A}) \in Cl(D)$ is the smallest of its kernel-equivalence class.*

Proof. Let $\mathfrak{A}$ be the reduced ideal contained in the ideal class $\mathcal{A}$. We have the relationship $N(\Psi(\mathcal{A})) = N(\mathfrak{A} \cap \mathcal{O}_D) = N(\mathfrak{A})$. Let $\mathcal{B}$ be an arbitrary class which is kernel-equivalent to $\Psi(\mathcal{A})$ in $Cl(D)$. Let $\mathfrak{b}$ be the reduced ideal contained in the ideal class $\mathcal{B}$. From the definition we have $\Phi(\mathcal{B}) = \mathcal{A}$. Here we have relationship $N(\mathfrak{b}\mathcal{O}_\Delta) = N(\mathfrak{b}) = N(\mathcal{B})$ and the ideal $\mathfrak{b}\mathcal{O}_\Delta$ is not generally reduced. If $\mathfrak{A}$ is a reduced ideal in $\mathcal{O}_\Delta$, then $N(\mathfrak{A}) \le N(\mathfrak{C})$ must hold for any ideal $\mathfrak{C}$ which is equivalent to $\mathfrak{A}$ in $\mathcal{O}_\Delta$. Especially choose $\mathfrak{C} = \mathfrak{B}$, then $N(\mathfrak{A}) \le N(\mathfrak{B})$ holds. Therefore, we proved $N(\Psi(\mathcal{A})) \le N(\mathcal{B})$.

From this we obtain the following unique representation of an ideal class:

Lemma 2. *Let $\mathcal{A}$ be an ideal class of $Cl(D)$. Then the ideal class $\mathcal{A}$ can be uniquely decomposed by*

$$\mathcal{A} = \mathcal{A}_0 \mathcal{Q}, \quad \mathcal{Q} = \mathcal{P}^r, \tag{3}$$

where $\mathcal{A}_0$ is the kernel-equivalent ideal with the smallest norm, $\mathcal{P}$ is a generator of $Ker(\Phi)$ and $0 \le r < q - (\Delta/q)$. Moreover, $\mathcal{A}_0$ can be computed by $\mathcal{A}_0 = \Psi(\Phi(\mathcal{A}))$.

Here we define the following two problems:

(3) Smallest Kernel-Equivalent Problem (SKEP): For a given discriminant $D = \Delta q^2$, a generator of $Ker(\Phi)$, and an ideal class $\mathcal{A} \in Cl(D)$ with $\Psi(\Phi(\mathcal{A})) \in \mathsf{SI}(D)$, find the smallest kernel-equivalent ideal class of $\mathcal{A}$.

(4) Decisional Kernel Problem (DKP): Given an discriminant $D = \Delta q^2$ and a generator of the kernel $Ker(\Phi)$, distinguish the following two distribution:

$$Rand = \{A_{Rand} \in_R Cl(D) | \Psi(\Phi(A_{Rand})) \in \mathsf{SI}(D)\}$$
$$Kernel = \{A_{Kernel} \in_R Ker(\Phi)\}.$$

Denote by A a distinguisher of the decisional kernel problem, which outputs either 0 or 1 for given $A_{Rand} \leftarrow Rand$ and $A_{Kernel} \leftarrow Kernel$. We define the advantage of the decisional kernel problem by algorithm A as follows:

$$Adv_{A,DKP}(k) = |Pr[A(A_{Rand}) = 1] - Pr[A(A_{Kernel}) = 1]|. \qquad (4)$$

If we have an algorithm A to solve the SKEP, then we can solve the DKP using algorithm A. Indeed, for a given ideal class $A \in Cl(D)$, we can compute the smallest kernel-equivalent ideal A_0 of A using algorithm A. Then we can distinguish whether $A \in Ker(\Phi)$ holds, by checking the equation $A_0 = \mathcal{O}_D$. Consequently we have the following reduction relationship.

Theorem 2. *Consider four number theoretic problems* QODLP, FP, SKEP, DKP, *then we have the following relationship among the problems:*

solving QODLP $\Rightarrow$ *solving* FP $\Rightarrow$ *solving* SKEP $\Rightarrow$ *solving* DKP,

where A $\Rightarrow$ B *means that algorithm* A, *which solves problem* A *can also solve problem* B *in polynomial time of* $\log|D|$.

4.1   Assumptions

The NICE cryptosystem depends on the two assumptions:

SKEP assumption: We say that an algorithm A is $(D, \mathcal{P}, t)$-SKEP algorithm with advantage ε, if A can solve SKEP with probability at least ε, in time t, for a given public-key $(D, \mathcal{P})$ of the NICE cryptosystem. We say that $(D, \mathcal{P}, t, \varepsilon)$-SKEP assumption holds, if there is no $(D, \mathcal{P}, t)$-SKEP algorithm with advantage ε.

DKP assumption: We say that an algorithm A is $(D, \mathcal{P}, t)$-DKP algorithm with advantage ε, if A can solve DKP with at least advantage ε, in time t, for a given public-key $(D, \mathcal{P})$ of the NICE cryptosystem. We say that $(D, \mathcal{P}, t, \varepsilon)$-DKP assumption holds, if there is no $(D, \mathcal{P}, t)$-DKP algorithm with advantage ε.

5   Security of the Original NICE Cryptosystem

In this section we prove the security of the original NICE cryptosystem. At first we prove that the encryption fucntion of the NICE cryptosystem is a one-way fucntion, if and only if the SKEP assumption is true. Then, we also prove that the NICE cryptosystem is secure against the chosen plaintext attack under the DKP assumption.

Theorem 3. *Let A be a one-wayness adversary of the NICE cryptosystem with advantage $\varepsilon = \mathsf{Succ}^{\mathsf{ow}}_{A,NICE}$ and time t. Then there is an algorithm for computing* SKEP *with the same advantage ε and the same time t.*

Proof. Let $\mathcal{M}$ be a message class of the NICE cryptosystem. The cipher-class of $\mathcal{M}$ is represented by $\mathcal{C} = \mathcal{M}\mathcal{Q}$, where $\mathcal{Q}$ is a random element in $Ker(\varPhi)$. Therefore the message class $\mathcal{M}$ and its cipher class $\mathcal{C}$ are kernel-equivalent. Let $\mathfrak{m}$ be the reduced ideal contained in the message class $\mathcal{M}$. We chose the message ideal with small norm such that $N(\mathcal{M}) = N(\mathfrak{m}) < |D|^{1/6}\sqrt{1/8} < \sqrt{|\varDelta|/4}$. Therefore $\mathfrak{m}\mathcal{O}_\varDelta$ is a reduced ideal and the norm of the ideal class $\varPsi([\mathfrak{m}\mathcal{O}_\varDelta]) = \varPsi(\varPhi(\mathcal{M}))$ is the smallest of its kernel-equivalent class. Moreover we have $\varPsi(\varPhi(\mathcal{M})) = \varPsi(\varPhi(\mathcal{C}))$ and the class $\mathcal{M}$ is the smallest kernel-equivalent class of $\mathcal{C}$. Thus the algorithm A can solve the SKEP.

As a statement against the chosen plaintext attack, we can prove the following theorem. The proof can be found in the appendix A.

Theorem 4. *Let A be a chosen plaintext attack algorithm against the NICE cryptosystem with advantage $\varepsilon = Adv^{\mathsf{IND-CPA}}_{A,NICE}$ and time t. Then there is an algorithm B for solving* DKP *with advantage $\varepsilon/2$ and time smaller than $t + T_\mathcal{E}$, where $T_\mathcal{E}$ is the encryption time of the NICE cryptosystem.*

Next we notice that the NICE cryptosystem is not secure against the chosen-ciphertext attack, namely Jaulmes and Joux proposed a chosen ciphertext attack [JJ00]. Moreover, the NICE cryptosystem is not non-malleable against the chosen plaintext attack, because we have the following relationship: Let $\mathcal{C} = \mathcal{M}\mathcal{P}^r$ and $\mathcal{C}' = \mathcal{M}'\mathcal{P}^{r'}$ be two ciphertexts of the NICE cryptosystem. If $\mathcal{M}\mathcal{M}' \in SI(D)$ holds, then the multiplication $\mathcal{C}\mathcal{C}'$ is the ciphertext of $\mathcal{M}\mathcal{M}'$. In theorem 4 we proved that the NICE cryptosystem is indistinguishable against the chosen plaintext attack. This observation is another example of the security gap between non-malleability and indistinguishability.

6   NICE-X Cryptosystem

We enhance the security of NICE cryptosystem in the following. Indeed, we present NICE-X, which is secure against the chosen ciphertext attack under the SKEP assumption in the random oracle model.

The construction and the security proof of the NICE-X is similar to those of the REACT [OP01b], which is based on the paper by Bellare and Rogaway [BR93]. When we apply the REACT to the NICE cryptosystem, then the REACT-NICE cryptosystem is IND-CCA2, relative to the Gap-SKEP [OP01a], which consists of solving the SKEP with access to the DKP. Indeed, the encryption of the NICE cryptosystem $\mathcal{M} \to \mathcal{C}$ is a probabilistic encryption because of $\mathcal{C} = \mathcal{M}\mathcal{P}^r$ in equation (1). For given $\mathcal{C}, \mathcal{M}$ we cannot check whether $\mathcal{C}$ is the ciphertext of $\mathcal{M}$, but we can do it by asking the decisional kernel oracle if $\mathcal{C}\mathcal{M}^{-1} \in Ker(\varPhi)$ holds. However the Gap-SKEP is stronger assumption than the SKEP.

We modify the REACT to be suitable for the NICE-X cryptosystem and we prove that the NICE-X cryptosystem is IND-CCA2 under only the SKEP. Let $\mathcal{E}(m, r)$ be a probabilistic encryption function of a message m and a random mask r. The REACT encrypts a message k as follows: $A = \mathcal{E}(m, r), B = k \oplus g(m), H = h(m, k, A, B)$, where g, h are some hash functions. However the NICE-X encrypts a message k using the random mask r instead of the message m, namely $B = k \oplus g(\mathcal{P}^r)$ instead of $B = k \oplus g(\mathcal{M})$. Here we consider the following problem: "For a given ciphertext c and a random mask r, distinguish whether r is the random mask of the ciphertext c." We call it the Random-mask-Checking Attack RCA. In the case of the ElGamal encryption, RCA is equivalent to solving the decisional Diffie-Hellman problem. Let (g^r, my^r) be a ciphertext of the ElGamal encryption, where $y = g^s$ is the public-key. Then the RCA is to check whether $t \in <g>$ is equivalent to y^r, which is to distinguish whether (g^r, g^s, t) is a Diffie-Hellman triple. Therefore there is no advantage to modify the ElGamal-REACT in this way. On the contrary, the NICE encryption is secure against the RCA, although it is a probabilistic encryption like the ElGamal encryption. We can distinguish whether an ideal class $\mathcal{P}^{r'} \in Kcr(\Phi)$ is the random mask of the cipher class $\mathcal{C}$ by checking $\mathcal{C}(\mathcal{P}^{r'})^{-1} \in \mathsf{SI}(D)$ from lemma 2. If yes, we have $r = r'$. Moreover if $\mathcal{P}^r$ is the random mask for the cipher class $\mathcal{C}$, then $\mathcal{C}(\mathcal{P}^r)^{-1}$ is the message class corresponding to $\mathcal{C}$. Therefore we can compute the message class $\mathcal{M}$ from given $\mathcal{C}, \mathcal{P}, r$, and the SKEP can be solved.

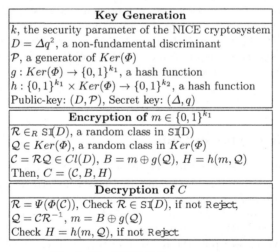

Key Generation
k, the security parameter of the NICE cryptosystem
$D = \Delta q^2$, a non-fundamental discriminant
$\mathcal{P}$, a generator of $Ker(\Phi)$
$g : Ker(\Phi) \to \{0,1\}^{k_1}$, a hash function
$h : \{0,1\}^{k_1} \times Ker(\Phi) \to \{0,1\}^{k_2}$, a hash function
Public-key: $(D, \mathcal{P})$, Secret key: (Δ, q)

Encryption of $m \in \{0,1\}^{k_1}$
$\mathcal{R} \in_R \mathsf{SI}(D)$, a random class in $\mathsf{SI}(D)$
$\mathcal{Q} \in Ker(\Phi)$, a random class in $Ker(\Phi)$
$\mathcal{C} = \mathcal{R}\mathcal{Q} \in Cl(D), B = m \oplus g(\mathcal{Q}), H = h(m, \mathcal{Q})$
Then, $C = (\mathcal{C}, B, H)$

Decryption of C
$\mathcal{R} = \Psi(\Phi(\mathcal{C}))$, Check $\mathcal{R} \in \mathsf{SI}(D)$, if not Reject,
$\mathcal{Q} = \mathcal{C}\mathcal{R}^{-1}, m = B \oplus g(\mathcal{Q})$
Check $H = h(m, \mathcal{Q})$, if not Reject

Fig. 2. The NICE-X Cryptosystem

The description of NICE-X is showed in figure 2. We use two hash functions $g : Ker(\Phi) \to \{0,1\}^{k_1}$ and $h : \{0,1\}^{k_1} \times Ker(\Phi) \to \{0,1\}^{k_2}$, which we see as a random oracle. A message m is chosen from $\{0,1\}^{k_1}$ and m is encrypted as follows: from the public-key $D, \mathcal{P}$ we can generate a random class $\mathcal{Q}$ in $Ker(\Phi)$

and a random class $\mathcal{R}$ in $\mathsf{SI}(D)$. Then the ciphertext C is the triple $(\mathcal{C}, B, H)$ where $\mathcal{C} = \mathcal{R}\mathcal{Q}$, $B = m \oplus g(\mathcal{Q})$, and $H = h(m, \mathcal{Q})$. We use $g(\mathcal{Q})$ instead of $g(\mathcal{R})$ to mask the message m, which is a different point from the REACT. The person who knows the secret key Δ, q can extract $\mathcal{R}$ using Ψ, Φ and computes $\mathcal{Q} = \mathcal{C}\mathcal{R}^{-1}$. The computation of $\mathcal{Q}$ is the extra cost from the decryption of the NICE cryptosystem. Here one checks if $\mathcal{R} \in \mathsf{SI}(D)$ to prevent the attack proposed by Joye, Quisquater, and Yung [JQY01]. Then he/she can decrypt $m = B \oplus g(\mathcal{Q})$. If $H = h(m, \mathcal{Q})$ holds, one accepts m as a valid ciphertext.

6.1   Security of NICE-X Cryptosystem

We change the security notations of section 3 to be suitable for the NICE-X cryptosystem. The messages class $\mathcal{M}_b$ and the challenge class $\mathcal{C}$ are replaced by the messages m_b and the challenge C, respectively. The advantage of the chosen ciphertext attack against the NICE-X cryptosystem is defined as follows:

$$Adv_{A,NICE-X}^{\mathsf{IND-CCA}}(k) = 2Pr[G, H \leftarrow \mathsf{Hash}; (m_0, m_1, st) \leftarrow A_1^{G,H,\mathcal{D_{NICE-X}}}(D, \mathcal{P});$$

$$b \leftarrow_R \{0,1\}; \mathcal{C} \leftarrow \mathcal{E}_{NICE-X}(m_b) : A_2^{G,H,\mathcal{D_{NICE-X}}}(st, C) = b] - 1.$$

We can prove the following theorem. The proof can be found in the appendix B. A difference from the statement of the REACT-RSA is the reduction time, namely our proposed reduction requires more $q_H T_{\mathcal{E}}$ than the REACT-RSA reduction, where q_H is the number of queries to oracle q_H and $T_{\mathcal{E}}$ is the encryption time of the NICE cryptosystem.

Theorem 5. *Let A be a (t, q_G, q_H, q_D)-chosen ciphertext attack algorithm of the NICE-X cryptosystem with advantage $\varepsilon = Adv_{A,NICE-X}^{\mathsf{IND-CCA}}$, with time t, after q_G, q_H, and q_D queries to the hash function G, H, and the decryption oracle, respectively. Then there is an algorithm for solving SKEP with a probability greater than $\frac{\varepsilon}{2}\left(1 - \frac{q_D}{2^{k_2}}\right)$ and time smaller than $t + (q_H + q_G)k + (q_H + q_G)T_{\mathcal{E}}$ in the random oracle model, where $T_{\mathcal{E}}$ is the encryption time of the NICE-X cryptosystem and k is the security parameter of the NICE cryptosystem.*

7   Comparison with Other Cryptosystems

We compare the NICE cryptosystem with the RSA and OAEP-RSA cryptosystem [BR94] and the REACT-RSA cryptosystem [OP01b]. The description of these schemes can be found the appendix C.

7.1   Efficiency Comparison

In order to compare these cryptosystems we assume that the same bit-lengths of the RSA-modulus n and the discriminant D are same. Indeed we choose 1024 bits and 2048 bits. In figure 3 we indicate the running time estimation for these algorithms. The numbers in the figure indicate the number of 1024-bit modular multiplications required for executing the encryption and the decryption of these algorithms. We do not add the time needed for computing a hash function, because it is much faster than the calculation time of a modular exponentiation. We assume that one multiplication of two 2048-bit integers is $2^2 = 4$ times slower than a 1024-bit multiplication.

Schemes	RSA	OAEP-RSA	REACT-RSA	NICE	NICE-X
Text size for a 1024-bit public key					
Plaintext	1024	496	1024	169	512
Ciphertext	1024	1024	2176	1024	1664
Running time for a 1024-bit public key					
Encryption	17	17	17	12800	12800
Decryption	384	384	384	26	49
Text size for a 2048-bit public key					
Plaintext	2048	992	2048	340	1024
Ciphertext	2048	2024	4352	2048	3328
Running time for a 2048-bit public key					
Encryption	68	68	68	102400	102400
Decryption	1536	1536	1536	102	194

Fig. 3. Comparison of the efficiency

The speed-up using the Chinese remainder theorem is used for the decryption of RSA and we estimate it 4 times faster than the original decryption. We need about $1.5x$ modular multiplications of $\mathbb{Z}/n\mathbb{Z}$ for computing a modular exponentiation $a^x \in \mathbb{Z}/n\mathbb{Z}$. Then the decryption of a 1024-bit RSA requires $1024 * 1.5/4 = 384$ multiplications, and the 2048-bit RSA requires $1024*4*1.5/4 = 1536$ multiplications. For the security parameter of the RSA-OAEP cryptosystem we choose $k_1 = 128, k_0 = 400$ for a 1024-bit n and $k_1 = 256, k_0 = 800$ for a 2048-bit n, respectively. For the security parameter of the REACT-RSA cryptosystem we choose $k_1 = 1024$ and $k_2 = 128$ for a 1024-bit n and $k_1 = 2048$ and $k_2 = 256$ for a 2048-bit n, respectively.

Next we discuss the running time of the NICE/NICE-X cryptosystem. We assume that the bit-length of the public key D of the NICE is equal to that of the RSA-modulus n. An ideal class $\mathcal{A} \in Cl(D)$ is represented by $\mathcal{A} = [(a, b)]$, where $0 < a < \sqrt{|D|/3}, b^2 \equiv D \bmod 4a$. The second coefficient b can be computed by a, D. A message ideal of the NICE is restricted $a < |D|^{1/6}/2$. For the security parameter of the NICE/NICE-X cryptosystem we choose $k_1 = 512$ and $k_2 = 128$ for a 1024-bit D and $k_2 = 1024$ and $k = 256$ for a 2048-bit D, respectively. From the timings of the implementation in software [HPT99] [PT00] we assume the timing of several operations of NICE as follows: A multiplication of $Cl(D)$ is about 23 times slower than that of the modular multiplication of $\mathbb{Z}/n\mathbb{Z}$. It takes about 2 times slower to generate a class idea of $\mathsf{SI}(D)$ than a modular multiplication of $\mathbb{Z}/n\mathbb{Z}$. The running time to compute $\Phi(\Psi(\cdot))$, which is used for the decryption of the NICE, is also around 1.5 times slower than the encryption of the RSA cryptosystem with $e = 2^{16} + 1$. The NICE-X cryptosystem has to compute $Q = \mathcal{CR}^{-1}$ in the decryption, which is the extra cost from the NICE cryptosystem. An inverse of an ideal $\mathcal{A} = [(a, b)]$ is computed by $\mathcal{A}^{-1} = [(a, -b)]$ and it is negligible. In this estimation we assume that n and D have the same bit-length.

The decryption of the NICE/NICE-X cryptosystem is much faster than that of the RSA, OAEP-RSA, and REACT-RSA. If we choose a longer public key, then the difference will become even larger because of the quadratic complexity $\mathcal{O}((\log |D|)^2)$ of the NICE decryption.

7.2 Security Comparison

In Figure 4 we show the comparison of the security of these cryptosystems. The original RSA cryptosystem is not IND-CPA. The security of the OAEP-RSA cryptosystem can be proven IND-CCA2 [FOPS01]. Let $\varepsilon_{OAEP}, t_{OAEP}$ be the advantage and the time of the chosen ciphertext attack adversary against RSA-OAEP after q_D, q_G, and q_H queries to the decryption oracle, and the hash functions g, h, respectively. Then the RSA problem can be solved with probability ε'_{OAEP}:

$$\varepsilon'_{OAEP} > \frac{\varepsilon^2_{OAEP}}{16}\left(1 - \frac{q_G}{2^{k_0-1}} - \frac{q_H}{2^{k-k_0-1}}\right)$$
$$-\varepsilon_{OAEP}\left(\frac{q_D}{2^{k_1+1}} + \frac{q_D(2q_G+1)}{2^{k_0+2}} + \frac{q_G}{2^{k+1}} + \frac{1}{2^{k-2k_0-4}}\right),$$

within a time bound $t'_{OAEP} \leq 2t_{OAEP} + q_H(q_H + q_G)\mathcal{O}(k^3)$. When we assume that $q_H = q_G = 2^{32}$ and that the running time of $\mathcal{O}(k^3)$ takes 1 ms for a 1024-bit RSA modulus, then we get $t'_{OAEP} \leq 2t_{OAEP} + T$, where T is 1169.9 million years. This time T is larger than the time to factor a 1024-bit RSA modulus [Sil00]. Similarly for $q_H = q_G = 2^{16}$, T is 99.4 days and it is meaningful. The advantage ε'_{OAEP} is about ε^2_{OAEP} for a practical size of the RSA modulus, i.e., 1024 bits. Indeed, for $k_1 = 128, k_0 < 400$ and $q_D, q_G < 2^{32}$, we have

$$\varepsilon'_{OAEP} \approx \frac{1}{16}\varepsilon^2_{OAEP} + \mathcal{O}(\frac{\varepsilon_{OAEP}}{2^l}), \quad l > 93.$$

For $\varepsilon_{OAEP} = 1/2^{64}$ we have $\varepsilon'_{OAEP} \approx 1/2^{132}$ and the advantage becomes negligible.

The security of the REACT cryptosystem can be also proved IND-CCA2 [OP01b]. Let $\varepsilon_{REACT}, t_{REACT}$ be the advantage and the time of the chosen ciphertext attack adversary against REACT-RSA after q_D, q_G, and q_H queries to the decryption oracle and the hash functions g, h, respectively. Then the RSA problem can be solved with probability ε'_{REACT}:

$$\varepsilon'_{REACT} > \frac{\varepsilon_{REACT}}{2} - \frac{q_D}{2^{k_2}},$$

within a time bound $t'_{REACT} \leq t_{REACT} + (q_H + q_G)k + q_G\mathcal{O}(k^3)$. The advantage and the time of this security reduction is very tight.

Security					
Schemes	RSA	OAEP-RSA	REACT-RSA	NICE	NICE-X
One-wayness	RSA	RSA	RSA	SKEP	SKEP
IND-CPA	–	RSA	RSA	DKP	SKEP
IND-CCA	–	RSA	RSA	–	SKEP
(advantage)	–	$\approx \varepsilon_{OAEP}^2$	$\approx \varepsilon_{REACT}$	–	$\approx \varepsilon_{NICE}$
(time)	–	$2t_{OAEP} + q_H($ $q_H + q_G)\mathcal{O}(k^3)$	$t_{REACT} + (q_H+$ $q_G)k + q_G\mathcal{O}(k^3)$	–	$t_{NICE} + (q_H + q_G$ $)k + (q_H + q_G)T_\varepsilon$

Fig. 4. Comparison of the security

Next, the NICE cryptosystem is not IND-CCA2 [JJ00] but IND-CPA (theorem 4). However the security of the NICE-X cryptosystem can be proved IND-CCA2 (theorem 5). Let $\varepsilon_{NCIE}, t_{NICE}$ be the advantage and the time of the chosen ciphertext attack adversary against NICE-X after q_D, q_G, and q_H queries to the decryption oracle and the hash functions g, h, respectively. Then the SKEP problem can be solved with probability ε'_{NICE}:

$$\varepsilon'_{NICE} > \frac{\varepsilon_{NICE}}{2}\left(1 - \frac{q_D}{2^k}\right) > \left(\frac{\varepsilon_{NICE}}{2} - \frac{q_D}{2^k}\right),$$

within a bounded time $t'_{NICE} \leq t_{NICE} + (q_H + q_G)k + (q_H + q_G)T_\varepsilon$, where T_ε is the encryption time of the NICE-X cryptosystem. The time t'_{NICE} and the advantage ε'_{NICE} of this security reduction are about the same with those of the REACT-RSA.

References

BLK00. J. Baek, B. Lee, and K. Kim "Provably secure length-saving public key encryption scheme under the computational Diffie-Hellmam assumption," ETRI J, Vol.22, No.4, (2000), pp.25-31.

BDPR98. M. Bellare, A. Desai, D. Pointcheval, and P. Rogaway, "Relations among notions of security for public-key encryption schemes," Advances in Cryptology – CRYPTO'98, LNCS 1462, (1998), pp.26-45.

BR93. M. Bellare and P. Rogaway, "Random oracles are practical: a paradigm for designing efficient protocols," First ACM Conference on Computer and Communications Security, (1993), pp.62-73.

BR94. M. Bellare and P. Rogaway, "Optimal asymmetric encryption - How to encrypt with RSA," Advances in Cryptology - EUROCRPT'94, LNCS 950, (1994), pp.92-111.

Ble98. D. Bleichenbacher, "A chosen ciphertext attack against protocols based on RSA encryption standard PKCS # 1," Advances in Cryptology – CRYPTO'98, LNCS 1462, (1998), pp.1-12.

Bon01. D. Boneh, "Simplified OAEP for the RSA and Rabin Functions," Advances in Cryptology – CRYPTO 2001, LNCS 2139, (2001), pp.275-291.

BW88. J. Buchmann and H. C. Williams, "A key-exchange system based on imaginary quadratic fields," Journal of Cryptology, 1, (1988), pp.107-118.

BST01. J. Buchmann, K. Sakurai, and T. Takagi, "An IND-CCA2 public-key cryptosystem with fast decryption," Darmstadt University of Technology, Technical Report No. TI-10/01, (2001). http://www.informatik.tu-darmstadt.de/TI/Veroeffentlichung/TR/Welcome.html

CGH98. R. Canetti, O. Goldreich, and S. Halevi, "The random oracle model, revisited," 30th Annual ACM Symposium on Theory of Computing, (1998).

CS98. R. Cramer and V. Shoup, "A practical public key cryptosystem provably secure against adaptive chosen ciphertext attack," Advances in Cryptology – CRYPTO'98, LNCS 1462, (1998), pp.13-25.

DDN00. D. Dolev, C. Dwork, and M. Naor, "Non-malleable cryptography," SIAM Journal of Computing, Vol. 30 (2), (2000), pp.391-437.

FO99a. E. Fujisaki and T. Okamoto, "How to enhance the security of public-key encryption at minimum cost," 1999 International Workshop on Practice and Theory in Public Key Cryptography, LNCS 1560, (1999), pp.53-68.

FO99b. E. Fujisaki and T. Okamoto, "Secure integration of asymmetric and symmetric encryption schemes," Advances in Cryptology – CRYPTO'99, LNCS 1666, (1999), pp.537-554.

FO01. E. Fujisaki and T. Okamoto, "A chosen-cipher secure encryption scheme tightly as secure as factoring," IEICE Trans. Fundamentals, Vol. E84-A, No.1, (2001), pp.179-187.

FOPS01. E. Fujisaki, T. Okamoto, D. Pointcheval, and J. Stern, "RSA-OAEP Is Secure under the RSA Assumption," Advances in Cryptology – CRYPTO 2001, LNCS 2139, (2001), pp.260-274.

HM89. J. L. Hafner and K. S. McCurley, "A rigorous subexponential algorithm for computation of class groups, " J. Amer. Math. Soc., 2, (1989), pp.837-850.

HPT99. M. Hartmann, S. Paulus, and T. Takagi, "NICE –New Ideal Coset Encryption–, " Conference of Hardware Embedding System (CHES), LNCS 1717, (1999).

HIME01. HIME, HITACHI Systems Development Laboratories, http://www.sdl.hitachi.co.jp/crypto/hime/, "Design and analysis of fast provably secure public-key cryptosystems based on a modular squaring" in these proceedings.

JJ00. E. Jaulmes and A. Joux; "A NICE cryptanalysis," Advances in Cryptology – EUROCRYPT'2000, LNCS 1807, (2000), pp.382-391.

JQY01. M. Joye, J.-J. Quisquater, and M. Yung, "On the power of misbehaving adversaries and security analysis of the original EPOC," In Proceedings of the Cryptographers' Track at RSA Conference '2001, LNCS 2020, (2001), pp.208-222.

KI01. K. Kobara and H. Imai, "Semantically secure McEliece public-key cryptosystems-conversions for McEliece PKC," 2001 International Workshop on Practice and Theory in Public Key Cryptography, LNCS 1992, (2001), pp.19-35.

KOMM01. K. Kurosawa, W. Ogata, T. Matsuo, and S. Makishima, "IND-CCA public key schemes equivalent to factoring n = pq," 2001 International Workshop on Practice and Theory in Public Key Cryptography, LNCS 1992, (2001), pp.36-47.

Len87. H. W. Lenstra, Jr., Factoring integers with elliptic curves, Annals of Mathematics, 126, (1987), pp.649-673.

LL91. A. K. Lenstra and H. W. Lenstra, Jr. (Eds.), The development of the number field sieve. Lecture Notes in Mathematics, 1554, Springer, (1991).

Mue01. S. Müller, "On the security of Williams based public key encryption scheme," 2001 International Workshop on Practice and Theory in Public Key Cryptography, LNCS 1992, (2001), pp.1-18.

OP01a. T. Okamoto and D. Pointcheval, "The Gap-Problems: a new class of problems fro the security of cryptographic schemes," 2001 International Workshop on Practice and Theory in Public Key Cryptography, LNCS 1992, (2001), pp.104-118.

OP01b. T. Okamoto and D. Pointcheval, "REACT: Rapid Enhanced-security Asymmetric Cryptosystem Transform," In Proceedings of the Cryptographers' Track at RSA Conference '2001, LNCS 2020, (2001), pp.159-175.

OUF98. T. Okamoto, S. Uchiyama, and E. Fujisaki, "EPOC: Efficient Probabilistic Public-Key Encryption," Submission to IEEE P1363a, (1998).

PP99. P. Paillier and D. Pointcheval, "Efficient public-key cryptosystem provably secure against active adversaries," Advances in Cryptology – ASIACRYPT'99, LNCS 1716, (1999), pp.165-179.

PT00. S. Paulus and T. Takagi, "A new public-key cryptosystem over quadratic orders with quadratic decryption time", Journal of Cryptology, 13, (2000), pp.263-272.

Poi99a. D. Pointcheval, "New public key cryptosystems based on the dependent-RSA problems," Advances in Cryptography - Eurocryt'99, LNCS 1592, (1999), pp. 239-254.

Poi99b. D. Pointcheval, "HD-RSA: Hybrid Dependent RSA – a New Public-Key Encryption Scheme," Submission to IEEE P1363a. October (1999).

Poi00. D. Pointcheval, "Chosen-ciphertext security for any one-way cryptosystem," 2000 International Workshop on Practice and Theory in Public Key Cryptography, LNCS 1751, (2000), pp.129-146.

RS91. C. Rackoff and D. Simon, "Noninteractive zero-knowledge proof of knowledge and chosen ciphertext attack," Advances in Cryptology – CRYPTO'91, LNCS 576, (1991), pp.433-444.

Sou01a. V. Shoup, "OAEP reconsidered," Advances in Cryptology – CRYPTO 2001, LNCS 2139, (2001), pp.239-259.

Sou01b. V. Shoup, "A proposal for an ISO standard for public key encryption," http://shoup.net/

SG98. V. Shoup and R. Gennaro, "Securing threshold cryptosystems against chosen ciphertext attack," Advanced in Cryptology – Eurocrypt'98, LNCS 1403, (1998), pp.1-16.

Sil00. R. Silverman, "A cost-based security analysis of symmetric and asymmetric key lengths," RSA Laboratories, Bulletin 13, (2000), pp.1-22.

TY98. Y. Tsiounis and M. Yung, "On the security of El Gamal based encryption," 1998 International Workshop on Practice and Theory in Public Key Cryptography, LNCS 1431, (1998), pp.117-134.

A   Proof of Theorem 4

Let $A = (A_1, A_2)$ be a chosen plaintext attack, which consists of the find-stage A_1 and the guess-stage A_2, with time t and an advantage $\varepsilon = Adv_{A,NICE}^{\mathsf{IND-CPA}}$. We will construct an adversary B who can solve the DKP with an advantage $\varepsilon/2$ and time $t + T_\varepsilon$.

Let $Rand$ and $Kernel$ be the two distributions defined in DKP of section 4. We sample variable $\mathcal{A}$ from random the distribution $Rand$ or $Kernel$. Then adversary B works as follows:

/* Adversary $B(\mathcal{A})$ */
1. B runs $A_1(D, \mathcal{P}, k)$ and obtains $\mathcal{M}_0, \mathcal{M}_1, st$.
2. B chooses $b \in_R \{0, 1\}$ and computes $\mathcal{B} = \mathcal{M}_b \mathcal{A} \in Cl(D)$.
   B then runs $A_2(st, \mathcal{M}_0, \mathcal{M}_1, \mathcal{B})$ and obtains c.
3. If $c = b$ Return 1, else Return 0.

We will study $Adv_{B,DKP}$ which is the advantage of adversary B for the decisional kernel problem. For the two distributions $Rand$ and $Kernel$ we will compare the probabilities of $b = c$.

When the variable $\mathcal{A}$ is sampled from the random distribution $Rand$, then the ideal class $\mathcal{B}$ is uniformly distributed in $Cl(D)$ and it is independent of b. Then we get

$$Pr[\mathcal{A} \leftarrow Rand : B(\mathcal{A}) = 1] = Pr[\mathcal{A} \leftarrow Rand : c = b] = 1/2.$$

On the other hand when $\mathcal{A}$ is sampled from the distribution $Kernel$, $\mathcal{B}$ is a valid ciphertext of $\mathcal{M}_b$. Then we have the following probability:

$$Pr[\mathcal{A} \leftarrow Kernel : B(\mathcal{A}) = 1] = Pr[\mathcal{A} \leftarrow Kernel : c = b]$$
$$= Pr_b[A_2(st, \mathcal{M}_0, \mathcal{M}_1, \mathcal{E}(\mathcal{M}_b)) = b]$$

In section 3 we defined the advantage of the chosen plaintext attack against the NICE cryptosystem:

$$Adv_{A,NICE}^{\mathsf{IND-CPA}} = 2Pr_b[A_2(st, \mathcal{M}_0, \mathcal{M}_1, \mathcal{E}(\mathcal{M}_b)) = b] - 1.$$

Therefore we get $Pr[\mathcal{A} \leftarrow Kernel : B(\mathcal{A}) = 1] = 1/2 + Adv_{A,NICE}^{\mathsf{IND-CPA}}/2$.

From equation (4) the advantage of B to distinguish the $Rand$ and $Kernel$ distribution is

$$Adv_{B,DKP} = |Pr[\mathcal{A} \leftarrow Rand : B(\mathcal{A}) = 1] - Pr[\mathcal{A} \leftarrow Kernel : B(\mathcal{A}) = 1]|$$
$$= Adv_{A,NICE}^{\mathsf{IND-CPA}}/2.$$

Consequently $Adv_{B,DKP}$ is $\varepsilon/2$. Adversary B has to compute $\mathcal{B} = \mathcal{M}_b \mathcal{A}$ in step 2, which is the extra cost from the time of adversary A. The time needed to compute $\mathcal{B} = \mathcal{M}_b \mathcal{A}$ is at most the encryption time of the NICE cryptosystem.

B Proof of Theorem 5

Denote by $A = (A_1, A_2)$ the chosen ciphertext adversary to the NICE-X cryptosystem. The indistinguishability of the NICE-X cryptosystem comes from the following statement: Any adversary, in order to have any information about the message m, has to ask the query $\mathcal{Q}$ to the random oracle g, or he has to ask the query $(-, \mathcal{Q}, -, -)$ to the random oracle h, where $\mathcal{Q} \in Ker(\Phi)$.

We describe a simulator S. The simulator would work as an adversary to solve the SKEP if it were in the real attack.

/* Simulator S */

1. S is given $C^* = (\mathcal{C}^*, B^*, H^*)$, where $\mathcal{C}^* \leftarrow \mathcal{R}^* \mathcal{Q}^*$, for $\mathcal{R}^* \leftarrow_R \mathsf{SI}(D)$, $\mathcal{Q}^* \leftarrow_R Ker(\varPhi)$, and $B^* \leftarrow_R \{0,1\}^{k1}$, $H^* \leftarrow_R \{0,1\}^{k2}$. Here S would like to compute the ideal class $\mathcal{R}^*$, which is smallest kernel-equivalent class of $\mathcal{C}^*$.

2. S runs $A_1(D, \mathcal{P}, k)$ and obtains m_0, m_1, st. Then S chooses $b \in_R \{0,1\}$ and outputs C^* as the ciphertext of m_b.

3. S runs $A_2(st, m_0, m_1, C^*)$ and obtains an answer c. Then S outputs the smallest kernel-equivalent class of $\mathcal{C}^*$, if one has been found among the queries asked to g and h (see below).

We have to simulate both the random oracle and the decryption oracle. The simulator S manages $LIST_h$ and $LIST_g$ which are initially empty.

g-**queries:** The simulator manages a list $LIST_g$ which consists of $(\gamma, g(\gamma))$, where $\gamma = \mathcal{Q} \in_R Ker(\varPhi)$. For any new query γ to the random oracle g, the simulator S returns a new k_1-bit random value $g(\gamma)$ and adds $(\gamma, g(\gamma))$ to the $LIST_g$. If $\gamma = \mathcal{Q}$ already appears as $(r, g(r))$ in the first coordinate of the list $LIST_g$, then we return $g(r) = g(\gamma)$.

h-**queries:** The simulator manages a list $LIST_h$ which consists of $(\delta, \beta, h(\delta))$, where $\delta = (m, \mathcal{Q}) \in_R \{0,1\}^{k1} \times Ker(\varPhi)$ and $\beta = m \oplus g(\mathcal{Q})$. For any new query $\delta = (m, \mathcal{Q})$ to the random oracle h, the simulator S returns a new k_2-bit random value $h(\delta)$, and computes $\beta = m \oplus g(\mathcal{Q})$ by asking the query $\mathcal{Q}$ to the random oracle g to obtain $g(\mathcal{Q})$. Note that β is well-defined. The simulator $\mathcal{S}$ then adds $(\delta, \beta, h(\delta))$ to the list $LIST_h$. If $\delta = (m, \mathcal{Q})$ already appears as $(s, h(s))$ in the first coordinate of the list $LIST_h$, then we return $h(s) = h(\delta)$.

Decryption queries: Let $C = (\mathcal{C}, B, H)$ be a ciphertext output by the adversary A, which is $C \neq C^*$. The simulator S decrypts C or rejects it as an invalid ciphertext. The simulator searches the tuples $(\delta, \beta, h(\delta))$ with $h(\delta) = H$ in the $LIST_h$. Note that if $\mathcal{C}\mathcal{Q}^{-1} \in \mathsf{SI}(D)$ holds, then $\mathcal{C}\mathcal{Q}^{-1}$ is the smallest kernel-equivalent class of $\mathcal{C}$ from lemma 2. Thus the simulator can check whether $\mathcal{Q}$ is the correct random masking of $\mathcal{C}$. Here for the above tuples $(\delta, \beta, h(\delta))$, the simulator checks whether $\beta = B$ and $\mathcal{C}\mathcal{Q}^{-1} \in \mathsf{SI}(D)$. If yes, the simulator found a tuple $(m, \mathcal{Q})$ such that $\mathcal{C} = \mathcal{R}\mathcal{Q}$ (for a $\mathcal{R} \in \mathsf{SI}(D)$), $B = m \oplus g(\mathcal{Q})$, and $H = h(m, \mathcal{Q})$. Then the simulator responds with m as the decryption of C. Otherwise, the simulator S considers it as an invalid ciphertext and returns the reject symbol Reject.

When we find $\mathcal{Q}^*$ in the final list $LIST_g$, we can obtain the smallest kernel-equivalent class of $\mathcal{C}^*$ by computing $\mathcal{R}^+ = \mathcal{C}^*(\mathcal{Q}^*)^{-1}$ and checking $\mathcal{R}^+ \in \mathsf{SI}(D)$ from lemma 2. Thus the simulator computes $\mathcal{C}^* \mathcal{Q}^{-1}$ and checks $\mathcal{C}^* \mathcal{Q}^{-1} \in \mathsf{SI}(D)$ for all $\mathcal{Q} \in LIST_g$. This completes the description of the simulator S.

We will estimate the probability that $\mathcal{Q}^*$ is contained in the final list of $LIST_g$ in the following. Our goal is to prove that $Pr^{sim}[\mathcal{Q}^* \in LIST_g]$ is non-negligible, where Pr^{sim} is the probability in the simulator game. The simulator S would work as an adversary to solve the SKEP if he were in the real attack. Denote by Pr^{real} the probability in the real attack. At first we will prove

$$Pr^{real}[\mathcal{Q}^* \in LIST_g] \geq \varepsilon/2.$$

The elements $Q^* \in LIST_g$ are generated by a query for $g(Q^*)$ or a query for $h(-, Q^*, -, -)$. Because of the randomness of the random oracle g, the attacker cannot obtain any advantage in the real game without $Q^* \in LIST_g$. Then the probability to guess the bit b is $1/2$.

$$Pr_b^{real}[(A_2(st, m_0, m_1, A_2(C^*) = b)|\neg(Q^* \in LIST_g)] = 1/2.$$

Next we have the following relationship:

$$Pr_b^{real}[A_2 = b] = 1/2 + Adv_{A,NICE-X}^{IND-CCA}/2$$
$$= Pr_b^{real}[(A_2 = b) \wedge \neg(Q^* \in LIST_g)] + Pr_b^{real}[(A_2 = b) \wedge (Q^* \in LIST_g)]$$
$$= Pr_b^{real}[(A_2 = b)|\neg(Q^* \in LIST_g)] \times Pr_b^{real}[\neg(Q^* \in LIST_g)]$$
$$+ Pr_b^{real}[(A_2 = b) \wedge (Q^* \in LIST_g)] \leq \frac{1}{2} + Pr_b^{real}[(Q^* \in LIST_g)]$$

Therefore $Pr^{real}[(Q^* \in LIST_g)] \geq \varepsilon/2$.

Next, we show that the adversary A cannot distinguish the simulation from the real attack until it issues a query for $g(Q^*)$. We would like to prove the following statement:

$$Pr^{sim}[(Q^* \in LIST_g)] \geq Pr^{real}[(Q^* \in LIST_g)](1 - \frac{q_D}{2^{k_2}}).$$

Here we assume that the following event happens during the simulation: Valid, "all valid ciphertexts issued by adversary A are decrypted during this simulation." In the case of Valid, all the decryption queries are answered correctly. Therefore, in the point of the adversary A's view, the distribution during the simulation is sampled from the same distribution during the real attack. We estimate the probability that event Valid does not happen. A valid ciphertext can be produced without asking the query (m, Q, C, B) to the random oracle h. An attacker guesses the values of $h(m, Q, C, B)$, but its probability is $1/2^{k_2}$. The probability which the attacker cannot guess H during the attack is $(1 - 1/2^{k_2})^{q_D} \geq 1 - q_D/2^{k_2}$. Therefore we proved the statement of the probability.

In the simulation we used at most $(q_G + q_H)$ operations for generating $Q, Q^* \in_R Ker(\Phi)$, computing ideal multiplications of $Cl(D)$, and computing two hash functions g, h. They can be computed in time $(q_G + q_H)T_{\mathcal{E}} + (q_G + q_H)k$, where $T_{\mathcal{E}}$ is the encryption time of the NICE-X cryptosystem, and k is the security parameter of the NICE cryptosystem.

C   RSA, OAEP-RSA, and REACT-RSA

RSA cryptosystem: Let $n = pq$ be the RSA modulus and let e be the encryption key of the RSA cryptosystem. The public key is (e, n). The secret key d is generated by equation $d \equiv e^{-1} \mod \varphi(n)$. A message m is chosen from $\{1, 2, 3, ..., n\}$ and it is encrypted by $c \equiv m^e \mod n$. Then c is the ciphertext of m. The message m is recovered by $m \equiv c^d \mod n$ using the secret key. Here we choose the encryption key $e = 2^{16} + 1$. The security of the RSA cryptosystem is based on the RSA problem, which finds m for given $n, e, c(\equiv m^e \mod n)$.

OAEP-RSA cryptosystem [BR94]: Let (n, e) and d be the public key and the secret key of the RSA, respectively. k is the bit-length of n. Let l, k_0, k_1 be positive integers which satisfy $k = l + k_0 + k_1$. Let g, h be two hash functions $g : \{0, 1\}^{k_0} \to \{0, 1\}^{k-k_0}$, $h : \{0, 1\}^{k-k_0} \to \{0, 1\}^{k_0}$. A message m is chosen from $\{0, 1\}^l$. For a random integer $r \leftarrow \{0, 1\}^{k_0}$, the message m is encrypted by

$$s = m||0^{k_1} \oplus g(r), \quad t = r \oplus h(s), \quad c \equiv (s||t)^e \bmod n.$$

Then c is the ciphertext of m. The message m is recovered as follows:

$$s||t \equiv c^d \bmod n, \quad r = t \oplus h(s), \quad M = s \oplus g(r),$$

if the k_1 least significant bits of M are zero, we return m as $M = m||0^{k_1}$, otherwise we return the reject symbol.

REACT-RSA cryptosystem [OP01b]: The REACT-RSA cryptosystem uses the same public key (e, n) and the same secret key d of the RSA cryptosystem. Let k_1, k_2 be positive integers. Let g, h be two hash functions $g : \mathbb{Z}/n\mathbb{Z} \to \{0, 1\}^{k_1}$, $h : \{0, 1\}^{k_1} \times \mathbb{Z}/n\mathbb{Z} \times \mathbb{Z}/n\mathbb{Z} \times \{0, 1\}^{k_2} \to \{0, 1\}^{k_2}$. A message m is chosen from $\{0, 1\}^{k_1}$. For a random integer $k \in_R \mathbb{Z}/n\mathbb{Z}$, the message m is encrypted by $A \equiv k^e \bmod$, $B = m \oplus g(k)$, $H = h(m, k, A, B)$, then $C = (A, B, H)$ is the ciphertext of m. The message m is recovered as follows:

$$k \equiv A^d \bmod n, \quad m = B \oplus g(k),$$

if $H = h(k, m, A, B)$ is satisfied, we return m as the message, otherwise we return the reject symbol. The security of the REACT-RSA is based on the RSA problem.

D   IND-CCA2 Public-Key Cryptosystems

In this appendix we list up several IND-CCA2 public-key cryptosystems. We classify them according to their reduced number-theoretic problems. We categorize them into four types of schemes: (1)the reduced problem is a decisional problem, (2)the reduced problem is a computational problem, (3)one-way permutation RSA type, and (4)the type without using the random oracle model. We also state their efficiency of the decryption process.

The schemes of type (1) are as follows: Pointcheval proposed the DRSA-1 scheme based on the decisional dependent-RSA problem [Poi99a]. The decryption of DRSA-1 is as fast as that of the RSA. Paillier and Pointcheval enhanced the Paillier cryptosystem to IND-CCA2 under the decisional composite residuosity and decisional partial discrete logarithms problems over $\mathbb{Z}/n^2\mathbb{Z}$ [PP99]. The decryption of Paillier and Pointcheval is slower than that of the RSA for a 1024-bit modulus. Tsiounis and Yung proved that the ElGamal cryptosystem is indistinguishable under the decisional Diffie-Hellman problem + the Schnorr signature problem [TY98]. Shoup and Gennaro also proposed an ElGamal based scheme under the decisional Diffie-Hellman problem [SG98]. Their decryption requires several exponentiations of the base group.

The schemes of type (2) are as follows: Pointcheval proposed the DRSA-2 scheme based on the computational dependent-RSA problem [Poi99a] and it is extended to the HD-RSA under the RSA problem [Poi99b]. Their decryptions are as efficient as the RSA's. Fujisaki and Okamoto [FO99a] [FO99b] and Pointcheval [Poi00] proposed a general conversion technique to make a public-key cryptosystem IND-CCA2. They proposed an ElGamal based cryptosystem under the computational Diffie-Hellman problem and an enhancement of the EPOC cryptosystem under the factoring problem of the form p^2q, where p, q are primes. Their decryptions need to compute the whole encryption process. Recently Okamoto and Pointcheval presented a conversion 'REACT' [OP01b]. The REACT is based on the paper by Bellare and Rogaway [BR93]. The REACT requires only two extra hash functions for its encryption and decryption. The REACT-RSA is IND-CCA2 under the RSA problem, but the REACT-ElGamal is IND-CCA2 under the Gap Diffie-Hellman problem and the REACT-EPOC is IND-CCA2 under the Gap high residuosity problem. The Gap problem is introduced by Okamoto and Pointcheval [OP01a]. Moreover there are two Diffie-Hellman based proposals to IEEE P1363a, the ECIES and the PSEC series, which are IND-CCA2 under the Gap Diffie-Hellman problem or on other assumption (there are more discussions about the schemes [Sou01b]). Their decryptions require 1 (or 2) exponentiations of the base group.

The schemes of type (3) are as follows: The decryption of all schemes of type (3) is as efficient as that of the RSA. Bellare and Rogaway proposed the OAEP-RSA [BR94]. Fujisaki et al proved that it is IND-CCA2 under the RSA assumption, although the security reduction is not tight [FOPS01]. As an enhancement of the OAEP-RSA, Shoup proposed the OAEP+, whose security reduction to the RSA problem is better than that of the OAEP-RSA [Sou01a]. Boneh enhanced the OAEP-RSA to the SAEP-RSA, which uses only a Feistel network, but whose security reduction is the same as in the OAEP-RSA [Bon01]. In the same paper, he also proposed the SAEP-Rabin, that is IND-CCA2 under the factoring problem and its security reduction is tight. There are three more IND-CCA2 schemes relative to the factoring problem. Kurosawa et al converted an encryption scheme using a reciprocal number to IND-CCA2 relative to the factoring problem [KOMM01]. Müller presented an IND-CCA2 Williams-type scheme [Mue01]. Hitachi proposed the HIME, which is an IND-CCA2 scheme relative to the factoring of the following composite integers $p^dq, (d > 1)$ or $\prod_{i=1}^{d} p_i, (d > 2)$, where p, q, p_i are prime numbers [HIME01].

The schemes of type (1),(2), and (3) are using the random oracle model, but the schemes of type (4) do not require the random oracle model. Cramer and Shoup proposed a discrete logarithm based cryptosystem under the decisional Diffie-Hellman problem for a subgroup of $\mathbb{Z}/p\mathbb{Z}$ [CS98]. Cramer-Shoup cryptosystem requires 3 (or 4) exponentiations of the base group.

Finally we note that Kobara et al enhanced the McEliece cryptosystem to an IND-CCA2 scheme [KI01].

Improvement of Probabilistic Public Key Cryptosystems Using Discrete Logarithm*

Dug-Hwan Choi[1], Seungbok Choi[2], and Dongho Won[3]

[1] Authentication Technology Research Center, Sungkyunkwan University
300 Chunchun-Dong, Jangan-Gu, Suwon
Kyunggi-Do 440-746, Korea
dchoi@dosan.skku.ac.kr
[2] Future Systems Incorporation, Koland Bldg. (8F), 1009-1
Daechi-Dong, Kangnam-Gu, Seoul 135-851, Korea
sbchoi@future.co.kr
[3] School of Electrical and Computer Engineering, Sungkyunkwan University
300 Chunchun-Dong, Jangan-Gu, Suwon
Kyunggi-Do 440-746, Korea
dhwon@simsan.skku.ac.kr

Abstract. We investigate two different probabilistic public key cryptosystems, one proposed by Okamoto and Uchiyama and the other by Paillier. Both of them are based on the discrete logarithmic function and the messages are calculated from the modular product of two those functions, one of which has a fixed value depending on a given public key. The improvements are achieved by a good choice for the public key so that it is possible to get efficient algorithms.

1 Introduction

Even though various public key cryptosystems have made after Diffe and Hellman [2] proposed that concept in 1976, nothing but RSA-Rabin and Diffe-Hellman propositions have been used pratically. Only Rabin scheme [8] and its variants are known to be provably secure against the passive adversary. However they do not guarantee that any partial information about the plaintext from the ciphertext could not be deduced. For this security level, *semantic security* was introduced and Probabilistic Public Key Encryption [3] was proposed by Goldwasser and Micali.

Okamoto and Uchiyama [5] suggested a new probabilistic encryption scheme using Logarithm function which is defined on p-Sylow subgroup of Z_n where $n = p^2 q$ with prime p and q. This is as secure as the intractability of factoring n. Also its semantic security under the *p-subgroup assumption* [5] is satisfied but the *non-malleability* [1] is not. If it is required, then we are able to convert the malleable scheme to the non-malleable one by EPOC-3. [4]

Paillier [6] also proposed new probabilistic encryption schemes under the assumption that deciding n-th residuosity is computationally hard. These schemes

---

* This work is supported by ITRC and COSEF(97-01-00-13-01-5).

K. Kim (Ed.): ICICS 2001, LNCS 2288, pp. 72–80, 2002.

are also proved to be one-way and semantically secure under appropriate assumption. These do not satisfy the non-malleability but it is possible to provide the non-malleable conversions by [7].

We found that, if a public key is chosen carefully, it is possible to exclude one modular multiplication in the original decryption without diminishing the security level given by original schemes. In case of the long messages, our schemes make the decryption speed faster than the original would do. For this purpose, we have to get some elements contained in the public key set for Okamoto-Uchiyama and Paillier schemes so that they are semantically secure. The proposed schemes in here also satisfy the one-wayness. Even though calculations are necessary to check the validity of randomly chosen public keys in the original schemes, we suggest a concrete method to get them for each of the improved schemes.

2   Okamoto-Uchiyama Probabilistic Public Key Cryptosystem

2.1   Review of Okamoto-Uchiyama Scheme

$\Gamma = \{\gamma \in Z_{p^2}^* | \gamma \equiv 1 \bmod p\}$ is the p-Sylow subgroup of $Z_{p^2}^*$ for an odd prime p. A function $L : (\Gamma, \cdot) \to (Z_p, +); x \mapsto \frac{x-1}{p}$ is an isomorphism satisfying

$$\forall a, b \in \Gamma, L(a \cdot b) \equiv L(a) + L(b) \bmod p .$$

If $y = x^m \bmod p^2$ and $L(x) \neq 0 \bmod p$ for $x \in \Gamma$ and $m \in Z_p$,

$$m = \frac{L(y)}{L(x)} \bmod p .$$

< Okamoto-Uchiyama Scheme >

Let $n = p^2 q$ with two large primes $p, q(|p| = |q| = k)$. $g \in Z_n^*$ is chosen randomly where the order of $g_p = g^{p-1} \bmod p^2$ is p. Let $h = g^n \bmod n$.
[Public Key] (n, g, h, k).
[Secret-Key] (p, q).
[Encryption] Let $m(0 < m < 2^{k-1})$ be a plaintext. Select $r \in Z_n$ uniformly.

$$C = g^m h^r \bmod n .$$

[Decryption] When $C_p = C^{p-1} \bmod p^2$,

$$m = \frac{L(C_p)}{L(g_p)} \bmod p .$$

2.2   Proposed Encryption Scheme

$(Z_{p^2}, +, \cdot)$ is a ring with identity. $(Z_{p^2}^*, \cdot)$ is a group with order $p(p-1)$ where $Z_{p^2}^*$ is a set $\{z \in Z_{p^2} | \gcd(z, p^2) = 1\}$.

It is clear that an element z in Z_{p^2} has an expression $\alpha p + \beta$ where α and β are in Z_p. Since $Z_{p^2}^*$ is a subset of Z_{p^2}, those elements can be expressed as the same form. Let $\Phi : Z_p \times Z_p^* \to Z_{p^2}^*$ be a map satisfying $\Phi((\alpha, \beta)) = \alpha p + \beta$.

Lemma 1. Φ *is bijective.*

Proof. Since $Z_p \times Z_p^*$ and $Z_{p^2}^*$ are multiplicative groups with same order $p(p-1)$, we need to prove that Φ is injective. It is easy to show the injectiveness of Φ. $\square$

Let $c \in Z_p^*$ be an element of a ring $(Z_{p^2}, +, \cdot)$. $c^{p-1} \bmod p^2$ is an element of Γ so that $(L(c^{p-1} \bmod p^2) - 1)c \bmod p$ is in Z_p. By Lemma 1, for every $c \in Z_p^*$, $bp + c$ with $b \equiv (L(c^{p-1} \bmod p^2) - 1)c \bmod p$ is in $Z_{p^2}^*$.

Theorem 1. *Let* $g = bp + c \in Z_{p^2}^*$ *where* b *and* c *are defined as above.* $g^{p-1} \equiv 1 + p \bmod p^2$.

Proof. By the assumption, c should be in Z_p^* and $b \in Z_p$ is dependent on the value of c. Then we get the followings:

$$(bp + c)^{p-1} \equiv c^{p-1} + (p-1)bc^{p-2}p \bmod p^2$$
$$\equiv 1 + (\tfrac{c^{p-1}-1}{p} \bmod p)p + (p-1)bc^{p-2}p \bmod p^2$$
$$\equiv 1 + (\tfrac{c^{p-1}-1}{p} - bc^{p-2} \bmod p)p \bmod p^2$$
$$\equiv 1 + p \bmod p^2$$

because

$$\tfrac{c^{p-1}-1}{p} - bc^{p-2} \equiv \tfrac{c^{p-1}-1}{p} - (\tfrac{c^{p-1}-1}{p} - 1)c^{p-1} \bmod p$$
$$\equiv \tfrac{c^{p-1}-1}{p} - \tfrac{c^{p-1}-1}{p}c^{p-1} + c^{p-1} \bmod p$$
$$\equiv \tfrac{c^{p-1}-1}{p}(1 - c^{p-1}) + c^{p-1} \bmod p$$
$$\equiv 1 \bmod p$$

since $\forall c \in Z_p^*, c^{p-1} \equiv 1 \bmod p$. $\square$

By Theorem 1, $L(g_p) = L(g^{p-1} \bmod p^2) = 1$. Thus for $0 < m < 2^{k-1}$,

$$m = L(1 + mp \bmod p^2) \bmod p$$
$$= L((g^m h^r)^{p-1} \bmod p^2) \bmod p \ .$$

< Improved Okamoto-Uchiyama Scheme >

Let $b \in Z_p$ and $c \in Z_p^*$ satisfy $b \equiv (\tfrac{c^{p-1}-1}{p} - 1)c \bmod p$. Choose $g \in Z_n^*$ with $g \equiv bp + c \bmod p^2$. Let $h \equiv g^n \bmod n$.
[**Public Key**] (n, g, h, k)
[**Secret-Key**] (p, q)
[**Encryption**] $0 < m < 2^{k-1}$. Select $r \in Z_n$ uniformly.

$$C \equiv g^m h^r \bmod n \ .$$

[**Decryption**] Let $C_p \equiv C^{p-1} \bmod p^2$.

$$m = L(C_p) \ .$$

2.3   Properties of Proposed Scheme

Our scheme satisfies the one-wayness and semantic security by the properties of Okamoto-Uchiyama scheme. Also the efficiency is increased when the original is compared with ours.

One-wayness. Our scheme is one-way if and only if factoring $n = p^2q$ is hard. Assume that there exists an oracle which inverts our scheme. Given the ciphertext $C = g^\eta \bmod n$ where $\eta \in Z_n$ and $\eta \geq 2^{k-1}$, this oracle provide the message $m < 2^{k-1}$. Then $\gcd(\eta - m, n)$ is either p, p^2, or pq because $\eta \equiv m \bmod p$. Therefore inverting our scheme allows factoring n. The other way is trivial because our scheme is broken in case of factoring $n = p^2q$. ([5])

Semantic Security. The public key for our scheme can behave like one for Okamoto-Uchiyama's. If our scheme does not satisfy the semantic security under the p-subgroup assumption, then the original does not because ours is one particular form of the original when our public key is applied. Therefore ours is semantically secure since the original is also.

Efficiency. The efficiency of ours is improved by the carefully chosen public key $g \in Z_n^*$ because the fact $L(g_p) \equiv 1 \bmod p$ makes the original decryption faster.

3   Paillier Probabilistic Public Key Cryptosystem

3.1   Review of Paillier Scheme

When p and q are large primes, we will denote $n = pq$, $\phi = (p-1)(q-1)$ and $\lambda = \mathrm{lcm}(p-1, q-1)$ where ϕ and λ are called as Euler's function and Carmichael's function on n. We need two facts $|Z_{n^2}^*| = n\phi$ and Carmichael's theorem which implies

$$\forall \omega \in Z_{n^2}^*, \begin{cases} \omega^\lambda = 1 \bmod n \\ \omega^{n\lambda} = 1 \bmod n^2 \end{cases}.$$

We use the notation $RSA[n,e]$ for the problem of extracting e-th roots modulo n where $n = pq$ is of unknown factorization.

Let $g \in Z_{n^2}^*$ and the function Ξ_g be defined by

$$Z_n \times Z_n^* \to Z_{n^2}^*; (x, y) \mapsto g^x y^n \bmod n^2.$$

By Lemma 3 in [6], Ξ_g is bijective when the order of $g \in Z_{n^2}^*$ is $n\mu$ with $\mu = 1, \ldots, \lambda$. Given $g \in Z_{n^2}^*$ with this condition, $\omega \in Z_{n^2}^*$ corresponds to n-th residuosity class of ω $[\![\omega]\!]_g \in Z_n$ and $y \in Z_n^*$ such that $\Xi_g([\![\omega]\!]_g, y) = \omega$. The Composite Residuosity Class Problem $Class\,[n]$ means that, given $\omega, g \in Z_{n^2}^*$ where $|g|$ is a nonzero multiple of n, compute $[\![\omega]\!]_g \in Z_{n^2}^*$. It is belived that $Class[n]$ is intractable. $D\text{-}Class[n]$ is the decisional problem associated to $Class[n]$ i.e.

Given ω, $g \in Z_{n^2}^*$ and $x \in Z_n$ with $n|\, |g| \neq 0$, decide whether $x = [\![\omega]\!]_g$.

The problem of factoring n is harder than *Class[n]* because the multiplicative subgroup

$$S_n = \{u \in Z_{n^2} | u = 1 + an \text{ for } a \in Z_n\}$$

gives the well-defined function

$$L : S_n \to Z_n; u \mapsto \frac{u - 1}{n}$$

so that, for any $\omega \in Z_{n^2}^*$ and $g \in Z_{n^2}^*$ with $n | |g| \neq 0$, we have

$$\frac{L(\omega^\lambda \bmod n^2)}{L(g^\lambda \bmod n^2)} = [\![\omega]\!]_g \bmod n \tag{1}$$

by Theorem 9 in [6].

To determine whether the order of a random selection $g \in Z_{n^2}^*$ is a nonzero multiple of n, we have an efficient way to check whether

$$\gcd(L(g^\lambda \bmod n^2), n) = 1 .$$

By Theorem 14 (Theorem 15) in [6], the following Main Scheme is one-way (semantically secure) if *Class[n]* (*D-Class[n]*) is intractable.

< Paillier Main Scheme >

Let $g \in Z_{n^2}^*$ have an order of nonzero multiple of n with $n = pq$.

[Public Key] (n, g).
[Secret-Key] λ.
[Encryption] Let $m \in Z_n$ be a plaintext. Select $r \in Z_n^*$ randomly.

$$C = g^m r^n \bmod n^2 .$$

[Decryption] When $C_\lambda = C^\lambda \bmod n^2$ and $g_\lambda = g^\lambda \bmod n^2$,

$$m = \frac{L(C_\lambda)}{L(g_\lambda)} \bmod n .$$

This main scheme can have a variant by (2) when the ciphertext space is restricted to the subgroup $< g >$ of $Z_{n^2}^*$ whose order is $n\alpha$ for $1 \leq \alpha \leq \lambda$. Under the same condition as the above, we can say the extention of (1) as the followings:

$$\forall \omega \in < g >, \ [\![\omega]\!]_g = \frac{L(\omega^\alpha \bmod n^2)}{L(g^\alpha \bmod n^2)} \bmod n . \tag{2}$$

When $h \in Z_{n^2}^*$ has an order $n\lambda$ and λ is divisible by α, let $g = h^{\lambda/\alpha} \bmod n^2$.

< **Paillier Subgroup Variant** >

Let $g \in Z_{n^2}^*$ have an order of nonzero multiple of n with $n = pq$.

[Public Key] (n, g).

[Secret-Key] α.

[Encryption] Let $m \in Z_n$ be a plaintext. Select $r \in Z_n$ randomly.

$$C = g^{m+rn} \bmod n^2 .$$

[Decryption] When $C_\alpha = C^\alpha \bmod n^2$ and $g_\alpha = g^\alpha \bmod n^2$,

$$m = \frac{L(C_\alpha)}{L(g_\alpha)} \bmod n .$$

(Decisional) Partial Discrete Logarithm Problem $PDL[n,g]$ ($D\text{-}PDL[n,g]$) are defined as follows:

Given $\omega \in\, <g>$ (and $x \in Z_n$), compute (decide whether) $[\![\omega]\!]_g$ ($= x$) .

By Theorem 19 (Theorem 20) in [6], this variant scheme is one-way (semantically secure) if $PDL[n,g]$ ($D\text{-}PDL[n,g]$) is hard.

3.2 Proposed Encryption Scheme

$(Z_{n^2}, +, \cdot)$ is a ring with identity. $(Z_{n^2}^*, \cdot)$ is a group with order $n\lambda$ where $Z_{n^2}^*$ is a set $\{z \in Z_{n^2} | \gcd(z, n^2) = 1\}$.

$\Gamma = \{\gamma \in Z_{n^2}^* | \gamma \equiv 1 \bmod n\}$ is the multiplicative cyclic subgroup $< 1 + n >$ of $Z_{n^2}^*$. The function $L : (\Gamma, \cdot) \to (Z_n, +); \gamma \mapsto \frac{\gamma - 1}{n}$ is well-defined on Γ in [6].

Lemma 2. *The function $L : (\Gamma, \cdot) \to (Z_n, +)$ is isomorphic.*

Proof. We want to show $L(\gamma\delta) = L(\gamma) + L(\delta)$.

$$
\begin{aligned}
L(\gamma\delta) &= \tfrac{\gamma\delta - 1}{n} \bmod n \\
&= \tfrac{(\gamma-1)(\delta-1)+(\gamma-1)+(\delta-1)}{n} \bmod n \\
&= \tfrac{\gamma-1}{n}(\delta-1) + \tfrac{\gamma-1}{n} + \tfrac{\delta-1}{n} \bmod n \\
&= \tfrac{\gamma-1}{n} + \tfrac{\delta-1}{n} \bmod n \\
&= L(\gamma) + L(\delta) \bmod n .
\end{aligned}
$$

Also it is clear that $\ker L = 1$. Therefore L is an isomorphism.   □

It is clear that an element z in Z_{n^2} has an expression $\mu n + \nu$ where μ and ν are in Z_n. Since $Z_{n^2}^*$ is a subset of Z_{n^2}, those elements have the same. Let $\Psi : Z_n \times Z_n^* \to Z_{n^2}^*$ be a map satisfying $\Psi((\mu, \nu)) = \mu n + \nu$.

Lemma 3. *Ψ is bijective.*

Proof. Since $Z_n \times Z_n^*$ and $Z_{n^2}^*$ are multiplicative groups with same order $n\phi$, we need to prove that Ψ is injective. It is easy to show the injectiveness of Ψ.   □

Since λ is in Z_n^*, there exists $\lambda^{-1} \in Z_n^*$ such that $1 = \lambda\lambda^{-1}$. Also for each $\nu \in Z_n^* \subset Z_{n^2}^*$, $\nu_\lambda = \nu^\lambda \bmod n^2$ is in Γ so that $L(\nu_\lambda)$ is in Z_n. Then we define $\mu \in Z_n$ as

$$\mu \equiv \lambda^{-1}\nu - \lambda^{-1}\nu\, L(\nu_\lambda) \bmod n \ . \tag{3}$$

Theorem 2. *Let $g = \mu n + \nu \in Z_{n^2}^*$ where $\mu \in Z_n$ and $\nu \in Z_n^*$ satisfy (3). $g^\lambda \bmod n^2$ is a generator of Γ i.e. $g^\lambda \equiv 1 + n \bmod n^2$.*

Proof. Because of the conditions of μ and ν, g should be in $Z_{n^2}^* \subset Z_{n^2}$ where $(Z_{n^2}, +, \cdot)$ is a ring.

$$\begin{aligned}
g^\lambda &\equiv (\mu n + \nu)^\lambda \bmod n^2 \\
&\equiv \nu^\lambda + \lambda\mu\nu^{\lambda-1}n \bmod n^2 \\
&\equiv 1 + L(\nu_\lambda)n + (\nu^\lambda - \nu^\lambda L(\nu_\lambda) \bmod n)n \bmod n^2 \\
&\equiv 1 + n \bmod n^2 \ .
\end{aligned}$$

The last equivalence is satisfied by Carmichael's theorem. □

By Theorem 2, $L(g_\lambda) = L(g^\lambda \bmod n^2) = 1$. Thus for $m \in Z_n$ and $r \in Z_n^*$,

$$\begin{aligned}
m &= L(1 + mn \bmod n^2) \bmod n \\
&= L((g^m r^n)^\lambda \bmod n^2) \bmod n \ .
\end{aligned}$$

< Improved Main Scheme >

Let $g = \mu n + \nu \in Z_{n^2}^*$ with $\mu \in Z_n$ and $\nu \in Z_n^*$ satisfying the relation

$$\mu \equiv \lambda^{-1}\nu - \lambda^{-1}\nu\, L(\nu_\lambda) \bmod n \ .$$

[Public Key] (n, g).
[Secret-Key] λ.
[Encryption] Let $m \in Z_n$ be a plaintext. Select $r \in Z_n^*$ randomly.

$$C = g^m r^n \bmod n^2 \ .$$

[Decryption] When $C_\lambda = C^\lambda \bmod n^2$,

$$m = L(C_\lambda) \ .$$

This improved scheme can have another variant also. When $h = \mu n + \nu \in Z_{n^2}^*$ has an order $n\lambda$ and λ is divisible by α, let $g = h^{\lambda/\alpha} \bmod n^2$. It is easy to show that

$$g^\alpha \equiv 1 + n \bmod n^2$$
$$\text{and}$$
$$\begin{aligned}
m &= L(1 + mn \bmod n^2) \bmod n \\
&= L((g^{m+rn})^\alpha \bmod n^2) \bmod n \ .
\end{aligned}$$

< Improved Subgroup Variant >

Let $h = \mu n + \nu \in Z_{n^2}^*$ with $\mu \in Z_n$ and $\nu \in Z_n^*$ satisfying the relation

$$\mu \equiv \lambda^{-1}\nu - \lambda^{-1}\nu\,L(\nu_\lambda) \bmod n \ .$$

Let $g \equiv h^{\lambda/\alpha} \bmod n^2$.
[Public Key] (n, g).
[Secret-Key] α.
[Encryption] Let $m \in Z_n$ be a plaintext. Select $r \in Z_n$ randomly.

$$C = g^{m+rn} \bmod n^2 \ .$$

[Decryption] When $C_\alpha = C^\alpha \bmod n^2$,

$$m = L(C_\alpha) \bmod n \ .$$

3.3   Properties of Proposed Schemes

Our improved versions satisfy the one-wayness and semantic security because the public key of ours can be one for the schemes proposed by P. Paillier and these are changed to ours when the public key suggested by ours are given to the original. Since we choose the public key $g \in Z_{n^2}^*$ satisfying $L(g_\lambda) \equiv 1 \bmod n$ or $L(g_\alpha) \equiv 1 \bmod n$, the decryption workload of the Paillier schemes is efficiently reduced. Thus each efficiency of the original schemes is improved by our propositions.

4   Conclusion

We have suggested a probabilistic encryption scheme which improves Okamoto-Uchiyama and Paillier schemes. The specific methods to get the public key sets for our propositions are given and these key sets are contained in the original ones. Also our key makes the calculation of each decryption reduced without diminishing the security level of the original schemes because these schemes are transformed to the ours by the choice of our key set. Therefore they are semantically secure and have a merit of better efficiency in the decryption than the original porpositions.

References

1. D. Dolev, C. Dwork, and M. Naor (1991) *Non-malleable Cryptography* Proc. of the 23rd STOC. ACM Press, New York.
2. W. Diffe and M. Hellman (1976), *New Directions in Cryptography* IEEE Tranctions on Information Theory, Vol. IT-22(6), pp. 644-654.
3. S. Goldwasser and S. Micali (1984), *Probabilistic Encryption* JCSS, 28, 2, pp. 270-299.
4. T. Okamoto and D. Pointcheval (2000), *Efficient Public Key Encryption (ver. 3)* Submission to P1363a, available on
   http://grouper.ieee.org/groups/1363/submission.html.

5. T. Okamoto and S. Uchiyama (1998), *A New Public Key Cryptosystem as secure as Factoring* Proc. of EUROCRYPTO'98, pp. 309-318.
6. P. Paillier (1999), *Public Key Cryptosystems Based on Composite Degree Rediduosity Classes* Proc. of EUROCRYPTO'99, LNCS 1592, pp. 223-238.
7. P. Paillier and D. Pointcheval (1999), *Efficient Public Key Cryptosystems Provably Secure Against Active Adversaries* Proc. of ASIACRYPT '99, LNCS 1716, pp. 165-179.
8. M. Rabin (1979), *Digitalized Signatures and Public Key Functions as Intractable as Factorization* MIT Laboratory for Computer Science TR-212.

Design and Analysis of
Fast Provably Secure Public-Key Cryptosystems
Based on a Modular Squaring

Mototsugu Nishioka[1], Hisayoshi Satoh[1], and Kouichi Sakurai[2]

[1] Systems Development Laboratory, Hitachi, Ltd. Japan
{nishioka,hisato}@sdl.hitachi.co.jp
[2] Department of Computer Science and Communication Engineering,
Kyushu University, Japan
sakurai@csce.kyushu-u.ac.jp

Abstract. We design a provably secure public-key encryption scheme based on modular squaring (Rabin's public-key encryption scheme [28]) over $\mathbb{Z}_N$, where $N = p^d q$ (p and q are prime integers, and $d > 1$), and we show that this scheme is extremely faster than the existing provably secure schemes. Security of our scheme is enhanced by the original OAEP padding scheme [3]. While Boneh presents two padding schemes that are simplified OAEP, and applies them to design provably secure Rabin-based schemes (Rabin-SAEP, Rabin-SAEP+), no previous works explores Rabin-OAEP. We gives the exact argument of security of our OAEP-based scheme. For speeding up our scheme, we develop a new technique of fast decryption, which is a modification of Takagi's method for RSA-type scheme with $N = p^d q$ [31]. Takagi's method uses Chinese Remainder Theorem (CRT), whereas our decryption requires no CRT-like computation. We also compare our scheme to existing factoring-based schemes including RSA-OAEP, Rabin-SAEP and Rabin-SAEP+. Furthermore, we consider the (future) hardness of the integer-factoring: $N = p^d q$ vs. $N = pq$ for large size of N.

1 Introduction

We design a public-key encryption scheme based on a modular squaring (Rabin's public-key encryption scheme [28]) over $\mathbb{Z}_N$, where $N = p^d q$ (p and q are prime integers, and $d > 1$). Our scheme has the following exceptional features:

- It is proven to be semantically secure against an adaptive chosen-ciphertext attack (IND-CCA2) in the random oracle model under the factoring assumption of N.
- It has a very fast encryption speed.
- The decryption speed (1536 bits) is about two-and-a-half times faster than that of RSA-OAEP (1024 bits) [3].
- The plaintext space is sufficiently large.
- The amount of computation for the encryption and decryption increases only slightly compared with previous schemes, even if the size of N increases in the future.

K. Kim (Ed.): ICICS 2001, LNCS 2288, pp. 81–102, 2002.
© Springer-Verlag Berlin Heidelberg 2002

1.1   What Is an Ideal Public-Key Cryptosystem?

We enumerate the conditions that an ideal public-key cryptosystem should satisfy and describe how our scheme satisfies these conditions. We believe that the following are the conditions that an ideal public-key encryption scheme should satisfy:

Security: It can be proven to be secure in the sense of IND-CCA2 under a weak assumption.

Efficiency:
(E-1) Both encryption and decryption speeds are fast.
(E-2) The ratio of a plaintext and a ciphertext "(Plaintext)/(Ciphertext)" is close to 1.
(E-3) The plaintext space is sufficiently large.
(E-4) It can be mounted with a small memory size (including public key and secret key sizes).

In terms of security, we believe that the factoring problem or the discrete logarithm problem are almost ideal as a number theoretic assumption of cryptosystems, because with sufficient study their computational intractability can be taken for granted [15,19,20]. Furthermore, there are two categories in number theoretic assumptions that are well utilized in the practical cryptosystems, i.e.:

Factoring Base: Factoring problem, RSA problem, Quadratic residue problem, etc,

Discrete Logarithm Base: Discrete logarithm problem, Computational Diffie-Hellman problem, Decisional Diffie-Hellman problem, etc.

Of these, the factoring problem and the discrete logarithm problem are the most intractable of the problems in these two categories.

1.2   Our Contribution

In constructing our scheme, we focused on the modular square function (Rabin's encryption function), because it is well known that inverting the encryption function on $\mathbb{Z}_N$ is as intractable as the factoring of N, where $N = pq$ (p and q are prime numbers). Another reason is that it has fast encryption speed. However, the following problems were encountered:

(P-1) The modular square function is not one-way trapdoor permutation, i.e., the decryption is not done uniquely.
(P-2) Rabin's scheme is not secure against a chosen-ciphertext attack.
(P-3) The decryption speed is not fast (i.e., it is as same as that of RSA).

In our scheme, we utilize OAEP [3] to solve the problems (P-1) and (P-2). OAEP is the padding scheme that can convert any one-way trapdoor permutation to a public-key encryption scheme secure in the sense of IND-CCA1 (Although at first it was believed that OAEP could convert such permutation to an

IND-CCA2 scheme, it has recently been pointed out that the converted scheme is not IND-CCA2 but IND-CCA1 [30]). Note, however, that OAEP has been not known to be applicable to more general one-way trapdoor functions including probabilistic functions and the modular square function. We shows that OAEP can be exceptionally applicable to the modular square function. The public-key encryption scheme that applies OAEP to Rabin-like one-way trapdoor permutation was presented [18]. Recently Boneh [6] presents two padding schemes that are simplified OAEP, and applies them to design provably secure Rabin-based schemes (Rabin-SAEP, Rabin-SAEP+). However, the padding method of OAEP is more complicated with comparing to that of SAEP(and SAEP+), then the argument of provable security of Rabin-OEAP is more sophisticated than that of SAEP(and SAEP+). Furthermore, we observe that the security of SAEP and SAEP+ depends on the random oracles more heavily than the security of OAEP and OAEP+ does, which suggests that reliability of the argument of provable security on the random oracle model is not uniform in the real world (cf. Section 3.3). Owing to OAEP, we can get the probabilistically uniqueness of the decryption and prove that it is secure in the sense of IND-CCA2 in the random oracle model by using Coppersmith's algorithm (cf. Section 3.4). We can also nearly clear the above conditions (E-2) and (E-3) by using OAEP. We think that the condition (E-3) is very important, because there are many protocols such as SET (Secure Electronic Transaction) in which additional information, such as identity information of users or information on cryptosystems, are attached with the data encryption key, even though the main purpose of public-key encryption schemes is to distribute the data encryption key of secret-key encryption schemes. For speeding up our scheme, we make $N = p^d q$ (p, q: prime numbers, $d > 1$) instead of $N = pq$ and we develop a new technique of fast decryption, which is a modification of Takagi's method for RSA-type scheme with $N = p^d q$ [31]. Takagi's method was done over $\mathbb{Z}_{p^d}$ after $\mathbb{Z}_N$ is divided into $\mathbb{Z}_{p^d}$ and $\mathbb{Z}_q$ by using the Chinese Remainder Theorem (CRT), and the calculated values on $\mathbb{Z}_{p^d}$ and $\mathbb{Z}_q$ were combined on $\mathbb{Z}_N$ by using CRT again. Our calculation method differs from this in that ours requires no CRT-like computation (cf. Section 2.1).

As a result, our method has the following advantages:

- It has less modular multiplications than the previous one (cf. Section 4.2).
- The actual decryption speed and mountaing size will be smaller than previous one because ours does not require Euclidean algorithm for CRT.

Although the difference of those decryption speeds is very small, it is expected that it will be non-negligible in smart card systems and in systems in which much decryption processing must be done at one time.

On the other hand, our scheme avoids the need for a hybrid scheme with a secret-key encryption scheme[1], meaning that solving (E-4) would require no secret-key encryption scheme to enable public-key encryption. Another problem with hybrid schemes is that they may require the use of two different secret-key cryptosystems in a single system, which would add to development costs.

---

[1] Recently, some hybrid schemes were proposed [8,25].

We also compare our scheme to existing factoring-based schemes including RSA-OAEP, Rabin-SAEP and Rabin-SAEP+. Furthermore, we consider the (future) hardness of the integer-factoring: $N = p^d q$ vs. $N = pq$ for large size of N. This is of its interest. The processing ability of computers is increasing rapidly, then the key length must also increase to stay ahead. This increase in key length impairs the efficiency of encryption schemes. However, our scheme can be used well into the future, because it can achieve efficient encryption and decryption processing even if the key length increases (cf. Section 3.1 and 4).

In this paper, the algorithm of our scheme is given in Section 2.1 and its security is described in Section 3. We evaluate our scheme by comparing it with previous schemes in Section 4.

2 Our Scheme

2.1 Algorithm of Our Scheme

Let k, k_0 and k_1 be security parameters with $2k_0 < k$ (we give the details of the length of each parameter k_0, k_1 and k in Section 4.1.). Set $n = k - k_0 - k_1 - 1$.

The random oracles G and H which are refered in encryption and decryption have input/output length of $G : \{0,1\}^{k_0} \to \{0,1\}^{k-k_0-1}$ and $H : \{0,1\}^{k-k_0-1} \to \{0,1\}^{k_0}$.

Key generation. Choose prime numbers p, q, such that $|p| = |q|$, $p \equiv 3 \pmod 4$, and $q \equiv 3 \pmod 4$, and choose an integer d with $d > 1$. And, let $N = p^d q$, where $|N| = k$, and $|x|$ denotes a binary length of x. Then, the private key is (p, q), and the public key is N. Note that $N/2 < 2^{k-1} < N < 2^k$.

Encryption. For a message $m \in \{0,1\}^n$, choose the random number $r \in \{0,1\}^{k_0}$, and compute

$$x = (m0^{k_1} \oplus G(r)) \| (r \oplus H(m0^{k_1} \oplus G(r))).$$

And compute $y = x^2 \bmod N$. Then, y is defined as a ciphertext of m.

Decryption. For the given ciphertext yC
(1) Check if y is a quadratic residue on $\mathbb{Z}_N$, namely check

$$y^{(p-1)/2} \equiv 1 \pmod p \quad \text{and} \quad y^{(q-1)/2} \equiv 1 \pmod q.$$

If y is not a quadratic residue, reject it.
(2) For $i, j \in \{0, 1\}$, compute

$$\gamma_0^{(i)} = (-1)^i y^{\frac{p+1}{4}} \bmod p, \qquad \gamma_1^{(j)} = \left((-1)^j y^{\frac{q+1}{4}} - x_0\right) p^{-1} \bmod q$$

and

$$\gamma_l^{(i,j)} = \frac{y - \Gamma_{l-1}^{(i,j)2} \bmod p^l q}{p^{l-1}q} \times (2\gamma_0^{(i)})^{-1} \bmod p \quad (2 \le l \le d),$$

where

$$\Gamma_1^{(i,j)} = \gamma_0^{(i)} + \gamma_1^{(j)}p \quad \text{and} \quad \Gamma_l^{(i,j)} = \Gamma_{l-1}^{(i,j)} + \gamma_l^{(i,j)}p^{l-1}q \quad (2 \leq l \leq d-1).$$

(3) For $i, j \in \{0, 1\}$, compute

$$x_{i,j} = \gamma_0^{(i)} + \gamma_1^{(j)}p + \sum_{l=2}^{d} \gamma_l^{(i,j)}p^{l-1}q.$$

Let those $x_{i,j}$ $(0 \leq i, j \leq 1)$ be x_1, x_2, x_3, x_4.

(4) For each x_i $(1 \leq i \leq 4)$, compute $s_i \in \{0, 1\}^{n+k_1}$ and $t_i \in \{0, 1\}^{k_0}$ such that $x_i = s_i || t_i$, if $x_i \in \{0, 1\}^{k-1}$. Otherwise, reject y (cf. *Remark 2*).

(5) For each s_i and t_i $(1 \leq i \leq 4)$, compute

$$r_i = H(s_i) \oplus t_i \qquad w_i = s_i \oplus G(r_i).$$

(6) For each w_i $(1 \leq i \leq 4)$, compute $m_i \in \{0, 1\}^n$ and $z_i \in \{0, 1\}^{k_1}$ such that $w_i = m_i || z_i$, and output

$$\begin{cases} m_i & \text{if } z_i = 0^{k_1}, \\ \text{``reject''} & \text{otherwise,} \end{cases}$$

as the decryption of the ciphertext y.

2.2   Soundness of Decryption

Theorem 1. In the above algorithm, the plaintext is correctly decoded from the valid ciphertext except with a negligible probability.

Proof. We show that x_i $(1 \leq i \leq 4)$ are all square roots of y in $\mathbb{Z}_N$. If it is shown, there are at most four square roots of y in $\{0, 1\}^{k-1}$. Therefore it is trivial that the probability that the decryption fails is less than $3/2^{k_1}$.

We show this by induction on d. Note that any element x in $\mathbb{Z}_N$ $(N = p^d q)$ can be written by

$$x = \gamma_0 + \gamma_1 p + \sum_{i=2}^{d} \gamma_i p^{i-1}q \quad (0 \leq \gamma_0, \gamma_2, \ldots, \gamma_{d-1} < p, \quad 0 \leq \gamma_1 < q),$$

and such γ_i is uniquely determined.

Let $d = 2$. Then, the element x in $\mathbb{Z}_{p^2q}$ can be written by $x = \gamma_0 + \gamma_1 p + \gamma_2 pq$ for some $\gamma_0, \gamma_1, \gamma_2 \in \mathbb{Z}$ $(0 \leq \gamma_0, \gamma_2 < p, 0 \leq \gamma_1 < q)$.

Suppose that $x^2 \equiv y \pmod{p^2 q}$. Then, we have

$$\begin{aligned} x^2 &\equiv (\gamma_0 + \gamma_1 p + \gamma_2 pq)^2 \\ &\equiv \gamma_0^2 + \gamma_1^2 p^2 + 2\gamma_0\gamma_1 p + 2\gamma_0\gamma_2 pq \equiv y \pmod{p^2 q}. \end{aligned} \tag{1}$$

And it follows that

$$\gamma_0{}^2 \equiv y \pmod{p} \quad \text{and} \quad (\gamma_0 + \gamma_1 p)^2 \equiv y \pmod{q}.$$

Since p and q are Blum numbers $C\gamma_0$ and γ_1 can be computed as follows (after testing if $y \bmod p$ and $y \bmod q$ are quadratic residue on $\mathbb{Z}_p$ and $\mathbb{Z}_p$ respectively):

$$\gamma_0 = y^{(p+1)/4} \bmod p \quad \text{or} \quad -y^{(p+1)/4} \bmod p,$$
$$\gamma_1 = (y^{(q+1)/4} - \gamma_0)p^{-1} \bmod q \quad \text{or} \quad (-y^{(q+1)/4} - \gamma_0)p^{-1} \bmod q.$$

Hence γ_0 is $\gamma_0^{(0)}$ or $\gamma_0^{(1)}$, and γ_1 is $\gamma_1^{(0)}$ or $\gamma_1^{(1)}$. Furthermore, γ_2 is induced from the equation (1) as follows:

$$\gamma_2 = \frac{y - (\gamma_0 + \gamma_1 p)^2 \bmod p^2 q}{pq} \times (2\gamma_0)^{-1} \bmod p.$$

Hence γ_2 is $\gamma_2^{(0,0)}$, $\gamma_2^{(1,0)}$, $\gamma_2^{(0,1)}$ or $\gamma_2^{(1,1)}$. Note that pq divides $y - (\gamma_0 + \gamma_1 p)^2 \bmod p^2 q$. We can also easily prove that y is a quadratic residue on $\mathbb{Z}_{p^2 q}$ if and only if $y \bmod p$ and $y \bmod q$ are respectively quadratic residue on $\mathbb{Z}_p$ and $\mathbb{Z}_q$.

From the above, it was shown that x_0, x_1, x_2, x_4 are all square roots of y in $\mathbb{Z}_{p^2 q}$.

Next, let $d > 2$. And assume that $\Gamma_{d-1}^{(i,j)} \ (= \gamma_0^{(i)} + \gamma_1^{(j)} p + \sum_{l=2}^{d-1} \gamma_l^{(i,j)} p^{l-1} q)$ are all square roots of y in $\mathbb{Z}_{p^{d-1}q}$ for $0 \le i, j \le 1$. Suppose that

$$x^2 \equiv y \pmod{p^d q}, \tag{2}$$

for some $x \in \mathbb{Z}_{p^d q}$. Then, from the assumption, x can be written by

$$x = \Gamma_{d-1}^{(i,j)} + \gamma_d^{(i,j)} p^{d-1} q,$$

for some $\gamma_d^{(i,j)} \in \mathbb{Z} \ (0 \le \gamma_d^{(i,j)} < p, \ 0 \le i, j \le 2)$. And we have

$$x^2 \equiv (\Gamma_{d-1}^{(i,j)} + \gamma_d^{(i,j)} p^{d-1} q)^2 \equiv \Gamma_{d-1}^{(i,j)}{}^2 + 2\Gamma_{d-1}^{(i,j)} \gamma_d^{(i,j)} p^{d-1} q \equiv y \pmod{p^d q},$$

from the equation (2). Hence $\gamma_d^{(i,j)}$ can be obtained by

$$\gamma_d^{(i,j)} = \frac{y - \Gamma_{d-1}^{(i,j)}{}^2 \bmod p^d q}{p^{d-1} q} \times (2\gamma_0^{(i)})^{-1} \bmod p.$$

Note that $p^{d-1} q$ divides $y - \Gamma_{d-1}^{(i,j)}{}^2 \bmod p^d q$. We can also easily prove that y is a quadratic residue mod $p^d q$ if and only if $y \bmod p$ and $y \bmod q$ are respectively quadratic residue on $\mathbb{Z}_p$ and $\mathbb{Z}_q$, by induction. □

Remark 1. We can send

$$\alpha = \begin{cases} 0 & \text{if } 0 < x < N/2, \\ 1 & \text{if } N/2 \le x < N, \end{cases}$$

or the Jacobi symbol $\beta = \left(\frac{x}{N}\right)$ with a ciphertext to support the decryption processing, where β is useful when d is even. Note that the security proof is not broken even if α and β are sent with the ciphertext.

Remark 2. The actual system must be implemented to be difficult to distinguish a failure in the integer-to-octets conversion from any subsequent failure, e.g. of the integrity check during OAEP-decoding, to prevent Manger's attack [21].

3   Security

3.1   Factoring Problem

In our scheme, (p, q, d) should be chosen to be intractable to factor $N = p^d q$. Although an efficient algorithm for factoring $N = p^d q$, when d is large ($d \approx \sqrt{\log p}$), is known [7], it is thought that the factoring of N will be intractable when d is small. Currently known prime factoring algorithms can be divided into two categories, namely those that depend on the size of the composite numbers and those that depend on those prime factors. The algorithms depending on prime factors, such as the ρ, $p - 1$, $p + 1$ methods, and the elliptic curve method([20], [27]) are known. The general number field sieve method depends on the size of the composite number.

Our scheme, in the case of $N = p^2 q$, uses 1300~1400-bit composite numbers which have 430~460-bit prime factors and 1500~1600-bit numbers in the case of $N = p^3 q$ in order to be as strong as 1024-bit RSA-type integers. For these composite numbers, the ρ method is ineffective, and if appropriate primes ($p - 1$ and $p + 1$ have large prime factors, etc.) are chosen in the key generation of our scheme, then the $p - 1$ and $p + 1$ methods are also ineffective. The elliptic curve method has an amount of calculation for finding a prime factor p of N $L_p[1/2, \sqrt{2}]$ ($L_p[a, b] = \exp((b + o(1))(\log p)^a (\log \log p)^{1-a})$), and an estimation of the amount of calculation for the number field sieve is $L_N[1/3, 1.901]$ ([9]), both of which are subexponential. In practice, depending on the implementation of the algorithm and the ability of computers, the size of the prime factors to be factorized is about $180 \sim 190$-bit by the elliptic curve method, and the size of the composite numbers to be factorized is about 512-bit by the number field sieve method. We can estimate the amount of calculation more precisely. Let $t_{EC}(p) = \log(L_p(1/2, \sqrt{(2)}))$ be the logarithm of the amount of calculation by the elliptic curve method and $t_{NFS}(N) = \log(L_N(1/3, 1.901))$ be that by the number field sieve method. Then, for 1024-bit RSA-type composite number $N = pq$ ($p, q : 512$ bits), the number field sieve method is more efficient than the elliptic curve method, and we have

$$\alpha := t_{NFS}(\text{1024-bit } N) = C_{NFS} + 59.42,$$

where C_{NFS} is the constant in the O-factors.

On the other hand, if we use 1344-bit composite number $N = p^2 q$ for our scheme ($p, q : 448$ bits), then the elliptic curve method is more efficient, and we have

$$\beta(448) := \alpha - t_{\mathrm{EC}}(\text{448-bit } p) = C - 0.28,$$

where C is some constant coming from the O-factors (In Figure 1, we show the graph of β.). This shows that factoring 1024-bit RSA-type integers is $e^{0.28} = 1.32$ times faster than that of 1344-bit p^2q-type integers. Hence we can say that the modulus of 1344-bit our scheme is stronger than that of 1024-bit RSA.

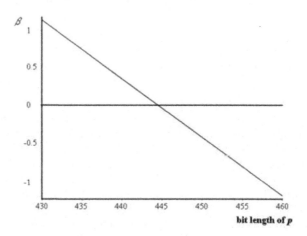

Fig. 1. The graph of β

In the case of $N = p^3q$, our scheme uses 1500~1600 bit N which have 375~400 bit prime factors. For example, when the bit length of N is 1536, we have

$$\beta(448) = C + 4.9.$$

We have $e^{4.9} = 134.3$ and we can say that the strength the integers of this type is comparable to that for 1024-bit RSA-type integers.

We will compare our scheme-type integers with 2048-bit and 4096-bit RSA-type integers. If we need such strength as 2048-bit RSA-type integers, then we use 2304-bit p^2q-type integers or 3072-bit p^3q-type integers, and for 4096-bit RSA-type integers, we use 4032-bit p^2q-type integers or 4928-bit p^3q-type integers (Figure 2). For p^2q-type integers, if the bit length is greater than 2700, then the number field sieve method is more efficient than the elliptic curve method. Thus we can use 4096-bit p^2q-type integers for our scheme. But we select 4032-bit integers so that the length of p and that of q are same. Then we have

$$t_{\mathrm{NFS}}(\text{4032-bit } N) = C_{\mathrm{NFS}} + 106.48,$$
$$t_{\mathrm{NFS}}(\text{4096-bit } N) = C_{\mathrm{NFS}} + 107.18.$$

The difference is $e^{(107.18-106.48)} = 2.01$. Hence we can say that the strength of these integers is nearly the same.

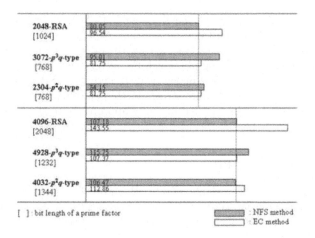

Fig. 2. Complexity for long modulus

3.2 Definitions

A *composite number generator* $\mathcal{G}$ is a probabilistic polynomial time (PPT) algorithm such that $\mathcal{G}(1^k)$ outputs a composite number N, where $|N| = k$.

Definition 1. Let $\mathcal{G}$ be a composite number generator. We say that algorithm M succeeds in (t, ϵ)-*factoring* $\mathcal{G}(1^k)$ if

$$\Pr\left[N \leftarrow \mathcal{G}(1^k) \; : \; M(N) = (p_1, p_2, \ldots, p_d)\right] \geq \epsilon,$$

where $N = \prod_{i=1}^{d} p_i$ (each p_i is prime numbers), and, moreover, in the experiment above, M runs in at most t steps.

We simply say that *the factoring $\mathcal{G}$ is intractable* if there is no polynomial time algorithm M which succeeds in (t, ϵ)-factoring $\mathcal{G}(1^k)$ for the non-negligible ϵ.

We quote the following definition from [2].

Definition 2 (IND-CPA, IND-CCA1, IND-CCA2 in the RO model).
Let $\Pi = (\mathcal{K}, \mathcal{E}, \mathcal{D})$ be an encryption scheme and let $A = (A_1, A_2)$ be an adversary. For atk $\in \{\text{cpa, cca1, cca2}\}$ and $k \in \mathbb{N}$, let

$$\mathbf{Adv}_{A,\Pi}^{\text{ind-atk}}(k) \overset{\text{def}}{=} 2 \cdot \Pr\left[H_1, H_2, \ldots, H_t \leftarrow \Omega; \quad (pk, sk) \leftarrow \mathcal{K}(1^k);\right.$$

$$(x_0, x_1, s) \leftarrow A_1^{H_1, H_2, \ldots, H_t, \, \mathcal{O}_1}(pk); \quad b \leftarrow \{0, 1\}; \quad y \leftarrow \mathcal{E}_{pk}^{H_1, H_2, \ldots, H_t}(x_b) :$$

$$\left. A_2^{H_1, H_2, \ldots, H_t, \, \mathcal{O}_2}(x_0, x_1, s, y) = b \right] - 1,$$

where

If atk=cpa	then	$\mathcal{O}_1(\cdot) = \varepsilon$	and	$\mathcal{O}_2(\cdot) = \varepsilon$
If atk=cca1	then	$\mathcal{O}_2(\cdot) = D_{sk}(\cdot)$	and	$\mathcal{O}_2(\cdot) = \varepsilon$
If atk=cca2	then	$\mathcal{O}_2(\cdot) = D_{sk}(\cdot)$	and	$\mathcal{O}_2(\cdot) = D_{sk}(\cdot)$.

We say that Π *is secure in the sense of* IND-ATK *in the random oracle model* if A being polynomial-time implies that $\mathbf{Adv}_{A,\Pi}^{\mathrm{ind-atk}}(\cdot)$ is negligible. For the above, we insist that A_1 outputs x_0, x_1 with $|x_0| = |x_1|$, and s denotes state information (possibly including pk) which the adversary wants to preserve. We also insist that the parameter t depends on Π and that Ω denotes a set of all maps from the set $\{0,1\}^*$ of finite strings to the set $\{0,1\}^\infty$ of infinite strings. The notation should be interpreted in a way appropriate to the context (see [2] for details).

3.3 OAEP versus SAEP

Recently, the simplified variants of OAEP and OAEP+, called SAEP and SAEP+, have been proposed [6]. SAEP apply to the RSA function, and SAEP+ apply to the RSA and Rabin functions. The padding methods of OAEP, OAEP+, SAEP and SAEP+ follow.

For OAEP and OAEP+:

$$\mathrm{OAEP}(m,r) = s||(r \oplus H(s)), \qquad \mathrm{OAEP+}(m,r) = s'||(r \oplus H(s')),$$

where

$$s = (m||0^{k_1}) \oplus G(r), \qquad s' = (m \oplus G(r))||H'(m||r),$$

H is a function from $\{0,1\}^{n+k_1}$ to $\{0,1\}^{k_0}$, G is a function from $\{0,1\}^{k_0}$ to $\{0,1\}^{n+k_1}$, H' is a function from $\{0,1\}^{n+k_0}$ to $\{0,1\}^{k_1}$, $m \in \{0,1\}^n$ and $r \in \{0,1\}^{k_0}$.

For SAEP and SAEP+:

$$\mathrm{SAEP}(m,r) = ((m||0^{s_0}) \oplus H(r))||r,$$
$$\mathrm{SAEP+}(m,r) = ((m||G(m||r)) \oplus H(r))||r,$$

where H is a function from $\{0,1\}^{s_1}$ to $\{0,1\}^{n+s_0}$, G is a function from $\{0,1\}^{n+s_1}$ to $\{0,1\}^{s_0}$, $m \in \{0,1\}^n$ and $r \in \{0,1\}^{s_1}$. Note that the functions H, H' and G are modeled as random oracle in the security analysis and that OAEP does not generally convert any one-way trapdoor permutation to a public-key encryption scheme secure in the sense of IND-CCA2 [30].

Rabin-SAEP and Rabin-SAEP+ can be proven to be IND-CCA2 in the random oracle model under the factoring assumption like our scheme. However, in real systems, the random oracles are represented by the functions made from practical hash functions, such as SHA [23], because no random oracles exist. Therefore, the proof of security in random oracle model cannot transfer to the real world, and it is important to analyze the security of provably secure schemes in the random oracle model in the real world. We believe that the security of SAEP and SAEP+ depends on the random oracles more heavily than the security of OAEP and OAEP+ does. For example, suppose that the adversary can compute the first m_1 bits and the last m_2 of $f^{-1}(y)$, where $m_1 + m_2 < |N|/2$, y

is a target ciphertext, and f is an encryption function of Rabin-SAEP, Rabin-SAEP+ or RSA-SAEP+. Then, it is impossible to apply Coppersmith's algorithm to compute the rest bits of $f^{-1}(y)$. Furthermore, suppose that the first m_1 bits of the output of the hash function H has bias in response to the last m_2 bits of the input. That is, the first m_1 bits of $H(x)$ can be computed with a probability of more than $1/2^{m_1}$ when the last m_2 bits of x are known. The adversary will then be able to guess a correct b with a probability of more than $1/2$ (cf. definition 2). However, we think that the possibility of this occurring in OAEP and OAEP+ is less than that in SAEP and SAEP+ because the plaintext is doubly protected by the two hash functions G and H.

From the above reason, we think that OAEP and OAEP+ are more secure padding methods than SAEP and SAEP+ in the real world. Therefore, we believe that our scheme has a higher security than Rabin-SAEP and Rabin-SAEP+ in the real world.

3.4   Coppersmith's Algorithm

In this section, we present the fact by Coppersmith [10].

[Coppersmith] Let N be a large composite integer of unknown factorization. Let

$$f(x) = x^k + a_{k-1}x^{k-1} + \cdots + a_2x^2 + a_1x + a_0 \in \mathbb{Z}[x]$$

be a monic polynomial of degree k. Then, there is an efficient algorithm to find all $x_0 \in \mathbb{Z}$ such that

$$f(x_0) = 0 \pmod{N} \qquad \text{and} \qquad |x_0| < N^{1/k}.$$

We denote by $T_C(N, k)$ the running time of Coppersmith's algorithm when finding roots of a polynomial $f \in \mathbb{Z}[x]$ of degree k.

3.5   Security of Our Scheme

We can obtain the following theorem.

Theorem 2. Our scheme is secure in the sense of IND-CCA2 in the random oracle model under the assumption of intractability of factoring $N (= p^d q)$.

Let $\mathcal{G}$ be a composite number generator such that $N (= p^d q) \leftarrow \mathcal{G}(1^k)$. We assume that the distribution of N is the same as that of N with our scheme.

The proof of Theorem 2 is immediately induced from the following theorem.

Theorem 3. Let $\Pi = (\mathcal{K}, \mathcal{E}, \mathcal{D})$ be the our encryption scheme with parameters k_0 and k_1, and let n be the associated plaintext length. Then, there exists an oracle machine U such that for each integer k the following is true. Suppose $A = (A_1, A_2)$ succeeds in $(t, q_D, q_G, q_H, \epsilon)$-breaking $\Pi(1^k)$ in the sense of IND-CCA2, namely,

$$2 \cdot \Pr[\, G, H \leftarrow \Omega \,; \, (pk, sk) \leftarrow \mathcal{K}(1^k) \,; \, (x_0, x_1, c) \leftarrow A_1^{G,H,\mathcal{D}_{sk}}(pk) \,;$$
$$b \leftarrow \{0, 1\} \,; \, y \leftarrow \mathcal{E}_{pk}^{G,H}(x_b) \,: \, A_2^{G,H,\mathcal{D}_{sk}}(y, x_0, x_1, c) = b \,] - 1 \geq \epsilon,$$

where A runs for at most t steps, makes at most q_D queries to the decryption oracle, makes at most q_G queries to G, and makes at most q_H queries to H.

Then, $M = U^A$ succeeds in (t', ϵ')-factoring $\mathcal{G}(1^k)$, where

$$t' = t + q_H T_C(N, 2) + q_G q_H T_S(k) + \tilde{T}(k) + \mathcal{O}(k)$$

$$\epsilon' = \frac{1}{3} \left(\epsilon - \frac{q_G}{2^{k_0}} \right) \left(1 - \frac{q_G}{2^{k_0}} \right) \left(1 - \frac{q_D}{2^{k_1-1}} - \frac{q_D}{2^{k_0}} \right).$$

Here, $T_S(k)$ denotes the running time of the encryption function $\mathcal{E}_{pk}(\cdot)$, and $\tilde{T}(k)$ denotes the running time of factoring N when an integer that has a commom factor with N is given.

The proof of Theorem 3 is in Appendix A.

4   Performance

In this section, we discuss our scheme's performance by comparing it with RSA-OAEP [3], RSA-OAEP+ [30], RSA-SAEP+ [6], Rabin-SAEP [6], Rabin-SAEP+ [6] and EPOC-2 [8]. Our reason for choosing these schemes is that we think these schemes have almost ideal features, described in Section 1.1, in the factoring base previous schemes (Although the security of RSA-OAEP+ and RSA-SAEP+ is based on RSA problem that is less intractable than the factoring problem and EPOC-2 is the hybrid scheme.).

4.1   Key Length

We first describe the length of each parameter k_0, k_1 and k in our scheme. We recommend taking $|k_0|, |k_1| \geq 128$ in the algorithm of our scheme like previous schemes. Table 1 compares the length of each modulus, namely $|N|$, of RSA-OAEP and of our scheme. Each modulus length is determined to make the intractability of the factoring almost the same when NFS and ECM are used (cf. Section 3.1). Note that the modulus lengths of RSA-OAEP, RSA-OAEP+, RSA-SAEP+, Rabin-SAEP and Rabin-SAEP+ are all the same.

Table 1. The length of modulus.

	Modulus length (bits)		
RSA-OAEP	1024	2048	4096
EPOC-2	1344	2304	4032
Our scheme ($N = p^2 q$)	1344	2304	4032
Our scheme ($N = p^3 q$)	1536	3072	4928

We show the key length of the secret and public keys of each scheme in Table 2: Here we set a standard to the 1024 bits modulus variant of RSA-OAEP.

Table 2. Public and secret key length.

	Public key (bits)	Secret key (bits)
RSA-OAEP	1026'2048	2048
Rabin-SAEP	1024	1024
EPOC-2	4032	1344
Our scheme ($N = p^2q$)	1344	896
Our scheme ($N = p^3q$)	1536	728

For convenience, we omit the bit length for the hash functions in each scheme. Note that the key lengths of RSA-OAEP, RSA-OAEP+ and RSA-SAEP+ are the same, and those of Rabin-SAEP and Rabin-SAEP+ are the same. Table 2 shows that the total key length of our scheme is shorter than that of the other schemes[2].

4.2   Encryption and Decryption Speeds

In Table 3, we give the cost for the modular multiplications of the encryption and the decryption in each scheme to evaluate the efficiency of the encryption and the decryption speeds: In each scheme, we make the plaintext length maximal. And, for fairness, we make all random numbers lengths 128 bits.

Table 3. Efficiency by the (converted) number of modular multiplications (1).

	Modulus length (bits)	Encryption	Decryption
RSA-OAEP	1024	2'1536	388
Rabin-SAEP	1024	1	388
EPOC-2	1344	386	1415
Our scheme ($N = p^2q$)	1344	2	260
Our scheme ($N = p^3q$)	1536	3	168

In RSA-OAEP, decryption by the Chinese remainder theorem is applied. We assume that a modular exponentiation a^x (x is k bit) requires $3k/2$ modular multiplications in the standard binary method, and $a^x b^y$ (x, y are k bit) requires $7k/4$ modular multiplications in the extended binary method [16]. Furthermore, we set a standard to the number of modular multiplications on a 1024-bit modulus. Hence we assume that an n-bit modular multiplication costs $(n/1024)^2 \times$ "the number of multiplications on an n-bit modulus". The reason for establishing the bit length of each encryption scheme was described in Section 3.1. Note that each number of modular multiplications of RSA-OAEP, RSA-OAEP+ and RSA-SAEP+ is the same, and that of Rabin-SAEP and Rabin-SAEP+ is the same.

---

[2] For convenience, we omit the size of the hash functions, security parameters and the secret-key cryptosystem (in EPOC-2).

Table 4. Efficiency by the (converted) number of modular multiplications (2).

	Modulus length (bits)	Encryption	Decryption
RSA-OAEP	2048	$8 \sim 12288$	3088
Rabin-SAEP	2048	4	3088
EPOC-2	2304	1134	6319
Our scheme ($N = p^2q$)	2304	6	1305
Our scheme ($N = p^3q$)	3072	9	1320
RSA-OAEP	4096	$32 \sim 98304$	24640
Rabin-SAEP	4096	16	24640
EPOC-2	4032	3473	30762
Our scheme ($N = p^2q$)	4032	16	6974
Our scheme ($N = p^3q$)	4928	23	5411

In our scheme, we utilize our calculation method for decryption (cf. Section 2.1). In the previous method [31], the average total number of modular multiplications T_1 is represented by $T_1 = \frac{|p|}{3} + \frac{25}{6}$, where we make $d = 2$ and set a standard to the number of modular multiplications on N. On the other hand, in our method, the average total number of modular multiplications T_2 is represented by $T_2 = \frac{|p|}{3} + \frac{11}{6}$. Although this difference is very small, it is expected that it will be non-negligible in smart card systems and in systems in which mauch decryption processing must be done at one time. For example, in a system in which 600 times decryptions are done at a single time on average, the difference in the modular multiplications between our method and the previous one amounts to 1400. Furthermore, the actual decryption speed and mountaing size will be smaller than previous one because ours does not require Euclidean algorithm for Chinese Remanider Theorem.

Let t_1 be the average time to do a modular multiplication, and let t_2 be the average time to calculate the hash values. From the above data, the average time for each encryption is: $2t_1 + 2t_2 \sim 1536t_1 + 2t_2$ (RSA-OAEP), $2t_1 + 3t_2 \sim 1536t_1 + 3t_2$ (RSA-OAEP+), $2t_1 + 2t_2 \sim 1536t_1 + 2t_2$ (RSA-SEAP+), $t_1 + t_2$ (Rabin-SAEP), t_1+2t_2 (Rabin-SAEP+), $386t_1+t_2+t_3$ (EPOC-2), $2t_1+2t_2$ (our scheme: $N = p^2q$), $3t_1+2t_2$ (our scheme: $N = p^3q$), where t_3 is the average time of encryption by the secret-key cryptosystem. Because t_2 is very small, we can say that our scheme is more efficient than RSA-OAEP and its transformations, and that our scheme is comparable with Rabin-SAEP and Rabin-SAEP+.

In Table 4, we give an estimation for 2048, 4096-bit RSA-OAEP and 2304, 4032-bit our scheme, and in the graph 3, we give an estimation for a bigger modulus. From these data, we can say that, for a big modulus, our scheme is much more efficient than the other schemes. And, this difference of the efficiency increases polynomially with $|N|$. This is owing to the fact that the efficiency of factoring $N = p^dq$ using the elliptic curve method is overtaken by that using the number sieve field method when $|N|$ is large (cf. Section 3.1).

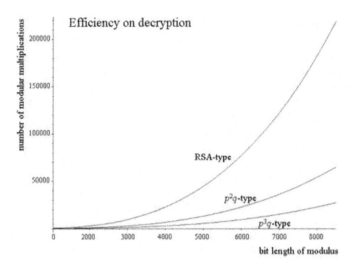

Fig. 3. Efficiency on decryption.

Table 5. Plaintext and ciphertext lengths

	Plaintext length (bits)	Ciphertext length (bits)
RSA-OAEP	768	1024
Rabin-SAEP	256	1024
RSA-SAEP+	384	1024
EPOC-2	Arbitrary	1344+α
Our scheme ($N = p^2 q$)	1088	1344

4.3 Plaintext and Ciphertext Lengths

We show the maximal plaintext and ciphertext lengths, in Table 5[3]. Here, we set a standard to the 1024 bits modulus variant of RSA-OAEP. And, we set $k_0 = 128$ and $k_1 = 128$ (cf. [3]) in RSA-OAEP and our scheme. Note that each pair of the plaintext and the ciphertext lengths of RSA-OAEP and RSA-OAEP+ is the same, and that of RSA-SAEP+ and Rabin-SAEP+ is the same.

The main purpose of the public key cryptosystems is to distribute the data encryption key of the secret key cryptosystems. However, to the best of our knowledge, just a few protocols send only the data encryption key. Many protocols including SET are used to send various kinds of information (e.g., information about the identity of users) along with the data encryption key. It is therefore important to choose a public key encryption scheme that can accommodate such practices, so. The plaintext space of our scheme is large enough to send the data encryption key with the attached information.

---

[3] α denotes the size of ciphertext by the secret-key encryption cryptosystem.

5   Conclusion

We presented a public-key encryption scheme that has almost ideal features. Our scheme is based on a modular squaring over $\mathbb{Z}_N$, where $N = p^d q$ (p and q are prime numbers, $d > 1$), and is secure in the sense of IND-CCA2 under the factoring assumption of N. We also showed our system's performance by comparing it with that of previous practical factoring base schemes, such as RSA-OAEP or Rabin-SAEP, etc. Our scheme could gain great profit by taking the modulo $N = p^d q$ instead of $N = pq$. In particular, its decryption speed becomes fast and its difference in efficiency in comparison with the above mentioned earlier schemes increases polynomially with $|N|$ (this is owing to the fact that the efficiency of factoring $N = p^d q$ using the elliptic curve method is overtaken by that using the number sieve field method when $|N|$ is large).

We believe that our scheme offers secure and practical public-key encryption that will be useful far into the future.

References

1. M. Bellare, A. Desai, D. Pointcheval and P. Rogaway.: Relations among notions of security for public-key encryption schemes, *Advances in Cryptology – Crypto'98*, LNCS 1462, Springer-Verlag, pp.26–45 (1998)
2. M. Bellare and P. Rogaway.: Random oracles are practical – a paradigm for designing efficient protocol, *First ACM Conference on Computer and Communications Security*, pp.62–73 (1993)
3. M. Bellare and P. Rogaway.: Optimal asymmetric encryption – How to encrypt with RSA, *Advances in Cryptology – Eurocrypt'94*, LNCS 950, Springer-Verlag, pp.92–111 (1994)
4. D. Bleichenbacher.: Chosen ciphertext attacks against protocols based on the RSA encryption standard PKCS#1, *Advances in Cryptology – Crypto'98*, LNCS 1462, Springer-Verlag, pp.1–12 (1998)
5. M. Blum and S. Goldwasser.: An efficient probabilistic public-key encryption scheme which hides all partial information, *Advances in Cryptology – Crypto'84*, LNCS 196, Springer-Verlag, pp.289-299 (1985)
6. D. Boneh.: Simplified OAEP for the RSA and Rabin functions, *Advances in Cryptology – Crypto2001*, LNCS 2139, Springer-Verlag, pp.275-291 (2001)
7. D. Boneh, G. Durfee and N. Howgrave-Graham.: Factoring $N = p^r q$ for large r, *Advances in Cryptology – Crypto'99*, LNCS 1666, Springer-Verlag, pp.326-337 (1999)
8. Call for Contributions on New Work Item Proposal on Encryption Algorithms, NTT, 2000-3-10.
9. D. Coppersmith.: Modifications to the number field sieve, *Journal of in Cryptology*, 6, 3, pp.169-180 (1993)
10. D. Coppersmith.: Finding a small root of a univariate modular equation, *Advances in Cryptology – Eurocrypt'96*, LNCS 1070, Springer-Verlag, pp.155-165 (1996)
11. R. Cramer and V. Shoup.: A practical public key cryptosystem provably secure against adaptive chosen ciphertext attack, *Advances in Cryptology – Crypto'98*, LNCS 1462, Springer-Verlag, pp.13-25 (1998)

12. D. Dolve, C. Dwork and M. Naor.: Non-malleable cryptography, *Proceedings of the 23rd Annual Symposium on Theory of Computing*, ACM, pp.542–552 (1991)
13. S. Goldwasser and M. Bellare.: *Lecture Notes on Cryptography*, http://www-cse.ucsd.edu/users/mihir/ (1997)
14. S. Goldwasser and S. Micali.: Probabilistic encryption, *Journal of Computer and System Sciences*, 28, 2, pp.270–299 (1984)
15. D.M. Gordon.: Designing and detecting trapdoors for discrete log cryptosystems, *Advances in Cryptology – Crypto'92*, LNCS 740, Springer-Verlag, pp.66-75 (1992)
16. D.E. Knuth.: *The Art of Computer Programming*, Addison-Wesley (1981)
17. N. Koblitz.: Elliptic curve cryptosystems, *Math. Comp.*, 48, 177, pp.203-209 (1987)
18. K. Kurosawa, W.Ogata, T.Matsuo and S.Makishima.: IND-CCA Public Key Schemes Equivalent to Factoring n=pq, *Proc. of PKC2001* (2001)
19. A.K. Lenstra and H.W. Lenstra,Jr.: *The Development of the Number Field Sieve*, Lect. Notes Math. 1554, Springer-Verlag (1993)
20. H.W. Lenstra,Jr.: Factoring integers with elliptic curves, *Annals of Math.*, 126, pp.649-673 (1987)
21. J. Manger.: A chosen ciphertext attack on RSA optimal asymmetric encryption padding (OAEP) as standardized in PKCS#1 v2.0, *Advances in Cryptology – Crypto2001*, LNCS 2139, Springer-Verlag, pp.230-238 (2001)
22. V.S. Miller.: Use of elliptic curves in cryptography, *Advances in Cryptology – Crypto'85*, LNCS 218, Springer-Verlag, pp.417-426 (1985)
23. National Institute of Standards, FIPS Publication 180, Secure Hash Standards (1993)
24. M. Naor and M. Yung.: Public-key cryptosystems provably secure against chosen ciphertext attacks, *Proceedings of the 22nd Annual Symposium on Theory of Computing*, ACM, pp.427–437 (1990)
25. T. Okamoto and D. Pointcheval.: EPOC-3: Efficient Probabilistic Public-Key Encryption-V3 (Submission to P1363a), May 2000
26. T. Okamoto and S. Uchiyama.: A new public-key cryptosystem as secure as factoring, *Advances in Cryptology – Eurocrypt'98*, LNCS 1403, Springer-Verlag, pp.308-318 (1998)
27. J.M. Pollard.: A Monte-Carlo method for factorization, BIT 15, pp.331-334 (1975)
28. M.O. Rabin.: Digital signatures and public-key encryptions as intractable as factorization, MIT, Technical Report, MIT/LCS/TR-212 (1979)
29. R.L. Rivest, A. Shamir and L. Adleman.: A method for obtaining digital signatures and public-key cryptosystems, *Communications of the ACM*, Vol.21, No.2, pp.120-126 (1978)
30. V. Shoup.: OAEP reconsidered, *Advances in Cryptology – Crypto2001*, LNCS 2139, Springer-Verlag, pp.239-259 (2001)
31. T. Takagi.: Fast RSA-type cryptosystem modulo $p^k q$, *Advances in Cryptology – Crypto'98* LNCS 1462, Springer-Verlag, pp.318-326 (1998)
32. H.C. Williams.: A modification of the RSA public key encryption procedure, *IEEE Trans. on Information Theory*, IT-26, 6, pp.726-729 (1980)
33. Y. Zheng and J. Seberry.: Practical approaches to attaining security against adaptive chosen Ciphertext Attacks, *Advances in Cryptology – Crypto'92*, LNCS 740, Springer-Verlag, pp.292-304 (1992)

A   Proof of Theorem 3

We first define the behavior of factoring algorithm M. M is given a composite number N $(= p^d q)$. It is trying to find the prime factor of N. The factoring algorithm M is defined as follows:

(0) An input to M is N, where $N \leftarrow \mathcal{G}(1^k)$.
(1) M chooses $w \in \{0,1\}^{k-1}$ at random. and computes $y = w^2 \bmod N$.
(2) M initializes two lists, called its G-list and its H-list, to empty. Then M picks $b \in \{0,1\}$ at random.

    Then, M simulates the two stages of $A = (A_1, A_2)$ as indicated in the next two steps.

(3) (*Simulation of the* **find**-*stage*) M runs A_1 on input pk, where pk denotes the public key of our scheme. M also provides A with fair random coins and simulates A's random oracles G and H as follows.
(3.1) When A_1 makes an oracle call $h \in \{0,1\}^{n+k_1}$ of H, M computes $x \in \{0,1\}^{k_0}$ such that $(x + 2^{k_1} h)^2 \equiv y \pmod{N}$ by using Coppersmith's algorithm if such x exists. Then, M sets $w^* = h||x$, and adds h to the H-list. Else M provides A_1 with a random string $H_h \in \{0,1\}^{k_0}$, and adds h to the H-list.
(3.2) When A_1 makes an oracle call g of G, M provides A with a random string $G_g \in \{0,1\}^{n+k_1}$, and adds g to the G-list.
    Let (x_0, x_1, c) be the output with which A_1 halts.

(4) (*Simulation of the* **guess**-*stage*) M runs A_2 on input (y, x_0, x_1, c). M responds to oracle queries as follows.
(4.1) Suppose A_2 makes H-query $h \in \{0,1\}^{n+k_1}$. M computes $x \in \{0,1\}^{k_0}$ such that $(x + 2^{k_1} h)^2 \equiv y \pmod{N}$ by using Coppersmith's algorithm if such x exists. Then, M sets $w^* = h||x$, and adds h to the H-list. Else M provides A_2 with a random string $H_h \in \{0,1\}^{k_0}$, and adds h to the H-list.
(4.2) Suppose A_2 makes G-query g. M provides A_2 with a random string $G_g \in \{0,1\}^{n+k_1}$, and adds g to the G-list.
(5) (*Simulation of the decryption oracle*) Suppose A makes a query y' to the decryption oracle. Here, we let (s_i, H_i) be a pair of the H-query s_i and its answer H_i for $i = 1, 2, \ldots, q_H$, and let (r_i, G_i) be a pair of the G-query r_i and its answer G_i for $i = 1, 2, \ldots, q_G$. Then, for $i = 1, 2, \ldots, q_H$ and $j = 1, 2, \ldots, q_G$, machine M
(5.1) Set $t_{i,j} = H_i \oplus r_j$.
(5.2) Compute $x'_{i,j} \in \{0,1\}^n$ and $z'_{i,j} \in \{0,1\}^{k_1}$ such that $x'_{i,j}||z'_{i,j} = s_i \oplus G_j$.
(5.3) Outputs

$$\begin{cases} x'_{i,j} & \text{if there is an } i,j \text{ such that } z'_{i,j} = 0^{k_1} \text{ and } y' = (s_i||t_{i,j})^2 \bmod N, \\ * & \text{otherwise.} \end{cases}$$

(6) M outputs w^* and halts the above simulations if this string was defined in the above experiment, and **fail** otherwise.

(7) M computes $\alpha = \gcd(w - w^*, N)$, and $l = (d+1)|\alpha|/k$. And, M outputs a pair of integers that is

$$(\sqrt[l]{\alpha}, \quad N/\sqrt[l]{\alpha^d}) \qquad \text{or} \qquad (\sqrt[d-l+1]{N/\alpha}, \quad N/\sqrt[d-l+1]{(N/\alpha)^d})$$

if $0 < \alpha < N$, and **fail** otherwise.

Note that the H-list and G-list include the queries of both the **find** and **guess** stages of A's execution and that M does not return the answer of h which defines w^* to A. In (3.1) and (3.2), it is possible to apply Coppersmith's algorithm to compute x, since $2k_0 < k$.

We consider the probability space given by the above experiment. The inputs N to M are drawn at random according to $\mathcal{G}(1^k)$. We call this "Game 1" and we let $\Pr_1[\cdot]$ denote the corresponding probability.

It is easy to verify that the amount of time t' to carry out Game 1 is

$$t' = t + q_H\, T_C(N, 2) + q_G q_H T_S(k) + \tilde{T}(k) + \mathcal{O}(k).$$

It is also easy to verify that there is a universal machine U such that the computation of M can be done by U^A.

We define the following event to consider the difference between the actual decryption oracle and the simulator of M:

FAIL is true if the output of the simulator is different from $\mathcal{D}_{sk}^{G,H}(y')$, where sk denotes the secret key of our scheme.

Then, we have the following lemma.

Lemma 1. The probability that the outputs of the actual decryption oracle and the simulator are different is upper-bounded by

$$\Pr_1[\textbf{FAIL}] \leq \frac{q_D}{2^{k_1-1}} + \frac{q_D}{2^{k_0}}.$$

Proof. If w^* is defined in the experiment, then M halts the simulation. Hence M simulates the decryption oracle until w^* is defined.

Let y be a target ciphertext, and let y' be the query to the decryption oracle. Let $s = x_b 0^{k_1} \oplus G(r)$, $t = r \oplus H(s)$, $(s||t)^2 \equiv y \pmod{N}$, $s' = x' 0^{k_1} \oplus G(r')$, $t' = r' \oplus H(s')$ and $(s'||t')^2 \equiv y' \pmod{N}$ for $x' \in \{0,1\}^n$ and $r, r' \in \{0,1\}^{k_0}$.

We consider the follwing event:

AskR' is true if r' is on the G-list.

AskS' is true if s' is on the H-list.

W'=AskR' $\wedge$ **AskS'**.

Then, it holds that $\Pr_1[\textbf{FAIL} \mid \textbf{W'}] = 0$. Hence we upper bound $\Pr_1[\textbf{FAIL}]$ by

$\Pr_1[\textbf{FAIL}]$

$$\begin{aligned}
&= \Pr_1[\textbf{FAIL} \mid \textbf{W'}] \cdot \Pr_1[\textbf{W'}] + \Pr_1[\textbf{FAIL} \mid \neg\textbf{AskS'}] \cdot \Pr_1[\neg\textbf{AskS'}] \\
&\quad + \Pr_1[\textbf{FAIL} \mid \textbf{AskS'} \wedge \neg\textbf{AskR'}] \cdot \Pr_1[\textbf{AskS'} \wedge \neg\textbf{AskR'}] \\
&\leq \Pr_1[\textbf{FAIL} \mid \neg\textbf{AskS'}] + \Pr_1[\textbf{FAIL} \mid \textbf{AskS'} \wedge \neg\textbf{AskR'}]. \quad (3)
\end{aligned}$$

Since $s \neq s'$, we have

$$\Pr_1[\mathbf{FAIL} \mid \neg\mathbf{AskS'}] \leq \frac{q_D}{2^{k_1}}. \tag{4}$$

Next, we consider $\Pr_1[\mathbf{FAIL} \mid \mathbf{AskS'} \wedge \neg\mathbf{AskR'}]$. We have

$$\begin{aligned}
\Pr_1[&\mathbf{FAIL} \mid \mathbf{AskS'} \wedge \neg\mathbf{AskR'}] \\
&= \Pr_1[\mathbf{FAIL} \mid \mathbf{AskS'} \wedge \neg\mathbf{AskR'} \mid r \neq r'] \cdot \Pr_1[r \neq r'] + \\
&\quad \Pr_1[\mathbf{FAIL} \mid \mathbf{AskS'} \wedge \neg\mathbf{AskR'} \mid r = r'] \cdot \Pr_1[r = r'] \\
&\leq \Pr_1[\mathbf{FAIL} \mid \mathbf{AskS'} \wedge \neg\mathbf{AskR'} \mid r \neq r'] + \Pr_1[r = r'].
\end{aligned}$$

Then, we have

$$\Pr_1[\mathbf{FAIL} \mid \mathbf{AskS'} \wedge \neg\mathbf{AskR'} \mid r \neq r'] \leq q_D/2^{k_1}.$$

It must be $t' = t \oplus H(s) \oplus H(s')$ for holding $r = r'$. However, H-query s is not made. Hence it follows that

$$\Pr_1[r = r'] \leq q_D/2^{k_0}. \tag{5}$$

From (3), (4) and (5), we have

$$\Pr_1[\mathbf{FAIL}] \leq \frac{q_D}{2^{k_1-1}} + \frac{q_D}{2^{k_0}}.$$

$\square$

Intuitively, Lemma 1 says that the advantage of M deriving new information from the decryption oracle is negligible.

We let $\Pr_2[\cdot] = \Pr_1[\cdot \mid \neg\mathbf{FAIL}]$ denote the probability distribution, in Game 1, conditioned on $\mathbf{FAIL}$ not being true, and call this "Game 2".

Let $s = x_b 0^{k_1} \oplus G(r)$, $t = r \oplus H(s)$ and $y = (s\|t)^2 \bmod N$ in Game 1. We consider the following events:

$\mathbf{BAD}$ is true if:
- G-oracle query r was made in the **find**-stage or the **guess**-stage, and
- $G_r \neq s \oplus x_b 0^{k_1}$.

$\mathbf{G} = \neg\mathbf{BAD}$.

Note that if A makes a H-query h ($\in \{0,1\}^{n+k_1}$) such that $(h\|x)^2 \equiv y$ $(\bmod\ N)$ for some $x \in \{0,1\}^{k_0}$, then M halts the simulation. Therefore, a G-query r should be made before such h is not on the H-list.

We let $\Pr_3[\cdot] = \Pr_2[\cdot \mid \mathbf{G}]$ denote the probability distribution, in Game 2, conditioned on $\mathbf{G}$ being true, and call this "Game 3".

Now, we consider the experiment which defines the advantage of A. Namely, first choose $N \leftarrow \mathcal{G}(1^k)$ and let $\mathcal{E}_*$ be the corresponding encryption function under our scheme. Then choose

$$\begin{aligned}
G_*, H_* &\leftarrow \Omega; \quad (x_0^*, x_1^*, c^*) \leftarrow A_1^{G_*, H_*, \mathcal{D}_*}(N, k_0, k_1); \\
b_* &\leftarrow \{0,1\}; \quad y^* \leftarrow \mathcal{E}_*^{G_*, H_*}(x_b^*),
\end{aligned}$$

and run $A_2^{G_*, H_*, \mathcal{D}_*}(y^*, x_0^*, x_1^*, c^*)$, where $\mathcal{D}_*$ is the corresponding decryption function under our scheme. Let $\mathrm{Pr}_1^*[\cdot]$ be the corresponding distribution and Game 1^* be the game.

We define Game 2^*, which is playing Game 1^* a little bit differently, as follows:

(1) Choose $N \leftarrow \mathcal{G}(1^k)$ and let $\mathcal{E}_*$ be the corresponding encryption function under our scheme.

(2) Choose y^* from the image of the encryption function uniformly at random first. Next, choose $H_* \leftarrow \Omega$ and $b_* \leftarrow \{0, 1\}$ uniformly at random.

(3) Choose $s^* \in \{0, 1\}^{n+k_1}$ and $t^* \in \{0, 1\}^{k_0}$ such that $(s^* || t^*)^2 \equiv y^* \pmod{N}$ at random. And, set $r^* = t^* \oplus H(s^*)$.

(4) We define G_*-oracle as follows:

$$G_*(x^*) = \begin{cases} s^* \oplus x_b^* 0^{k_1} & \text{if } x^* = r^*, \\ g_* \in_R \{0, 1\}^{n+k_1} & \text{if } x^* \neq r^*. \end{cases}$$

(5) Run $A_2^{G_*, H_*, \mathcal{D}_*}(y^*, x_0^*, x_1^*, c^*)$ for $(x_0^*, x_1^*, c^*) \leftarrow A_1^{G_*, H_*, \mathcal{D}_*}(N, k_0, k_1)$.

We claim that Game 3 and Game 2^* are identical in the sense that the view of A at any point in these two games is the same before the H-query h ($\in \{0, 1\}^{n+k_1}$) such that $(h || x)^2 \equiv y \pmod{N}$ for some $x \in \{0, 1\}^{k_0}$ is made. Indeed, we have chosen the event $\mathbf{G}$ so that the oracle queries we are returning in Game 2 will mimic Game 2^* as long as $\mathbf{G}$ remains true.

Let us introduce the following additional events (of Game 1):

AskH is true if at the end of the **guess**-stage, h ($\in \{0, 1\}^{n+k_1}$) such that $(h || x)^2 \equiv y \pmod{N}$ for some $x \in \{0, 1\}^{k_0}$ is on the H-list.

AskR is true if at the end of the **guess**-stage, r is on the G-list.

AskS is true if at the end of the **guess**-stage, s is on the H-list.

$\mathbf{W} = \mathbf{AskR} \wedge \mathbf{AskS}$.

Lemma 2. The probability that the good event fails is upper-bounded by

$$\mathrm{Pr}_2[\neg \mathbf{G}] \leq \frac{q_G}{2^{k_0}}.$$

Proof. It is immediately induced by the following:

$$\mathrm{Pr}_2[\neg \mathbf{G}] = \mathrm{Pr}_2[\neg \mathbf{G} \mid \neg \mathbf{AskH}] \leq \mathrm{Pr}_2[\neg \mathbf{G} \mid \neg \mathbf{AskS}]$$
$$\leq \mathrm{Pr}_2[\mathbf{AskR} \mid \neg \mathbf{AskS}] \leq \frac{q_G}{2^{k_0}}.$$

$\square$

The following lemma gives a relation among $\mathrm{Pr}_3[A = b]$ and $\mathrm{Pr}_3[\mathbf{W}]$.

Lemma 3. The winning probability in Game 3 is bounded below by

$$\mathrm{Pr}_3[\mathbf{W}] \geq 2\mathrm{Pr}_3[A = b] - 1 - \frac{q_G}{2^{k_0}},$$

where "$A = b$" denotes the event that A is successful in predicting bit b.

Proof. We upper bound $\Pr_3[A = b]$ by

$$
\begin{aligned}
\Pr_3&[A = b] \\
&= \Pr_3[A = b \mid \mathbf{W}] \cdot \Pr_3[\mathbf{W}] + \Pr_3[A = b \mid \neg\mathbf{AskR}] \cdot \Pr_3[\neg\mathbf{AskR}] \\
&\quad + \Pr_3[A = b \mid \mathbf{AskR} \wedge \neg\mathbf{AskS}] \cdot \Pr_3[\mathbf{AskR} \wedge \neg\mathbf{AskS}] \\
&\leq \Pr_3[\mathbf{W}] + \Pr_3[A = b \mid \neg\mathbf{AskR}] \cdot \Pr_3[\neg\mathbf{AskR}] + \Pr_3[\mathbf{AskR} \wedge \neg\mathbf{AskS}] \\
&= \Pr_3[\mathbf{W}] + \Pr_3[A = b \mid \neg\mathbf{AskR}] \cdot (1 - \Pr_3[\mathbf{W}] - \Pr_3[\mathbf{AskR} \wedge \neg\mathbf{AskS}]) \\
&\quad + \Pr_3[\mathbf{AskR} \wedge \neg\mathbf{AskS}]. \qquad\qquad\qquad\qquad\qquad\qquad\qquad (6)
\end{aligned}
$$

Now observe that if $\neg\mathbf{AskR}$ is true then A has no advantage in predicting b:

$$
\Pr_3[A = b \mid \neg\mathbf{AskR}] = \frac{1}{2}. \qquad (7)
$$

We also have

$$
\Pr_3[\mathbf{AskR} \wedge \neg\mathbf{AskS}] \leq \frac{q_G}{2^{k_0}}. \qquad (8)
$$

Therefore, it follows

$$
\Pr_3[\mathbf{W}] \geq 2\Pr_3[A = b] - 1 - \frac{q_G}{2^{k_0}},
$$

from (6), (7) and (8). $\qquad\qquad\qquad\qquad\qquad\qquad\qquad\qquad\qquad\qquad\square$

From Lemma 3, we have

$$
\Pr_2[\mathbf{W}] \geq \epsilon - \frac{q_G}{2^{k_0}}. \qquad (9)
$$

From Lemma 1, Lemma 2 and (9), we have

$$
\begin{aligned}
\Pr_1[\mathbf{AskH}] &\geq \Pr_1[\mathbf{AskS}] \geq \Pr_2[\mathbf{AskS}] \cdot \Pr_1[\mathbf{FAIL}] \\
&\geq \Pr_2[\mathbf{AskS} \mid \mathbf{G}] \cdot \Pr_2[\mathbf{G}] \cdot \Pr_1[\mathbf{FAIL}] \\
&\geq \Pr_3[\mathbf{AskS}] \cdot \Pr_2[\mathbf{G}] \cdot \Pr_1[\mathbf{FAIL}] \\
&\geq \Pr_3[\mathbf{W}] \cdot \Pr_2[\mathbf{G}] \cdot \Pr_1[\mathbf{FAIL}] \\
&\geq \left(\epsilon - \frac{q_G}{2^{k_0}}\right) \left(1 - \frac{q_G}{2^{k_0}}\right) \left(1 - \frac{q_D}{2^{k_1 - 1}} - \frac{q_D}{2^{k_0}}\right).
\end{aligned}
$$

The equation $X^2 \equiv y \pmod{N}$ has four solutions in $\mathbb{Z}_N$ as indicated in the proof of Theorem 1, and two of those are less than $N/2$. Note that $N/2 \in \{0, 1\}^{k-1}$. Since $w, w^* \in \{0, 1\}^{k-1}$, the probability that it holds $1 < \alpha < N$ is more than $1/3$, where $\alpha = \gcd(w - w^*, N)$.

Therefore, we have

$$
\begin{aligned}
\Pr_1&[N \leftarrow \mathcal{G}(1^k) : \; M(N) = (p, q)] \\
&\geq \frac{1}{3}\left(\epsilon - \frac{q_G}{2^{k_0}}\right) \left(1 - \frac{q_G}{2^{k_0}}\right) \left(1 - \frac{q_D}{2^{k_1 - 1}} - \frac{q_D}{2^{k_0}}\right).
\end{aligned}
$$

Concrete Security Analysis
of CTR-OFB and CTR-CFB Modes of Operation

Jaechul Sung[1], Sangjin Lee[1], Jongin Lim[1], Wonil Lee[1], and Okyeon Yi[2]

[1] Center for Information Security Technologies(CIST),
Korea University, Anam Dong, Sungbuk Gu,
Seoul, Korea
{sjames,sangjin,jilim,nice}@cist.korea.ac.kr
[2] Information Security Technology Division, ETRI, Taejon, Korea
oyyi@etri.re.kr

Abstract. In [1], they gave the notions of security for the symmetric encryption and provided a concrete security analysis of the XOR, CTR, and CBC schemes. Among the three schemes, the CTR scheme achieves the best concrete security in their analysis. In this paper, we propose the new schemes, CTR-OFB and CTR-CFB, which have the security as same as that of the CTR scheme on the point of the concrete security analysis and achieve higher resistance against some practical attacks than the CTR scheme.

Keywords: Modes of Operation, Concrete Security, Pseudorandom Function Family, Symmetric Encryption Schemes.

1 Introduction

The DES has four modes of operation in [11]. The four modes are the ECB, CBC, CFB, and OFB. Since DES modes of operation were introduced, many other modes of operation for block ciphers have been suggested and analyzed. Moreover modes of operations for block ciphers received much attention lately, partly due to an announcement by NIST that they are considering an update to their list of standardized. In [12] several modes of operation were suggested, such as the ABC, CTR, IACBC, ICPM, OCB, XCBC, and etc.

S.Goldwasser and S.Micali was the first to introduce the formal notions of security for encryption [7]. They presented two notions of security for asymmetric encryption, semantic security and polynomial security, and proved them equivalent with respect to polynomial-time reductions. M.Bellare et al.[1] presented the four notions of security for symmetric encryption in the framework of concrete security under the attack assumptions of chosen-plaintext attack(CPA) and chosen-ciphertext attack(CCA):

1. Left-or-Right indistinguishability (LOR)
2. Real-or-Random indistinguishability (ROR)
3. Find-then-Guess security (FTG)
4. Semantic security (SEM)

K. Kim (Ed.): ICICS 2001, LNCS 2288, pp. 103–113, 2002.

They proved the first two notions are security preserving under the any attack assumption and the last two notions are also security preserving. Furthermore the first two notions are the stronger notions of security than the last two notions. Therefore showing an encryption scheme LOR or ROR secure implies tight reductions to all other notions but showing an encryption scheme FTG secure or SEM secure does not. So, if the bounds are equal, it is better to demonstrate security with respect to one of the first two notions, since that immediately translates into equally good bounds for the other notions. With the notions of security, especially left-or-right indistinguishability(LOR), they proved the concrete security analysis of the XOR, CTR, and CBC Schemes.

The counter(CTR) mode was originally introduced by W.Diffie and M. Hellman in 1979 [6]. Recently H.Lipmaa, P.Rogaway, and D. Wagner suggested the CTR mode in standardizing AES modes of operation [12]. The CTR mode has significant efficiency advantages, which can be preprocessed because of the independence of message blocks and easy to random-access. Furthermore the CTR mode gives the better concrete security than the XOR and CBC schemes [1].

In this paper we define new modes of operation for block ciphers. The new modes of operation are counter-based-OFB(CTR-OFB) mode and counter-based-CFB mode(CTR-CFB), which provide the concrete security as same as that of CTR mode. Although the CTR scheme changes the input bits of the underlying function serially, our scheme can randomize some input bits. So this can have more resistant against the SQUARE-type attacks [5,8] and the conventional differential attack with low hamming weight differential than the CTR scheme. Also our new schemes provide the better concrete security than the OFB and CFB schemes, which achieve the same concrete security as the CBC scheme does.

This paper is organized as follows. In Section 2 we give some preliminary definitions, the notions of security, and some results of [1]. In Section 3 and 4 we propose the new modes of operations and prove the concrete security. In Section 5 we summarize our conclusions.

2   Preliminaries

In this section we describe some relevant definitions. Our treatment follows that of M.Bellare, K.Kilian, P.Rogaway [2], and M.Bellare, A.Desai, E.JokiPii, and P.Rogaway [1].

In [1], they considered the four definitions of security for symmetric encryption under the two attack assumptions of chosen-plaintext attack(CPA) and chosen-ciphertext attack(CCA). Here we will only consider the notion of the left-or-right indistinguishablity(LOR) under the CPA model, which gives the other three notions with comparable bounds.

We define that $a \leftarrow A(x_1, x_2, \ldots)$ denote the experiment of running A on inputs $x_1, x_2, \ldots$ if $A(\cdot, \cdot, \ldots)$ is any probabilistic algorithm. Let $\Pi = (K, E, D)$ be an encryption scheme, where algorithm K is the key generator, E is the encryption algorithm, and D is the decryption algorithm.

The approach to concrete security is via parameterization of the resources of the adversary A. Let t be A's running time, q_e be the number of encryption oracle queries, and μ_e be the amount of the ciphertext A sees in response to its oracle queries.

In the LOR sense adversary is allowed queries of the form (x_0, x_1) where x_0, x_1 are equal length messages. Consider the two different games. In the first, each query is responded to by encrypting the left message. In the second, it is right message. In formal definition, the left-or-right oracle is defined by $E_k(LR(\cdot, \cdot, b))$, where $b \in \{0, 1\}$, to take input (x_0, x_1) and do the following:

If $b = 0$, it computes $C \leftarrow E_K(x_0)$ and return C.
If $b = 1$, it computes $C \leftarrow E_K(x_1)$ and return C.

Now we can define the LOR-CPA as the following.

Definition 1. *[1] Let* $\mathsf{SE} = (\mathsf{K}, \mathsf{E}, \mathsf{D})$ *be a symmetric encryption scheme. Let* $b \in \{0, 1\}$ *and* $k \in N$. *Let* A_{cpa} *be an adversary has access to the oracle* $E_K(LR(\cdot, \cdot, b))$. *Consider the following experiment:*

$$\text{Experiment } \mathbf{Exp}_{\mathsf{SE}, A_{cpa}}^{lor-cpa-b}(k)$$
$$K \leftarrow \mathsf{K}(k)$$
$$d \leftarrow A_{cpa}^{E_k(LR(\cdot, \cdot, b))}(k)$$
$$\textbf{Return } d$$

Define the advantages of the adversary via

$$\mathbf{Adv}_{\mathsf{SE}, A_{cpa}}^{lor-cpa}(k) = \Pr[\mathbf{Exp}_{\mathsf{SE}, A_{cpa}}^{lor-cpa-1}(k) = 1] - \Pr[\mathbf{Exp}_{\mathsf{SE}, A_{cpa}}^{lor-cpa-0}(k) = 1].$$

Define the advantage functions of the scheme as follows. For any t, q_e, μ_e,

$$\mathbf{Adv}_{\mathsf{SE}}^{lor-cpa}(k, t, q_e, \mu_e) = \max_{A_{cpa}}\{\mathbf{Adv}_{\mathsf{SE}, A_{cpa}}^{lor-cpa}(k)\}.$$

If a reasonable adversary cannot obtain the significant advantage, we consider an encryption scheme to be good. In the similar way we can define the LOR-CCA with the decryption oracle. For details, see [1].

We will consider the symmetric encryption schemes based on finite pseudo-random functions(PRFs) or permutations(PRPs) [2]. Let $Rand^{l \rightarrow L}$ be the family of all functions from $\{0, 1\}^l$ to $\{0, 1\}^L$ and $Perm^l$ be the family of all permutations on $\{0, 1\}^l$. We will not define PRFs and PRPs for detail. The following implies the relation of the advantage between PRFs and PRPs.

Proposition 1. *[1] For any permutation family* P *with length* l,

$$\mathbf{Adv}_P^{prf}(t, q) \leq \mathbf{Adv}_P^{prp}(t, q) + \frac{q^2}{2^{l+1}}.$$

Now we will see the concrete security of the symmetric encryption schemes, i.e., the XOR, CTR, and CBC schemes, using RFs(random functions), RPs (random permutations), PRFs, and PRPs. Let a function family F be input length l, output length L, and key-length k. To specify the function we will use

$f = F_K$. The followings are specified the XOR, CTR, and CBC schemes respectively. The message x to be encrypted is regarded as a sequence of l-bit blocks, $x = x_1 \cdots x_n$, and let r be the nonce and addition is modulo 2^l.

- **The XOR scheme: XOR$[F]$ = (K-XOR, E-XOR, D-XOR)**
  The key generation algorithm K-XOR just outputs a random k-bit key K for the underling function family F, thereby specifying a function $f = F_K$ of l-bits to L-bits. Define E-XOR$_K(x)$ = E-XOR$^{F_K}(x)$ and D-XOR$_K(z)$ = D-XOR$^{F_K}(z)$, where:

function E-XOR$^f(x)$	**function D-XOR$^f(z)$**
$\quad r \leftarrow \{0,1\}^l$	$\quad$ Parse z as $r\|\|y_1 \cdots y_n$
$\quad$ for $i = 1, \cdots, n$	$\quad$ for $i = 1, \cdots, n$
$\quad\quad$ do $y_i = f(r+i) \oplus x_i$	$\quad\quad$ do $x_i = f(r+i) \oplus y_i$
$\quad$ return $r\|\|y_1 y_2 \cdots y_n$	$\quad$ return $x = x_1 x_2 \cdots x_n$

- **The CTR scheme: CTR$[F]$ = (K-CTR, E-CTR, D-CTR)**
  The key generation algorithm K-CTR is the same as the XOR scheme, meaning just outputs a random k-bit key K for the underling function family F. Define E-CTR$_K(x, ctr)$ = E-CTR$^{F_K}(x, ctr)$ and D-CTR$_K(z)$ = D-CTR$^{F_K}(z)$, where:

function E-CTR$^f(x, ctr)$	**function D-CTR$^f(z)$**
$\quad$ for $i = 1, \cdots, n$	$\quad$ Parse z as $ctr\|\|y_1 \cdots y_n$
$\quad\quad$ do $y_i = f(ctr+i) \oplus x_i$	$\quad$ for $i = 1, \cdots, n$
$\quad ctr \leftarrow ctr + n$	$\quad\quad$ do $x_i = f(ctr+i) \oplus y_i$
$\quad$ return $(ctr, ctr\|\|y_1 y_2 \cdots y_n)$	$\quad$ return $x = x_1 x_2 \cdots x_n$

- **The CBC Scheme: CBC$[F]$ = (K-CBC, E-CBC, D-CBC)**
  The key generation algorithm K-CBC is the same as the XOR scheme, meaning just outputs a random k-bit key K for the underling permutation family F(The CBC Scheme is required that $l = L$) . Define E-CBC$_K(x)$ = E-CBC$^{F_K}(x)$ and D-CBC$_K(z)$ = D-CBC$^{F_K}(z)$, where:

function E-CBC$^f(x)$	**function D-CBC$^f(z)$**
$\quad y_0 \leftarrow \{0,1\}^l$	$\quad$ Parse z as $y_0\|\|y_1 \cdots y_n$
$\quad$ for $i = 1, \cdots, n$	$\quad$ for $i = 1, \cdots, n$
$\quad\quad$ do $y_i = f(y_{i-1} \oplus x_i)$	$\quad\quad$ do $x_i = f^{-1}(y_i) \oplus y_{i-1}$
$\quad$ return $y_0\|\|y_1 y_2 \cdots y_n$	$\quad$ return $x = x_1 x_2 \cdots x_n$

Let us see the concrete security of the schemes. We first summarize the security of the XOR scheme.

Theorem 1. *[1]* **[The Concrete Security of the XOR Scheme]**

(i) **(The Lower Bound on Insecurity of XOR using a RF)**
   Let $R = Rand^{l \to L}$. Then, for any t, q_e, and μ_e, such that $\mu_e q_e / L \leq 2^l$,

$$\mathbf{Adv}^{lor-cpa}_{XOR[R]}(\cdot, t, q_e, \mu_e) \geq 0.316 \cdot \frac{\mu_e \cdot (q_e - 1)}{L \cdot 2^l}.$$

(ii) **(The Upper Bound on Insecurity of XOR using a RF)**
Let $R = Rand^{l \to L}$. Then, for any t, q_e, μ_e,

$$\mathbf{Adv}_{XOR[R]}^{lor-cpa}(\cdot, t, q_e, \mu_e) \leq \frac{\mu_e \cdot (q_e - 1)}{L \cdot 2^l}.$$

(iii) **(Security of XOR using a PRF)**
Suppose F be a PRF family with input-length l and output-length L. Then, for any t, q_e, and $\mu_e = qL$,

$$\mathbf{Adv}_{XOR[F]}^{lor-cpa}(\cdot, t, q_e, \mu_e) \leq 2 \cdot \mathbf{Adv}_F^{prf}(t, q) + \frac{\mu_e \cdot (q_e - 1)}{L \cdot 2^l}.$$

The CTR scheme is the stateful version of the XOR scheme. This scheme achieves the better security than that of the XOR. The adversary has no advantage in the ideal case.

Theorem 2. [1] [**The Concrete Security of the CTR Scheme**]

(i) **(Security of CTR using a RF)**
Let $R = Rand^{l \to L}$. Then, for any t, q_e, and $\mu_e \leq L2^l$,

$$\mathbf{Adv}_{CTR[R]}^{lor-cpa}(\cdot, t, q_e, \mu_e) = 0.$$

(ii) **(Security of CTR using a PRF)**
Suppose F be a PRF family with input-length l and output-length L. Then, for any t, q_e, and $\mu_e = min(qL, L2^l)$,

$$\mathbf{Adv}_{CTR[F]}^{lor-cpa}(\cdot, t, q_e, \mu_e) \leq 2 \cdot \mathbf{Adv}_F^{prf}(t, q).$$

Although in the CBC scheme $l = L$ is required and each F_k should be a permutation, we will still consider the case that F is a pseudorandom function family($l = L$). Also we will see the case that F is a pseudorandom permutation family.

Theorem 3. [1] [**The Concrete Security of the CBC Scheme**]

(i) **(The Lower Bound on Insecurity of CBC using a RF)**
Let $R = Rand^{l \to l}$. Then, for any t, q_e, and μ_e, such that $\mu_e \leq l2^{\frac{l}{2}}$,

$$\mathbf{Adv}_{CBC[R]}^{lor-cpa}(\cdot, t, q_e, \mu_e) \geq 0.316 \cdot \left(1 - \frac{2}{2^{l/2}}\right) \cdot \left(\frac{\mu^2}{l^2} - \frac{\mu}{l}\right) \cdot \frac{1}{2^l}.$$

(ii) **(The Upper Bound on Insecurity of CBC using a RF)**
Let $R = Rand^{l \to l}$. Then, for any t, q_e, and μ_e,

$$\mathbf{Adv}_{CBC[R]}^{lor-cpa}(\cdot, t, q_e, \mu_e) \leq \left(\frac{\mu^2}{l^2} - \frac{\mu}{l}\right) \cdot \frac{1}{2^l}.$$

(*iii*) **(Security of CBC using a PRF)**

   Suppose F be a PRF family with input-length l and output-length l. Then, for any t, q_e, and $\mu_e = ql$,

$$\mathbf{Adv}_{CBC[F]}^{lor-cpa}(\cdot, t, q_e, \mu_e) \leq 2 \cdot \mathbf{Adv}_F^{prf}(t, q) + \left(\frac{\mu^2}{l^2} - \frac{\mu}{l} \right) \cdot \frac{1}{2^l}.$$

(*iv*) **(The Lower Bound on Insecurity of CBC using a RP)**

   Let $RP = Perm^l$. Then, for any $t, q_e(= \mu_e/l)$, and $\mu_e(\leq l2^{\frac{l}{2}})$,

$$\mathbf{Adv}_{CBC[RP]}^{lor-cpa}(\cdot, t, q_e, \mu_e) \geq 0.316 \cdot \left(\frac{\mu^2}{l^2} - \frac{\mu}{l} \right) \cdot \frac{1}{2^l}.$$

(*v*) **(The Upper Bound on Insecurity of CBC using a RP)**

   Let $RP = Perm^l$. Then, for any t, q_e, and μ_e,

$$\mathbf{Adv}_{CBC[RP]}^{lor-cpa}(\cdot, t, q_e, \mu_e) \leq \left(\frac{\mu^2}{l^2} - \frac{\mu}{l} \right) \cdot \frac{1}{2^l}.$$

(*vi*) **(Security of CBC using a PRP)**

   Suppose F be a PRP family with length l. Then, for any t, q_e, and $\mu_e = ql$,

$$\mathbf{Adv}_{CBC[F]}^{lor-cpa}(\cdot, t, q_e, \mu_e) \leq 2 \cdot \left(\mathbf{Adv}_F^{prp}(t, q) + \frac{q^2}{2^{l+1}} \right) + \left(\frac{\mu^2}{l^2} - \frac{\mu}{l} \right) \cdot \frac{1}{2^l}.$$

In the above theorems we can see that the CTR scheme has the best security in a random function model. This also gives the best concrete security in a pseudorandom function model. The CTR model has no collision on the inputs of the function f. Since the function f is in a random function, the attacker have no information to distinguish in the LOR sense. However the XOR and CBC scheme may have an collision on input of f by the birthday paradox, this can leak some information to distinguish. This motivates our schemes. Our scheme pursue the CTR scheme to achieve the perfect concrete security in the random function model under the LOR sense.

3   The CTR-OFB and CTR-CFB Schemes

In the previous section we considered that the CTR mode has the best concrete security in the LOR-CPA sense. This comes from the collision-freeness on the input of the function f. Here we propose the new schemes, the counter-based OFB scheme and CFB scheme, which we call the CTR-OFB scheme and CTR-CFB scheme respectively.

Now we define our schemes. Let a function family F be input length l, output length L, and key-length k. To specify the function we will use $f = F_K$. The message x to be encrypted is regarded as a sequence of l-bit blocks, $x = x_1 \cdots x_n$. Let r be the nonce with v-bit, addition is modulo 2^{l-v}, and ctr be the $(l-v)$-bit integer. The notation $a||b$ means the concatenation of a and b, and $lsb_j(a)$ takes j bits of a from 0 to $j-1$ bit position.

- **The CTR-OFB scheme:**
  CTR-OFB$[F]$ = (K-CTR-OFB, E-CTR-OFB, D-CTR-OFB)
  The key generation algorithm K-CTR-OFB is the same as the XOR scheme, meaning just outputs a random k-bit key K for the underling function family F. Define E-CTR-OFB$_K(x, ctr)$ = E-CTR-OFB$^{F_K}(x, ctr)$ and D-CTR-OFB$_K(z)$ = D-CTR-OFB$^{F_K}(z)$, where:

function E-CTR-OFB$^f(x, ctr)$	function D-CTR-OFB$^f(z)$
$v_0 = r \leftarrow \{0,1\}^v$ and $y_0 = v_0 \| ctr$	Parse z as $y_0 \| y_1 \cdots y_n$
for $i = 1, \cdots, n$	Parse y_0 as $v_0 \| ctr$
do $y_i = f(v_{i-1} \| ctr + i) \oplus x_i$	**for** $i = 1, \cdots, n$
$v_i = lsb_v(f(v_{i-1} \| ctr + i))$	**do** $x_i = f(v_{i-1} \| ctr + i) \oplus y_i$
$ctr \leftarrow ctr + n$	$v_i = lsb_v(f(v_{i-1} \| ctr + i))$
return $(ctr, y_0 \| y_1 y_2 \cdots y_n)$	**return** $x = x_1 x_2 \cdots x_n$

- **The CTR-CFB scheme:**
  CTR-CFB$[F]$ = (K-CTR-CFB, E-CTR-CFB, D-CTR-CFB)
  The key generation algorithm K-CTR-CFB is the same as the XOR scheme, meaning just outputs a random k-bit key K for the underling function family F. Define E-CTR-CFB$_K(x, ctr)$ = E-CTR-CFB$^{F_K}(x, ctr)$ and D-CTR-CFB$_K(z)$ = D-CTR-CFB$^{F_K}(z)$, where:

function E-CTR-CFB$^f(x, ctr)$	function D-CTR-CFB$^f(z)$
$r \leftarrow \{0,1\}^v$ and $y_0 = r \| ctr$	Parse z as $y_0 \| y_1 \cdots y_n$
for $i = 1, \cdots, n$	Parse y_0 as $r \| ctr$
do $y_i = f(lsb_v(y_{i-1}) \| ctr + i) \oplus x_i$	**for** $i = 1, \cdots, n$
$ctr \leftarrow ctr + n$	**do** $x_i = f(lsb_v(y_{i-1}) \| ctr + i) \oplus y_i$
return $(ctr, y_0 \| y_1 y_2 \cdots y_n)$	**return** $x = x_1 x_2 \cdots x_n$

The above schemes are conter-based and give the concrete security as same as the CTR scheme. We will see this in the following section. Also the CTR-OFB scheme can be preprocessed because of the independence of the message block. So we can see that our scheme have the same security of the CTR scheme on the concrete security point of view.

The CTR-OFB scheme is similar to the OFB scheme and the CTR-CFB scheme is also similar to the CFB scheme. However, since the OFB and CFB schemes achieve the same concrete security as the CBC scheme does, our new schemes have the better concrete security than the OFB and CFB schemes.

The modes of operation for symmetric encryption are generally using block ciphers. It is well known that block ciphers are difficult to be constructed to attain PRFs or PRPs. So we should see a scheme not only on the theoretical point of view but also on the practical point of view.

For the most powerful known attacks on block ciphers are Differential Cryptanalysis(DC) [3,4] and Linear Cryptanalysis(LC) [9,10]. For the CTR scheme we know that inputs of f are using serially. This may give an easy way to construct to the chosen plaintext pairs with the low hamming weight differential. If the underlying block ciphers have crucial weakness in this attack, the CTR scheme is easy to attack. However for our schemes inputs of f are the concatenation of randomized v bits and $l - v$ bit counter. This make difficult to construct the plaintext pairs having low hamming weight. Also the randomization of the input bits of the underlying function in the CTR-OFB and CTR-CFB schemes also give the higher resistance against the SQUARE-type attacks [5,8] than that of the CTR scheme.

4   Security Analysis of the CTR-OFB and CTR-CFB Schemes

Here we will see the concrete security of the our proposed schemes. We will use the same notations in section 2. For our scheme the function family F is with the input length l, output length L, and key length k. The following theorem give the concrete security of the CTR-OFB schemes.

Theorem 4. [The Concrete Security of the CTR-OFB Scheme]

(i) **(Security of CTR-OFB using a RF)** Let $R = Rand^{l \to L}$. Then, for any t, q_e, and $\mu_e \leq L2^{l-v}$,

$$\mathbf{Adv}_{CTR-OFB[R]}^{lor-cpa}(\cdot, t, q_e, \mu_e) = 0.$$

(ii) **(Security of CTR-OFB using a PRF)** Suppose F be a PRF family with input-length l and output-length L. Then, for any t, q_e, and $\mu_e = min(qL, L2^{l-v})$,

$$\mathbf{Adv}_{CTR-OFB[F]}^{lor-cpa}(\cdot, t, q_e, \mu_e) \leq 2 \cdot \mathbf{Adv}_F^{prf}(t, q).$$

Proof. The proof of (ii) can be achived as the same way of [1]. So we need only to prove (i).

Let (P_i, Q_i) be the oracle queries of the adversary A, each consisting of a pair of equal length messages. Let n_i be the number of blocks in the i-th query. We denote $P_i = p_1^i \cdots p_{n_i}^i$ and $Q_i = q_1^i \cdots q_{n_i}^i$. Let $r_i \in \{0,1\}^v$ be the nonce associated to (P_i, Q_i) as chosen at random by the oracle, for $i = 1, \cdots, q_e$. Let be the orcle answers such that $(r_i||ctr, y_1^i, \cdots, y_{n_i}^i) \leftarrow O(P_i, Q_i)$, $i = 1, \cdots, q_e$.

In answering the i-th query, the oracle applies the underlying function f to the n_i strings either $r_i||ctr+1$, $lsb_v(p_1^i \oplus y_1^i)||ctr+2$, $\cdots$, $lsb_v(p_{n_i-1}^i \oplus y_{n_i-1}^i)||ctr+n_i$ or $r_i||ctr + 1$, $lsb_v(q_1^i \oplus y_1^i)||ctr + 2$, $\cdots$, $lsb_v(q_{n_i-1}^i \oplus y_{n_i-1}^i)||ctr + n_i$.

Let D be the following event, defined for either game: $r_i||ctr + 1$, $lsb_v(p_1^i \oplus y_1^i)||ctr+2$, $\cdots$, $lsb_v(p_{n_i-1}^i \oplus y_{n_i-1}^i)||ctr+n_i$ and $r_i||ctr+1$, $lsb_v(q_1^i \oplus y_1^i)||ctr+2$, $\cdots$, $lsb_v(q_{n_i-1}^i \oplus y_{n_i-1}^i)||ctr + n_i$ have no same value for $i = 1, \cdots, q_e$. We define $Pr_0[\cdot]$ to be the probability of an event in game 0 and $Pr_1[\cdot]$ to be the probability of an event in game 1.

Claim 1. $Pr_0[D] = Pr_1[D] = 1$ for $\mu_e \leq L2^{l-v}$

Proof: In any case we know that the input string does not have the same value since the values of counter are different. So the probability of each game is 1.

Claim 2. $Pr_0[A = 1|D] = Pr_1[A = 1|D]$

Proof: Given the event D, we have that, in either game, the function f is evaluated at a new point each time. Thus the output is randomly and uniformly distributed over $\{0,1\}^L$ and each block is a message block XORed with a random value. So we have $Pr_0[A = 1|D] = Pr_1[A = 1|D]$.

Now we compute the advantage of A as follows:

$$\mathbf{Adv}_{CTR-OFB[R]}^{lor-cpa}(\cdot, t, q_e, \mu_e) = Pr_1[A=1] - Pr_0[A = 1]$$
$$= Pr_1[A=1|D] \cdot Pr_1[D] + Pr_1[A = 1|\bar{D}] \cdot Pr_1[\bar{D}] -$$
$$Pr_0[A=1|D] \cdot Pr_0[D] - Pr_0[A = 1|\bar{D}] \cdot Pr_0[\bar{D}]$$

By Claim 1 and 2, we have $\mathbf{Adv}_{CTR-OFB[R]}^{lor-cpa}(\cdot, t, q_e, \mu_e) = 0$.

In the similar way, we can prove the following theorem, which gives the concrete security of the CTR-CFB scheme.

Theorem 5. [The Concrete Security of the CTR-CFB Scheme]

(*i*) **(Security of CTR-CFB using a RF)**
Let $R = Rand^{l \to L}$. Then, for any t, q_e, and $\mu_e \leq L2^{l-v}$,

$$\mathbf{Adv}_{CTR-CFB[R]}^{lor-cpa}(\cdot, t, q_e, \mu_e) = 0.$$

(*ii*) **(Security of CTR-CFB using a PRF)**
Suppose F be a PRF family with input-length l and output-length L. Then, for any t, q_e, and $\mu_e = min(qL, L2^{l-v})$,

$$\mathbf{Adv}_{CTR-CFB[F]}^{lor-cpa}(\cdot, t, q_e, \mu_e) \leq 2 \cdot \mathbf{Adv}_F^{prf}(t, q).$$

Proof. This proof is as same as the proof of Theorem 4.

5   Conclusion

In this paper we propose the new modes of operation, the CTR-OFB and CTR-CFB scheme. Each scheme have the perfect concrete security on the sense of [1] as same as the CTR mode do. The CTR scheme have the inputs of f serially. However our schemes can randomize some input bits of the f. We may think that this makes the attack difficult to analyze the scheme on the practical attack point of view, for example, the differential cryptanalysis with low hamming weight differential and the SQUARE-type attacks.

The CTR mode can be preprocessed because of the independence of message blocks and are easy to random-access. The CTR-OFB scheme also can be preprocessed. But it does not permit random-access.

References

1. M. Bellare, A. Desai, E. JokiPii, and P. Rogaway, *A Concrete Security Treetment of Symmetric Encryption: Analysis of the DES Modes of Operation*, Proceedings of the 38th Symposium on Foundations of Computer Science, IEEE, 1997. The revised version is available at *http://www-cse.ucsd.edu/users/mihir*.
2. M. Bellare, J. Kilian, and P. Rogaway, *The Security of the Cipher Block Chaining Message Authentication Code*, Advanced in Cryptology - CRYPTO'94, LNCS 839, pp. 341–358, Springer-Verlag, 1994.
3. E. Biham and A. Shamir, *Differential cryptanalysis of DES-like cryptosystems*, Advances in Cryptology - CRYPTO'90, LNCS 537, pp. 2–21, Springer-Verlag, 1991.
4. E. Biham and A. Shamir, *Differential cryptanalysis of the full 16-round DES*, Advances in Cryptology - CRYPTO'92, LNCS 740, pp. 487–496, Springer-Verlag, 1992.
5. J. Daeman, L. Knudsen, and V. Rijmen, *The Block Cipher Square*, Fast Software Encryption 1997, LNCS 1636, pp. 46–59, Springer-Verlag, 1997.
6. W. Diffie and M. Hellman, *Privacy and Authentication: An introduction to Cryptography*, Proceedings of the IEEE, 67(1979), pp. 397–427, 1979.
7. S. Goldwasser and S. Micali, *Probabilistic Encryption*, Journal of Computer and System Sciences, Vol.28, pp. 270–279, April 1984.
8. Stefan Lucks, *The Saturation Attack - a Bait for Twofish*, Fast Software Encrption 2001, 2001, to appear.
9. M. Matsui, *Linear cryptanalysis method for DES cipher*, Advances in Cryptology - EUROCRYPT'93, LNCS 765, pp. 386–397, Springer-Verlag, 1994.
10. M. Matsui, *The first experimental cryptanalysis of the Data Encryption Standard*, Advances in Cryptology - CRYPTO'94, LNCS 839, pp. 1–11, Springer-Verlag, 1994.
11. National Bureau of Standards, *DES modes of operation*, FIPS-Pub.46, National Bureau of Standards, U.S. Department of Commerce, Washington D.C., December 1980.
12. National Institute of Standards and Technology, *AES Mode of Operation Development Effort, http://csrc.nist.gov/encryption/modes*.

Appendix:
The Figures of the CTR-OFB and CTR-CFB Schemes

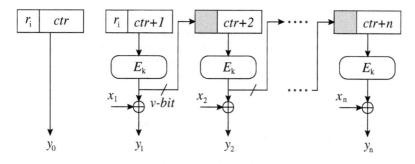

Fig. 1. The CTR-OFB Scheme

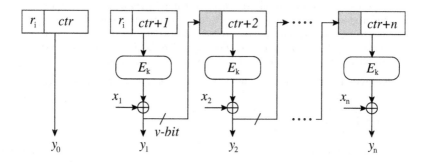

Fig. 2. The CTR-CFB Scheme

Decentralized Event Correlation
for Intrusion Detection

Christopher Krügel, Thomas Toth, and Clemens Kerer

Distributed Systems Group
Technical University Vienna
Argentinierstrasse 8, A-1040 Vienna, Austria
{chris,ttoth,C.Kerer}@infosys.tuwien.ac.at

Abstract. Evidence of attacks against a network and its resources is often scattered over several hosts. Intrusion detection systems (IDS) which
attempt to detect such attacks therefore have to collect and correlate information from different sources. We propose a completely decentralized
approach to solve the task of event correlation and information fusing of
data gathered from multiple points within the network.
Our system models an intrusion as a pattern of events that can occur at
different hosts and consists of collaborating sensors deployed at various
locations throughout the protected network installation.
We present a specification language to define intrusions as distributed
patterns and a mechanism to specify their simple building blocks. The
peer-to-peer algorithm to detect these patterns and its prototype implementation, called *Quicksand*, are described. Problems and their solutions
involved in the management of such a system are discussed.

Keywords: Intrusion Detection, Event Correlation, Network Security

1   Introduction

Intrusion detection systems (IDS) are network security tools that process local
audit data or monitor network traffic to search for specific patterns (misuse
based) or certain deviations from expected behavior (anomaly based) which
indicate malicious activities against the protected network. The traces of a simple
attack are often visible in a single log-file or can be monitored at a single network
interface card. However, sophisticated attackers do not concentrate on a single
host alone but try to disguise their activities by distributing them over several
machines. Each single action considered for itself looks innocent but in their
entirety they constitute an attack. This makes it necessary to collect and relate
audit data from different sources (a process called *event correlation*).

As Zamboni [1] pointed out, most existing IDSs perform their data processing
centrally, despite their distributed data collection [5,12]. This causes limitations
in their scalability, ease of configuration and fault tolerance. The failure of the
central unit completely deactivates the correlation process and effectively blinds
the IDS. The processing capacity of this node also limits the number of events it
can handle in a certain amount of time. When too many sensors are forwarding

K. Kim (Ed.): ICICS 2001, LNCS 2288, pp. 114–131, 2002.

their messages to the central host the resulting backlog increases the reaction time of the system or might cause data loss.

To circumvent such shortcomings, hierarchical designs have been introduced. Systems like Emerald [10], GrIDS [13], AAFID [2,1] or NetSTAT [14] have a layered structure where data is locally preprocessed and filtered. Only events that might be part of a distributed attack scenario are forwarded to a higher level entity. Emerald [10,9] uses a publish/subscribe system to disseminate relevant data between hosts. Nevertheless, these systems use dedicated machines that act as central points for collecting data from remote sensors.

Although hierarchical structures and filtering at low levels allow better scalability, the systems are still vulnerable to faults and overloading of nodes that are close to the root of the hierarchy. Top level nodes still limit scalability and their failure can cut off large parts of the IDS (the whole subtree below).

We attack the inherent problems of centralized, dedicated nodes by proposing a completely decentralized approach where the detection of an intrusion is restricted to those nodes where parts of the attack are directly observable. Sensors are deployed at every (possible) host of the protected network and at potentially interesting network spots (e.g. router, switch). These sensors collaborate and exchange information in a peer-to-peer fashion without a centralized coordinator.

Only a few systems have already attempted to use a similar approach. The best known is CSM (Cooperating Security Managers) [16], a design which distinguishes between a local ID component and an information forwarding unit at each node. The forwarding unit allows to exchange information between nodes along the login chain of users. While this system was the first to show the possibility of distributed cooperation in principal, its applicability is limited by the fact that cooperation is only done along user's login chains. Another system called Micael [3] proposes mobile agents to accomplish the task of distributed event correlation without a central entity but its current status is unclear.

The following paper describes Quicksand, the prototype implementation of our decentralized intrusion detection approach. First, we define a specification language to define basic events and distributed patterns, which are composed of such events. Then, we discuss Quicksand's system design and management as well as the underlying algorithm to disseminate information between peers. Finally, the performance of our prototype is evaluated and compared to intrusion detection systems that attempt to correlate data in a hierarchical or centralized fashion.

2   System Overview

We define an intrusion as a pattern of basic events that occur at multiple hosts. A basic event is characterized as the occurrence of something of interest that could be the sign of an intrusion (e.g. the receipt of a certain IP packet, a failed authentication or a password file access). Such events can either stem from a local misuse or an anomaly incident. In addition to our sensors, we plan to integrate a number of existing third-party systems to perform the local detection and feed their event data into our correlation algorithm.

By relating events from multiple nodes, one can detect a number of attacks that would remain unnoticed by only focusing on local activity. One type of attacks can be found by checking for suspicious signatures of network connections between hosts as described in GrIDS [13]. This includes worms spreading in a network as well as telnet chains (a number of consecutive telnet logins by an intruder to hide his tracks). The opening of a connection between two computers looks harmless when monitored locally but when the whole network is considered, suspicious chain or tree patterns might emerge.

Relating events between different hosts might also increase the chance for anomaly detection sensors to notice an attack. Consider a port scan from a certain machine that is scanning each host on a subnet for an open port 80 and 8080. Whenever the firewall (or port scan detector) monitors such an activity, it forwards the information about the scan and its source to the web server. When the web server later receives abnormal traffic exactly from that source an attack is assumed. Sensors on different hosts can usefully exchange information to enhance their detection rate. The aim of this information flow is that each local node gets a better understanding of what happens inside the whole network. The combination of local information from different places leverages the understanding of the network traffic each participating node monitors.

Quicksand can be used to describe and detect situations where a sequence of events (like port scans and then abnormal packets) occurs on multiple hosts. This is used to modify the reaction of sensors at nodes that get aware of emerging, hostile patterns. Quicksand implements a flexible mechanism that allows to specify response functions for each step of an intrusion pattern. In contrast to traditional systems which usually only react (e.g. generate warnings or harden the firewall) after an incident has been detected, our mechanism allows fine-grain control over potential counter activities while threatening situations emerge.

The detection process finds patterns by sending messages between sensors running on nodes where interesting events occur. Similar to the approach presented in [15], the installation is covered by a *web* of sensors that needs to be managed. As novel intrusions are discovered, signature databases of all sensors have to be consistently updated. Quicksand uses a central management station that stores the current configurations of all installed ID probes. We have implemented a mechanism that allows to transfer event and pattern descriptions as well as response functions over a (SSL encrypted) connection to sensors. A protected management channel is utilized to load and unload these building blocks and perform reconfiguration on-the-fly.

In addition, the detection of distributed patterns demands the knowledge of the correct temporal order of events. We have designed two different solutions to this well-known issue in distributed systems that trade additional overhead for improved accuracy.

3   Pattern Specification

The design of our pattern specification language is guided by two conflicting goals. The first goal states that the language should be as expressive as possible. It would be desirable to allow the description of complex relationships between events on different hosts using regular or tree grammars. As our system relies on peer-to-peer message passing between hosts without a central coordination point, arbitrary complex patterns might cause the amount of data that needs to be exchanged to explode. In the worst case each host has to send all its data to every other node. This conflicts with the second goal, which demands that the amount of transferred data between hosts should be minimal. Therefore we have to impose limitations on the expressiveness of our pattern language.

An event is the smallest entity of a pattern and defined as the occurrence of something that might be part of an intrusion (e.g. an IP packet from a certain source address, the invocation of a certain process). In addition, we introduce the notion of *artificial events*. An artificial event is not related to an actual activity in the environment but is created during a response by Quicksand itself. It is piped back into the sensor's event input queue and can be utilized to satisfy constraints of different (or the same) pattern. Artificial events are a useful mechanism to exchange information between attack scenarios or to model timeouts. The output of a certain attack pattern can be used as input for another pattern to build hierarchical structures or to implement scenarios that count the number of times a certain basic pattern has occurred. A timeout can be implemented by starting a timer in a response that creates an artificial event when it expires.

Basic event objects consist of a list of attribute-value pairs that are filled by the sensor from the actual observed event instance. We define these objects as annotated C **structures** where each member, which can be a basic C **type** or a reference (pointer) to another structure, represents a certain attribute. This allows to assemble more complex event descriptions as compounds of simpler objects. The annotation is used to specify the attribute that determines the target of *send events* (which are explained below).

A sensor that scans its input for the occurrence of a certain event needs to be compiled against this event's corresponding C **struct**. Therefore, we cannot add new basic types to a certain sensor while the system is running. This is no real limitation, as the basic events themselves usually do not change frequently. It is more important to be able to update and modify pattern descriptions which represent the actual attacks, an operation supported by Quicksand.

A pattern describes activities on individual hosts as well as interaction between machines. The basic building block of a pattern is a sequence of basic events that happens locally on one machine (called *host sequence*). One can specify a list of events at a local host by enumerating them and imposing certain constraints on their attributes. We distinguish between constraints which relate single event attributes with constant values and constraints which relate different attributes of events using variables. A connection (context) between event sequences on different hosts is established by *send events*.

Definition: *A pattern* P, *relating events that occur at n distinct hosts, consists of n sequences of events, one for each node (an event sequence at a single node is called* host sequence*).*
A set of events S_A *at host* A *is linked to a set of events* S_B *at host* B, *iff* S_A *contains a send event to host* S_B. *Any event that refers to a remote host (e.g. a packet sent to a host, the reception of a packet from a host) might be used as send event.*

The detection of a port scan that looks for an open port 80 can be used as a send event when the web server (the referred host in this example) is known to the host that detects the scan. It is only required that the *target* B of the send event can be determined locally at A (usually by using the event data). The first event of S_B has to be the next event to occur after the send event in S_A. It is required that the send event is the last event in S_A. As mentioned above, basic events are annotated to specify the attribute which is used to derive the target.

Definition: *Pattern* P *is valid, iff the following properties hold.*

1. *Each set of events is at least linked to one other set.*
2. *Every set except one (called the* root set*) contains exactly one send event as the last event of the host sequence. The root set contains no send event.*
3. *The connection graph contains no cycles. The connection graph is built by considering each event set as a vertex and each link between two sets as an edge between the corresponding vertices.*

These definitions only allow tree-like pattern structures (i.e. the connection graph is a tree), where the node with the root set is the root of the tree. Although this restriction seems limiting at first glance, most desirable situations can still be described. Usually, activity at a target host depends on events that have occurred earlier at several other hosts. This situation can be easily described by our tree patterns where connection links from those hosts end at the root set.

3.1   Attack Specification Language

This section describes the syntax and semantics of our pattern description language (called Attack Specification Language - ASL).

A pattern definition is written as follows

```
ATTACK "Scenario Name" [ nodes ] pattern
```

The *nodes* section is used to assign an identifier to each node that is later referred in the pattern definition.

The *pattern* section specifies the pattern. It consists of a list of event sets, one for each node that appears in the node section. The event set is a list of identifiers, each describing an event. A predefined label called send is used to identify the target nodes of send events.

Each event can optionally be defined more precisely by constraints on its attribute values. These attribute values can be related to constant values or to variables by a number of operators (=, !=, <, >, >= and <=) or to constant values by a **range** or an **in** operator. The argument of **range** is a pair of values specifying the upper and lower bound of a valid range of values while the argument of **in** is a list that enumerates all valid values.

A variable is defined the first time it is used. One must assign a value (bind an attribute value) to each defined variable exactly once while it may be used arbitrarily often as a right argument in constraint definitions. The scope of variables is global and its type is inherited from the defining attribute.

For each event, an optional response function can be specified. This function is invoked whenever the corresponding event description is fulfilled and it can take the values of already bound variables or constants as arguments. A response function can be used to generate alerts for the system administrator or to perform active counter measures against an intruder (e.g. harden the firewall). As these functions are invoked locally at the node where the corresponding event description has been detected, our system possess no central response component that can be taken out easily. In addition, response functions are used to create artificial event objects that are fed back into the detection process. This allows information exchange between different scenarios or the elegant implementation of timeouts.

With these explanations, we introduce the (incomplete) syntax (in BNF) of the pattern section (all identifiers represent strings).

```
pattern    : {event set}+
event set  : node-id '{' {event}+ '}'
event      : ['send('target-id'):'] event-id
             '[' {constraint ';' }* ']'
constraint : assignment | [label] relation [ response ]
assignment : '$'variable-id '='
             ( attribute-id | constant )
relation   : attribute-id operator
             ['('] {value ',' }* value [')']
response   : '<' function-id'(' arg-list ')' '>'
value      : constant | '$'variable-id
arg-list   : { arg-id ',' }* arg-id | e
```

The following example shows a classical telnet chain scenario.

```
ATTACK "Telnet Chain" [ Node1, Node2 ]
Node1 {
 send(Node2): tcp_connect [DstPort = 23;]
}
Node2 {
 tcp_connect [DstPort = 23;] <report('telnet-chain');>
}
```

It describes a connection from Node1 to port 23 at Node2 and from there to port 23 of another remote machine. Node2 describes the root set as it has no outgoing send event. The target of the send event can easily be extracted from the tcp_connect event as the destination IP address as specified in the annotation of a tcp-connect basic event. Whenever a telnet-chain has been detected at the root node, a report function is invoked.

4   System Design

4.1   Architecture

Our system consists of a set of hosts which are connected by a network. Each node runs a Quicksand sensor which is made up of several local intrusion detection units and an event correlation component.

Local Intrusion Detection Unit. The local intrusion detection components are responsible for gathering audit data. They can be implemented as host based systems which collect data from the operating system and user processes or as network based systems that monitor packets via the machine's interface cards. Local ID units can apply misuse or anomaly based methods to extract interesting information (i.e. events) from the local data stream.

For our first prototype, we have built a network based sensor that operates similar to Snort [11] and an anomaly sensor that scans HTTP and DNS traffic [7] for suspicious (i.e. abnormal) content. Events which are relevant for distributed attacks and therefore specified as interesting basic events in ASL are passed to the event correlation unit.

Event Correlation Unit. The event correlation unit receives a prefiltered stream of events from the local intrusion detection units as well as messages containing relevant data from other nodes and executes the distributed misuse detection algorithm (which is detailed in Section 4.2).

In order to be able to integrate third-party probes, the interface between the local and the event correlation units has to be designed carefully. Information needs to be passed in both directions as the local sensors have to send interesting events to the correlator and receive configuration and setup information.

We have defined generic events as well as configuration directives based on the intrusion detection message exchange format (idmef) [4] defined by the IETF intrusion detection working group (idwg). When the local system does not support that format, specific plug-in modules are inserted to translate between each system's dialect and the standard format. Instead of creating our own message format we attempt to rely upon existing standards.

The correlation unit is also responsible to react upon detected security violations. The ASL definition of an attack contains a description of counter measures which should be invoked. A possible active response is the reconfiguration of firewalls or the interruption of open network connections. Such activities have to be performed fast and reliable, therefore no central entity is involved. In addition, a passive response can be executed by transmitting an alarm message to a dedicated node running a control unit.

Control Unit. At least one host needs an installed control unit. This central module is utilized by the system administer to configure the system. Its task includes the processing of attack specifications written in ASL and the configuration of the local intrusion detection and event correlation units that are distributed over the protected network.

Pattern specifications that are written in our Attack Specification Language need to be translated into data structures and functions suitable for our detection process. This is done by an ASL parser that compiles attack patterns into C source files which are then compiled into object files. Response functions are specified as C functions and compiled into separate object files.

Object files are merged into shared libraries which may be shipped to sensors over a secure connection (SSL). These libraries are automatically installed and can thereupon be utilized by the sensor. The control unit is used to remotely load and unload attack patterns. Whenever a certain pattern is loaded into the correlation engine a dedicated thread is started to execute it. Instead of using a generic interpreter to process pattern specifications, the detection algorithm is directly run as compiled code. To remain flexible and to be able to integrate new patterns on-the-fly (i.e. without changing the code of the event correlation unit), the algorithm's code is divided into a pattern dependent and a pattern independent part. The independent part is implemented as shared code in the correlation unit. Similar to an operating system, this part provides basic services that can be accessed by the threads running the attack scenario dependent code. The pattern related code is stored in the libraries produced by the ASL parser and the response libraries. These libraries are dynamically loaded into the sensors and execute the pattern dependent parts.

The control unit uses a local database to store information about the libraries that have already been installed at each host. This database holds a list of symbols exported by each object file as well as a list of installed object files at each host. That information is used to prevent the installation of libraries that depend on response functions in different object files which have not been installed yet. Currently, we only issue a warning message that enumerates the symbols that cannot be resolved. The problem has to be handled by the system administrator who has to perform the necessary installations manually. In the future, we plan an automatic support to perform this update automatically.

4.2   Pattern Detection

The purpose of the pattern detection process is to identify actual events that satisfy an attack scenario (written in ASL). When a set of events fulfills the temporal and content based constraints of a scenario an alert is raised. We are aware of the fact that a distributed system design might result in tremendous message overhead. This potential danger is addressed at the level of pattern specification (as only tree-shaped patterns are allowed) as well as in the pattern detection algorithm.

Pattern Graph. In order to be able to process an attack description, it has to be translated from ASL into a data structure suitable for our system. This is done by transforming a scenario into an acyclic, directed graph (called *pattern graph*). An attack scenario describes sequences of events located at different hosts that are connected by send events. Each single event specified by an ASL scenario is represented as a node of the resulting graph. The nodes of each host sequence are connected by directed edges. An edge leads from a node representing a certain event to the node which represents the immediate successor of that event in the ASL pattern description. Send events require a little different treatment as they are the last event in their host sequence and therefore do not have an immediate successor. In this case a directed edge to the first node of the host sequence to which the send event points is inserted. Additionally, nodes are assigned unique identification numbers (ID).

Messages. A message is a compact, more suitable representation of an event. Most attack descriptions rely only on a small subset of the event's attributes for correlation (e.g. only IP addresses instead of the complete IP header). In ASL, only attributes that are assigned or compared to variables are of interest to the further detection process. Therefore there is no need to operate on the complete event objects. Whenever an event matches an event description, a message is generated and the appropriate values are copied into it. Then the message is forwarded to the node corresponding to the event description for further processing by the detection algorithm.

Each message is written as a triple <ID, timestamp, list of (attribute,value)>. The ID of the message is set to the identification of the node where the message is sent to (i.e. the node whose event description has matched). The timestamp denotes the occurrence of the original event and the attribute/value list holds the values of the relevant event attributes (which have been copied from the original event). The ID of a message defines its type. Different actual message instances with identical IDs are considered to be of the same message type.

Constraints. An attack description in ASL imposes a number of different constraints on the events that must be taken into account by the detection algorithm.

Temporal constraints between events are introduced by send events. The first event in the host sequence at the target of each send event has to occur after the send event itself. In addition, the events of a host sequence have to occur in the same order as they are defined in the ASL description.

An event description that relates an event attribute to a constant value creates a *static constraint* while the use of variables that relate attributes of different events introduce *dynamic constraints*. It is obvious that it cannot always be immediately determined whether an event satisfies its dynamic constraints while all static constraints of a certain event can be resolved immediately. Actual event instances that fulfill the static constraints of a certain event description are transformed into messages which are forwarded to the actual detection process. Its task is to resolve all dynamic and temporal constraints.

Detection Process. The idea is that each node of the pattern graph can be considered as the root of a subtree of the complete tree pattern. There are **node constraints** assigned to each node of the graph such that if there are messages which satisfy the node constraints, there are events that fulfill the dynamic and temporal constraints of the complete subtree above that node. Whenever the node constraints of a node are satisfied, certain messages may be moved one step closer to the root node, hence they are pushed over the node's outgoing edge to its neighbor node below (as we have a tree shaped graph, there is at most one outgoing edge for each node). Then they are processed at the destination node. This allows to successively satisfy subtrees of the complete pattern and move messages closer to the root node of the pattern graph. Whenever events at the root node fulfill the constraints there, the pattern has been detected.

The advantage of this approach is the fact that only local information is necessary to decide which messages should be forwarded. This allows to actually distribute nodes of the pattern graph over several hosts and have each node making local decisions without a central coordination point. Different host sequences may (and usually do) occur at different hosts. The node constraints have to make sure that all events described by the subtree pattern have occurred, that their temporal order is correct and that all dynamic constraints (which can be resolved up to this point) are met.

The node constraints consist of

1. the set of temporal constraints between the event that is associated with the node and all events that are associated with the node's predecessors in the pattern graph **and**
2. all dynamic constraints that can be resolved at this node. Dynamic constraints (i.e. a variable definition at one node and its use at another one) are inserted into the pattern graph at the earliest node possible. The earliest possible node is determined by finding the first common node in the paths from each of the constraint operands to the pattern graph's root node. When one node is on the path of the other one, the constraint is inserted directly there, otherwise it is inserted at the node where both paths merge.

Whenever messages fulfill the constraints of a certain node, it is guaranteed that events have occurred that satisfy all constraints (static, dynamic and temporal) of the subtree above this node. An example of a pattern graph with dynamic constraints is shown in Figure 1.

We define the *event pool* for each node as the place that stores messages that can potentially be used to satisfy the node constraints. Message instances that are not needed to satisfy the node constraints of this node (but of nodes closer to the root node) are stored in the *bypass pool*.

Detection Algorithm. Having determined the node constraints and the event and bypass pools for each node the algorithm to actually move event objects between nodes can be explained. Each arriving message is checked to determine whether it can be used to satisfy node constraints (i.e. belongs to the event pool).

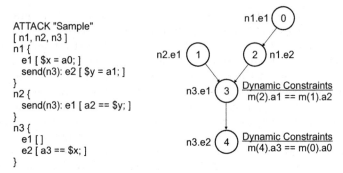

```
ATTACK "Sample"
[ n1, n2, n3 ]
n1 {
    e1 [ $x = a0; ]
    send(n3): e2 [ $y = a1; ]
}
n2 {
    send(n3): e1 [ a2 == $y; ]
}
n3 {
    e1 []
    e2 [ a3 == $x; ]
}
```

The occurrence of event n1.e2 results in the creation of message
<2, time of occurrence, (a1, value of a1)>

Fig. 1. Pattern Graph and Node Constraints

If this is the case, the algorithm attempts to find a tuple of messages of *different* type (i.e. all referring to different event descriptions with a different ID) that match *all* the node constraints. The tuple has to include one actual message for each type of the event pool and the new message has to be part of the tuple as well.

Consider a potential tuple for node 3 in Figure 1 with its event pool {m1, m2, m3}. The event pool consists of messages with IDs 1 and 2 as messages of these types are used in the node's single dynamic constraint. In addition, both messages are used to resolve the temporal constraints between messages of node 3 and its predecessors, namely node 1 and node 2. Therefore the tuple must consist of messages with ID 1, 2 and 3. When such a tuple (or tuples, when more than one set of events match the node constraints) can be identified, the detection process has found events that match the subtree pattern starting at the local node. The tuple's messages have to be moved over the outgoing link to the next node. As messages in the event pool might be needed later to satisfy the node constraints together with newly arriving instances, the original ones remain in the pool and only copies are forwarded. Whenever a matching tuple is found all messages in the bypass pool are automatically moved over the outgoing link as well, as they are needed at nodes closer to the root. In order to prevent the system from being flooded by duplicate messages each event pool entry is only copied and forwarded to the next node once.

The situation is slightly different for send nodes. As a send node can have different next neighbor nodes at different hosts (depending on the target of the send event) the copying and moving of pool entries must be handled differently. The send node has to keep track which pool entries have already been copied to the destinations of the send events for each different destination.

We use timers to remove elements from the event pools after a certain, configurable time span because elements cannot be kept infinitely long as memory is a limited resource. This means that patterns which evolve over a long time might remain undetected. Note that this is not a limitation of our approach but a problem that affects all systems that operate online and have to keep state.

The following example in Figure 2 shows the step-by-step operation of the detection process to detect the distributed pattern which is described by the scenario in Figure 1.

Dotted arrows indicate the copying of event messages to the next neighbor. Associated with each node are two sets enclosed in brackets. The first holds the node's current event pool entries, the second its bypass pool.

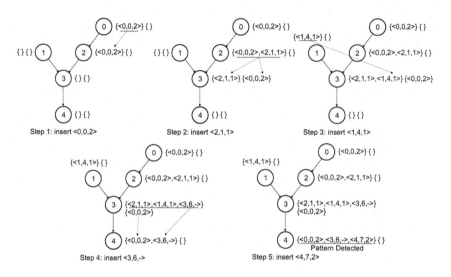

Fig. 2. Pattern Detection

4.3   Time Issues

The use of patterns that can specify temporal relationships between events occurring at different hosts introduces the following challenge. The relative order of distributed events needs to be known locally at the node that has to determine whether certain events satisfy the time constraints of a distributed pattern.

In case of centrally collected events, order is usually established by timestamping messages when they arrive from remote sensors. A system without a central unit can solve this task in two different ways.

One variant uses physical timestamps to mark local events (and the resulting messages) and a synchronization mechanism to keep all physical clocks within acceptable accuracy bounds. This allows the receiver of a message to establish a relative order between remote and local events by comparing the messages' timestamps. Unfortunately, this model is very difficult to implement in large, heterogenous network environments as it requires external means of clock synchronization.

The other variant utilizes logical clocks. Lamport's happened-before relation [8] defines the notion of *causal relationship* to describe sequential events. Adapted to our system, two events are sequential when one happens-before the other in Lamport's sense. Logical clocks are used by having each host marking every sent packet with a logical timestamp. The sender can then refer to this logical timestamp when it detects that a certain packet has contained the send event of a certain pattern. When the receiver later gets the message indicating that a certain received packet has carried a send event it is easy to determine which local events have occurred before and after the reception of that packet (with a known logical timestamp). We have implemented a prototype of such a time service for Quicksand which is presented in detail in [6] but it requires a modification of the TCP/IP stack of the underlying operating system which has only been done for Linux.

For our tests (that involved different operating systems) a simpler approach has been chosen. We assume a constant delay between the actual send event and the message containing the information that this send event has occurred. At the receiving node, the processing of local messages is simply delayed for that constant time span to get a correct temporal order between remote and local events. We are aware of the fact that this might introduce incorrect temporal dependencies. When messages arrive too early, our system might report intrusions that have not taken place (false positives) while we would miss intrusions (false negatives) when message take longer than expected. By using a reasonable long delay (e.g. one second), false negatives can be practically prevented (especially in local networks). In addition, we have never experienced any false positive as the incidentally occurrence of events specified by a certain pattern within this short time span is very unlikely.

5   System Evaluation

The aim of this section is to show that the proposed detection process operates as efficient as current solutions while providing superior fault tolerance and scalability properties. This makes it necessary to define the evaluation criteria that we use to measure these properties.

We measure *fault tolerance* as the percentage of nodes of the complete network which have their events correlated after a single machine running an IDS part (sensor or correlator) fails (or is taken out). This indicates the percentage of distributed patterns that can still be detected. When a node failure partitions the set of hosts into several subsets where events are still related within each of these subsets, the highest percentage among all of them is chosen. When a correlator that is responsible only for a subset of all nodes fails, the remaining system may still perform event correlation on a reduced set of hosts. The fault tolerance measures exactly that fraction of nodes.

The *scalability* of distributed intrusion detection systems is characterized by two values. One indicates the total network traffic between all nodes (total) while the other measures the maximum network traffic at a single node (peak).

We compare Quicksand (representing a completely decentralized system) to a design that deploys sensors at every host and centrally collects their data (centralized approach) and to one that introduces several layers of processing nodes (hierarchical approach) on top of its sensors.

5.1 Theoretical Results

For our theoretical discussion, we assume a network with n hosts and the occurrence of n*e interesting events during a time interval of length Δ. The interval Δ also specifies when messages representing events 'time out' and are removed from the detection process. While the number of events in the whole system is assumed to be proportional to the number of nodes, the number of events at each single host may not exceed a certain threshold τ. This is reasonable as it allows a certain variance of the distribution of events within the system (i.e. modelling local hot spots like WWW or NFS servers in very large networks) without allowing a single node from having to deal with arbitrary many events as the number of nodes grows larger.

The coverage of a network (in %) after a single node failure is given below for the different systems. We assume that the hierarchical system uses $l = \lfloor \log_m((m-1)*n) \rfloor$ layers with m^k nodes (k ... 0 to l-1) in each layer, where m specifies the number of children for each node.

System	Type of Node	Coverage
Centralized	Sensor	$\frac{n-1}{n}$
	Correlator	0
Hierarchical	Sensor	$\frac{n-1}{n}$
	Correlator (at layer k)	$n - \frac{m^{l-k}-1}{m-1}$
	Root	$\frac{n-1}{m}$
Decentralized	Node	$\frac{n-1}{n}$

The loss of a node at layer k in the hierarchical model stops correlation of the complete subtree with $\frac{m^{l-k}-1}{m-1}$ nodes. When the root is lost, each subtree with $\frac{n-1}{m}$ nodes can still do correlation. Not surprisingly, this shows that centralized and hierarchical system are more vulnerable especially to the loss of important nodes (i.e. nodes in top layers positions or the root) than our completely distributed design. We do correlation only at nodes where the relevant events are actually observable, therefore a loss of some hosts cannot influence the detection capability of the remaining system.

The theoretical scalability values of our system depend on assumptions about the used patterns and the number of send events to different targets during the time interval Δ.

As explained in Section 4.2, all messages at the send node have to be copied to each new target of a send event. This results in message traffic which is proportional to the number of send events with different targets during Δ. The average number of send events at a single node during Δ is indicated as ω. Depending on the used patterns, different amounts of messages have to be copied over send event links. In the optimal case, only one message instance representing the send event itself has to be transmitted. When the attack scenario contains dynamic constraints between events that are separated by one or more send event link, additional messages have to be moved to the target host as well. The situation worsens when a message has to move over several consecutive send links as it gets copied to each target at every step (yielding potential exponential growth of the number of messages). Therefore the depth δ of a pattern (defined as the maximal number of consecutive send links a message has to traverse) is an important factor to determine scalability of our system. Usually, not all event patterns define or use variables and messages created from those events do not need to be forwarded. v denotes the fraction of pattern descriptions of an attack scenario that actually do define or use variables and result in messages that might need to be transferred over the net. When e events occur at a single node, only e*v of them might need to be sent over send links.

As each message only contains part of the data of the complete event object we save bandwidth in comparison to systems that have to send the whole event itself (because they do not know which information is important at a higher level). The ratio between the event object size and the message size (including ID and timestamp) is written as r.

The explanation (and notation) given above allows to formulate the estimate of the total network traffic as

$$\frac{n*(e*v)*\omega^\delta}{r}$$

As each node equally participates in the detection process, the peak traffic is equal to the traffic at a single node which results in

$$\frac{n*(e*v)*\omega^\delta}{n*r} = \frac{(e*v)*\omega^\delta}{r}$$

Although the formula shows the potential for an exponential explosion of the created network traffic, the next section will show that δ is usually very small for our area of application (i.e. less than 2 in almost all real cases). It is interesting to notice that the peak traffic does not depend on the total number of nodes in the system which indicates good scalability properties. Additionally, the factors v and r help to keep the total bandwidth utilization reasonably low.

The following table shows the total and peak traffic values for a centralized and a hierarchical solution. We assume that each hierarchy layer is capable of reducing the events it forwards to a higher level node by a constant factor c.

System	Total Traffic	Peak Traffic
Centralized	$e*n$	$e*n$
Hierarchical	$e*n*\sum_{i=1}^{i<l} c^i$	$e*n*c^{l-1}$

The total and peak traffic values for the centralized solution reflect the fact that all event data is sent to a single location. The traffic in the hierarchical system comes from the data that is forwarded by nodes to their higher layer parents. As a fraction (determined by c) of the event data is forwarded over several levels the total traffic consists of the sum of the traffic volumes between each layer. The peak traffic occurs at the root node (i = l-1). Although it is significantly smaller than in the centralized case, it still depends on the number of nodes in the system.

In both the decentralized and the hierarchical system, the total traffic volume increases when compared to a centralized design. Nevertheless, the peak traffic indicates that they scale much better than a centralized one.

5.2   Experimental Results

Our system is designed to provide a scalable solution for enterprise sized networks. Unfortunately, we do not have the resources to perform scalability tests on such a scale. Therefore, we had to restrict ourselves to our department's network and installed Quicksand on our web server, the DNS server, our firewall and 6 additional hosts. These machines are running Linux 2.2.14 and SunOS 5.5.1 on different Pentium II, Athlon and Sparc hosts. The idea is to gather experimental data that can be compared to values we would expect from the theoretical considerations.

We use our network based sensor (which is similar to Snort [11]) to collect data from the network and have basic event definitions (C structs) for common network traffic packets (i.e. TCP, UDP, IP and Ethernet). These are the basic building blocks for attack scenarios written in ASL (Attack Specification Language). We use 16 different distributed patterns that aim to detect distributed signatures and anomalies (as explained in Section 1) with the following properties.

Property	Average	Maximum	Minimum
Pattern depth (δ)	1.19	2.00	1.00
Events with variables (v)	0.83	1.00	0.50

The given numbers are based on a week of real data collected in our network during which we processed 16374 events. We used a time interval (Δ) of 24 hours.

Property	Average	Maximum	Minimum
Events per Δ	2340	3818	1732
Send event targets for single node (during Δ)	1.62	5	0
Total traffic (in messages)	3922	7536	3159
Peak traffic (in messages)	1011	2722	744

As expected our used patterns did not result in a message explosion and the total number of messages never exceeded twice the number of actual events. The unexpected high peak traffic values resulted from many scans for port 80 that the firewall reported to the web server. In our setup, a high fraction of the messages concentrated on a few machines (web server, DNS server) while regular nodes transmitted fewer messages. However, an increase of nodes in our local network would not raise the load at these machines significantly (as the port scan messages were caused by machines on the Internet anyway) while producing more total traffic inside the network. In such a case (as with large intranets) we expect that the ratio between the messages at these servers and messages at regular nodes decreases.

6   Conclusion

We present Quicksand, an intrusion detection system that implements a distributed pattern detection scheme to relate events that occur at different hosts. This can be used to detect distributed signatures (like telnet chains) and anomalies.

The system design and implementation details have been described and a pattern language introduced. This specification language has to be restricted in order to prevent a message explosion in our peer-to-peer architecture. The consequential decentralized algorithm to find events that satisfy such patterns exhibits good scalability and fault tolerance properties.

References

1. Jai Sundar Balasubramaniyan, Jose Omar Garcia-Fernandez, David Isacoff, Eugene Spafford, and Diego Zamboni. An Architecture for Intrusion Detection using Autonomous Agents. In *14th IEEE Computer Security Applications Conference*, December 1998.
2. Marc Crosbie and Eugene Spafford. Defending a computer system using autonomous agents. In *Proceedings of the 18th National Information Systems Security Conference*, October 1995.
3. Jose Duarte de Queiroz, Luiz Fernando Rust da Costa Carmo, and Luci Pirmez. Micael: An autonomous mobile agent system to protect new generation networked applications. In *2nd Annual Workshop on Recent Advances in Intrusion Detection*, September 1999.
4. IETF Intrusion Detection Working Group. Intrusion Detection Message Exchange Format. http://www.ietf.org/html.charters/idwg-charter.html.
5. Judith Hochberg, Kathleen Jackson, Cathy Stallins, J. F. McClary, David DuBois, and Josephine Ford. NADIR: An automated system for detecting network intrusion and misuse. *Computer and Security*, 12(3):235–248, May 1993.
6. Christopher Krügel and Thomas Toth. An efficient, IP based solution to the 'Logical Timestamp Wrapping' problem. In *6th International Conference on Telecommunications*, 2001.
7. Christopher Krügel, Thomas Toth, and Engin Kirda. Service Specific Anomaly Detection for Intrusion Detection. In *ACM Symposium on Applied Computing (to appear)*, 2002.
8. L. Lamport. Time, clocks and the ordering of events in a distributed system. *Comms. ACM*, 21(7):558–65, 1978.

9. Peter G. Neumann and Phillip A. Porras. Experience with EMERALD to date. In *1st USENIX Workshop on Intrusion Detection and Network Monitoring*, pages 73–80, Santa Clara, California, USA, April 1999.
10. Phillip A. Porras and Peter G. Neumann. EMERALD: Event Monitoring Enabling Responses to Anomalous Live Disturbances. In *Proceedings of the 20th NIS Security Conference*, October 1997.
11. Martin Roesch. Snort - lightweight intrusion detection for networks. In *USENIX Lisa 99*, 1999.
12. S. R. Snapp, J. Brentano, G. V. Dias, T. L. Goan, L. T. Heberlein, C. Ho, K. N. Levitt, B. Mukherjee, S. E. Smaha, T. Grance, D. M. Teal, and D. Mansur. DIDS (Distributed Intrusion Detection System) - Motivation, Architecture and an early Prototype. In *14th National Security Conference*, pages 167–176, October 1991.
13. S. Staniford-Chen, S. Cheung, R. Crawford, M. Dilger, J. Frank, J. Hoagland, K. Levitt, C. Wee, R. Yip, and D. Zerkle. GrIDS - A Graph based Intrusion Detection System for large networks. In *Proceedings of the 20th National Information Systems Security Conference*, volume 1, pages 361–370, October 1996.
14. G. Vigna and R. Kemmerer. NetSTAT: A network-based intrusion detection system. In *Proceedings of the 14th Annual Computer Security Applications Conference*, December 1998.
15. Giovanni Vigna, Richard A. Kemmerer, and Per Blix. Designing a Web of highly-configurable Intrusion Detection Sensors. In *Recent Advances in Intrusion Detection*. Springer Lecture Notes in Computer Science, 2001.
16. Gregory B. White, Eric A. Fisch, and Udo W. Pooch. Cooperating Security Managers: A peer-based intrusion detection system. *IEEE Network*, pages 20–23, January/February 1996.

Enhancing the Security of Cookies

Vorapranee Khu-smith and Chris Mitchell

Information Security Group, Royal Holloway, University of London,
Egham, Surrey, TW20 0EX, United Kingdom
{V.Khu-Smith,C.Mitchell}@rhul.ac.uk

Abstract. Cookies are pieces of information generated by a Web server
to be stored in a user's machine. The information in cookies can range
from selected items in a user's shopping cart to authentication information used for accessing restricted pages. While cookies are clearly very
useful, they can also be abused. In this paper, security threats that cookies can pose to a user are identified, as are the security requirements
necessary to defeat them. Various options to meet the security requirements are then examined. Proposed user-controlled approaches and their
implementations are presented and compared with a server-controlled
approach, particularly the 'Secure Cookies' method, to illustrate the relative advantages and disadvantages of the two approaches.

Keywords: Cookies; Internet security; Web security

1 Introduction

Today, browsing and on-line shopping are becoming increasingly convenient. A
user can personalise a Web page, have his/her own shopping cart, and be automatically authenticated without repeatedly entering username and password.
However, the stateless nature of the HTTP protocol does not support such features. Therefore, cookies were introduced to enable Web servers to maintain
current session state and recognise individual users.

Cookies are pieces of information generated by a Web server to be stored in a
user's machine. The information in cookies can be, for example, selected items in
a user's shopping cart, authentication information used for accessing restricted
pages, or account details. The first time a browser contacts a Web server, a cookie
is sent from the latter to the former. It is then stored in the user's PC in a file
called either cookies.txt, Cookies, or MagicCookie in the Windows, UNIX, and
Macintosh operating systems respectively. The next time the browser requests a
Web page from the Web server, it sends the corresponding cookies.

A cookie consists of six elements, namely Name, Expiration Date, Domain
name, Path, Secure, and String Data. The first part is the name of the cookie.
The expiry date defines the cookie's lifetime. Domain name and path are used
when a browser searches for a cookie corresponding to the host of the requested
URL. The path attribute specifies the subset of the URLs to which the cookie
belongs. The secure attribute indicates whether the cookie is transmitted in
secure mode such as TLS and HTTPS. The String Data field is where all other

K. Kim (Ed.): ICICS 2001, LNCS 2288, pp. 132–145, 2002.

information of the server's choice is stored. Detailed cookie specifications can be found in [6] and [8].

While cookies are clearly very useful, they can also be abused to impersonate a user, compromise user privacy and, in some cases, reveal confidential user information. Although a number of papers, e.g. [2,3,7,11], point out potential security threats, most of them focus on facts about cookies, such as what they are and how they are used, and do not appear to provide a satisfactory security analysis and solution.

In [9] the 'Secure Cookies' method is proposed, in which security measures are applied to cookies by a Web server. Consequently, this approach allows Web servers to control what, when and how the security procedures will be performed. Whilst this may be appropriate for many applications, in some environments users will wish to control the security of their own cookies. This paper, therefore, examines possible user-controlled approaches to enhance cookie security.

In this paper, the threats that cookies can pose to a user are identified, as are the security requirements necessary to defeat them. Various options available to meet the security requirements are examined. A server-controlled approach is then outlined followed by proposed user-controlled approaches and their implementations. A comparison between the server-managed and user-managed approaches is subsequently given. The final section summarises and concludes the paper.

2   Cookie-Related Security Threats

In this section, security threats that cookies can pose to a user are identified. The threats can be divided into three main categories namely confidentiality of cookie contents, monitoring user behaviour using cookies, and malicious cookies.

2.1   Confidentiality of Cookie Contents

In order to allow users to browse among restricted pages without repeatedly identifying themselves, cookies are used to store authentication information such as username and password. Consequently, it is important to ensure the confidentiality and authenticity of cookies storing such information. Otherwise, anyone who can access such cookies can potentially impersonate the user. Although, in many cases, authentication information stored in cookies is in a server-specific format, and hence the contents are not immediately obvious to the reader, it is still possible for an attacker to simply replay an intercepted cookie and impersonate a user.

Implementation flaws, particularly in Web browsers, can also bring about security risks to a user. For example, a vulnerability in Internet Explorer Version 4 and 5 for Windows 95, 98, NT, and 2000 allowed any sites to see the content of other sites' cookies [4]. This is because the browser confused a site having a long URL ending with the domain name of another server with that other server. Consequently, it was possible for a malicious web site to give itself a long URL ending with a sequence of characters identical to the URL of another site, and the malicious site was then able to access cookies stored by that other site. If a cookie contains personal information, e.g. confidential data such as account details, then the consequences of such a vulnerability can be significant. Although the Web browser flaw has been fixed, it is possible that there are other undiscovered vulnerabilities that can pose security threats to users.

Another example of an implementation flaw lies in the way that some users include a Netscape Navigator folder in a publicly accessible directory. In an environment where there are not as many computers as users, it is not unusual to provide public spaces for users to access their data from any computers within the environment. Such spaces are accessible to any users with a valid username and password. For example, a college can provide a drive for students to store their home pages. In this drive, each student has his/her own directory. Any students with a valid username and password can access the drive, and hence other students' directories. An empirical study [5] showed that a number of cookie files could be found in this publicly accessible drive. This is because some users had stored their Netscape Navigator folder in such a publicly accessible directory. Since Netscape Navigator stores user cookie files in the user's Netscape folder, this means that user cookies will also be stored in a publicly accessible location, i.e. accessible by any other users with a legitimate username and password. The study showed that it was possible to use this weakness to obtain personal details, ranging from user names to full user details including contact addresses and telephone numbers.

2.2   Monitoring User Behaviour Using Cookies

A particularly controversial issue concerning cookies is that they can be used as tracking devices to follow user movements across the Internet. Web-advertising agencies such as DoubleClick, Focalink, Globaltrack, and ADSmart run advertisement banners on various sites. Their clients add an <IMG> tag to the client HTML page, pointing to a URL on the advertising agency's server. When a Web browser sees this <IMG> tag, it contacts the advertising agency server to retrieve the graphic. The first time the graphic is downloaded from the site, the user browser will receive a cookie containing a random ID. From then on, every time the browser connects to a site containing the agency's advertisement banners, it sends the cookie (the random ID) along with the URL of the page that is being read [11].

After a period of time, the advertising agency will be able to generate a user profile, revealing user browsing habits and interests. This might be used to improve advertising campaigns, to target advertisements to user interests, and to avoid repeatedly showing the same advertisements to a user. The ability to track users is a potential violation of user privacy. It is also possible that advertising servers might share such information without user consent although, at the time of writing, there is no strong evidence of such behaviour.

Even though using cookies as a tracking device may not reveal the actual identity of a user, the fact that an advertising agency server can maintain a list of URLs that a user has viewed can lead to a possible leak of user personal details. In particular, if a Web server uses the GET method to input data from an HTML form to a CGI script, the information will be sent as a part of the URL. Therefore, anyone who can read the URL will be able to obtain the information in the form.

2.3   Malicious Cookies

It is often rather misleadingly stated that cookies are just text files, and hence are harmless. Although cookies are application data files, cookies can include special tags that can introduce executable code, just as Microsoft Office application files can contain Macro viruses.

In HTML, in order to distinguish text from 'markup' symbols, a set of characters such as '<', which typically indicates the beginning of an HTML tag, are defined as special. Tags can either affect the formatting of a Web page or introduce a program that will be executed by Web browsers. For example, a <SCRIPT> tag introduces code from a variety of scripting languages [10]. Many Web servers use information stored in cookies to create dynamic pages. Therefore, if a cookie includes those special tags, when a page incorporating this cookie is displayed, a malicious program can be called and executed. The security effects of such a program can range from alterations of the submitted form to bypassing an authentication process. However, what exactly can be done by the called program depends on the language in which it is written as well as the Web server's security context configuration.

3   Security Requirements

In this section, a number of security requirements are identified to deal with the security threats discussed earlier. Note that these security requirements do not address the threat of cookies being used to monitor user behaviour. Such threats are typically tackled by using tools specially designed to monitor the activity of cookies.

3.1   Cookie Confidentiality

As stated in section 2.1, cookies can be used to store authentication data and personal details such as mailing addresses and credit card numbers. Therefore, it is important to provide confidentiality for such information. Information in cookies might be revealed in two major ways. Firstly, a cookie can be intercepted while it is transmitted, and, secondly, a cookie can be disclosed when it is stored in a user's machine. In general, it is a good practice to provide confidentiality for both scenarios.

3.2   Cookie Integrity

In order to prevent attacks such as cross-site scripting, i.e. where special tags are inserted into cookies as described in Section 2.3, maintaining the integrity of cookie data is vital. Moreover, if a cookie is used for user authentication and the content of the cookie is changed, then the authorisation process will fail. An attacker could modify such a cookie, and hence prevent a legitimate user from accessing a service. Domain name and Path in cookies are also important. If it is possible to make changes to these elements, cookies can be sent to an entity who is not a legitimate owner.

3.3   Cookie Authentication

Although the content of a cookie may be encrypted and protected from unauthorised modifications, there remains the possibility of an attack where one entity 'presents' a user authentication cookie copied from another party. Consequently, it is important to be able to verify if the entity that is supplying a cookie is the owner of that cookie.

4   Meeting the Security Requirements

There are various ways of meeting the security requirements stated above. This section examines the possible options in more detail, and considers their advantages and disadvantages.

4.1   Secure Channels

The first approach is to protect the cookie transmission channel, i.e. the link between a Web browser and a Web server, by using secure protocols such as TLS and HTTPS. This provides confidentiality and integrity by use of an encrypted channel and MACs respectively. However, a secure channel only protects the information against eavesdropping and modifications en route. Once the cookie reaches its destination it is stored in clear text, and hence this option provides only partial protection. Anyone who has access to the stored cookie file will be able to read, change, or replay it.

An advantage of this option is that it is transparent to a user. Moreover, it makes use of existing capabilities, and therefore does not require modifications to Web browsers. However, in order to send a cookie securely, the cookie's 'Secure' attribute must be set. Since cookies are generated by servers, it is completely in the server's hands whether the 'Secure' attribute is set. Given that most users are not aware whether or not cookies are sent via a secure channel, many Web servers send them in clear.

Secure channels can provide user authentication; however this does not guarantee the origin of a cookie. In order to provide cookie authentication, there should exist some means to link the user authentication used in the secure channel establishment process with the cookie itself.

4.2   Access Control for User PCs

Another way of providing security for cookies is to protect user PCs against unauthorised access. A user authentication technique, e.g. using passwords, can restrict access to a PC, thereby protecting stored cookies against unauthorised reading and modification. The main advantage of such an approach is its simplicity. It is relatively easy to implement since most users are accustomed to using passwords. There is also no need to modify Web browsers.

A disadvantage of this mechanism is that it only protects cookies while they are in a PC. Therefore, it may have to be employed with other mechanisms to enhance security. Furthermore, it does not provide cookie authentication, since a malicious user can still reuse a stolen cookie. The use of passwords may also require additional management, such as password tests to prevent dictionary attacks.

4.3   Cryptographic Protection within Cookie Files

A combination of cryptographic techniques can be used to meet all three security requirements. Cookie encryption can be used for confidentiality. Both cookie integrity and cookie authentication can be provided by a MAC or digital signature. As part of cookie authentication, cookie replay protection can be achieved by incorporating cookie transfer into an authentication protocol (e.g. using a time stamp).

Cookie encryption and integrity protection can protect a cookie both when it is stored and when it is transmitted, as opposed to the use of secure channels, which do not protect stored cookies (see Section 4.1). There is thus no need for additional access control to user PCs. However, a disadvantage of using cryptographic techniques is that keys are required. Key management issues, such as how to securely exchange the keys, where they should be stored, and who should keep them, have to be taken into account. Additionally, there is a possibility that Web browser modifications or additional software will be required.

4.4   Summary

The three techniques discussed above can be effective, and, in practice, a system can combine some or all of these methods to enhance security. However, the last technique, i.e. applying cryptographic protection to the cookie file itself, appears to offer the widest range of security services. For this reason, this approach is the focus of the remainder of the paper.

Applying cryptographic measures to cookie files can be performed by Web servers or Web browsers. Whoever does so will have control over what, when and how the measures are performed. In the remainder of this paper, we examine the two approaches, and consider their respective merits.

5   Server-Managed Cookie Encryption

Server-controlled cookie encryption has the major benefit of user transparency. If implemented appropriately, no changes to Web browsers will be required. A disadvantage of this approach is obviously that users will have little control over the protection of their own cookies. An example of this approach is the Microsoft Passport scheme [1] which was introduced to provide an online user-authentication service. It employs encrypted cookies as a means for exchanging user-authentication information between a Microsoft Passport server and participating web sites.

Another example of server-managed cookie encryption is the 'Secure Cookie' scheme proposed by Park and Sandhu [9]. Although these two schemes are similar in the way that they both use encrypted cookies, the latter is more general, in that it is a means of protecting all user cookies, and not just those cookies generated by a single application. As a result, we use the Park and Sandhu 'Secure Cookie' scheme as the basis for a comparison with the new cookie security scheme proposed in Section 6. In the remainder of this section, we provide a brief overview of the Park and Sandhu scheme.

In this approach, Web servers are required to use 'Secure Cookies' of specific kinds, each with predefined types of content and protection. Examples include Name Cookies, Life Cookies, Key Cookies, and Seal Cookies. A Name Cookie, for example, contains a user name that can be used for user authentication. A Key Cookie contains an encryption key. The integrity of all cookies is protected by a Seal Cookie that holds either a MAC or a signed hash of the other cookies.

In order to have a set of Secure Cookies, a Web browser needs to contact another server called the Cookie Issuer, which generates the Secure Cookies. The Web browser then sends the cookies to the Web server, which will verify or decrypt them as appropriate. Examples of Secure Cookies are listed in Table 1.

Table 1. Secure Cookie Components

Domain	Flag	Path	Cookie_Name	Cookie_Value	Secure	Date
acme.com	True	/	Name_cookie	Alice	False	12/31/2000
⋮	⋮	⋮	⋮	⋮	⋮	⋮
acme.com	True	/	Life_cookie	12/31/99	False	12/31/2000
acme.com	True	/	Pswd_cookie	Hashed password	False	12/31/2000
acme.com	True	/	Key_cookie	Encrypted key	False	12/31/2000
acme.com	True	/	Seal_cookie	Signed Message Digest of MAC	False	12/31/2000

This approach satisfies the security requirements of confidentiality, integrity, and user authentication by using encryption, a signed message digest or MAC, and a digital signature respectively. However, it does not provide protection against replay attacks. A stolen Secure Cookie can be submitted to the Web server.

The Key Cookie, as stated earlier, stores a session key that is used to encrypt and decrypt other cookies. The session key can be encrypted using either a server public key or a server secret key. In either case, Web servers are responsible for key management.

While this approach may be appropriate in many applications, in some circumstances users may wish to read their own cookies and control their security. Consequently, in the next section we examine possible approaches that give users more control.

6   User-Managed Cookie Encryption

With the user-managed approach, users obviously have the benefit of control over what, when and how the security mechanisms should be applied. However, a special Web browser or additional software is required in order to enable users to perform the security procedures. The client may also have to store cryptographic keys, which could be a security threat in some circumstances. As a result, there is a need for a key management system to support the use of cryptography.

In this section, two possible approaches, using symmetric and asymmetric cryptography, are described. In order to provide cookie confidentiality, integrity and authentication, the schemes use encryption, MACs, signatures, and time stamps. The security mechanisms described below will be applied only to the cookie value, to minimise the complexity of the protocols.

6.1   Using Symmetric Cryptography

In this approach, a user selects cookie encryption by sending a request for cookie encryption to the Web server. This will trigger a key establishment protocol. If a user chooses to encrypt cookies, he/she will be required to authenticate him/herself to prevent unauthorised users, who may have access to the user's PC, from activating the security procedure and using cookies. This can be achieved by using simple schemes such as passwords. Key establishment can be performed by sending the key via a secure channel or using other key distribution techniques such as those involving a trusted third party. After successful key establishment, the user and the server then share a symmetric cookie key and the server will encrypt cookies with the key and send them to the user. The encrypted cookies are then stored locally in the user's PC.

The next time the Web browser requests a Web page from the site, it looks for corresponding encrypted cookies. A time stamp is generated and concatenated with the cookies, and a MAC is computed on this data. A page request, the time stamp, the encrypted cookies and the MAC are then sent to the server. On receipt of the request, the server verifies the MAC and checks whether the time stamp is within the acceptance window. The server can change information in the cookies whenever the user requests a page, and the new encrypted cookie will be sent back to the user with the requested page.

A time stamp and a MAC are included in order to prevent an intercepted cookie from being replayed. Without knowing the secret cookie key, an adversary will not be able to create a valid MAC. There are no cryptographic requirements for time stamp generation — the time stamp only needs to be within the acceptance window.

Given that symmetric cryptographic operations are typically simple to compute, the encryption operation will not significantly increase the server workload. Users only need to decrypt a received cookie if they want to see the content. However, this approach needs a secure mean to distribute the symmetric key the first time the user and server communicate. Moreover, users and servers need to maintain the shared secret key. As the number of users (n) grows, the number of keys increases approximately to n^2, and the task of key management will therefore become increasingly complicated over time.

Since the security of this approach depends on the secrecy of the shared key, it is vital to store the key securely. A user can store the keys in a smart card or in his/her PC (password protected).

6.2   Using Asymmetric Cryptography

In this approach, users are required to have a certificate and an asymmetric key pair. If a user wishes to have his/her cookies encrypted, the first time the user requests a Web page his/her certificate and the page message will be signed and sent. The server then generates a secret key, encrypts the cookies with this secret key, encrypts the secret key with the user's public key, and sends the encrypted secret key with the encrypted cookies and the requested page. There is no need for the user to decrypt the cookie unless he/she wants to know the cookie content. The next time the user contacts the server, the encrypted cookie, a time stamp, and a MAC is sent with a Web page request.

As for the symmetric technique, this approach allows users to decide if they want to encrypt the cookie or not. If the user sends a certificate, the server will know that the cookie must be sent encrypted. The user will also be required to enter a password for user authentication, since his/her private key may be stored locally in the PC. However, if it is stored in another more secure way, e.g. on a smart card, user authenticity verification may not be required. As for the symmetric cryptography approach, the time stamp is used to prevent replay of an intercepted cookie.

A drawback of this technique is that certificates and key pairs are required for the client. A Public Key Infrastructure (PKI) will also be needed to create and manage public key certificates.

7   Protocols for User-Managed Cookie Protection

In this section, two protocols for user-managed cookie security are described in detail, building on the general approaches described in the previous section.

7.1   Cookie Encryption Using Symmetric Cryptography

In this approach, a secure channel is employed to distribute a cookie key. How this secure channel is established is outside the scope of this paper, but it could, for example, be provided using protocols such as TLS and HTTPS.

The main advantage of this method is convenience. It makes use of existing security protocols to distribute the cookie key. However, doing so requires the establishment of a secure channel. Another drawback is that key management is relatively complicated, since there will be a large number of cookie keys for users to manage and store securely.

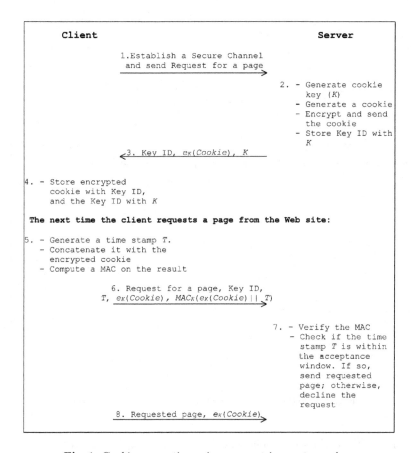

Fig. 1. Cookie encryption using symmetric cryptography.

The protocol is specified in Figure 1. In this figure:

- 'Client' can represent additional software, a modified Web browser, a plug-in, or an applet which performs security procedures for users,
- 'Server' represents a Web server,
- $X||Y$ denotes the concatenation of data items X and Y,
- K denotes a secret key used to encrypt cookies (the 'cookie key'),
- $e_K(M)$ denotes message M encrypted (using symmetric encryption) with key K,

Fig. 2. Cookie encryption using asymmetric cryptography.

- 'Key ID' identifies the cookie key K used to encrypt the cookie,
- T denotes a time stamp, and
- $MAC_K(M)$ denotes a MAC on message M using a variant of key K (note that it is important that the key used to compute the MAC is not precisely the same as the key used for encryption, particularly if the MAC is a CBC-MAC based on the same block cipher used to perform cookie encryption).

7.2   Cookie Encryption Using Asymmetric Cryptography

The public key based scheme is presented in Figure 2. In Figure 2, the following notation is employed (in addition to that used in Figure 1):

- $E_P(X)$ denotes data X encrypted (using asymmetric encryption) with public key P,
- $S_C(M)$ denotes the signature of the client on message M (computed using the client private key), and
- P_C denotes the public key of the client.

An advantage of this approach is that cookie keys are stored encrypted. The only key a user has to keep secret is the private key with which cookie keys are decrypted. Therefore, the key management task is not so complicated. There is also less difficulty in key distribution than in the system based on symmetric cryptography.

In order to allow servers to detect an attack where a malicious user deletes the request for cookie encryption (the user certificate), a signature is required on the request message indicating whether cookie encryption is enabled. Otherwise, an attacker can just delete the certificate. If an attack is detected, the server can send a message to inform the user and ask if the user wants to try again. This, however, introduces a risk of denial of service where a malicious user keeps modifying the message causing the page request to fail. Moreover, the additional computation of signing process can increase the user machine's workload and possibly slow the page request process.

8   Secure Cookies versus User-Managed Cookie Security

In this section, a detailed comparison between the server-controlled and user-controlled approaches is provided.

8.1   User Authentication

The Secure Cookies method [9] offers three choices for user authentication namely by IP address, by signature, and by password. The first alternative is not always appropriate since in some environments IP addresses are assigned dynamically. Moreover, it is prone to an IP spoofing attack.

The second user authentication mechanism is signature-based, and requires users to have a public key pair in order to sign a time stamp in a cookie. However, the signing process can increase the user workload, and hence possibly slow the web-browsing process.

Both Secure cookies and user-managed cookies employ user passwords. However, in the Secure Cookies technique the password must be sent via a secure channel to a Web server for verification. Therefore, in addition to the cookie security process, a secure channel may be needed. In the user-managed methods, on the other hand, the password is verified locally, which lessens its exposure and avoids the need for secure channel establishment.

8.2   Integrity and Confidentiality

The Secure Cookies scheme uses a signed message digest of cookies and MACs. The user-managed cookie security schemes also use either a signature or a MAC. In both cases, an additional computation for encryption process is required.

Both techniques offer a choice between symmetric and asymmetric encryption to provide cookie content confidentiality. However, one disadvantage of the user-managed scheme is that the required cryptographic keys will have to be stored by the clients.

8.3   Cookie Authentication

In the Secure Cookies approach, there is no mechanism for replay protection. The user-managed cookie security schemes, however, use a time stamp to provide protection against replay attacks. However, a MAC must be included in order to prevent a malicious user from replaying a stolen cookie with a new time stamp, potentially increasing the user's workload.

8.4   Other Issues

The main advantage of the user-managed approach over Secure Cookies is probably the fact that it allows users to have more control over the security of their own cookies. By contrast, with Secure Cookies, Web servers have full control over cookie encryption. Although this may be appropriate for many applications, in some environments users may wish to have more control over their security — indeed, if users wish to read their own cookies, e.g. to deal with the threat of user tracking, then a user-controlled approach is essential. Moreover, the Secure Cookies method may be more complex since a set of cookies is required for each Web site. There is also a need for an Issuing Cookie server to generate the Secure Cookies.

9   Summary and Conclusions

Although cookies are a very useful mechanism for maintaining session state, as discussed earlier they can pose a number of security threats. As a result, three cookie security requirements, namely confidentiality, integrity, and authentication, can be identified.

Secure Cookies [9] is a server-controlled approach for cookie protection that satisfies some of these security requirements. However, in some environments users may want to control their own cookie security, especially if the user tracking threat is to be effectively combated. In this paper, two user-controlled approaches, using symmetric and asymmetric encryption of cookies, are proposed. The main differences between how security services are provided in the server-managed 'Secure Cookies' and the user-managed approaches are summarised in Table 2.

The 'Secure Cookies' approach is potentially more flexible since it offers various options to provide each security requirement. It can also be made transparent to the user's Web browser. However, it is relatively complex because it requires an additional server to generate Secure Cookies. Moreover, in most cases, it requires a number of cookies, while in the user-managed approaches a variety of

Table 2. Comparisons

Security Requirements	User-Managed Cookie Encryption	Server-Managed Secure Cookies
User Authentication	Password-based	IP Address, Password, Signature
Integrity	Signature, MAC	Signed hash, MAC
Confidentiality	Encryption	Encryption
Replay Protection	Time Stamp	N/A

information can be stored in a single cookie. Finally, the Secure Cookies scheme has the potentially major disadvantage that it prevents users examining the contents of cookies stored in their own machine.

If a password scheme is used, the Secure Cookies method requires it to be sent via a secure channel. On the other hand, the user-managed techniques verify a password locally, and hence minimise its exposure. The user-managed approaches also provide additional protection against replay, while the Secure Cookies scheme does not.

References

1. Microsoft Passport scheme. Available at http://www.passport.com/.
2. S. Garfinkel, and G. Spafford. *Web Security & Commerce.* O'Reilly, 1997.
3. B. Hancock. Security views: some cookies are not tasty. *Computers & Security*, **17**(5):374–376, 1998.
4. B. Haselton and J. McCarthy. Internet Explorer open cookie jar. http://www.peacefire.org/security/iecookies/, May 2000.
5. V. Khu-smith. An implementation flaw concerning Netscape Navigator and cookies. January 2001.
6. D. Kristol and L. Montulli. *HTTP State Management Mechanism — RFC2109.* IETF, 1997.
7. S. Laurent. *Cookies.* McGraw Hill, 1998.
8. Netscape. *Persistent Client State HTTP Cookies*, 1996.
9. J. Park and R. Sandhu. Secure cookies on the web, *IEEE Internet Computing*, **4**(4):36–44, 2000.
10. D. Ross, I. Brugiolo, J. Coates, and M. Roe. Cross-site scripting overview. http://www.microsoft.com/technet/security/, Febuary 2000.
11. D. Stein. *Web Security.* Addison Wesley, 1998.

A New Stack Buffer Overflow Hacking Defense Technique with Memory Address Confirmation

Yang-Seo Choi[1], Dong-il Seo[1], and Sung-Won Sohn[2]

[1] Anti-CyberTerror Team, ETRI
161 Kajong-Dong, Yusong-Gu, Taejon, 305-350, Korea
{yschoi92,bluesea}@etri.re.kr
[2] Network Security Department, ETRI
161 Kajong-Dong, Yusong-Gu, Taejon, 305-350, Korea
swsohn@etri.re.kr

Abstract. Stack buffer overflow hacking became generally known due to the Morris' Internet Worm in 1988. Since then buffer overflow hacking has been used to attack systems and servers by hackers very frequently. Recently, many researches tried to prevent it, and several solutions were developed such as Libsafe and StackGuard; however, these solutions have a few problems. In this paper we present a new stack buffer overflow attack prevention technique that uses the system call monitoring mechanism and memory address where the system call is made. Because of its detection mechanism this system can be used for unknown attack detection, too.

1 Introduction

The buffer overflow hacking technique which became famous due to the Morris's Internet Worm[1] in 1988 is still being used. In fact, for the past 10 years, buffer overflow hacking has been the most famous attack. Actually before the Internet Worm had been known to the world, people had thought that it was impossible to control the program flow by buffer overflow, but after the Worm, they learned that it is possible. In 1996, AlephOne presented the article "Smashing the stack with fun and profit" in Phrack magazine[2]. This article gave very detail explanation about the stack buffer overflow attack. The buffer overflow hacking would be more easier and even the novice hacker could try the buffer overflow attack.

Based on the data from the Securityfocus[3], and Securiteam[4], the most famous hacking technique was the buffer overflow hacking method. Almost every operating system had the buffer overflow vulnerabilities. Even if no vulnerability has been found yet, it is expected that some buffer overflow vulnerabilities will be found [3, 4]. Fig. 1 shows how often buffer overflow attack have been tried for the past 2 years in Korea. Fig. 1 shows the ratio between the number of total attacks and buffer overflow attack: the buffer overflow attacks make up 40.6% of the total number of attacks. If 1999's data is included the, the percentage of buffer overflow attacks rises to 47.5%[5].

K. Kim (Ed.): ICICS 2001, LNCS 2288, pp. 146–159, 2002.

Recently, due to the buffer overflow defense techniques, the buffer overflow attack is beginning to decrease, but buffer overflow hacking is still being used frequently.

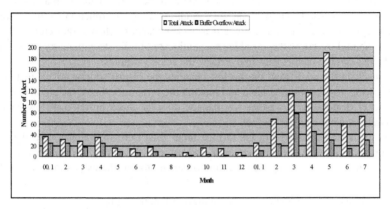

Fig. 1. The Number of Buffer Overflow Attack[5]

In fact, employing the buffer overflow defense technique has several problems because some of them could be cracked by expert hackers[26], several defense systems need the vulnerable program's source code, and other systems take too much overhead. In this paper we present a new buffer overflow attack defense technique that uses a system call monitoring method that does not need the program source code, and can potentially be used for an unknown attack.

This paper is composed of 5 chapters. In chapter 2, we explain the buffer overflow hacking technique, and in chapter 3 we introduce related works. In chapter 4, we describe our technique to prevent buffer overflow hacking, and the conclusion will be in the chapter 5.

Every testing in this paper was tested on RedHat Linux[28] system version 7.1 installed on an Intel Pentium III 800MHz CPU.

2 Stack Smashing Attack[2, 27]

The buffer overflow attack is done as follows. At first, a hacker tries to overflow the stack variable by inserting the well-designed attack codes that consist of return address and some instructions. Certainly the codes were hard coded by complex processing[2]. With these codes, the hacker tries to modify a function's return address and makes the program flow move to the injected codes. The instructions in the injected codes are the instructions that the hacker wants to execute, i.e. shell program execution, modifying the system configuration files, opening a port, etc.

There are 2 kinds of buffer overflow hacking methods. The first is buffer overflow hacking that uses the heap memory area, the second is uses the stack memory area. Generally buffer overflow means the stack buffer overflow, because only the stack buffer overflow can change the program flow and cause the program to be controlled by the hacker.

All the local function variables are allocated in the stack memory area, and almost every variable is used as a local variable. Additionally every function's return address is stored in the stack area. Therefore if a hacker wants to change the return address, then he should overflow the variable that is allocated in the stack memory area. These are the reasons the hacker tries to overflow the stack variables.

In stack buffer overflow, the hacker changes the return address of a function by overflowing any function variable and making the program execute the commands that the hacker desires. In this paper we manage only the stack buffer overflow hacking technique as we mentioned before.

The stack acts as a type of LIFO(Last input First Out) mechanism, and the stack memory area is used for function calls and returns. Additionally the followings are stored in the stack memory area, and the order of storing things is shown in the Fig. 2[6].

– Various kinds of parameters which are used in function calls
– Return Address : the address for next instruction(Instruction Pointer)
– Frame Pointer
– Stack local variables used in a called function

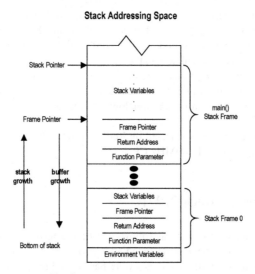

Fig. 2. Stack Memory Area Lay Out

The return address is stored before the stack variables, and the stack operates using the LILO mechanism. So the stack local variables have lower memory address than the return address because the stack growth direction is to the lower memory address, and, the variable's buffer growth direction is opposite to the stack growth direction. Because of the difference of between the memory growth directions, if the stack variable overflows enough, then the return address could be overwritten[2].

All kinds of the stored information in the stack area can be different from other systems because of the difference of operating system, CPU types and amount of memory, but generally the information storing sequence is as Fig. 2.

Usually if the buffer overflow has occurred then the "Segmentation Fault" or "Bus Error" message is printed and the program halted because the program instruction pointer has moved to an invalid memory address or invalid instruction. But if the instruction pointer moves to a valid memory area and valid instruction, then the program executes the instruction without any error. If this program is owned to root and has been set the suid bit then the instruction would be executed as a root.

With Example 1 we can confirm how the buffer overflow hacking technique works. The program source in Example 1 is obtained from [2].

Example from AlephOne.(1996) Smashing the Stack for Fun and Profit. Phrack 49[2]

Example 1. Hacking Exploit Code

```
char shellcode[] =
"\xeb\x1f\x5e\x89\x76\x08\x31\xc0\x88\x46\x07\x89"
"\x46\x0c\xb0\x0b\x89\xf3\x8d\x4e\x08\x8d\x56\x0c"
"\xcd\x80\x31\xdb\x89\xd8\x40\xcd\x80\xe8\xdc\xff"
"\xff\xff/bin/sh";
char large_string[128];
void main() {
    char buffer[96];
    int i;
    long *long_ptr = (long *) large_string;
    for (i = 0; i < 32; i++)
        *(long_ptr + i) = (int) buffer;
    for (i = 0; i < strlen(shellcode); i++)
        large_string[i] = shellcode[i];
    strcpy(buffer,large_string);
    return;
}
```

If Example 1 is compiled and executed then we get a shell with Example 1's owner's right. So if Example 1 is owned to root user then we get the root shell. Fig. 3 explains how Example 1 is executed.

Fig. 3 a) shows the stack lay out and program flow before overflow. This program is in a normal state. In Fig. 3 b), the stack overflow has been caused by the function strcpy(). The array buffer[96] is overwritten by the array large_string [128]. The return address has changed to the memory address of the array buffer[96], and the attack code has been inserted. But the attack code is not executed yet. In Fig. 3 c), the instruction pointer moved to attack code because of changed return address, and the attack code is being executed. The shell program("/bin/sh") is executed.

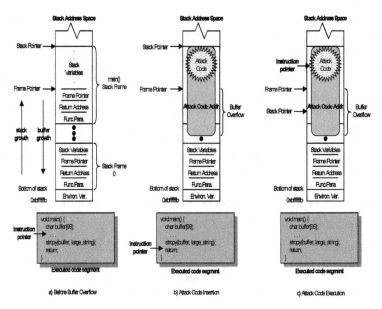

Fig 3. Stack Buffer Overflow Process

3 Related Work

This section describes related works in protecting systems against stack smashing attacks. Section 3.1 and Section 3.2 describe techniques specifically designed to protect a system against stack smashing attacks, while the remaining sections describe programming techniques to reduce or eliminate the vulnerabilities that enable stack smashing attacks.

3.1 StackGuard

StackGuard[7] is developed by a part of the Immunix Project[8]; It is a type of compiler extension. It prevents the return address from modifying in the stack area. It uses 2 kinds of defense methods.

First, it uses the canary word. The canary word is inserted to the stack just before, the return address when a function has been called, and when the function returns, StackGuard checks the canary word. If the canary word has been changed then it stops the function returns. If the canary word has not been changed, then the function progresses normally. In this method, the StackGuard assumes that if the return address has been changed by the stack buffer overflow then the canary word must be changed.

If the canary word has determined constantly and the position of the canary word is discovered by a hacker, then the hacker can skip the canary word without changing. To prevent this, the canary word determined when the program is executed. This StackGuard can be used with gcc[32] version 2.7.2.2.

The second, the StackGuard uses the MemGuard[9, 14] debugging tool. The Synthetix project [10, 11, 12, 13] introduced a notion called "quasi-invariants". Quasi-invariants are state properties that hold true for a while but may change without notice. Quasi-invariants are used to specify optimistic specializations: code optimizations that are valid only while the quasi-invariants hold. This work was extended to treat return addresses on the stack as quasi-invariant during the activation lifetime of the function. The return address is read-only (invariant) while the function is active, thus preventing effective buffer overflow against the stack. MemGuard is a tool developed to help debug optimistic specializations by locating code statements that change quasi-invariant values. MemGuard provides fine-grained memory protection: individual words of memory (quasi-invariant terms) can be designated as read-only, except when explicitly written via the MemGuard API.

3.2 Libsafe and Libverify[15]

In contrast to previous methods and without requiring access to a program's source code, it can transparently protect processes against stack smashing attacks, even on a system-wide basis.

The Libsafe method intercepts all calls to library functions that are known to be vulnerable. A substitute version of the corresponding function implements the original functionality, but in a manner that ensures that any buffer overflows are contained within the current stack frame. This method has been implemented as a dynamically loadable library called Libsafe.

The Libverify method uses binary rewriting of the process memory to force verification of critical elements of stacks before use. This method has also been implemented as a dynamically loadable library called Libverify.

The key idea behind Libsafe is the ability to estimate a safe upper limit on the size of buffers automatically. This estimation cannot be performed at compile time because the size of the buffer may not be known at that time. Thus, the calculation of the buffer size must be made after the start of the function in which the buffer is accessed. Our method is able to determine the maximum buffer size by realizing that such local buffers cannot extend beyond the end of the current stack frame. This realization allows the substitute version of the function to limit buffer writes within the estimated buffer size. Thus, the return address from that function, which is located on the stack, cannot be overwritten, and control of the process cannot be commandeered.

The Libverify library relies on verification of a function's return address before use, a scheme similar to that found in StackGuard. The difference is the manner of implementation. Whereas StackGuard introduces the verification code during compilation, Libverify injects the verification code at the start of the process execution via a binary rewrite of the process memory. Furthermore, Libverify uses the actual return address for verification instead of a "canary" value representing the return address. Thus, in contrast to StackGuard, Libverify can protect pre-compiled executables.

3.3 Non-Executable Stack

Casper Dik and "Solar Designer" have developed Solaris and Linux patches, respectively, that make the stack non-executable [16, 17], precisely to address the stack smashing problem. These patches simply make the stack portion of a user process's virtual address space non-executable, so that attack code injected onto the stack cannot be executed. They offer the advantages of *zero* performance penalty, and that programs work and are protected without re-compilation. However, they do necessitate running a specially-patched kernel, unless this approach is adopted as a standard; Solaris 2.6 incorporates a non-executable stack as a configuration option in /etc/system. This technique is non-trivial and not obvious, for the following reasons:

– gcc uses executable stacks for function trampolines for nested functions.
– The Linux kernel (for example) uses executable user stacks for signal delivery.
– Functional programming languages, and some other programs, rely on executable stacks for runtime code generation.

Solar Designer's Linux patch addresses the trampoline problem and other use of executable stacks by detecting such usage and permanently enabling an executable stack for that process. The patch deals with signal handlers by dynamically enabling an executable stack only for the duration of the signal handler. Both of these compromises offer potential opportunities for intrusion(e.g. a buffer overflow vulnerability in a signal handler.)

3.4 Array Bounds Checking for C

Richard Jones and Paul Kelly developed a gcc patch [18] that does a full array of bounds checking for C programs. Compiled programs are compatible with other gcc modules because they have not changed the representation of pointers. Rather, they derive a "base" pointer from each pointer expression and check the attributes of that pointer to determine whether the expression is within bounds. The performance costs are substantial: a pointer-intensive program (ijk matrix multiply) experiences 30´ slowdown. Since slowdown is proportionate to pointer usage, which is quite common in privileged programs, this performance penalty is particularly unfortunate. The compiler is also not mature; complex programs such as elm fail to execute when compiled with this compiler.

3.5 Memory Access Checking

Purify[19] is a memory usage debugging tool for C programs. Purify uses "object code insertion" to instrument all memory accesses. The approach is similar to StackGuard, in that it does integrity checking of memory, but it does so on each memory access, rather than on each function call return. As a result, Purify is generally slower than StackGuard, imposing a slowdown of 2 to 5 times the execution time of optimized code, making Purify more suitable for debugging software. StackGuard, in contrast, is intended to be left on for production use of the compiled code. Thus for high assurance, protective measures such as StackGuard should be employed unless one is very sure that all potential buffer overflow vulnerabilities have been eliminated.

3.6 Type-Safe Languages

All of the vulnerabilities described here result from the lack of type safety in C. If only type-safe operations can be performed on a given variable, then it is not possible to use creative input applied to variable foo to make arbitrary changes to the variable bar. Type-safety is one of the foundations of the Java security model. Consequently, errors in the Java type checking system are one of the ways that Java programs and Java virtual machines can be attacked [20, 21]. If the correctness of the type checking system is in question, then programs depending on that type checking system for security get the same benefit as type-unsafe programs. Applying StackGuard techniques to Java programs and JVMs may yield beneficial results.

4 A Defense Technique for the Stack Smashing Attack with System Call Monitoring

In this paper we present a new buffer overflow attack defense technique that uses the system call monitoring and memory address confirmation. Actually there are several kinds of system call monitoring mechanism. Any kind of system call monitoring method could be used to out technique. The main idea of this paper is not the system call monitoring but the memory address confirmation. Actually no other application uses the memory addresses to detect hacking trials. But we use the memory address where the system call has been called. This memory address is very important because it will be used to decide whether the actual hacking occurred or not.

4.1 Main Idea

The main ideas of this technique are as follows.
(1) The entire stack smashing attack occurred in the stack memory area
(2) The inserted instructions by hackers are executed in the stack area, too. Actually all the instruction should be executed in the text or heap area.
(3) We do know which instructions are executed in the stack area normally.
(4) The critical instructions that can be thought of as hacking are never executed in the stack area.
(5) We do know which instructions can be identified as a hacking such as shell program execution or modifying important system configuration file, and so on[23, 24].
(6) So if a dangerous instruction is executed in the stack area then we can decide that the instruction is hacking.
(7) Therefore if we can confirm the address of an executed instruction is in the stack memory area, then we can detect the stack smashing attack trial even if the attack is not known.

As we saw in the chapter 2, (1) and (2) are clear, and (3) is explained in section 3.3. On the analogy of (3), the (4) is clear. From [23, 24] (5) is clear. Therefore we can say that (6), (7).

Actually, research on the vulnerable system activities already has been done by many researchers for intrusion detection system. So we can use this research [23, 24].

```
-------------------------------------------------------------------------
08048000-08054000 r-xp 00000000 03:06 354265      /usr/sbin/in.telnetd
08054000-08056000 rw-p 0000b000 03:06 354265      /usr/sbin/in.telnetd
08056000-0806e000 rwxp 00000000 00:00 0
40000000-40013000 r-xp 00000000 03:01 342722      /lib/ld-2.1.3.so
40013000-40014000 rw-p 00012000 03:01 342722      /lib/ld-2.1.3.so
40014000-40015000 rw-p 00000000 00:00 0
. . .
40286000-40287000 rw-p 0000b000 03:01 342770      /lib/libresolv-2.1.3.so
40287000-40289000 rw-p 00000000 00:00 0
bfffd000-c0000000 rwxp ffffe000 00:00 0
-------------------------------------------------------------------------
```

Fig. 4. in.telnetd program's map file

How can we confirm the memory address? We have confirmed the memory address in the linux system by the `maps` files for every running process. Their directories are /proc/<process id>. Fig. 4 shows the `in.telnetd` program's `maps` file.

In Fig. 4, we can confirm where the text, heap, and stack areas are. In Fig. 4, the text area is the memory address 08048000-08054000. The address 08048000-08054000 is hexa value. The heap area is from 40000000 to 4028900, and the stack area is from bfffd000 to c0000000. As we mentioned before, if any instruction is executed in the stack area (bfffd000-c0000000) then the instruction must be a hacking trial.

At last, we should confirm the executed instruction's memory address i.e. the memory address of requested system call. The memory address of requested system call could be identified by a `strace` command. Fig. 5 shows the system calls for a /bin/ps program using the `strace` command.

```
-------------------------------------------------------------------------
. . .
[400cd6cd] stat("/dev/pts/3", {st_mode=S_IFCHR|0620, st_rdev=makedev(136,
           3), ...}) = 0
[400ceb14] write(1,"5208pts/300:00:00strace\n",315208pts/3 00:00:00 strace)
[400cd6cd] stat("/proc/5209", {st_mode=S_IFDIR|0555, st_size=0, ...}) = 0
[400cea24] open("/proc/5209/stat", O_RDONLY) = 7
[400cead4] read(7, "5209 (ps) R 5208 5208 5176 34819"..., 511) = 182
[400cea8d] close(7)                = 0
[400cea24] open("/proc/5209/statm", O_RDONLY) = 7
[400cead4] read(7, "168 168 139 15 0 153 28\n", 511) = 24
[400cea8d] close(7)                = 0
[400cea24] open("/proc/5209/status", O_RDONLY) = 7
[400cead4] read(7, "Name:\tps\nState:\tR (running)\nPid:"..., 511) = 414
[400cea8d] close(7)                = 0
[400cea24] open("/proc/5209/cmdline", O_RDONLY) = 7
[400cead4] read(7, "ps", 2047)     = 2
. . .
-------------------------------------------------------------------------
```

Fig. 5. System calls of program /bin/ps

Fig. 6 shows the system calls for the program in Example 1. We can identify the instruction for executing the /bin/sh program is executed in the stack area(bffffa52).

```
-------------------------------------------------------------------------
. . .
[4000f8d4]  mprotect(0x4001c000, 970752, PROT_READ| PROT_WRITE)=0
[4000f8d4]  mprotect(0x4001c000, 970752, PROT_READ| PROT_EXEC)=0
[4000f891]  munmap(0x40015000, 26022)       = 0
[400ce7fd]  personality(PER_LINUX)          = 0
[400ae257]  getpid()                         = 5084
[bffffa52]  execve("/bin/sh",["/bin/sh"],[/* 0 vars */]) = 0
[4000f78c]  brk(0)                           = 0x809a304
[4000f84d]  old_mmap(NULL, 4096, PROT_READ | PROT_WRITE, MAP_PRIVATE |
MAP_ANONYMOUS, -1, 0) = 0x40014000
[4000ee54]  open("/etc/ld.so.preload", O_RDONLY) = -1 ENOENT(No such file or
directory)
[4000ee54]  open("/etc/ld.so.cache", O_RDONLY) = 3
[4000ed5d]  fstat(3, {st_mode=S_IFREG|0644, st_size=26022, ...}) = 0
[4000f84d]  old_mmap(NULL, 26022, PROT_READ, MAP_PRIVATE, 3, 0) = 0x40015000
. . .
-------------------------------------------------------------------------
```

Fig. 6. System calls of the Example 1 program

4.2 System Contribution & Execution

The memory address confirmation mechanism can be implemented similar to any kind of intrusion detection system that uses the system call monitoring mechanism[22]. We present a basic defense system briefly.

To detect stack smashing attacks the follows should be included in the system.

– P   : The list of programs with possible vulnerabilities.
– M   : Memory map of P
– S   : System call list that could be decided as a hacking
– System Call Monitoring Infrastructure with system call interceptor

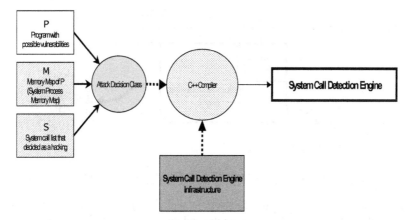

Fig. 7. Offline system for production of detection engines in our approach

First, the list(P) of programs with possible vulnerabilities. In our suggestion, only the selected programs are monitored. These programs are may be daemon programs such as `telnetd`, `ftpd`, `httpd`, etc.

Second, the memory maps for every selected program. These memory maps are used to decide whether the system call is called in the stack area or not.

Third, system calls list that could be a hacking trial. Actually it is a series of commands. These system calls are vulnerable system calls such as execution a shell program, opening a port, changing the critical system configuration files etc.

Fourth, the system call monitoring infrastructure with the system call interceptor. As we mentioned before our technique can be used with any kind of system call monitoring mechanism and system call interceptor. But we recommend the system call interceptor within the operating system kernel. Other alternative implementations include interception of system calls as they pass through the system call library, libc, or using the system call tracing and process control facilities of many UNIX variants. However, these approaches do not offer the same level of security as our kernel-based approach since they can be easily bypassed. It is also doubtful that either approach can be made as efficient as the kernel approach since the kernel approach alone allows interception and modification without process context switching.

The offline system generates detection engines based on the four elements. The four elements are merged and compiled. This stack buffer overflow hacking defense technique makes the attack decision class with the vulnerable program list P, the memory maps M for each P, and the system calls S that can be identified as a hacking attack. The attack decision class would be combined with system call detection engine infrastructure and become the system call detection engine. This system call detection engine would be installed in the kernel area.

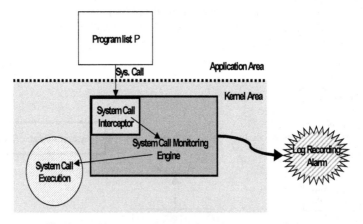

Fig. 8. Runtime system for execution of detection engine

Fig. 8 can be used to explain online system approach. The system call detection engine checks only one system call at a time and does not record the sequences of system calls requested by P. The system call detection engine only checks whether the system call was called from the text area or from the stack area. If the system call is made from the stack area and the system call is defined as hacking then the system call detection engine decides whether stop the process or not. Additionally this detection engine can record the detection logs and alarm the system administrator.

4.3 System Characteristics

The special characteristics of this stack buffer overflow hacking defense system are as follows.

First, because of detection mechanism, this system can detect an unknown attack. This system's detection mechanism uses only the defined system calls and the memory addresses, not the attack patterns; so even if a new stack buffer overflow attack is attempted, the entire attack can be detected by this system.

Second, this system does not need the program source code and any program can execute a normal instruction in stack area.

Third, this system doesn't have to remember the sequences of the system calls.

Fourth, there is a chance to pass this detection mechanism by using the OMEGA Project[33] type attack. But it is very difficult to use OMEGA Project type attack.

The attack of OMEGA Project type tries to hack a system by using a normal function in the code area. If the attack is successful, the commands are executed in the code area. So the system calls would be called in the normal code area. But if an attacker wants to use this method, he(or she) should confirm the absolute memory address where does the normal function is loaded before attacks a system. If the attacker identified the memory address, it's not same all the time. Even if a program has executed in the same system, the memory address of a loaded function can be changed by execution environment. If a program executed in a different system, then the address is almost always changed.

One can calculate a function's memory address dynamically. But to do this, attacker should execute special libraries at the stack area. Then our system can detect it.

5 Conclusion

We have presented a stack buffer overflow attack defense technique. The stack buffer overflow attack is the most famous hacking technique in the past 10 years. Our approach uses memory maps of the vulnerable programs. It does not need the program source code and can detect a new buffer overflow attacks.

This stack buffer overflow hacking defense technique makes the attack decision class with the vulnerable program list P, the memory maps M for each P, and the system calls S that can be identified as a hacking attack. The attack decision class would be combined with system call detection engine infrastructure and become the system call detection engine. This system call detection engine would be installed in the kernel area.

The contributed system catches all the system calls that are made from the vulnerable programs and checks whether or not the system call has been made from the stack memory area. Then, the system call can be identified as a hacking attempt.

All the stack smashing attacks occur in the stack area, and we can confirm the system call's memory address; therefore we can detect every stack smashing attack. This technique could be used for unknown attack detection, but more research for the unknown attack detection is needed.

There is a chance to bypass our mechanism but it is too difficult to do. But we should research about this and make further study about this.

References

[1] Larry Boettger, "The Morris Worm : How it Affected Compter Security and Lessons Learnd by it", SANS Institute White Paper, Dec. 2000.
[2] Aleph One, "Smashing The Stack For Fun And Profit", Phrack 49th Ed. File 14th of 16, Phrack.org, Nov. 1996
[3] Securityfocus.com, http://www.securityfocus.com/
[4] Beyond Security Ltd. http://www.securiteam.com/
[5] CERTCC-KR, "Hacking Statistics", Korea Information Security Agency. 2001.
[6] "IA-32 Intel Architecture Software Developer's Manual. Volume 1-3", Intel Corporation, 2000
[7] Crispin Cowan, Steve Beattie, Ryan Finnin Day, Calton Pu, Perry Wagle, and Erik Walthinsen, "Protecting Systems from Stack Smashing Attacks with StackGuard", Linux Expo, Raleigh, NC, May 1999.
[8] Immunix project, http://immunix.org/
[9] Qian Zhang, "The Synthetix MemGuard Kernel Programmer's Interface", June 1997
[10] Crispin Cowan, Tito Autrey, Charles Krasic, Cal-ton Pu, and Jonathan Walpole. Fast Concurrent Dynamic Linking for an Adaptive Operating System. In International Conference on Configurable Distributed Systems (ICCDS'96), Annapolis, MD, May 1996.
[11] Crispin Cowan, Andrew Black, Charles Krasic, Calton Pu, Jonathan Walpole, Charles Consel, and Eugen-Nicolae Volanschi. Specialization Classes: An Object Framework for Specialization. In Proceedings of the Fifth International Workshop on Object-Orientation in Operating Systems (IWOOOS '96), Seattle, WA, October 27-28 1996.
[12] Eugen N. Volanschi, Charles Consel, Gilles Muller, and Crispin Cowan. Declarative Specialization of Object-Oriented Programs. In Proceedings of the Conference on Object-Oriented Programming Systems, Languages, and Applications (OOPSLA'97), Atlanta, GA, October 1997.
[13] Calton Pu, Tito Autrey, Andrew Black, Charles Consel, Crispin Cowan, Jon Inouye, Lakshmi Kethana, Jonathan Walpole, and Ke Zhang. Optimistic Incremental Specialization: Streamlining a Commercial Operating System. In Symposium on Operating Systems Principles (SOSP), Copper Mountain, Colorado, December 1995.
[14] Crispin Cowan, Dylan McNamee, Andrew Black, Calton Pu, Jonathan Walpole, Charles Krasic, Renaud Marlet, and Qian Zhang. A Toolkit for Specializing Production Operating SystemCode. Technical Report CSE-97-004, Dept. of Computer Science and Engineering, Oregon Graduate Institute,March 1997.

[15] Arash Baratloo, Timothy Tsai, and Navjot Singh, "Transparent Run-Time Defense Against Stack Smashing Attacks", Proceedings of the USENIX Annual Technical Conference, June 2000.

[16] "Solar Designer". Non-Executable User Stack.
http://www.false.com/security/linux-stack/.

[17] Casper Dik. Non-Executable Stack for Solaris. Posting to comp.security.unix, January 2 1997.

[18] Richard Jones and Paul Kelly. Bounds Checking for C.
http://www-ala.doc.ic.ac.uk/ phjk/BoundsChecking.html, July 1995.

[19] Reed Hastings and Bob Joyce. Purify: Fast Detection of Memory Leaks and Access Errors. In Proceedings of the Winter USENIX Conference, 1992. Also available at http://www.rational.com/support/techpapers/fast_detection/.

[20] Drew Dean, Edward W. Felten, and Dan S. Wallach. Java Security: From HotJava to Netscape and Beyond. In Proceedings of the IEEE Symposium on Security and Privacy, Oakland, CA, 1996.
http://www.cs.princeton.edu/sip/pub/secure96.html.

[21] Jim Roskind. Panel: Security of Downloadable Executable Content. NDSS (Network and Distributed System Security), February 1997.

[22] R. Sekar, T.Bowen, M.Segal "On Preventing Intrusions by Process Behavior Monitoring", USENIX, Proceedings of the Workshop on Intrusion Detection and Network Monitoring, April 9-12, 1999

[23] A. Kosoresow, "Intrusion detection via system call traces", IEEE Software '97.

[24] Rebecca Gurley Bace, "Intrusion Detection", Macmillan Technical Publishing, 2000

[25] Maccabe, "Computer Systems : Architecture, Organization and Programming", pp159-171, 1993

[26] Bulba, Kil3r, "Bypassing Stackguard and Stackshield", Phrack 56th Ed. File 5th of 16, Phrack.org, May. 2000

[27] mudge, "Compromised - Buffer-Overflows, from Intel to SPARC Version 8", lopht.com, Oct. 1996

[28] Linus Torvalds et al. Linux Operating System. http://www.linux.org/.

[29] Nathan P. Smith. Stack Smashing vulnerabilities in the UNIX Operating System. http://millcomm.com/nate/machines/security/stack-smashing/nate-buffer.ps, 1997.

[30] Alexander Snarskii. FreeBSD Stack Integrity Patch.
ftp://ftp.lucky.net/pub/unix/local/libc-letter, 1997.

[31] E. Spafford. The Internet Worm Program: Analysis. Computer Communication Review, January 1989.

[32] Richard M. Stallman. Using and Porting GNU C. Free Software Foundation, Inc., Cambridge, MA.

[33] Lamagra, The Omega Project,
http://packetstorm.decepticons.org/papers/unix/omega.txt, 2001.

Efficient Revocation Schemes for Secure Multicast

Hartono Kurnio, Rei Safavi-Naini, and Huaxiong Wang

Centre for Computer Security Research
School of Information Technology and Computer Science
University of Wollongong
Wollongong 2522, Australia
{hk22,rei,huaxiong}@uow.edu.au

Abstract. Multicast communication is the main mode of communication for a wide range of Internet services such as video broadcasting and multi-party teleconferencing where there are multiple recipients. A secure multicast system allows a group initiator (or a centre) to send message over a multicast channel to a *dynamically* changing group of users. The main challenge in secure multicasting is efficient group key management. We propose new schemes for user revocation that can be used to establish a common key among subgroups of users. The schemes can be used with static or dynamic group initiator and allow temporary and permanent revocation of users. We also give a method of adding authentication to the proposed schemes. We prove security and compare efficiency of the new schemes.

1 Introduction

Multicasting is widely used for sending data to a group of users in various applications such as news feeds and pay-TV, and collaborative applications including shared white-boards and teleconferencing. Multicasting is attractive because in comparison to unicasting, it reduces network traffic. This is because to send data to multiple recipients, multicasting sends a single copy of data through most of the path, while in unicasting a different copy is sent to each receiver from the start. Hence unicasting becomes inefficient when the number of recipients is large.

Multicasting is used in scenarios that can be broadly divided into two categories: dynamic and static group initiator. In *dynamic group initiator* systems, a group controller sets up the system and after that any user in the group can form a subgroup. The system is suitable for applications such as teleconferencing that requires collaboration among users. It also has the advantage of being flexible and not having a single point of failure. In *static group initiator* systems only a fixed group controller has the ability to form a subgroup. Applications in this category are news feeds, stock quotes and pay-TV, that require one-way communication from a single source (group controller) to the receivers.

Providing security is essential in many group applications such as pay-TV where only customers who have paid the charges must be able to receive the

K. Kim (Ed.): ICICS 2001, LNCS 2288, pp. 160–177, 2002.

broadcast. *Secure multicast key distribution* systems establish a *group key*, also called a *session key*, among authorised receivers that allows them to have private communications among themselves.

In most cases, groups are dynamic: some members may be revoked or some new members may join the group. When a user's membership is revoked, it is crucial that he/she can not read future communication. That is, the system needs to provide *forward secrecy*. To conceal future communications from the revoked users, the group key must be updated through a *rekeying* process.

Rekeying with the aim of removing a subset of users is also studied in the context of *blacklisting* [13]. The main attack in a rekeying system is from a colluding group of revoked users. The revoked members may belong to multiple rounds of the rekeying protocol. A rekeying system provides *t-resistance* if collusion of at most t revoked users cannot break forward secrecy property of the system.

Multicast groups may include a very large number of users and so the rekeying systems must be efficient. The main efficiency measures of a rekeying system are *rekeying bandwidth* that measures the communication cost of rekeying, and *storage size* of the group controller and the users. Reducing storage is an important requirement of applications such as the decoder unit in a pay-TV system or smart cards, that have restricted memory.

1.1   Our Work

We consider efficient revocation schemes for multicast environments with dynamic and static group initiator (GI). Our main construction is inspired by [20] and [1] and uses a tree structure in conjunction with Shamir secret sharing scheme [18], and Diffie-Hellman key agreement [7], resulting in a scheme with a number of advantages. The main advantage of our system over the scheme in [20] is that our scheme is applicable to dynamic GI and static GI scenario both, while the scheme in [20] and its variants are only applicable to the static GI scenario. Moreover our scheme requires less bandwidth for rekeying (see section 5.2 for more detail). Compared to the schemes in [1,17] our proposed scheme provides higher level of collusion resistance: that is it provides $(n-1)$-resistance while the schemes in [1,17] provide t_{max}-resistance and when t_{max} is large, say $t_{max} = n-1$, it becomes very inefficient. This is because the required bandwidth of the systems for any $t \leq t_{max}$ users is $O(t_{max})$. We prove security of the proposed schemes and give methods of reducing storage and communication cost of the schemes.

Although the main focus of this paper is not on authenticated communication, we will show how to extend our scheme to provide this property.

1.2   Related Work

Establishing secure group communication is studied in different contexts. In the following we outline some of these approaches and their relationship to this paper.

Dynamic Conferences. A dynamic conference system consists of a number of users who can establish subgroups. The subgroup key may be calculated without any other message exchange [2], or might require participants to broadcast messages [3]. The system in [2] allows pre-determined sizes for the subgroup while in conference systems such as [3] all participants must broadcast.

Broadcast Encryption. Fiat and Naor [8] introduced broadcast encryption system for pay-TV applications. In this system, after an initial key setup stage, the transmitter can send a message that can only be decrypted by an authorised group such that collusion of up to t users outside the authorised group cannot decrypt the transmission.

The transmitter requires $O(t^2 \log n \log^2 t)$ bandwidth and each user has to store $O(t \log^2 n \log t)$ keys for t-resistance scheme, where n is the total number of users in the group. User revocation systems using broadcast encryption were proposed in [19,13]. These systems are for static group initiator and are without any computational assumption. Fiat and Naor's broadcast encryption schemes [8] provide security against collusion of up to t receivers, and can cater for both private and public keys settings. Our proposed schemes are computationally secure. Our basic scheme provides security against any number of colluders and its variant provides security for t-collusion.

Secure Multicast. Secure multicast key distribution systems have received much attention in recent years [20,21,15,4,5,6,14,10,1,17]. Early surveys of multicast security issues are in [11] and [12]. One of the main contributions in this area has been the introduction of hierarchical key tree by Wallner *et al* [20] and independently by Wong *et al* [21]. In these systems a d-ary logical tree is used to allocate keys to the users. In this scheme for a group of size n, the group controller and the users have to store $\frac{dn-1}{d-1}$ and $\log_d n + 1$ keys, respectively. The required bandwidth for revocation of a single user is $d \log_d n$ keys. The scheme allows revocation of any number of users and provides protection against collusion of arbitrary number of revoked users.

A number of authors have proposed variants of hierarchical key tree to improve the efficiency. McGrew and Sherman [15] applied one-way function to binary tree $(d = 2)$ and reduced the bandwidth to $\log_2 n$ keys. Canetti *et al* [4,5] explored the trade-off between the bandwidth, user storage, and group controller storage and showed how to use pseudo-random number generators to achieve bandwidth of $(d - 1) \log_d n$ keys. Chang *et al* [6] used Boolean function minimisation techniques on binary tree and reduced the total number of group controller keys to $\log_2 n$ keys, while retaining the same order of bandwidth and user keys. The main drawback of Chang *et al.* scheme is that collusion of two revoked users may reveal all keys in the system (1-resistance). Safavi-Naini and Wang extended Chang *et al* scheme to increase the threshold parameter for resistance (i.e., schemes with t-resistance), their approach is combinatorial and uses perfect hash families in the construction. However, their scheme is efficient only if the threshold parameter t is much smaller than n. In all the above work, some of them are extremely efficient with respect to storage and communication cost, assume static group initiator schemes.

Dynamic group initiator schemes are studied in [1,14,10]. The tree approach in Kurnio *et al* [14] is similar to the work of Kim *et al* [10] and is basically an extension of the static group initiator scenario in [20], using Diffie-Hellman key exchange on binary trees. In this model, there is no group controller and the basic rekeying protocol is for single user revocation. This means that for multiple user revocation, multiple rounds of the basic process must be executed which results in a complex system. User storage is $\log_2 n$ keys and the bandwidth for revocation of a single user is $\log_2 n$ keys.

A completely different approach to revocation was proposed by Anzai *et al* [1]. Their solution uses a $(t_{max} + 1, n + t_{max})$-threshold secret sharing scheme together with Diffie-Hellman key agreement and allows revocation of up to t_{max} users (i.e., the scheme has t_{max}-resistance). The scheme requires a user to store 1 key and the group controller to store a polynomial of degree at most t_{max} over $GF(q)$. However, the scheme needs $n + t_{max}$ public keys and has the bandwidth of t_{max} keys for revocation of t, $t \leq t_{max}$, users. The scheme is only practical for small t_{max}.

The paper is organised as follows. In section 2, we propose a dynamic GI scheme and show how to transform the scheme into a static GI scheme. In section 3, we propose a more efficient method of system setup, while keeping the revocation operation unchanged. In section 4, we describe extensions to the schemes. In section 5, we evaluate and compare our schemes with other schemes and finally conclude our work in section 6.

2   A Dynamic GI Scheme

Let $\mathcal{U} = \{U_1, \ldots, U_n\}$ be a set of n users, and GC denote a group controller who initialises the group. GC generates a set $\mathcal{K}$ of secret keys and constructs the set $\mathcal{Y}$ of the corresponding public keys. For all $1 \leq j \leq n$, GC sends a subset $\mathcal{K}_j$ of secret keys, $\mathcal{K}_j \subset \mathcal{K}$, to U_j via a private channel. The public keys are published on a public bulletin-board. Moreover, GC may also publish other necessary public information.

At a later time, a user, also called the *group initiator* GI, can form a subgroup by evicting some users from the group. This is done only by GI multicasting a single message which allows legitimate users to calculate a group key while revoked users cannot.

2.1   System Setup

The group controller GC is in charge of the system setup and does the following.

1. Generates two large primes p (around 1024 bits) and q (around 160 bits), where $q|p - 1$, and a generator g of the multiplicative group of $GF(p)$. He then publishes p, q and g.

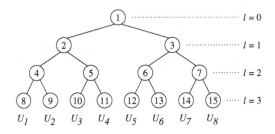

Fig. 1. Tree structure for $n = 8$ and $d = 2$

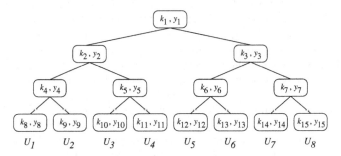

Fig. 2. Logical key tree for figure 1

2. Builds a tree of degree d with n leaves. Every node in the tree is either a leaf or a parent with d child nodes. Let $\mathcal{N}$ denote the set of nodes in the tree and $m = |\mathcal{N}|$ be the total number of nodes in the tree; so, for balanced trees $m = \frac{dn-1}{d-1}$ [1]. Each node is labelled by a unique number i where $i \neq 0$. For $1 \leq j \leq n$, GC logically associates user U_j with a leaf of the tree. Knowledge of the tree structure together with node labels and users' association with leaves are public. Figure 1 is an example of a tree structure. The nodes are labelled by i, for $1 \leq i \leq m$, starting from root node to leaf nodes, and from left to right directions.

3. Generates a set of secret keys, $\mathcal{K} = \{k_i \mid i \in \mathcal{N}, k_i \in GF(q)\}$, and a set $\mathcal{Y} = \{y_i \mid i \in \mathcal{N}, y_i = g^{k_i} \bmod p\}$ of public keys. For all $i \in \mathcal{N}$, node i is associated with a pair of secret key k_i and public key y_i.

4. Publishes all the public keys and securely sends to user U_j the set of secret keys, $\mathcal{K}_j \subset \mathcal{K}$, from U_j's leaf to the root. U_j keeps these keys as his secret keys. (For example, the logical key tree of figure 1 is shown in figure 2 and secret keys for U_1 is $\mathcal{K}_1 = \{k_1, k_2, k_4, k_8\}$.)

The system setup above requires GC to generate all $\frac{dn-1}{d-1}$ secret keys and to publish all $\frac{dn-1}{d-1}$ public keys. A user has to store secret keys from a leaf to the root (height of tree), which is $h + 1$, where $h = \log_d n$ keys. We then have,

---

[1] Although the tree may be unbalanced, we assume a balanced tree in our efficiency analysis.

Proposition 1. *In the above scheme, the storage size for* GC *and users are* $\frac{dn-1}{d-1}$ *and* $\log_d n + 1$ *secret keys, respectively. There are* $\frac{dn-1}{d-1}$ *public keys.*

2.2   User Revocation

Suppose a group initiator GI wants to form a subgroup $\mathcal{U}_L \subset \mathcal{U}$. This can be achieved by revoking the users $\mathcal{U}_R = \mathcal{U} \setminus \mathcal{U}_L$ and forming a new group key for $\mathcal{U}_L$, as follows.

Let $\mathcal{N}_{U_j}$ be a set of nodes from U_j's leaf to the root. Assume $|\mathcal{M}| = s$.

1. GI randomly chooses an element r of $GF(q)$ and multicasts $Y = g^r \bmod p$.
2. GI uses the algorithm in section 2.3 to find a set $\mathcal{M}$ of nodes that satisfy the following conditions:
   (i) $\forall U_j \in \mathcal{U}_L, \mathcal{M} \cap \mathcal{N}_{U_j} \neq \emptyset$;
   (ii) $\forall U_j \in \mathcal{U}_R, \mathcal{M} \cap \mathcal{N}_{U_j} = \emptyset$.
   (The algorithm guarantees that $\forall U_j \in \mathcal{U}_L, |\mathcal{M} \cap \mathcal{N}_{U_j}| = 1$, which means each user in $\mathcal{U}_L$ has exactly one node in $\mathcal{M}$.)
3. GI chooses a set $\mathcal{I}$ of $s-1$ distinct elements of $GF(q)$ such that $0 \notin \mathcal{M}$ and $\mathcal{M} \cap \mathcal{I} = \emptyset$. Then for all $c \in \mathcal{I}$, GI calculates,

$$y_c = \prod_{a \in \mathcal{M}} (y_a)^{L(\mathcal{M}, a, c)} \bmod p,$$

where

$$L(\mathcal{M}, a, c) = \prod_{b \in \mathcal{M}, b \neq a} \frac{c - b}{a - b} \bmod q.$$

Finally, GI multicasts $Y_c \parallel c$, where $Y_c = (y_c)^r \bmod p$ and $\parallel$ denotes concatenation.
4. A user $U_j \in \mathcal{U}_L$ uses a secret key $k_e \in \mathcal{K}_j$, where $e \in \mathcal{M} \cap \mathcal{N}_{U_j}$, and the multicasted data to calculate the group key (i.e., session key) GK as follows.

$$GK = (Y^{k_e})^{L(\mathcal{I} \cup \{e\}, e, 0)} \times \prod_{c \in \mathcal{I}} (Y_c)^{L(\mathcal{I} \cup \{e\}, c, 0)} \bmod p.$$

The DH (Diffie-Hellman) problem which is the basis of our revocation protocol can be stated as follows.

Given a generator g of the multiplicative group of $GF(p)$, and inputs $y = g^x \bmod p$ and $y' = g^{x'} \bmod p$, compute $DH(g; y, y') = g^{xx'} \bmod p$.

The DH problem is believed to be hard [7].

Theorem 1. *In the above scheme,*

(i) *each user in* $\mathcal{U}_L$ *is able to calculate a group key* GK;
(ii) *assuming that the Diffie-Hellman (DH) problem is hard, collusion of any number of users in* $\mathcal{U}_R$ *cannot find* GK.

Proof. (i) We show that all users in $\mathcal{U}_L$ can compute a common key GK based on their secret keys and the multicasted data.

Without loss of generality, assume $\mathcal{M} = \{1, 2, \ldots, s\}$. Let the secret keys and public keys associated with $\mathcal{M}$ be $\{k_a \mid a \in \mathcal{M}\}$ and $\{y_a = g^{k_a} \bmod p \mid a \in \mathcal{M}\}$, respectively. Notice that there implicitly exists a unique polynomial $f(x)$ of degree at most $s - 1$ such that $g^{f(a)} = g^{k_a}$, for all $a \in \mathcal{M}$. Using the public keys of $\mathcal{M}$, that is, $\{g^{k_a} \bmod p \mid a \in \mathcal{M}\}$, one can calculate $y_c = g^{f(c)}, \forall c \in \mathcal{I}$, as follows,

$$
\begin{aligned}
y_c &= g^{f(c)}, \\
&= g^{\sum_{a \in \mathcal{M}} (k_a) \times L(\mathcal{M}, a, c)} \bmod p, \\
&= \prod_{a \in \mathcal{M}} (y_a)^{L(\mathcal{M}, a, c)} \bmod p.
\end{aligned}
$$

So, each user in $\mathcal{U}_L$ computes the group key as

$$
\begin{aligned}
GK &= (Y^{k_e})^{L(\mathcal{I} \cup \{e\}, e, 0)} \times \prod_{c \in \mathcal{I}} (Y_c)^{L(\mathcal{I} \cup \{e\}, c, 0)} \bmod p, \\
&= (g^{r \times f(e)})^{L(\mathcal{I} \cup \{e\}, e, 0)} \times \prod_{c \in \mathcal{I}} (g^{r \times f(c)})^{L(\mathcal{I} \cup \{e\}, c, 0)} \bmod p, \\
&= \prod_{c \in (\mathcal{I} \cup \{e\})} (g^{r \times f(c)})^{L(\mathcal{I} \cup \{e\}, c, 0)} \bmod p, \\
&= g^{r(\sum_{c \in (\mathcal{I} \cup \{e\})} f(c) \times L(\mathcal{I} \cup \{e\}, c, 0))} \bmod p, \\
&= g^{r \times f(0)} \bmod p.
\end{aligned}
$$

(ii) We show that the collusion of any subgroup of $\mathcal{U}_R$ is not able to find the secret group key GK. Our proof uses a reduction argument. That is we show that, if there exists an oracle (probabilistic polynomial-time) G that can compute GK using all the information known to $\mathcal{U}_R$, then the same oracle can be used to solve the DH problem.

It is sufficient to show that if there exists a probabilistic polynomial-time algorithm G that on input $g^r, g^{rf(c)}, \forall c \in \mathcal{I}$ and $g^{f(a)}, \forall a \in \mathcal{M}$, outputs $g^{rf(0)}$ with a non-negligible probability, where $f(x)$ is a polynomial of degree at most $s - 1$, then G can be used to solve the DH problem. That is, given g^{x_1} and g^{x_2}, where x_1 and x_2 are two randomly chosen elements of $GF(q)$, G can be used to find $g^{x_1 x_2}$. Let $\mathcal{I} = \{c_1, \ldots, c_{s-1}\}$. Choose $s - 1$ randomly chosen elements $a_1, \ldots, a_{s-1} \in GF(p)$ and construct a unique polynomial $h(x)$ of degree at most $s - 1$ such that $h(c_i) = a_i, \forall i, 1 \leq i \leq s - 1$, and $g^{h(0)} = g^{x_2}$. This can be used to calculate $g^{h(c_i)} = g^{a_i}, 1 \leq i \leq s - 1$, and also $g^{h(\alpha)}$ for all $\alpha \in GF(q)$, and hence $g^{h(a)}, \forall a \in M$. Furthermore, since we know a_i we can compute

$$
(g^{x_1})^{a_i} = (g^{x_1})^{h(c_i)}, \quad c_i \in \mathcal{I}.
$$

Now if G is given the input, g^{x_1}; $(g^{x_1})^{a_i}, i = 1, \ldots, s - 1$; and $g^{h(a)}, a \in M$, it will output $g^{x_1 h(0)} = g^{x_1 x_2}$. This means G can solve the DH problem and so contradicts the hardness assumption of the DH problem. $\qquad\square$

The above scheme can be used multiple times (rounds) with a different r for each round. The proof for multiple round can be found in appendix.

2.3   An Algorithm for Finding $\mathcal{M}$

Let the root of the tree be considered as level zero and leaves be considered at level h [2] (similar to figure 1). The algorithm for finding $\mathcal{M}$ is as follows.

Let $\mathcal{N}$ denote the set of nodes in the tree and $\mathcal{N}^{(l)}$ be the set of all nodes on level l. That is, $\mathcal{N} = \{\cup \mathcal{N}^{(l)} \mid 0 \leq l \leq h\}$. Also, let $\mathcal{N}_{U_j}$ be the set of nodes from U_j's leaf to the root and let $N_{U_j}^{(l)}$ be the node at level l in $\mathcal{N}_{U_j}$. Assume $\mathcal{N}_R = \{\cup \mathcal{N}_{U_j} \mid U_j \in \mathcal{U}_R\}$ is the set of nodes of the revoked users and let $\mathcal{N}_R^{(l)}$ denote the set of nodes at level l of $\mathcal{N}_R$. That is, $\mathcal{N}_R = \{\cup \mathcal{N}_R^{(l)} \mid 0 \leq l \leq h\}$ and $\mathcal{N}_R^{(l)} \subseteq \mathcal{N}^{(l)}$. The algorithm is as follows.

(1)	$\mathcal{M} = \emptyset, \mathcal{U}_{left} = \mathcal{U}_L$
(2)	For $l = 0$ to h do
(3)	$\mathcal{N}_{temp1} = \mathcal{N}^{(l)} \setminus \mathcal{N}_R^{(l)}$
(4)	$\mathcal{U}_{temp} = \{U_j \mid U_j \in \mathcal{U}_{left}, N_{U_j}^{(l)} \in \mathcal{N}_{temp1}\}$
(5)	$\mathcal{N}_{temp2} = \{i \mid i = N_{U_j}^{(l)}, U_j \in \mathcal{U}_{temp}\}$
(6)	$\mathcal{M} = \mathcal{M} \cup \mathcal{N}_{temp2}$
(7)	$\mathcal{U}_{left} = \mathcal{U}_{left} \setminus \mathcal{U}_{temp}$
(8)	If $\mathcal{U}_{left} = \emptyset$ do
(9)	Stop
(10)	End
(11)	End

Intuitively, the algorithm works as follows. It starts from the root and visits all nodes in each level, before moving to the next level down. A node is put in $\mathcal{M}$ if the following conditions are satisfied. (i) the node does not belong to a revoked user, and (ii) no other node on its path to the root is in $\mathcal{M}$.

Let $\mathcal{U}_{left}$ be the set of users who do not have a node in $\mathcal{M}$. Step (1) initialises $\mathcal{M} = \emptyset$ and $\mathcal{U}_{left} = \mathcal{U}_L$. Step (2) repeats step (3) to step (10) for each level from $l = 0$ to $l = h$. Step (3) puts up all nodes in level l, except those belonging to the revoked users, in $\mathcal{N}_{temp1}$. Step (4) looks for users in $\mathcal{U}_{left}$ who have at least one node in $\mathcal{N}_{temp1}$. These users are kept in $\mathcal{U}_{temp}$. It is possible that nodes in $\mathcal{N}_{temp1}$ do not belong to any user in $\mathcal{U}_{temp}$. Step (5) will select nodes in $\mathcal{N}_{temp1}$ that belong to at least one user in $\mathcal{U}_{temp}$. These nodes are stored in $\mathcal{N}_{temp2}$. Step (6) adds $\mathcal{N}_{temp2}$ to $\mathcal{M}$ and step (7) subtracts $\mathcal{U}_{temp}$ from $\mathcal{U}_{left}$. Step (8) checks if $\mathcal{U}_{left}$ is empty, that is if all users in $\mathcal{U}_L$ have at least a node in $\mathcal{M}$. If this is the case, then the algorithm stops; otherwise it goes to the next level.

Theorem 2. *The output set $\mathcal{M}$ of the above algorithm satisfies the following properties*

---

[2] For an unbalanced tree, level h is located at the *lowest* leaves of the tree where $h + 1$ is height of the unbalanced tree.

(i) $\forall U_j \in \mathcal{U}_L, \mathcal{M} \cap \mathcal{N}_{U_j} \neq \emptyset$, and
(ii) $\forall U_j \in \mathcal{U}_R, \mathcal{M} \cap \mathcal{N}_{U_j} = \emptyset$.

In fact the algorithm guarantees that $\forall U_j \in \mathcal{U}_L, |\mathcal{M} \cap \mathcal{N}_{U_j}| = 1$ and $\mathcal{M}$ is a minimal set.

Proof (sketch). Step (3) excludes all nodes belonging to the revoked users and fulfils condition (ii). We note that a node on a lower level belongs to more users and a node on the highest level (a leaf node, $l = h$) belongs to a single user. In step (2), the algorithm runs from the lowest level ($l = 0$) to the highest level ($l = h$), and constructs $\mathcal{M}$ from the most common nodes to the least common nodes. Together with steps (4), (5) and (7), the algorithm ensures that once the node at level l of the user U_j, $N_{U_j}^{(l)}$, is in $\mathcal{M}$, nodes on the higher levels belonging to the same user will not be in $\mathcal{M}$. This guarantees that $\forall U_j \in \mathcal{U}_L, |\mathcal{M} \cap \mathcal{N}_{U_j}| = 1$ and results $\mathcal{M}$ to be minimal. The algorithm may terminate at level $l, l < h$, if condition (i) is satisfied. Otherwise, it will proceed to level $l = h$ to guarantee this condition.   □

Theorem 3. *In the above scheme, revocation of one user requires the bandwidth $(d - 1) \log_d n - 1$ keys. Revoking t users, $t > 1$, in the best case requires zero bandwidth (no multicast), and in the worst case $(1 - \frac{1}{d})n - 1$ bandwidth.*

Proof. Revoking one user always has $|\mathcal{M}| = (d - 1) \log_d n$ and needs the bandwidth of $(d - 1) \log_d n - 1$ keys. For $t > 1$ users, the best case is when the first common ancestor (the ancestor that is highest in the tree) of the leaves associated with the $|\mathcal{U}_L| = d^a, 1 \leq a \leq h - 1$, remaining users is not an ancestor of the revoked users. In this case, $|\mathcal{M}| = 1$ and so the required bandwidth is zero. All users in $\mathcal{U}_L$ have a common ancestor and are able to use the secret key corresponding to this common ancestor for secure communication. The worst case is when $t = d^{h-1}$ and the leaves associated with the revoked users have different parents in which case $|\mathcal{M}| = (1 - \frac{1}{d})n$ and the required bandwidth is $(1 - \frac{1}{d})n - 1$ keys.   □

2.4   An Example

Consider figure 1 and let $\mathcal{U}_L = \{U_1, U_2, U_5, U_6, U_7, U_8\}$. Suppose U_1 wants to form a subgroup by revoking $\mathcal{U}_R = \{U_3, U_4\}$. U_1 does the following.

1. generates r and multicasts $Y = g^r \bmod p$.
2. sets $\mathcal{N}_R = \{1, 2, 5, 10, 11\}$ and executes the algorithm. The result is $\mathcal{M} = \{3, 4\}$.
3. uses y_3 and y_4 to calculate $y_c, \forall c \in \mathcal{I}$. Suppose U_1 chooses $\mathcal{I} = \{16\}$. He calculates $y_{16}, Y_{16} = (y_{16})^r \bmod p$ and multicasts $Y_{16} \parallel 16$.
4. U_1 and U_2 use k_4, and $U_5, \ldots, U_8$ use k_3 to calculate GK.
5. U_3 and U_4 are not able to calculate GK since they do not have k_3 or k_4.

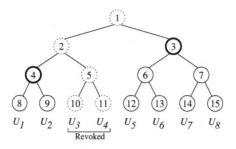

Fig. 3. Dashed nodes are nodes belong to $\mathcal{U}_R$. Bold nodes are nodes of the minimal set $\mathcal{M}$

2.5   A Static GI Scheme

In a static GI scheme, there is a single group controller who initially sets up the group and at a later stage revokes the membership as required by the new subgroup to be formed.

The dynamic GI scheme of section 2 can be easily converted into a static GI scheme. The two systems have similar performance and system setup except in the static case public keys are not required.

The user revocation works as follows.

1. GC chooses a random number r in $GF(q)$ and multicasts $Y = g^r \bmod p$.
2. GC finds $\mathcal{M}$ and uses Lagrange Interpolation to generate a polynomial $f(x)$ of degree $s - 1$ such that $f(a) = k_a$ for all $a \in \mathcal{M}$,

$$f(x) = \sum_{a \in \mathcal{M}} k_a \times L(\mathcal{M}, a, x) \bmod q.$$

3. GC chooses $\mathcal{I}$ and calculates $k_c = f(c), \forall c \in \mathcal{I}$. He then multicasts $Y_c \parallel c$, where $Y_c = Y^{k_c} \bmod p$, for all $c \in \mathcal{I}$.
4. Each user $U_j \in \mathcal{U}_L$ uses his secret key $k_e \in \mathcal{K}_j$, where $e = \mathcal{M} \cap \mathcal{N}_{U_j}$, to calculate group key GK, as shown below.

$$GK = (Y^{k_e})^{L(\mathcal{I} \cup \{e\}, e, 0)} \times \prod_{c \in \mathcal{I}} (Y_c)^{L(\mathcal{I} \cup \{e\}, c, 0)} \bmod p,$$

$$= (g^{r \times f(e)})^{L(\mathcal{I} \cup \{e\}, e, 0)} \times \prod_{c \in \mathcal{I}} (g^{r \times f(c)})^{L(\mathcal{I} \cup \{e\}, c, 0)} \bmod p,$$

$$= \prod_{c \in (\mathcal{I} \cup \{e\})} (g^{r \times f(c)})^{L(\mathcal{I} \cup \{e\}, c, 0)} \bmod p,$$

$$= g^{r(\sum_{c \in (\mathcal{I} \cup \{e\})} f(c) \times L(\mathcal{I} \cup \{e\}, c, 0))} \bmod p,$$

$$= g^{r \times f(0)} \bmod p.$$

Alternatively, GC can choose a value for the group key GK, encrypt GK with the secret key $k_a, \forall a \in \mathcal{M}$, and multicast the result. Each user $U_j \in \mathcal{U}_L$ decrypts the encrypted multicast using his secret key k_e to find GK. Security of the above scheme follows from Theorem 1.

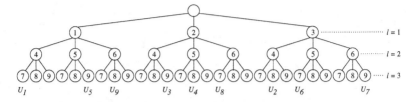

Fig. 4. Tree structure for the example

3   A Variant of Key Generation and Allocation

In the scheme described above, the total number of keys in the system is m. In this section we propose a variant of the scheme that reduces the number of system keys while maintaining security. Reducing the number of system keys has the advantage of reducing the storage required by GC and the amount of published information.

The reduction in the number of system keys is at the cost of reducing collusion resistance of the system. That is, the modified system provides collusion resistance for up to t_{max} colluders where t_{max} is a pre-determined threshold parameter. This may be a limitation for some applications where we cannot bound the collusion size beforehand, while in other situations it may be a reasonable assumption.

The basic idea is as follows. For a d-ary tree of height h, we choose dh keys for the system and allocate a key to each node in such a way that, d distinct keys are assigned to the nodes in the ith level, $1 \le i \le h$, such that the d children of the same parent have distinct keys. Each user is associated with a leaf but not all leaves are assigned to users. The keys of a user are his leaf key together with all the node keys along the path to the root. The leaves corresponding to the users are chosen in such a way that for any set of t_{max} users, $\{U_{i_1}, \dots, U_{i_{t_{max}}}\}$, and a user $U_j \notin \{U_{i_1}, \dots, U_{i_{t_{max}}}\}$, there exists at least one key which belongs to user U_j, but does not belong to a user in $\{U_{i_1}, \dots, U_{i_{t_{max}}}\}$. In other words, d and h must be chosen such that $n < d^h$ (Note that there are d^h leaves of the tree). Figure 4 illustrates the leaf assignment for 9 users in a 3-ary tree of height 3. The scheme requires only 9 keys.

In the following we give a construction for this approach using polynomials over finite fields.

GC does the following:

1. Generates p, q and g, similar to the basic system setup, and publishes them.
2. Selects t_{max}, the required level of collusion resistance. Then he chooses a prime d and computes $u = \lceil \log_d n \rceil$. Next he chooses the tree depth, h, such that $h > t_{max} \times (u - 1)$ and $h \le d$.
3. Forms a set of polynomials $\mathcal{F}_d[x]_u = \{f(x) \in \mathcal{F}_d[x] \mid deg\ (f(x)) \le u - 1\}$ and associates a polynomial $f_j(x) \in \mathcal{F}_d[x]_u$ to a user U_j for $1 \le j \le n$. Note that $|\mathcal{F}_d[x]_u| = d^u \ge n$.

4. Chooses a set of h distinct elements of $GF(d)$, $\mathcal{L} = \{\alpha^{(1)}, \ldots, \alpha^{(h)}\}$, each associated with one level of the tree. To each user $U_j, 1 \leq j \leq n$, GC assigns an identity vector $V_j = (f_j(\alpha^{(1)}), \ldots, f_j(\alpha^{(h)})) = (v_j^{(1)}, \ldots, v_j^{(h)})$ over $GF(d)$.

5. Generates a set of secret keys, $\mathcal{K} = \{k_a^{(l)} \mid k_a^{(l)} \in GF(q), 0 \leq a \leq d - 1, 1 \leq l \leq h\}$ and a set of public keys, $\mathcal{Y} = \{y_a^{(l)} = g^{k_a^{(l)}} \bmod p \mid 0 \leq a \leq d - 1, 1 \leq l \leq h\}$.

6. All public keys are published. For $1 \leq j \leq n$, GC secretly sends a set of secret keys, $\mathcal{K}_j = \{k_{v_j^{(b)}}^{(b)} \mid 1 \leq b \leq h\} \subset \mathcal{K}$ to user U_j.

We observe that the underlying structure above is a balanced tree of degree d with d^h leaves. Since $n < d^h$, only some of the leaves are associated with the users.

The system setup requires GC to generate $d \times h = d(t_{max}(u - 1) + 1) \approx d(t_{max}(\log_d n - 1) + 1) \approx dt_{max} \log_d n$ secret keys, and to publish $d \times h \approx dt_{max} \log_d n$ public keys. A user has to keep $h = t_{max}(u-1)+1 \approx t_{max}(\log_d n - 1) + 1 \approx t_{max} \log_d n$ secret keys. We then have,

Proposition 2. *For the above scheme, storage of GC is $dt_{max} \log_d n$ secret keys, storage of a user is $t_{max} \log_d n$ secret keys and there are $dt_{max} \log_d n$ public keys.*

Revocation process is the same as described in section 2.2 and security of user revocation follows from Theorem 1. However, we note that bandwidth requirement of this scheme is different from that of the basic one.

Theorem 4. *In the above scheme, the bandwidth for t user revocation is at most $(d - 1)t_{max} \log_d n$ keys.*

Proof. The maximum value of $|\mathcal{M}|$ corresponds to $t = 1$ in which case $|\mathcal{M}| = dt_{max} \log_d n - t_{max} \log_d n = (d-1)t_{max} \log_d n$. This is the worst case. Revocation of $t > 1$ users has smaller $|\mathcal{M}|$ and so the required bandwidth for revocation of any t users is less than $(d - 1)t_{max} \log_d n$ keys. ☐

We need to show that with the above key allocation, $\mathcal{M}$ as defined in section 2.2, is not an empty set. This is true because each user has a subset of h keys that corresponds to a polynomial of degree at most $u - 1$. It follows that the number of common keys of any two users is at most $u - 1$. This is because if two users, U_i and U_j, have a common key at level p, it means that $f_i(\alpha^{(p)}) = f_j(\alpha^{(p)})$. The condition $h > t_{max} \times (u - 1)$ yields that a set $\mathcal{M}$ satisfying the required conditions can be found. Since the cardinality of $\mathcal{M}$ determines the communication bandwidth, we would like the size of $\mathcal{M}$ to be as small as possible. To find $\mathcal{M}$, we can use the same algorithm as in section 2.3. However, the resulting $\mathcal{M}$ is not necessarily minimal.

An Example. Let $n = 9, t_{max} = 2$, and $d = 3$. It follows that $u = 2, h > 2$, i.e., $h = 3$, and $\mathcal{F}_3[x]_2 = \{f(x) \in \mathcal{F}_3[x] \mid deg\ (f(x)) \leq 1\}$. Polynomials for users are the following.

$$f_1(x) = 0 \bmod 3 \quad f_2(x) = x \bmod 3 \quad f_3(x) = 2x \bmod 3$$
$$f_4(x) = 1 \bmod 3 \quad f_5(x) = 1 + x \bmod 3 \quad f_6(x) = 1 + 2x \bmod 3$$
$$f_7(x) = 2 \bmod 3 \quad f_8(x) = 2 + x \bmod 3 \quad f_9(x) = 2 + 2x \bmod 3.$$

Let $\mathcal{L} = \{2, 0, 1\}$, then vectors for the users are the following.

$$V_1 = (0, 0, 0) \quad V_2 = (2, 0, 1) \quad V_3 = (1, 0, 2)$$
$$V_4 = (1, 1, 1) \quad V_5 = (0, 1, 2) \quad V_6 = (2, 1, 0)$$
$$V_7 = (2, 2, 2) \quad V_8 = (1, 2, 0) \quad V_9 = (0, 2, 1).$$

Moreover, sets of secret keys for the users are the following.

$$\mathcal{K}_1 = \{k_0^{(1)}, k_0^{(2)}, k_0^{(3)}\} \quad \mathcal{K}_2 = \{k_2^{(1)}, k_0^{(2)}, k_1^{(3)}\} \quad \mathcal{K}_3 = \{k_1^{(1)}, k_0^{(2)}, k_2^{(3)}\}$$
$$\mathcal{K}_4 = \{k_1^{(1)}, k_1^{(2)}, k_1^{(3)}\} \quad \mathcal{K}_5 = \{k_0^{(1)}, k_1^{(2)}, k_2^{(3)}\} \quad \mathcal{K}_6 = \{k_2^{(1)}, k_1^{(2)}, k_0^{(3)}\}$$
$$\mathcal{K}_7 = \{k_2^{(1)}, k_2^{(2)}, k_2^{(3)}\} \quad \mathcal{K}_8 = \{k_1^{(1)}, k_2^{(2)}, k_0^{(3)}\} \quad \mathcal{K}_9 = \{k_0^{(1)}, k_2^{(2)}, k_1^{(3)}\}.$$

Figure 4 is a tree structure for the example above. Note that entries of vectors are mapped to unique integers, that is an entry $a^{(l)}$ is mapped to an integer $a + (l - 1)d + 1$.

The above polynomial based construction can be generalised to constructions from error-correcting codes by interpreting Reed-Solomon codes as polynomials over finite fields.

4   Extensions

In this section, we propose extensions to the proposed system with the aim of making it more practical. A similar approach has been previously used by [1].

4.1   Temporary and Permanent Revocation

User revocation to form new groups, as described above, is on a temporary basis. That is a revoked user in one session may become an authorised user in the next session. However it might be necessary to permanently revoke a user, for example when his keys are compromised. In this case it is necessary to update the system keys.

Permanent revocation needs the assistance of the group controller GC. Below we show a method to revoke users permanently. Suppose users in $\mathcal{U}_R$ must be permanently revoked. That is the system keys must be updated such that the key information known to the revoked group has to be changed. In the following, we show how GC can update the keys.

1. GC uses the temporary revocation algorithm to revoke the users in $\mathcal{U}_R$ and obtains a group key GK.
2. GC and all users in $\mathcal{U}_L$ update their secret keys as $k_i' = k_i \times GK \bmod q$.
3. GC replaces public value g with $g' = g^{\frac{1}{GK}} \bmod p$.

Using this method, the system's public keys remain unchanged as $y_i = (g')^{k_i} \bmod p$.

It is possible to generate the system keys using a pseudo-random function f_k [9], in which case GC only needs to hold a single secret key k which is the index to a pseudo-random function family. The system's secret keys are obtained as $k_i = f_k(i) \bmod q$. In this case GC computes $k'_i = f_k(i) \times GK \bmod q$. For every permanent revocation, GC only needs to update GK without the need to change $f_k(i)$.

4.2   Authentication

In multicast applications, authenticity of data and the sender is very important. Below we describe an adaptation of a technique in [16] that can be used to prevent modifying or forging multicasted data, and to identify the sender in a dynamic GI scheme.

Suppose a sender U_j wants to multicast a message msg.

1. U_j uses the secret key, k_a, corresponding to his leaf node as his identity id, that is $id = k_a$. He then generates a random number r over $GF(q)$ and computes $g^r \bmod p$.
2. U_j calculates $hash = f_{owh}(msg \parallel j \parallel g^r) \bmod q$, where f_{owh} is a publicly known one way hash function, and calculates a signature $sign = (-hash \times r) + id \bmod q$.
3. U_j multicasts the signed data $M = msg \parallel j \parallel g^r \parallel sign$.

Receivers verify as follows.

1. Compute $hash' = f_{owh}(msg \parallel j \parallel g^r) \bmod q$ from M and assign $id' = y_a$, where y_a is the public key of k_a.
2. Check $id' \stackrel{?}{=} g^{sign} \times (g^r)^{hash'} \bmod p$. If they are equal, then receivers can be sure of the integrity of the data and authenticity of the sender. Otherwise, either data or the sender, or both, are tampered with.

The above authentication system can be used to authenticate multicasted data $Y = g^r \bmod p$ and $Y_c \parallel c, \forall c \in \mathcal{I}$, sent during the revocation stage, as described below.

Let U_z be the GI. GI computes $msg = Y \parallel Y_c \parallel c, \forall c \in \mathcal{I}$ (we combine all data into one message). Then GI calculates $hash = f_{owh}(msg \parallel z)$ and $sign = (-hash \times r) + id$ over $GF(q)$. Finally GI multicasts $M = msg \parallel z \parallel sign$. Receivers will calculate $hash'$ and id', and check $id' \stackrel{?}{=} g^{sign} \times Y^{hash'} \bmod p$.

The system described above is secure for the basic scheme. This is because the secret key of the user's leaf node k_a is unique. The system will not work if the variant described in section 3 is used for the system setup as in this case k_a is known by more than one user. However using $\sum_{b \in \mathcal{N}_{U_j}} k_b \bmod q$ for his identity, will give a unique identity to the sender U_j. During verification, the receivers compute $id' = \prod_{b \in \mathcal{N}_{U_j}} y_b \bmod p$ from public keys.

Table 1. Performance comparison for dynamic GIschemes

	Current Scheme	Our Schemes	
	[1]	Basic	Variant
Storage			
GC	1 polynomial	$\frac{dn-1}{d-1}$	$dt_{max}h$
User	1	$h+1$	$t_{max}h$
Bandwidth			
Revoke $t = 1$	t_{max}	$(d-1)h-1$	$< (d-1)t_{max}h$
$t > 1$ Max	t_{max}	$(1-\frac{1}{d})n-1$	$< (d-1)t_{max}h$
Min	t_{max}	zero	$< (d-1)t_{max}h$
Resistance	t_{max}	$n-1$	t_{max}
Public Keys	$n + t_{max}$	$\frac{dn-1}{d-1}$	$dt_{max}h$

$h = \log_d n$, where d is tree degree, n is total users.

5   Performance Evaluation and Comparison

In this section we evaluate and compare costs of rekeying (revocation) in different schemes including the schemes proposed in this paper. We consider the following parameters: bandwidth, GC storage, user storage, level of collusion resistance, and the number of public keys. We assume all system keys are of the same size, although different for public and secret case, and use the key length as the unit to measure all the above parameters, except collusion resistance which is measured in terms of the number of users. For schemes based on tree structure, the required bandwidth for multiple user revocation depends on the number and the position of the revoked users. So, we consider the maximum and minimum bandwidth cost.

5.1   Dynamic GI Scheme

The scheme in [1] can only be used for revocation of up to t_{max} users. Our proposed basic scheme does not have this limitation. Our scheme has a slightly higher user storage but it is still in the order of $\log_d n$ which is acceptable for most applications. The bandwidth requirement of our scheme depends on the number of revoked users t and their positions on the tree varying between zero and $\frac{n}{2} - 1$ keys for $d = 2$. The scheme in [1] requires a constant bandwidth t_{max} keys. The scheme is not efficient when t_{max} is large since for any $t \leq t_{max}$ users, the required bandwidth is t_{max} keys. GC storage and the required public keys in our scheme and its variant are comparable to those of the scheme in [1] since the polynomial is determined by t_{max} coefficients.

5.2   Static GI Scheme

GC storage in the proposed $n - 1$-resistance scheme is as efficiency as other $(n - 1)$-resistance schemes. The scheme in [6] requires $2h$ keys for GC storage,

Table 2. Performance comparison for static GI schemes

	Current Schemes			Our Schemes	
	[20,21]	[15]* [4,5]	[6]*	Basic	Variant
Storage					
GC	$\frac{dn-1}{d-1}$	$\frac{dn-1}{d-1}$	$2h$	$\frac{dn-1}{d-1}$	$dt_{max}h$
User	$h+1$	$h+1$	h	$h+1$	$t_{max}h$
Bandwidth					
Revoke $t=1$	dh	$(d-1)h$	$(d-1)h$	$(d-1)h-1$	$< (d-1)t_{max}h$
$t>1$ Max	n	n	$\frac{n}{2}$	$(1-\frac{1}{d})n-1$	$< (d-1)t_{max}h$
Min	zero	zero	zero	zero	$< (d-1)t_{max}h$
Resistance	$n-1$	$n-1$	1	$n-1$	t_{max}

$h = \log_d n$, where d is tree degree, n is total users.
* This scheme is only for binary tree, i.e., $d = 2$.

however, it only provides 1-resistance. The user storage for all schemes are the same.

The bandwidth of revocation for one user in our scheme is slightly less than that of other existing schemes, with the schemes in [20,21] being the most expensive ones. For revocation of t, $t > 1$, users, the worst case is when $t = d^{h-1}$ and the leaves associated with the revoked users have different parents. In this case the required bandwidth is maximum. The best case is when the first common ancestor (the ancestor that is highest in the tree) of the leaves associated with the $|\mathcal{U}_L| = d^a, 1 \leq a \leq h - 1$, remaining users is not an ancestor of the revoked users. In this case the required bandwidth is minimum. The maximum required bandwidth in [20,21,15,4,5] cannot be lower than n. Our scheme has lower cost and the lowest cost is when $d = 2$. All schemes have the same minimum cost. Also, we note that while our schemes support temporary and permanent removal, current schemes only provide permanent removal.

6   Conclusion

We considered the problem of user revocation in multicast environment, with dynamic GI and static GI. We proposed a construction for dynamic GI and proved its security. We showed that the scheme can be easily modified to a static GI scheme, while maintaining its security. We also proposed extensions to this basic scheme to reduce the number of system keys, and then showed how users can be permanently removed from the group. We also described a method of adding authentication to the scheme. We compared the parameters of our schemes with those of existing schemes and showed their superior performance.

References

1. J. Anzai, N. Matsuzaki and T. Matsumoto. A Quick Group Key Distribution Scheme with "Entity Revocation". *Advances in Cryptology – ASIACRYPT '99, Lecture Notes in Computer Science 1716*, pages 333-347, 1999.

2. C. Blundo, A. De Santis, A. Herzberg, S. Kutten, U. Vaccaro and M. Yung. Perfectly Secure Key Distribution for Dynamic Conferences. *Advances in Cryptology – CRYPTO'92, Lecture Notes in Computer Science 740*, pages 471-486, 1993.
3. M. Burmester and Y. Desmedt. A Secure and Efficient Conference Key Distribution System. *Advances in Cryptology – EUROCRYPT '94, Lecture Notes in Computer Science 950*, pages 275-286, 1995.
4. R. Canetti, J. Garay, G. Itkis, D. Micciancio, M. Naor and B. Pinkas. Issues in Multicast Security: A Taxonomy and Efficient Constructions. *Proceedings of INFOCOM '99*, pages 708-716, 1999.
5. R. Canetti, T. Malkin and K. Nissim. Efficient Communication-Storage Tradeoffs for Multicast Encryption. *Advances in Cryptology – EUROCRYPT '99, Lecture Notes in Computer Science 1592*, pages 459-474, 1999.
6. I. Chang, R. Engel, D. Kandlur, D. Pendarakis and D. Saha. Key Management for Secure Internet Multicast Using Boolean Function minimisation Techniques. *Proceedings of INFOCOM '99*, pages 689-698, 1999.
7. W. Diffie and M. Hellman. New Directions in Cryptography. *IEEE Trans. Inform. Theory* **22**, pages 644-654, 1976.
8. A. Fiat and M. Naor. Broadcast Encryption. *Advances in Cryptology – CRYPTO '93, Lecture Notes in Computer Science 773*, pages 480-491, 1994.
9. O. Goldreich, S. Goldwasser and S. Micali. How to Construct Random Functions. *JACM*, Vol. 33, No. 4, pages 792-807, 1986.
10. Y. Kim, A. Perrig and G. Tsudik. Simple and Fault-Tolerance Key Agreement for Dynamic Collaborative Groups. *Proceedings of CCS '00*, pages 235-244, 2000.
11. P. S. Kruus. A Survey of Multicast Security Issues and Architectures. *21st National Information Systems Security Conferences*, 1998.
12. P. S. Kruus and J. P. Macker. Techniques and Issues in Multicast Security. *MILCOM '98*, 1998.
13. R. Kumar, S. Rajagopalan and A. Sahai. Coding Constructions for Blacklisting Problems Without Computational Assumptions. *Advances in Cryptology – CRYPTO '99, Lecture Notes in Computer Science 1666*, pages 609-623, 1999.
14. H. Kurnio, R. Safavi-Naini, W. Susilo and H. Wang. Key Management for Secure Multicast with Dynamic Controller. *Information Security and Privacy – ACISP 2000, Lecture Notes in Computer Science 1841*, pages 178-190, 2000.
15. D. A. McGrew and A. T. Sherman. Key Establishment in Large Dynamic Groups Using One-Way Function Trees. *Manuscript*, 1998.
16. K. Nyberg, R.A. Rueppel. Message Recovery for Signature Schemes Based on the Discrete Logarithm Problem. *Advances in Cryptology – EUROCRYPT '94, Lecture Notes in Computer Science 950*, pages 182-193, 1995.
17. R. Safavi-Naini and H. Wang. New Constructions of secure multicast re-keying schemes using perfect hash families. *7th ACM Conference on Computer and Communication Security*, ACM Press, 2000, 228-234.
18. A. Shamir. How to Share a Secret. *Communications of the ACM* **22**, pages 612-613, 1979.
19. D. R. Stinson. On Some Methods for Unconditionally Secure Key Distribution and Broadcast Encryption. *Designs, Codes and Cryptography* **12**, pages 215-243, 1997.
20. D. M. Wallner, E. J. Harder and R. C. Agee. Key Management for Multicast: Issues and Architectures. *Internet Draft (draft-wallner-key-arch-01.txt)*, ftp://ftp.ietf.org/internet-drafts/draft-wallner-key-arch-01.txt.
21. C. K. Wong, M. Gouda and S. S. Lam. Secure Group Communication Using Key Graphs. *Proceedings of SIGCOMM '98*, pages 68-79, 1998.

A   Security Proof for Multiple Rounds

For simplicity, we assume that the scheme is run twice (two rounds) for the same $\mathcal{M}$ and we show that an adversary who can collude with the users in $\mathcal{U}_R$, after seeing all the broadcast (and even the group key for the first round) is not able to compute the group key of the second round. We may further assume that both $\mathcal{M}$ and $\mathcal{I}$ are the same for two rounds and so the polynomial f will be the same. The only different values in the two runs are the random values r_1 and r_2, respectively. We will again employ the "reducibility arguments" for the proof. Assume that G be a probabilistic polynomial-time algorithm that on input of $g^{r_1}, g^{r_2}, g^{r_1 f(c)}, g^{r_2 f(c)}, \forall c \in \mathcal{I}, g^{f(a)}, \forall a \in \mathcal{M}$ and $g^{r_1 f(0)}$ it outputs $g^{r_2 f(0)}$ with a non-negligible probability. We show that we can use G to solve the DH problem. Let g^{x_1}, g^{x_2} be two elements with x_1, x_2 randomly chosen from $GF(q)$. As before, we choose $s - 1$ random elements $a_1, \ldots, a_{s-1} \in GF(p)$. There exists a unique polynomial $h(x)$ of degree at most $s - 1$ such that $h(c_i) = a_i, \forall i, 1 \le i \le s - 1$, and $g^{h(0)} = g^{x_2}$. We also randomly choose r_1 and compute $g^{r_1 h(0)}$. We can feed G with the following data: (1) $g^{r_1}, g^{r_1 h(c)}, \forall c \in \mathcal{I}, g^{h(a)}, \forall a \in \mathcal{M}$ and $g^{r_1 h(0)}$ (i.e., all the information obtained by the adversary from the first round); (2) $g^{x_1}, (g^{x_1})^{a_i}, i = 1, \ldots, s - 1$ (the public information from the second round). By the assumption of G, it outputs $g^{x_1 h(0)} = g^{x_1 x_2}$, which shows that G can solve the DH problem and we obtain a contradiction, and therefore the desired result follows.

Binary Codes
for Collusion-Secure Fingerprinting

Gérard Cohen[1], Simon Litsyn[2], and Gilles Zémor[1]

[1] École nationale supérieure des télécommunications,
Dept. of Computer Science and Networks,
46 rue Barrault, 75 634 Paris 13, France
{cohen,zemor}@infres.enst.fr
[2] Tel-Aviv University, Dept. of Electrical Engineering-Systems, Ramat Aviv 69978
litsyn@eng.tau.ac.il

Abstract. We examine the problem of Collusion-Secure Fingerprinting in the case when marks are binary and coalitions are of size 2. We are motivated by two considerations, the pirates' probablity of success (which must be non-zero, as was shown by Boneh and Shaw) on one hand, and decoding complexity on the other. We show how to minimize the pirates' probability of success: but the associated decoding complexity is $O(M^2)$, where M is the number of users. Next we analyze the Boneh and Shaw replication strategy which features a higher probability of success for the pirates but a lower decoding complexity. There are two variations. In the case when the fingerprinting code is linear we show that the best codes are *linear intersecting codes* and that the decoding complexity drops to $O(\log^2 M)$. In the case when the fingerprinting code is allowed to be non-linear, finding the best code amounts to finding the largest B_2-*sequence* of binary vectors, an old combinatorial problem. In that case decoding complexity is intermediate, namely $O(M)$.

Keywords: Fingerprinting, watermarking, Collusion-resistant code, intersecting code, B_2-sequence.

1 Introduction

Suppose a *Distributor* wishes to create and distribute a large number of copies of a large binary file $\Phi \in \{0,1\}^N$. In order to trace illegal copies he will *mark* each copy of Φ. The marking process of some copy of Φ consists of changing the bits of Φ belonging to some subset of a privileged set $L \subset \{1, \ldots N\}$ of coordinates called *marks*. The subset of marks associated to a copy of Φ is called a *fingerprint* and can be seen as a binary vector of length $\ell = |L|$. The set of marks L is supposed to be unknown to anyone but the distributor. Furthermore, the set of marks is usually supposed to be a small subset of $\{1, \ldots N\}$, so that modifying a fingerprint by randomly changing bits of a copy of Φ implies changing many bits of the original file and damaging the data significantly.

The problem of *collusion* occurs when a coalition of c pirate users compare their fingerprinted copies: whenever their set of copies differ on some coordinate

K. Kim (Ed.): ICICS 2001, LNCS 2288, pp. 178–185, 2002.
© Springer-Verlag Berlin Heidelberg 2002

they will know it is a mark. They can then produce an illegal copy by changing at will bits on the subset of marks they have found out. Following previous work, we shall always suppose that they cannot access the other marks.

In this paper we shall concern ourselves with the case when the size of the coalition is $c = 2$. To state the problem more formally, let Q be an alphabet with q elements and let $x, y \in Q^{\ell}$. The *envelope* of x and y is the set $e(x, y)$ of vectors $z \in Q^{\ell}$ such that $z_i \in \{x_i, y_i\}$ for $i = 1 \ldots \ell$. We see that when $q = 2$ and $Q = \{0, 1\}$, and when x and y are two fingerprints, the envelope $e(x, y)$ is exactly the set of corrupted fingerprints that can be produced from the copies identified by x and y.

The distributor would like to use a set of binary fingerprints, or code $\mathcal{C}$, with the property that any corrupted fingerprint of any envelope $e(x, y)$ always identifies unambiguously either x or y. Unfortunately, as was pointed out by Boneh and Shaw [2], such sets of fingerprints do not exist for $|\mathcal{C}| > 2$.

The problem of devising codes such that any $s \in e(x, y)$ identifies at least one of its "parents" x or y is therefore, strictly speaking, meaningful only when the alphabet size q is at least 3. This problem was taken up by Hollman et al. in [6], and for larger c in [1,14,15,16].

In the binary case, one must fall back on fingerprinting schemes that *almost* always identify collusions of two (or more) users. This is the approach of Boneh and Shaw which we pursue here by devising improved coding schemes for the case $c = 2$.

In section 2 we address the problem of finding identifying codes with the best possible rate. We shall see that *linear 3-wise intersecting codes* are near-optimal candidates. However there does not seem to be any associated decoding scheme, short of exhaustive search, to recover the "guilty" codewords from a corrupted fingerprint. Therefore in the following sections we are willing to trade coding efficiency for decoding complexity. In section 3 we examine a replication strategy introduced by Boneh and Shaw: we shall see that linear codes can be decoded very efficiently when replication is used, and that in this case the linear codes needed for identification are exactly *linear intersecting codes*. Finally, we drop linearity in section 5. When replication is used, we shall see that the property we need for identification is exactly that of long-studied combinatorial objects called B_2-sequences of vectors. If the best-known B_2-sequences are used, decoding complexity is intermediate.

2   3-Wise Intersecting Codes

Let $s \in \{0, 1\}^{\ell}$, and define G_s to be the set of pairs $\{x, y\} \subset \mathcal{C}$ such that $s \in e(x, y)$. It is convenient to view G_s as the set of edges of a graph. Following the terminology of [6], let us say that $s \in \{0, 1\}^{\ell}$ has an *identifiable parent* if the graph G_s is a *star*, i.e. if there exists a vertex x incident to every edge of G_s. Equivalently, this means that at least one of the members of the guilty coalition that produced s can be identified unambiguously.

We shall need the following definition:

Definition 1. *The code C is said to be $(2,2)$-separating if, for any 4-tuple of codewords $(a, b, c, d) \in C^4$, there exists a coordinate i where either $(a_i, b_i, c_i, d_i) = (1, 1, 0, 0)$ or $(a_i, b_i, c_i, d_i) = (0, 0, 1, 1)$.*

This property, originally introduced in [11], is relevant to idenfication because of the following:

Proposition 1. *The code C is $(2,2)$-separating if and only if, for any $s \in \{0, 1\}^\ell$, the graph G_s does not contain any two disjoint edges.*

Proof. If the 4-tuple (a, b, c, d) cannot be separated then s can be constructed such that for all i, $s_i \in \{a_i, b_i\}$ and $s_i \in \{c_i, d_i\}$: we then have $s \in e(a, b)$ and $s \in e(c, d)$. Conversely, if, for some i, $(a_i, b_i, c_i, d_i) = (1, 1, 0, 0)$, then for all s either $s \notin e(a, b)$ or $s \notin e(c, d)$. □

In a way, this is the best we can get, because there is just one graph, any two edges of which intersect, and which is not a star: the triangle. The triangle cannot be forbidden because, whatever the code C, it is always possible for any two pirates with codewords x and y to create a hybrid s such that G_s contains a triangle. Indeed, let z be a third codeword and let s be such that for every i, s_i is the binary symbol that occurs the most often among x_i, y_i, z_i. Then the triangle made up of $\{x, y\}$, $\{y, z\}$, $\{z, x\}$ is in G_s.

Probability of hitting a triangle. Let d be the Hamming distance between x and y. Since the pirates do not know the original file Φ, the only thing they have identified is the set $supp(x + y)$ of coordinates where x and y differ. They can therefore do little more than choose at random one of the 2^d hybrids. They will therefore succeed if they manage to choose an s which coincides with a codeword z on $supp(x + y)$: this can happen with probability at most $|C|/2^d$.

To get a good identifying code C we need therefore only choose a $(2,2)$ separating code with a large minimum distance. One way of doing this is by choosing a linear code.

Definition 2. *A binary linear code is r-wise intersecting ([5]) if the supports of any r linearly independent codewords intersect. For $r = 2$, we simply say that the code is intersecting.*

Proposition 2. *A linear binary code is 3-wise intersecting if and only if it is $(2,2)$-separating.*

Proof. Let C be a 3-wise intersecting code. By linearity, it is enough to prove that the code separates all pairs of non intersecting coalitions containing $\mathbf{0}$. Consider two such coalitions $\Xi = \{a, b\}$, $\Xi' = \{\mathbf{0}, c\}$. We deal with two cases, according to the rank of $\{a, b, c\}$

If the rank is 2, then any coordinate in $a \setminus c$ separates. (Note that such a coordinate i exists since the code is intersecting). Indeed, since $a + b + c = \mathbf{0}$, we get $a_i = b_i = 1$ and $(a, b, \mathbf{0}, c)_i = (1100)$.

If $\{a, b, c\}$ are independent, so are $\{a + c, a, b\}$; then the 3-wise intersection yields a coordinate where $a_i = b_i = (a + c)_i = 1$ and the required separation.

Conversely, assume that $\mathcal{C}$ is $(2, 2)$-separating. Consider any three independent codewords a, b, c. By hypothesis, since $a + c \neq b$, the two pairs $\{0, a + c\}$ and $\{a, b\}$ are separated on some coordinate i. Thus $(0, a + c, a, b)_i = (0011)$ and $a_i = b_i = c_i = 1$. $\qquad\square$

The largest known dimension k of 3-wise intersecting codes is asymptotically

$$k = \left(1 - \frac{1}{3} \log_2 7 + o(1)\right) \ell \approx 0.064\ell \qquad (1)$$

see [5] or [12]. Furthermore, any 3-wise intersecting code must have a minimum distance d such that $d/\ell > k/\ell + \lambda$ for a fixed $\lambda > 0$. Therefore, 3-wise intersecting codes can accomodate a number $M = |\mathcal{C}|$ of users exponential in the codelength ℓ with an exponentially small probability $2^{-\lambda\ell}$ for pirate coalitions to escape identification.

Remark. When the 3-wise intersecting codes of asymptotic dimension given by (1) are used, it can be shown by applying the linear programming bound of coding theory [10] that $\lambda \geq 0.16 + o(1)$.

The problem, however, with the use of 3-wise identifying codes is the lack of any apparent decoding algorithm, short of exhaustive search through all possible $\binom{M}{2}$ coalitions of 2 among M users. The coding strategy we will now look at deteriorates the maximum number of users to just below exponential in ℓ, but admits greatly simplified decoding schemes.

3   The Boneh and Shaw Replication Strategy

Boneh and Shaw use the following *replication* trick: the set of fingerprints $\mathcal{C} \subset \{0, 1\}^\ell$ is actually constructed from a code $C \subset \{0, 1\}^n$ where $\ell = tn$. The code C is therefore of a smaller length than ℓ and t is the *replication parameter*. A fingerprint $X \in \mathcal{F}$ is constructed from a codeword $x \in C$ simply by replicating each symbol t times, i.e. changing 0 to $00 \cdots 0$ and 1 to $11 \cdots 1$. Let us call the set of t coordinate positions of X that stem from a single coordinate of x a *block*. The point of the replication strategy is that the partition of the set L of marks into blocks is kept secret by the distributor. In other words, when two pirate users compare their fingerprinted copies corresponding to fingerprints X and Y, they will have no way of deciding whether two uncovered marks belong to the same block or not. The consequence is that whenever the pirates decide to change a fraction p of the set of uncovered marks, they will, on average, change a fraction p of the marks belonging to any single block.

3.1   Pirate Strategies and Decoding

Suppose two colluding pirates are in possession of two legal copies fingerprinted by X and Y and that X and Y originate from $x, y \in C$. We see that, because they have no way of differentiating between the set of uncovered marks, the pirates have essentially the following type of strategy: they can pick one of the copies, fingerprinted by X say, and change randomly and independently with probability p every coordinate of the set of uncovered marks. Their only degree of freedom is the choice of p.

On the other hand, when confronted with this illegal copy of Φ the distributor will try to trace one of the legal copies it was constructed from, i.e. reconstruct x or y. The distributor has two strategies:

1. He can associate to every block a binary symbol by majority decision. If there are more "0" bits than "1" bits in block j decide $z_j = 0$: otherwise decide $z_j = 1$. If p is close to 0, we see that with high probability (depending on the size of the replication parameter t) $z_j = x_j$. If p is close to 1 we will have $z_j = y_j$ with high probability. In those cases, if t is big enough ($t > \log n$), the vector $z \in \{0, 1\}^n$ thus constructed will simply be equal to x or to y, and the pirates are foiled. However, if $p \approx 1/2$ we may have alternatively $z_j = x_j$ or $z_j = y_j$ and the resulting scrambled vector z may be undecipherable.

2. The alternative strategy is to associate to any corrupted block which contains both zeros and ones a third *erased* symbol, say ε. This strategy yields a ternary vector $\zeta \in \{0, 1, \varepsilon\}^n$.

What replication ensures is that the second strategy need be applied only when the first has failed to produce a legitimate codeword $z \in C$. With high probability this will happen only when the pirates have chosen p sufficiently separated from 0 and 1. In that case the second strategy will yield with high probability the ternary vector

$$\zeta = \zeta(x, y)$$

defined by $\zeta_j = x_j = y_j$ when $x_j = y_j$ and by $\zeta_j = \varepsilon$ when $x_j \neq y_j$.

If the code C has the property that $\zeta(x, y)$ always identifies $\{x, y\}$ for any pair of codewords $\{x, y\}$, then, for any sufficiently big replication parameter t, (this means that t should be sufficiently larger than $\log n$) one of the two decoding strategies will almost always identify x or y.

Let us say that code C is a ζ-*code* if for any two distinct pairs of codewords $\{x, y\}$ and $\{x', y'\}$, we have $\zeta(x, y) \neq \zeta(x', y')$. Note that in this case, the second (erasure) decoding strategy has the advantage, when it succeeds, of always identifying *both* pirates x and y, while the first decoding strategy will identify only one of the two.

The problem of devising fingerprinting schemes that resist coalitions of two pirate users is now reduced to finding ζ-codes. Because the distributor wants to accomodate the largest possible number of users we want to know what is the largest possible ζ-code of length n.

In the sequel it will be implicit that the distributor uses a replicated code expanded from a ζ-code.

3.2   Optimal Strategy for the Pirates

In light of the preceding discussion, the optimal stategy for the pirates is to choose the parameter p which equates probabilities of success of the distributor with both strategies.

With the first one (majority decision), the probability of non-detection of a corrupted symbol with a replication factor of t is:

$$P = \Sigma_{i \geq t/2} \binom{t}{i} p^i (1-p)^{t-i} \approx \binom{t}{t/2} (p(1-p))^{t/2} \approx (4p(1-p))^{t/2}.$$

With the second strategy (erasure), the probability of non-detection of a given erased symbol is

$$P' = \max\{p^t, (1-p)^t\}.$$

Setting $P = P'$ gives $4p(1-p) = (1-p)^2$ or $4p(1-p) = p^2$, which gives the non-intuitive $p = 1/5$ or $p = 4/5$ as the optimal strategy for the pirates. In this case the probability that the pirates succeed in foiling both decoding strategies is upper bounded by

$$n \left(\frac{4}{5} \right)^t.$$

4   Linear ζ-Codes and Intersecting Codes

Suppose now that a *linear* ζ-code is used. What is the decoding complexity ? For the first strategy, decoding amounts to a majority decision on every block and is trivial. The second decoding strategy really amounts to decoding erasures by solving a system of linear equations. Specifically, the distributor is faced with $\zeta \in \{0, 1, \varepsilon\}^n$ and looks for $\{x, y\}$ such that $\zeta = \zeta(x, y)$. This means finding (x, y) such that:

$$x_j = y_j = 0 \text{ if } \zeta_j = 0 \tag{2}$$
$$x_j = y_j = 1 \text{ if } \zeta_j = 1 \tag{3}$$
$$x_j + y_j = 1 \text{ if } \zeta_j = \varepsilon \tag{4}$$

Suppose the linear code C is generated by the rows of a $k \times n$ generator matrix (g_{ij}), so that $x = (x_j)$ and $y = (y_j)$ are specified by $\alpha = (\alpha_i)_{i=1...k}$ and $\beta = (\beta_i)_{i=1...k}$ where

$$x_j = \sum_{i=1}^{k} \alpha_i g_{ij} \qquad y_j = \sum_{i=1}^{k} \beta_i g_{ij}. \tag{5}$$

We see that finding all possible α and β amounts to solving the linear system in $2k$ variables $(\alpha_i, \beta_i)_{i=1...k}$ and $\#\{j, \zeta_j = \varepsilon\} + 2\#\{j, \zeta_j = 0\} + 2\#\{j, \zeta_j = 1\}$ equations given by $(2), (3), (4)$ and (5). That C is a ζ-code means that the affine space of solutions is always of dimension 1 and that the corresponding two solutions (x, y) and (y, x) determine completly the "guilty" pair $\{x, y\}$.

Curiously, *linear* ζ-codes turn out to be exactly the same objects as linear *intersecting* codes (see definition 2). We now prove this.

Proposition 3. *A linear code is intersecting if and only if it is a ζ-code.*

Proof. If C is not intersecting, there exist two codewords, say c, c', with empty support intersection; let $c'' = c + c'$, we get $\zeta(c, c') = \zeta(c'', \mathbf{0})$. Thus C is not a ζ-code.

Conversely, assume that C is not a ζ-code. Consider four codewords a, b, a', b' such that $\zeta(a, b) = \zeta(a', b')$. It is easy to check that the codewords $a + a'$ and $b + a'$ have non-intersecting supports so that C is not intersecting. □

Binary intersecting codes have been studied in different contexts, with a variety of applications [3,5,7,9,13], and this is a new one. The largest-known dimension of an intersecting code of length n is asymptotically

$$k = (1 - \frac{1}{2} \log_2 3 + o(1))n \approx 0.207n.$$

The preceding discussion shows that if a linear intersecting code is replicated and used for identification, decoding can be accomplished with a complexity in $O(n^2)$, i.e. $O(\log^2 M)$ where M is the number of users.

5 ζ-Codes and B_2-Sequences

In this section we drop linearity. Let x and y be two codewords of a code $C \subset \{0,1\}^n$. Notice that $\zeta(x, y)$ is obtained from the real sum $x + y$ by changing every 2 coordinate into a 1, every 1 coordinate into ε and leaving 0's unchanged. We see therefore that C is a ζ-code if and only if the real sums $x + y$ are all different for every pair $\{x, y\}$ of codewords. Such a code is an old combinatorial object, known as a B_2-*sequence* of binary vectors and was investigated by Lindström in [8].

Let $|C_n|$ be the maximal size of a ζ-code of length n. Denote the *maximum possible rate* R_ζ of ζ-codes by

$$R_\zeta = \limsup_{n \to \infty} \frac{1}{n} \log_2 |C_n|.$$

For a long time, the best known bounds on R were due to B. Lindström [8] and read:

$$0.5 \leq R_\zeta \leq 0.6 \tag{6}$$

The lower bound is obtained in the following way. For any even $n = 2m$ pick the set of $2^m - 1$ columns that make up the parity-check matrix of a double-error correcting primitive BCH code. As is easy to check, for an arbitrary choice of four codewords there exists at least one coordinate in which their (modulo-two, *a fortiori* real) sum is odd, hence all the corresponding pairwise vector sums are different.

The upper bound was recently improved in [4] to:

Theorem 1. *The maximum possible rate R of ζ-codes satisfies*

$$R_\zeta \leq 0.5752....$$

Decoding. Suppose one of the best known ζ-codes is used, i.e. is made up of the set of columns of 2-error correcting BCH code. When faced with the ternary vector $\zeta = \zeta(x, y)$, the distributor can create the binary vector $\sigma \in \{0, 1\}^n$ such that $\sigma_j = 1$ if $\zeta_j = \varepsilon$ and $\sigma_j = 0$ otherwise. Clearly, $\sigma = x + y \bmod 2$, and recovering $\{x, y\}$ from σ is exactly the syndrome-decoding problem which can be accomplished for the BCH code with a complexity linear in its length (e.g. with the Berlekamp-Massey algorithm). The length of the BCH code is the number M of codewords of the ζ-code C so that the decoding complexity of the latter is $O(M)$.

References

1. Barg, A., Cohen, G., Encheva, S., Kabatiansky, G., Zémor, G.: A hypergraph approach to the identifying parent property. SIAM J. on Discrete Math. **14** (2001) 423–431
2. Boneh, D., Shaw, J.: Collusion-Secure Fingerprinting for Digital Data. IEEE Trans. Inf. Theory. **IT-44** (1998) 1897–1905 (preliminary version in Crypto'95).
3. Cohen, G., Lempel, A.: Linear intersecting codes. Discrete Math. **56** (1985) 35–43
4. Cohen, G., Litsyn, S., Zémor, G.: Binary B_2-Sequences: a new upper bound. J. Combinatorial Theory, Series A. **94** (2001) 152–155
5. Cohen, G., Zémor, G.: Intersecting codes and independent families. IEEE Trans. Inf. Theory. **IT-40** (1994) 1872–1881
6. Hollmann, H.D., van Lint, J.H., Linnartz, J.-P., Tolhuizen, L.M.: On codes with the identifiable parent property. J. Combinatorial Theory, Series A. **82** (1998) 121–133
7. Katona, G., Srivastava, J.: Minimal 2-coverings of a finite affine space based on $GF(2)$. J. Stat. Planning and Inference. **8** (1983) 375–388
8. Lindström, B.: On B_2-Sequences of Vectors. J. of Number Theory. **4** (1972) 261–265
9. Miklós, D.: Linear binary codes with intersection properties. Discrete Appl. Math. **9** (1984) 187–196
10. McEliece, R. J., Rodemich, E. R., Rumsey, H. C. Welch, L. R.: New upper bounds on the rate of a code via the Delsarte-MacWilliams inequalities. IEEE Trans. Inf. Theory. **IT-23** (1977) 157–166
11. Sagalovitch, Yu. L.: Methods of introducing redundancy for raising the reliability of a finite automaton. Problems of Information Transmission. **4** (1968) 62–72
12. Sagalovitch, Yu. L.: Separating systems. Problems of Information Transmission. **30** (1994) 105-123
13. Sloane, N. J. A.: Covering arrays and intersecting codes. J. Combinator. Designs. **1** (1993) 51–63
14. Staddon, J. N., Stinson, D. R., Wei, R.: Combinatorial properties of frameproof and traceability codes. IEEE Trans. Inf. Theory. **IT-47** (2001) 1042–1049
15. Stinson, D.R., Tran Van Trung, Wei, R.: Secure Frameproof Codes, Key Distribution Patterns, Group Testing Algorithms and Related Structures. J. Stat. Planning and Inference. **86** (2000) 595-617
16. Stinson, D.R., Wei, R.: Combinatorial properties and constructions of traceability schemes and frameproof codes. SIAM J. Discrete Math. **11** (1998) 41-53.

Copyright Protection
of Object-Oriented Software

Jarek Pastuszak[1], Darek Michałek[1], and Josef Pieprzyk[2]

[1] Systems Research Institute, Polish Academy of Sciences
Warsaw, Poland
jarek.pastuszak@bsb.com.pl
[2] Department of Computing, Macquarie University
Sydney, NSW 2109, Australia
josef@ics.mq.edu.au

Abstract. The work deals with copyright protection for software written in object-oriented language (such as C++ and Java). The model used allows to insert watermarks on three "orthogonal" levels. For the first level, watermarks are injected into objects. The second level watermarking is used to select proper variants of the source code. The third level uses transition function that can be used to generate copies with different functionalities. Generic watermarking schemes were presented and their security discussed.

1 Introduction

Copyright protection is an universally accessible right of the author to release their original work to the public restricting, however, the use of the work and keeping the ownership of the work to themselves. In effect, the authors claiming copyright to their works prevent others from commercialisation of the works, leaving this opportunity for themselves. Publishing of multimedia works on the Internet allows authors to reach the world-wide audience crossing boundaries of different legal systems. Relying on the legal protection only seems to be no longer reasonable.

Now more than ever, it is obvious that copyright protection must be supported by technology that allows the owners to prove their rights even if the works have been manipulated. Copyrighted works cover all kinds of human creativity, including music, painting, writing, etc. Note that originality of "traditional" pieces of art can be relatively easily established by testing the physical media that carry the work. For instance, the age of a canvas together with chemical analysis of paint detects less sophisticated painting forgeries.

As far as electronic media are concerned, copying of multimedia data creates identical copies of the original. The original and copies are indistinguishable. A typical solution is to introduce to multimedia documents (i.e. to their electronic versions) small inconspicuous changes. For instance, a small noise is added to music CDs, a small changes of background colours are made to video, etc. The string added to the original multimedia document is called watermark and typ-

K. Kim (Ed.): ICICS 2001, LNCS 2288, pp. 186–199, 2002.
© Springer-Verlag Berlin Heidelberg 2002

ically encodes name or identity or logo or other piece of information uniquely identifying the author or the copyright holder of the document.

Copyright protection of multimedia documents applies a variety of techniques from information hiding, cryptography, coding, and signal processing. Software can be seen as a just another kind of multimedia documents. Unfortunately, watermarking techniques developed for multimedia documents (such as voice, graphics, video, image) are not applicable for software protection. At the same time, software trade over the Internet seems to be one of the main fully electronic e-commerce applications with an enormous potential.

This work addresses software protection written in an object-oriented language (such as C++ or Java).

The remainder of the paper is structured as follows. In Section 2 we briefly overview the literature. Section 3 contains the model and specifies four basic algorithms necessary to handle watermarks. Security evaluation of the scheme is discussed in Section 4. Generic schemes are described in Section 5. Experiments which have been done are briefly summarised in Section 6. Section 7 concludes the work.

2   Literature Review

In recent years, research effort has been concentrated mainly on multimedia watermarking. Geometric distortions (spatial domain) and their applicability for watermarking were discussed in [8,12]. An alternative approach based on signal distortion (frequency domain or spread spectrum) was discussed in numerous papers including [2,7,11]. Watermarking of textual documents is investigated in [3,5,13]. Most promising watermarking techniques seem to be based on geometric distortions. The text is simply converted into an image. The image now can be seen as a collection of lines, words or characters. Each of these objects can be spatially manipulated [3,4]. For instance, a line can be positioned on three different levels: up, down or can stay unchanged. Words can be shifted horizontally and characters can be distorted by shortening or lengthening their parts.

Classical techniques for software protection are presented in [9]. A watermark can be inserted into the source or executable code. The document is, therefore, treated as a passive entity. Watermarks, which can be hidden inside program's data structure (static watermarks) or built up in real time by the program during its execution (dynamic watermarks), were studies in [6]. The software can also be divided into two functionally dependent modules. One module contains the bulk of the code while the second module is placed into a tamper resistant device. The software works correctly only if two matching modules are used together (see [1]). This idea can be extended when the tamper resistant device is accessible via the Internet to solve the problem of charging for software usage.

3   Model

Object oriented programming encapsulates objects that apart from necessary data structure, possess also a well defined collection of methods defining suitable functionalities. Note that the inheritance mechanism allows to create a hierarchy of objects with modified functionalities.

Given an object o and its functionality f. This fact is denoted as $\varphi(o) = f$ where φ is an operator that for a given object extracts its functionality. Two objects o_1, o_2 are considered mutants of each other if they are different but they share the same functionality, i.e. $o_1 \neq o_2$ and $\varphi(o_1) = \varphi(o_2)$.

The space of mutant objects over the set $\mathcal{F}_o$ is denoted by

$$\mathbf{O} = \bigcup_i O_i$$

where $O_i = \{o : \varphi(o) = f_i\}$ for $i = 1, 2, \ldots$ and $\mathcal{F}_o = \{f_1, f_2, \ldots\}$ is the set of all functionalities.

The space of all effective programs over the set of objects $\mathbf{O}$ with n elements is

$$\mathbf{S} = \{S = (o_1, \ldots, o_k) : o_i \in \mathbf{O}; i = 1, \ldots, k; k = 0, 1, \ldots, n\}.$$

The corresponding space of functionalities is $\mathbf{F} - \{F = \varphi(S) : S \in \mathbf{S}\}$. Given the set $\mathbf{W}$ of all watermarking values. The transition function $\delta : \mathbf{O} \times \mathbf{W} \to \mathbf{S}$ transforms a pair (object, watermark) into an effective program. The set of all transition functions is $\Delta = \{\delta_1, \delta_2, \ldots\}$.

Definition 1. *The functional watermarking is a six-tuple* $FW = \langle \mathcal{F}_o, \mathbf{O}, \mathbf{W}, \Delta, \mathbf{S}, \mathbf{F} \rangle$, *where* $\mathcal{F}_o$ *is the set of all functionalities of objects,* $\mathbf{O}$ *- the space of objects,* $\mathbf{W}$ *- the set of watermarks,* Δ *- the set of transition functions,* $\mathbf{S}$ *- the space of effective programs, and* $\mathbf{F}$ *- the set of functionalities of effective programs.*

To handle watermarks in copyrighted software, one would need to define the following four basic algorithms:

- G – *generation* of watermarks,
- I – *insertion* of watermarks,
- E – *extraction* of watermarks,
- R – *identification* of software.

The algorithm G can be defined as follows

$$G : \mathbf{S}_M \times \mathbf{U} \times \mathbf{U}_{special} \to \mathbf{W}$$

where $\mathbf{S}_M$ is the collection of the master programs, $\mathbf{U}$ is the set of all users to whom the software can be sold and $\mathbf{U}_{special}$ is the set of copyright holders (either the owner or the creator).

Watermarks can be inserted into

- objects (static insertion) or/and
- programs (dynamic insertion).

1. Static watermark insertion algorithm I_{static}, which consists of two steps:

   (a) $I_{static}^{(1)} : \mathbf{O} \times \mathbf{W} \to \mathbf{O}$. This is a procedure of embedding watermarks (secrets) into every mutant object. This procedure is very time consuming and it is recommended to be used very rarely (for example once a year or for a new version of the software),

   (b) $I_{static}^{(2)} : \mathbf{S}_w \times \mathbf{O} \to \mathbf{S}'_w$. In this step, we replace mutant objects by their watermarked equivalents.

2. Dynamic insertion algorithm $I_{dynamic}$, which consists of the following three steps:

   (a) $I_{dynamic}^{(1)} : \mathbf{S}'_{expected} \times \mathbf{W} \to \mathbf{S}' \subset \mathbf{S}$ (optional step). In this step we embed a watermark into an expected program (or programs). Note, that usually, the watermark is actually a fingerprint (i.e. is unique for a particular *End-User* or a particular software).

   (b) $I_{dynamic}^{(2)} : \mathbf{S}' \times \mathbf{W} \to \mathbf{O}' \subset \mathbf{O}$ (required step). In this step, a subset $\mathbf{O}'$ of the mutant objects is generated. There is an expected software hidden into this set. Next, the set is distributed to the *End-User*. Note, that in some cases, the expected software $S_{expected}$ could be a permutation of the set $\mathbf{O}'$. If $I_{dynamic}^{(2)}$ does not depend on the set of watermarks $\mathbf{W}$, then it has the form: $I_{dynamic}^{(2)} : \mathbf{S}' \to \mathbf{O}'$. In this case the embedding (or hiding) of effective programs $\mathbf{S}'$ into the set of mutant objects $\mathbf{O}'$ needs no additional information.

   (c) $I_{dynamic}^{(3)} : \mathbf{O}' \times \mathbf{W} \to \Delta' \subset \Delta$ (required step). Transition functions are generated from the given sets of mutant objects and watermarks.

As the result of executing the above algorithms, we obtain some instance of the functional watermarking, which can be defined as follows:

Definition 2. *An instance of functional watermarking is a six-tuple $IFW = \langle \mathcal{F}'_o, \mathbf{O}', \mathbf{W}', \Delta', \mathbf{S}', \mathbf{F}' \rangle$,where $\mathcal{F}'_o \subseteq \mathcal{F}_o$, $\mathbf{O}' \subseteq \mathbf{O}$, $\mathbf{W}' \subseteq \mathbf{W}$, $\Delta' \subseteq \Delta$, $\mathbf{S}' \subseteq \mathbf{S}$, $\mathbf{F}' \subseteq \mathbf{F}$ and for all $x_i \in \mathbf{X}'$, there exists $\delta(\mathbf{O}', w_i) = S' \in \mathbf{S}'$ such that $\varphi(S') \in \mathbf{F}'$.*

Now we are ready to define *watermark extraction* algorithm E (called *recogniser* or *extractor*), which reconstructs watermarks from the given instance of the functional watermarking. Note, that *recogniser* can only observe the sequence of mutant objects that is delivered to the *End-User*, and the watermarked program.

There are two types of watermark extraction algorithms.

1. *Static extractor $E_{static} : \mathbf{O}' \to \mathbf{W}_{static} \subseteq \mathbf{W}$, which recreates static watermark (embedded by watermark insertion algorithm I_1).*

2. *Dynamic extractor $E_{dynamic} : \mathbf{S}' \to \mathbf{W}_{dynamic} \subseteq \mathbf{W}$, which reconstructs dynamic watermark by scanning watermarked programs during its execution.*

Consider the *identification* algorithm R defined as

$$R : \mathbf{W}_{static} \times \mathbf{W}_{dynamic} \times \mathbf{U} \to \{accept, reject\}$$

Note, that the algorithm R is based on two different watermarks. The watermarks can be identical or alternatively one may depend on the other. Typically, it is expected that the watermarks can be generated by the copyright holders from their secret keys and verified by any body from the public information (for instance, using certificates of corresponding public keys).

Clearly, generation of watermarks influences its robustness. Generation algorithms for watermarking can produce:

- *fixed strings* (representing the name of the copyright holder),
- *random strings* whose elements were drawn randomly according to some probability distribution,
- *pseudo-random strings* generated cryptographically,
- *digital signatures* generated from the document, secret key of the copyright holder and other identification information.

Fingerprints are a special type of watermarks. From now on a fingerprint is a watermark that uniquely identifies the copy of a document and the copyright holder.

Consider the way in which watermark strings can be recognised by extractor. In general, we have

1. *static watermarks* that can only be extracted by examination of software, which is not currently working, either from a binary code of the program or from a source code of the program. It is usually used to assert the source of the program (the copyright holder, who created the software), or
2. *dynamic watermark* that can only be extracted by scanning the software during its execution. In general, there are two types of *dynamic watermarks*. The first one is connected with the functionality of the software (to be more precisely, it reflects the form of building blocks used). The second one defines the order of blocks and their structure that are in the software. We also call them *structure watermark*, because it depends on the structure of the software delivered to the *End-User*.

4   Security Evaluation of the Scheme

In a sense, watermarking provides a new identity to a piece of software. The insertion algorithm I takes the master copy (whose identity is publicly known) and embeds the new identity by "adding" the watermark. The identity of the software can hold the following properties:

- *uniqueness* – each copy of the software is obtained using a unique data typically known to the copyright holder only (this normally includes their secret key),
- *invisibility* – secret keys are never given explicitly in a copy of software. This also means that the recovery of secrets from watermarks extracted from the software is computationally difficult,

- *verifiability* – any one can check the authenticity of the software, i.e. the watermark extracted from a software is publicly verifiable,
- *non-repudiation* – the creator of the software cannot disavow their authorship,
- *resistance against partial manipulation* – the identity survives a partial modification of the software or in other words, the valid identity can be retrieved even after the watermark has been partially distorted.
- *unforgeability* – an attacker is unable (due to its limited resources) to modify the software so that it assumes a new valid (but forged) identity (watermark). A weaker notion would require that any modification made to the software would destroy the identity (watermark) leading to invalid identity (watermark).

Given an instance of functional watermarking $FWI = \langle \mathcal{F}'_o, \mathbf{O}', \mathbf{W}', \Delta', \mathbf{S}', \mathbf{F}' \rangle$ and the corresponding collection of algorithms $\{G, I, E, R\}$. An attack on this scheme is the probabilistic algorithm that runs in polynomial time, takes the instance FWI as the input and returns another instance of the functional watermarking $FWI_{forge} = \langle \mathcal{F}'_o, \mathbf{O}", \mathbf{W}', \Delta", \mathbf{S}", \mathbf{F}' \rangle$ such that the following three conditions are satisfied:

1. the functionality of the space of mutant object is preserved, or

$$\bigwedge_{o" \in \mathbf{O}"} \bigvee_{o' \in \mathbf{O}' \wedge f \in \mathcal{F}'_o} f(o") = f(o').$$

2. the functionality of the space of effective programs is the same, or

$$\bigwedge_{S" \in \mathbf{S}"} \bigvee_{S' \in \mathbf{S}' \wedge F \in \mathbf{F}'} F(S") = F(S')$$

and

$$\bigwedge_{\delta \in \Delta"} \bigvee_{S" \in \mathbf{S}"} \delta(\mathbf{O}", \mathbf{W}') = S"$$

3. the identification algorithm R fails to detect forgery, or

$$(R_{FWI} = accept) \implies (R_{FWI_{forge}} = accept)$$

Note, that there is another possible definition of attack when the third condition holds only. In this scenario, the attacker wants to modify a given (watermarked) software in such a way that it preserves its identity with wrong functionality. Obviously, the attacker wants to damage the copyright holder reputation.

The functional watermarking scheme is called *unconditionally secure*, when there is no polynomial time algorithm that can produce a new instance of the scheme whose watermarks are accepted by the identification algorithm.

The functional watermarking scheme is called *practically secure*, when there is a polynomial time algorithm that can break the scheme, but the attack is not useful in practice in terms of:

- the time necessary to produce a forged copy of software (if time necessary to successfully forge a software is longer than the time needed to design a software of the same or similar functionality),
- the amount of necessary modifications (if the number of modifications is bigger than the number of instructions in the software),
- the required resources and expertise of the adversary (if the attacker must be an expert equipped with large and expensive human and computing resources).

5   Generic Watermarking Schemes

For the whole section we assume, that the following assumptions are true. The producer holds its unique secret key k_p. We further assume that there is a *Public Key Infrastructure* with a trusted *Certification Authority* that holds the matching public key K_p in the form of publicly available certificates. There is also a cryptographically strong collision-free hash function $H : \Sigma^* \to \Sigma^m$ with public description. The function takes a message of arbitrary length $M \in \Sigma^*$ and yields an m-bit digest $d = H(M) \in \Sigma^m$. The producer also applies a cryptographic signature $SG_p : \Sigma^* \to \Sigma^r$ by $SG_p(M) = D_{k_p}(H(M)) \in \Sigma^r$. In other words, we say that $\{k_p, K_p, H, SG_p\} \in U_{special}$.

5.1   Fingerprints with Mutant Objects

Given the set of mutant objects O, a mutant object $o \in O$, *label function* $V(o)$, *id function* $id_o(o)$ and the collection of copyright holders (creators of the software) $U_{special}$ and set of watermarks **W**.

Watermarking with mutant object can be defined as follows.

1. The producer creates and announces a *public object table*, that contains records of the form:
   - $V(o)$, i.e. the label (index) of the mutant object,
   - *object*, i.e. information about object o,
   - $D_{k_p}(H(o))$, i.e. the digital signature of the corresponding object (in fact, the body of the object).
2. The producer generates so called *common secret comSEC* according to function $G : U_{special} \to$ **W**.
3. The producer computes $h = H(object||V(o)||comSEC)$ and calculates the signature $SG_p = D_{k_p}(h)$.
4. Finally the *id function* becomes $id_o(o) = V(o)||SG_p$.

Note, that there are two possibilities, when we are thinking about *object*:

1. *object* $= H(o)$ – in this option, the copyright owner reveals only partial information about the object. Note, that there is no further need to publish the *object* field in the *public object table*. Unfortunately, in this scenario, there is a potential possibility of attack, in which one could exchange the correct object's identity while the body of the objects would remain unchanged. This attack, however, works only in the situation when the identity of the object is not calculated during the program execution,

2. $object = o$ – in this option, the copyright holder publishes the whole information about the object (i.e. the source code of the object) and the attack mentioned above doesn't work any more. Of course, the publication of the source codes is in conflict with the copyright holder interest. Perhaps, the easiest and the most straightforward solution of this problem can be a deposit of source codes with a trusted authority.

5.2   Object-Oriented Fingerprints

Programs written in an object-oriented languages are typically shorter due to re-usability of objects. A large part of the functionality of software is hidden in objects.

Definition 3. *Given a software $S = (o_1, o_2, \ldots, o_n)$ where $o \in \mathbf{O}$. The identity sequence of a software S is defined as follows:*

$$id_S = (V(o_1), \ldots, V(o_n))$$

Definition 4. *Given a software $S = (o_1, o_2, \ldots, o_n) \in \mathbf{S}$, the copyright holder keys (K_p, k_p), cryptographically strong hash function H, a cryptographic signature scheme and a common secret $comSEC$. Then the object-oriented fingerprint of the software S is defined as follows:*

$$OFP_S = D_{k_p}(H(H(S_M)||comSEC||id_p||\text{edition date}||\text{vendor id}))$$

such that $id_S = OFP_S$ where S_M is the master copy of the software which is used to produce all fingerprints and id_p is the identity of the producer.

Note, that our fingerprint is just a digital signature. The embedding fingerprint procedure, identity verification by a user and the identity verification by Copyright Holder are defined as in [10].

5.3   Fixed Functionality Schemes

Several effective programs (or multiple copies of the same *master program*) with the same functionality are generated using a single transition function. Given an instance of functional watermarking $\langle \mathcal{F}'_o, \mathbf{O}', \mathbf{W}', \Delta', \mathbf{S}', \mathbf{F}' \rangle$.

Assumptions:

1. $\Delta' = \{\delta\}$ or $|\Delta'| = 1$ – we consider a single transition function only,
2. $\mathbf{F}' = \{F\}$ or $|\mathbf{F}'| = 1$ – we assume, that all effective programs have the same functionality F.
3. $comSEC$ – we have some *common secret* previously obtained from a generation function $G : U_{special} \rightarrow \mathbf{W}$

Algorithms:

1. *Watermark generation G*:
   (a) Generation of the collection of *object watermarks* $w^{(o)}$ in order to insert them into appropriate mutant objects.
   (b) Generation of the watermark $w^{(fing)}$ for some effective program according to a user (buyer of software) and the copyright holder information.
   (c) Generation of the transition watermark $w^{(\delta)}$.
2. *Watermark insertion I*:
   (a) Creation of the new instances of mutant objects with their identities $id_o(o) = w^{(o)}$.
   (b) Embedding of the fingerprint into an effective program S_{fing}.
   (c) Creation of the *End-User* subset of the mutant objects:

   $$\mathbf{O'} = \{o : \bigwedge_{S \in \mathbf{S'}} (o \in S) \implies (o \in \mathbf{O'})\}$$

   (d) Generation of the transition function δ:

   $$\delta(\mathbf{O'}, w^{(\delta)}) = S_{fing} \wedge \bigwedge_{x \neq w^{(\delta)}, x \in \mathbf{W'}} \delta(\mathbf{O'}, x) \neq S_{fing}$$

3. *Watermark extraction E*:
   (a) dynamic extractor first computes $w^{(\delta)}_{candidate}$ that is used to find $S_{candidate}$ $= E_{dynamic}(w^{(\delta)}_{candidate})$ and get $w_{dynamic}$,
   (b) static extractor retrieves

   $$\mathbf{W_{static}} = \{w_{static} : \bigwedge_{o \in \mathbf{O'}} w_{static} = id_o(o)\}$$

4. *Identification R*:
   $R(\mathbf{W_{static}}, w_{dynamic}, \mathbf{U}) = reject$ if and only if both the static and dynamic watermarks are wrong.

The *Identification Algorithm* needs some further elaboration. Note, that identification fails only when both watermarks are invalid. But if some parts of watermarks fail and some do not, then identification can be still considered successful. In this situation the copyright holder (and sometimes the *End-User* too) knows that, although unsuccessful, the attack has occurred.

5.4   Fixed Program Scheme

In this scheme one effective program (which is the *fingerprinted* version of some *master program*) is generated using a single transition function. First, we define a bijective function (mapping) called *Hiding Function H*.

Let (n_k) means the k-elements sequence of integers given form set $\mathcal{Z}_l$, where $k = 1, 2, \ldots, l$. Let $N^{(k)}$ denotes the set of all possible sequences (n_k) for fixed k, or:

$$N^{(k)} = \{(n_k) : \bigwedge_k n_k \in \mathcal{Z}_l \wedge k = 1, 2, \ldots, l\}$$

Let r be some large positive integer.

Definition 5. *Given the set N and the number r which are defined above. Then the Hiding Function H is defined as follows:*

$$H^{(k)} : \{0,1\}^r \to N^{(k)}$$

and the reverse function:

$$[H^{(k)}]^{-1} : N^{(k)} \to \{0,1\}^r$$

So the *Hiding Function* for the given integer $n \in \{0,1\}^r$ produces the unique sequence of integers $(n_1, n_2, \ldots, n_k)$ for some fixed $1 \le k \le l$.

Given the *Hiding Function* $H^{(k)}$ defined in Definition 5 and the integer $n \in \{0,1\}^r$, then the transition function $\delta \in \Delta'$ could be denoted as follows:

$$\delta := H^{(k)}(n)$$

Now we are ready to describe the scheme. Given an instance of functional watermarking $\langle \mathcal{F}'_o, \mathbf{O}', \mathbf{X}', \Delta', \mathbf{S}', \mathbf{F}' \rangle$.

<u>Assumptions:</u>

1. $\mathbf{S}' = \{S\}$ or $|\mathbf{S}'| = 1$ – there is a single effective program only, this means that there is a single functionality F or $\mathbf{F}' = \{F\}$ and $|\mathbf{F}'| = 1$,
2. $\Delta' = \{\delta\}$ or $|\Delta'| = 1$ – there is a single transition function only,
3. $comSEC$ – a *common secret* is generated according to a function $G : U_{special} \to \mathbf{W}$,
4. $H^{(n)}$ – a well-defined *Hiding Function*, where $n = |\mathbf{O}'|$ and:

$$H^{(n)} : \{0,1\}^r \to \{(n_k) : \bigwedge_k (n_k \in \mathcal{Z}_n \wedge o_{n_k} \in \mathbf{O}') \wedge k = 1, 2, \ldots, n\}$$

<u>Algorithms:</u>

1. *Watermark generation G:*
   (a) Generation of the collection of *object watermarks* $w^{(o)}$ in order to insert them into mutant objects.
   (b) Generation of the watermark $w^{(fingerprint)}$ of some effective program for a user and for copyright holder information.
   (c) Generation of the transition watermark $w^{(\delta)}$.
2. *Watermark insertion I:*
   (a) Creation of the new instances of the space of mutant objects with their identities $id_o(o) = w^{(o)}$.
   (b) Embedding of the fingerprint $w^{(fingerprint)}$ into the effective program S.
   (c) Creating of the *End-User* subset of the mutant objects:

$$\mathbf{O}' = \{o_i^{u_k} : (o_i^{u_k} \in S) \Longleftrightarrow (o_i^{u_k} \in \mathbf{O}')\}$$

   where the lower index denotes the position of the given mutant object in the set of the mutant objects while the upper index denotes the position of the object into the effective program S and

$$H^{(n)}(w^{(\delta)}) = (u_1, u_2, \ldots, u_n)$$

   In the other words, program S is some permutation of the $\mathbf{O}'$ generating according to *Hiding Function*.

(d) Generation of the transition function δ:

$$\bigwedge_{x \in \mathbf{W}'} \delta(\mathbf{O}', x) = H^{(n)}(w^{(\delta)})$$

3. *Watermark extraction E*:

   (a) *Dynamic Extractor* $E_{dynamic} = \{E^{(\delta)}, E^{(fingerprint)}\}$ retrieves the *fingerprint* and the *transition* watermark.
   (b) *Static Extractor* recovers the set of *object watermarks*:

$$\mathbf{W_{static}} = \{w_{static} : \bigwedge_{o \in \mathbf{O}'} w_{static} = id_o(o)\}$$

4. *Identification R*:
   $R(\mathbf{W_{static}}, \{w_{candidate}^{(\delta)}, w_{candidate}^{(fingerprint)}\}, \mathbf{U}) = reject$ if and only if all three watermarks $w_{candidate}^{(\delta)}$, $w_{candidate}^{(fingerprint)}$ and w_{static} are invalid.

6   Experiments

Experiments were conducted to verify the concept and the implementation of watermarking schemes. Watermarking was implemented for source codes with extensive use of the RTTI (Run Time Type Information) mechanism. This mechanism is available in most object-oriented programming languages (such as C++, Java, Smalltalk, etc.). The object identity can be encoded into the name of the class. This requires to create a separate class for each object. This solution separates identity of the object from its functionality.

6.1   Watermark Embedding

Given a watermark. The watermark can be embedded into a piece of software using

- conversion of watermarks (identities of objects) to the form accepted by compiler. As an object identity may be of arbitrary sequence of bits, the sequence has to be converted into the form accepted as the valid name. Most compilers accept any alpha-numerical sequences that start from a letter. For instance, we can fix the first letter adding a long enough string taken from the watermark,
- generation of source code versions containing object identities. Depending on the language and compiler in hand, it is possible to use substitutions defined by macros or other mechanisms allowing fast manipulations.

Now consider an example in C++ language.

Example 1. **Coding Example.**

```
// Header file with encrypted class names
#include "encode.h"

// Original source-code
class TMyClass { public:
  TMyClass( void );
  virtual ~TMyClass( void );
  void Execute( void );
};
```

The *encode.h* file could look like this:

```
#define TMyClass T34F49A00B53F219DAA6B2A169834
```

As one can see, the *encode.h* file has very simple form, so it can be generated automatically. As a result of preprocessor work, we obtain:

```
class T34F49A00B53F219DAA6B2A169834 { public:
  T34F49A00B53F219DAA6B2A169834( void );
  virtual ~T34F49A00B53F219DAA6B2A169834( void );
};
```

6.2   Watermark Extraction

There are two cases:

- recovery of class names of objects used in the watermarked software – this depends on the pair: language and compiler that gives a suitable mechanism. For instance in C++, to identify the object type, one would need to use operator `typeid` that gives, among many other information, also the class name. So the function call

    ```
    typeid(MyObject).name()
    ```

    generates

    ```
    T34F49A00B53F219DAA6B2A169834
    ```

    that is encoded identity of the class. This operation is done for each object according to the transition function,
- reconstruction of object identities – we use the decoding function.

The RTTI mechanism seems to be very effective for watermark embedding. An attacker who would like to forge the software, would need to modify the compiler (and the RTTI mechanism) so it will generate names required by the attacker leaving the functionalities and other names untouched. Although this seems to be possible, it will be a tedious and resource consuming exercise requiring a very high level of expertise.

Generation of source code versions is easy to implement and can be done very efficiently. It suffices to create a file `encode.h` to put the necessary substitution function.

A drawback of the solution is the existence of a separate class for each object. This does not have any impact if the program applies different objects. It can be a problem if majority (or perhaps all) objects are from the same class.

7   Conclusions

The work introduces a general model for copyright protection of source code software written in object-oriented languages such as C++ or Java. It turns out that watermarks can be embedded into programs using different techniques. In the first, watermarks are hidden inside objects used in the program. We actually discussed one way in which the watermarks can be embedded into object names. Variants of source code instructions can be also chosen by watermarks. Finally, watermarks can also be used to modify transition function that can be used to generate copies of slightly different functionalities.

Acknowledgement

This work was partially supported by the Australian Research Council grant A00103078.

References

1. D. Aucsmith. Tamper resistant software. In R. Anderson, editor, *Information Hiding, First International Workshop, Cambridge, 1996*, pages 317 – 334. Springer-Verlag, 1996. Lecture Notes in Computer Science No. 1174.
2. M. Barni, F. Bartolini, V. Cappellini, and A. Piva. Copyright protection of digital images by embedding unperceivable marks. *Image and Vision Computing*, 16:997–906, 1998.
3. H. Bergal. Watermarking cyberspace. *Communications of ACM*, 41:19–24, 1998.
4. J. Brassil and L. O'Gorman. Watermarking document images with bounding box expansion. In R. Anderson, editor, *Information Hiding, First International Workshop, Cambridge, 1996*, pages 227–235. Springer-Verlag, 1996. Lecture Notes in Computer Science No. 1174.
5. J.T. Brassil, S. Low, N.F. Maxemchuk, and L. O'Gorman. Electronic masking and identification techniques to discourage document copying. *IEEE Journal on Selected Areas in Communications*, 13(8), 1995.
6. C. Collberg and C. Thomborson. On the limits of software watermarking. Technical Report 164, Department of Computer Science, the University of Auckland, Auckland, New Zealand, 1998.

7. I. Cox, J. Kilian, T. Leighton, and T. Shamoon. A secure, robust watermark for multimedia. In R. Anderson, editor, *Information Hiding, First International Workshop, Cambridge, 1996*, pages 183 – 206. Springer-Verlag, 1996. Lecture Notes in Computer Science No. 1174.

8. O. Faugeras. *Three Dimensional Computer Vision: A Geometric Viewpoint*. MIT Press, 1993.

9. D. Glover. *The protection of computer software*. Cambridge University, 2nd Edition, 1992.

10. J. Pieprzyk. Fingerprints for copyright software protection. In M. Mambo and Y. Zheng, editors, *Information Security, Second International Workshop, ISW'99*, pages 178–190. Springer-Verlag, 1999. Lecture Notes in Computer Science No. 1729.

11. C.I. Podilchuk and W. Zeng. Image-adaptive watermarking using visual models. *IEEE Journal on Selected Areas in Communications*, 16:525–539, 1998.

12. R.G. van Schyndel, A.Z. Tirkel, and C.F. Osborne. A digital watermark. In *Proceedings IEEE International Conference on Image Processing*, volume 2, pages 86–90, 1994.

13. P. Wayner. *Digital copyright protection*. Academic Press, Boston, 1997.

Off-Line Authentication Using Watermarks

Hyejoung Yoo[1], Kwangsoo Lee[2], Sangjin Lee[2], and Jongin Lim[2]

[1] sparxcom corporation,
Anam Dong 5ga, Sungbuk Gu, Seoul, Korea
hjyoo@sparxcom.com
[2] Center for Information Security Technologies(CIST),
Korea University, Anam Dong, Sungbuk Gu,
Seoul, Korea
{kslee,sangjin,jilim}@cist.korea.ac.kr

Abstract. In this paper, we propose a new method for the secure and practical watermarking system. This method is designed to use printed images in the off-line authentication, where one determines whether a printed image has been altered intentionally since the first printing time, perhaps by a malicious party. This system is based on the fragile watermarking technologies to detect the transforms such as PS(printing and scanning) distortions, slight rotations and croppings, etc. With the human visual system, this watermarking system can detect the watermark in a image with random noises and can be applied for the practical use.

Keywords: off-line authentication, e-commerce, PS distortion, PSP distortion.

1 Introduction

Digital watermarking is a technology to mark a discriminating information into digital images for the purposes of the ownership verification or authentication. This technology is becoming important in the electronic commerce because the digital data can be splitted on the developing network systems such as the World Wide Web.

There are two types of watermarking systems. One is the robust watermarking system that is designed to resist attacks that attempt to remove or destroy the mark. Such attacks include lossy compression, filtering, and geometric scaling. The other is the fragile watermarking system that is designed to detect slight changes to the watermarked image with high probability. The fragile watermarking system is mainly used for the image authentication because it ensures the genuineness of images with high probability.

In the off-line authentication, the watermark extracted on the scanning process determines the genuineness of the printed image of a digital image downloaded from network systems. The off-line authentication is important due to the usage of digital images on the internet and in electronic commerce such as on-line ticketing.

There are some problems in the off-line authentication of the printed image with the information of the genuineness. On the process that the digital image is printed and the printed image is scanned, there is possibility that the image is

K. Kim (Ed.): ICICS 2001, LNCS 2288, pp. 200–213, 2002.
© Springer-Verlag Berlin Heidelberg 2002

modified by random noises. We call this the PS(Printing and Scanning) distortion. For the correctness of the off-line authentication, the PS distortion must be removed.

In this paper, we design the fragile watermarking system for the off-line authentication that resists some problems such as the PS distortion. The principal property in the scheme is the correctness and validity of the off-line authentication. With the human visual system, this watermarking system can detect the watermark in the image with random noises.

Although invisibility is less important in the off-line authentication, our system has a regulation factor so that one can adjust the visibility of the mark in compliance with the various kinds of applications. If we decrease the value of the regulation factor, we can embed the watermark invisibly into the digital image and authenticate its printed image more securely.

This paper is organized as follows. Section 2 is a brief review of the general flow of watermarking technologies. In section 3 we consider the basic properties of the watermark used in the offline authentication. In section 4 we present new watermarking method for off-line authentication. Section 5 shows the security of the method. We suppose various attack scenarios and explain them by the experimental results that our method is secure against these attacks. Conclusion and suggestion for the further research are offered in section 6.

2   General Flow of Watermarking Technologies

Digital watermarking is a method to insert a digital information into an image so that the information can be extracted for the purposes of authorship or ownership verification. This technologies have been proposed as solutions for fundamental problems in digital communication.

Early watermarking methods are no more than incrementing or decrementing an image component to encode binary '1' or '0', respectively[4]. Tirkel[6] and van Schyndel[7] applied the properties of m-sequences to produce oblivious watermarks resistant to filtering and cropping. It is reasonably robust to cryptographic attacks. Tirkel and Osborne[6] were the first to apply spread spectrum techniques to digital image watermarking. Spread spectrum method is cryptographically secure and is capable of achieving error free transmission of the watermarked image at the limits given by the maximum channel capacity[8].

A fragile watermark is a mark that is readily altered or destroyed when the host image is modified through a linear or non-linear transformation[9]. Since digital images are easy to modify, a secure authentication system is useful in proving that no tampering has occurred. The sensitivity of fragile watermarks to modification leads to their use in image authentication. That is, it may be of interest for parties to verify that an image has not been edited, damaged, or altered since it was marked. Image authentication systems can be adapted in law, commerce, defense, and journalism, etc.

Schneider and Chang[10] presented a methodology for designing content based digital signatures which could be used to authenticate digital images.

The goal of this method was to develop the way to be able to prove some form of authenticity, while still allowing desired forms of manipulation such as lossy compression. Lin and Chang[11] described an effective technique for image authentication which could prevent malicious manipulations but allowed JPEG lossy compression. The authentication signature was based on the invariance of the relationship between DCT coefficients at the same position in the separate blocks of an image. Lu, Mark Liao, and Sze[12] proposed a combined watermarking scheme which could simultaneously achieve the copyright protection and the content authentication by hiding watermarks only once. In this method, they proposed to quantize the selected wavelet coefficients into masking threshold units. Then, the watermark was embedded by modulating the quantization result to either a right or left masking threshold unit using the cocktail watermarking[13].

Wang and Knox[14] presented the watermarking method which was an invisible watermark based on a digital halftoning screen. When the image to be protected was halftoned in preparation for printing, the digital watermark, in the form of a logo-style image, was embedded into the correlation patterns of the halftoned image.

Even if one of the main distribution methods of the digital data is a printed media, however, there are few watermarking methods that are robust to printing and scanning noises, i.e, PS distortion. We propose a new technique to be robust to PS distortion . This is the first proposal of the off-line authentication scheme using watermarks.

3   Basic Properties

There are two kinds of attacks. One is an intentional attack such as the duplication and the other is an unintentional attack such as the compression and the PS distortion through the flow of the scheme.

In case of an intentional attack, attackers may try to duplicate a printed document or a printed ticket to fool the system without triggering alarm to enter a theater or a stadium with the copy of that.

Some noises are added to the watermarked image. D/A and A/D transform noises would make trouble to authenticate the data. And a slight rotation may happen in scanning processes of the image for the authentication of a ticket.

In the following, we discuss the seven basic properties required for off-line authentication.

1. **robustness to the PS distortions:** In the off-line authentication scheme, since printing and scanning are indispensable procedure, the PS distortion is added essentially to the watermarked image. Therefore, the robustness to the PS distortion is one of the most important properties of the off-line authentication scheme.
2. **robustness to slight rotations:** When the image is scanned, a slight rotation may occur. Therefore, the watermark must be robust to slight rotations.

3. **robustness to the cropping:** This is an essential property of all sorts of watermarking schemes. The watermark must be detected until the damage does not affect the value of the documents or the tickets.
4. **robustness to folding and crumpling:** Watermark is inserted in a document or in a ticket which the user carries at his pleasure. So there is a possibility that the watermarked image is folded or crumpled.
5. **fragility to copying:** The fragility to copy is the most important property for off-line authentication scheme such as on-line issuing of documents and on-line ticketing. If the scheme is not fragile to copy, it is not applicable to the real life and will be very useless.
6. **fragility to the PSP distortions:** Reprinting after scanning of the printed ticket is one way of copying. A ticket is easily copied in high quality by this method. The watermark is fragile to the PSP distortion.
7. **propriety of the time requirement:** In order to apply to the real life, the propriety of the time requirement is very important. If embedding and detecting processes are not efficient, the off-line authentication scheme cannot be realized. Especially, the detecting time of the on-line ticketing scheme is more important. The detecting time must be reasonble enough to enter a theater or a stadium without any delay.

4   The Scheme

In this section, we present the watermarking scheme for the off-line authentication which discriminates the genuineness of the printed image. To our knowledge, this is the first practical proposal about the authentication of the printed data. Here, we handle images which is used for the off-line authentication in practical application such as on-line issuing of documents and on-line ticketing.

4.1   Off-Line Authentication

Digital data can be manipulated and distributed easily by using computers. The easy manipulation of digital data causes a real threat for the data issuers. Issuers of on-line data have had no means to authenticate their printed data practically up to now.

Lin and Chang[15] showed several techniques for extracting invariants from the original and rescanned image. But they did not consider the fragility to copy and to psp distortion. They proposed a hypothetical model of the pixel value distortions and the parameters of this model were approximated by experiments. There are some problems in their analysis of the geometirc distortions of RSC(rotation, scaling, and cropping) to apply to ps model of off-line authentication. First, the image is not rotated around the same center, that is, the central point of each rotation may be different in ps process. The rotations are not closed under the composite operation($R_1 R_2 \neq R$), and, if the image is rotated more than once, the new kind of noise is added to that, not a general rotation error. Second, rotation and cropping are not commutative($RC \neq CR$) unless the rotation degree equals zero. Scaling and cropping are also not commutative ($SC \neq CS$) unless the scaling factors are all equal.

In order to construct the off-line authentication scheme, the printed materials which are issued via internet must preserve the contents in their integrity. Data integrity verification is equivalent to the functionality of the authentication.

There are two types of watermarking schemes for authentication. The first type is sensitive even to the slightest change of the digital medium, whereas the other type notifies an authentication violation only when significant modifications of the visual content occur. The former is useful for channel authentication and the latter is useful for tracing the changes of the contents of the documents. In our case the traceability of the changes is much more important.

One of the general properties of the watermarking schemes for the digital data authentication is perceptual transparency. An embedded watermark should not be visible under normal observation and should not interfere with the functionality of the image. In most cases, this refers to preserving the aesthetic qualities of an image and increasing the security. In our case, we see that authentication of the contents is more important than aesthetic qualities of the image. The invisibility is less important where the images in the documents or tickets are not for the commercial purpose. But the degree of visibility must be decided with prudence, since it affects the security of the scheme. Our method has the regulation factor of the visibility. As this regulation factor is decreased, the watermark becomes more invisible.

4.2   Embedding

Off-line authentication scheme contains PS steps. Printing and scanning add random noises to the data. These noises are resulted from the luminance and contrast variations, the blurring of adjacent pixels, and the geometric distortions. Therefore, even if we print/scan the same image by the same printer/scanner, the results are different. Thus, the watermark must be robust to random PS noises. At the same time, it should be fragile to the PSP distortion and the copy distortion.

We model the variation of the gray values of pixels before and after the scanning process. The variance is usually larger on dark and bright pixels. Table 1 shows the repetitive experimental results about variation of the pixel values which are caused by scanning process.

We describe the embedding approach which intensifies the regularity of the watermark and the irregularity of the random noises caused by scanning process.

First, we must randomize the watermark information. This process prevents an attacker who knows the embedded watermark from re-embedding it into the scanned tickets and re-authenticating them. But we omit this procedure, since we embed the logo-style watermark in order to authenticate whether the printed data are issued by the determined issuer or not.

The algorithm that we present below is guaranteed to produce the watermark which is pertinent to the off-line authentication scheme.

Table 1. Scanners CCD error variance

OV (Original Value)	MV (Error Mean Value)	MV−OV	ED (Error Deviation)
250	253	+3	5
200 ∼ 210	206	+6	−3 ∼ 16
190 ∼ 200	193	+3	−6 ∼ 12
179 ∼ 180	169	−1	−10 ∼ 11
130 ∼ 140	129	−1	−10 ∼ 8
80 ∼ 90	84	+4	−3 ∼ 14
60 ∼ 70	69	+9	0 ∼ 19
50 ∼ 60	70	+20	13 ∼ 28
30 ∼ 40	48	+18	20 ∼ 28
0 ∼ 10	35	+35	30 ∼ 48

1. Decide the information rate of the watermark W to be embedded.
2. Find the parts I_s of the original image I of which pixel values are contained in $[b_1, b_2]$, where the pixel values in $[b_1, b_2]$ have the small changing variances by the scanning process, so $I(i,j) \in I_s$ is more robust to PS distortions. In our experiments, we let $b_1 = 80$ and $b_2 = 180$.
3. Divide I by the size of W in order to disperse the watermark message so as to be robust to PS distortions. For example, let the size of I be $M \times N$ and the size of the logo be $a \times b$, then the number of the division parts of I is l, where $l = \lceil M/a \rceil \times \lceil N/b \rceil$. We denote these division parts by D_i ,$1 \le i \le l$. In case of $M/a \notin \mathbb{Z}$ or $N/b \notin \mathbb{Z}$, we append $'0'$ to the remainder parts in order to equalize the sizes of the remainders with the size of W. Thus, $size(D_i) = size(W)$ for all i, $1 \le i \le l$.
4. For all i, check that D_i is contained the part of I_s for some s. If more than half of D_i is contained in I_s, we say that I_s contains D_i and denote $D_i \subset I_s$.
5. Each division part D_i is assigned to a different value of the regulation factor r_i;

$$r_i \le 0.05 \quad if \; D_i \subset I_s \; for \; some \; s$$
$$r_i > 0.05 \quad if \; \nexists \; s \; such \; that \; D_i \subset I_s$$

and then calculate $I^* = \sum_{i=1}^{l} I_i^*$, where $I_i^* = D_i + r_i W \; (1 \le i \le l)$, where $I_i^*(j,k) = D_i(j,k) + r_i W(j,k)$.
Depending on the value of r_i , the different amount of partial information of W is embedded in each division part. The higher the value of r_i is, the more the partial information of W is embedded into D_i. Here, all regualation factors can be assigned to the same, that is, $r_i = r$ for all i. The visibility of W is decided by the regulation factor. The more visible the mark is, the better detected it is. By this step the interference of the PS distortions decreases and the embedded watermark gets to be more invisible.
6. The watermarked image, Iem, is lined with $I^*(j,k)$ depending on i.

We show the logo-style watermark, the original image, and the watermarked image in Figure 1.

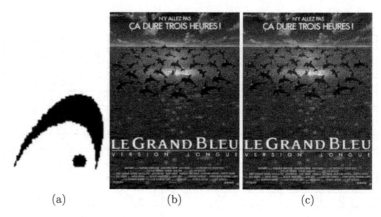

(a)   (b)   (c)

Fig. 1. (a) watermark: sparxcom co. logo (100×100); (b) original image: grand-bleu (600×800); (c) watermarked image with $r = 0.035$, $l = 48$, $b_1 = 80$, and $b_2 = 180$

One of the practical applications of our watermarking scheme is on-line ticketing. On-line ticketing is a kind of e-commerce system all sorts of which users can purchase via internet. After the user chooses the ticket to buy, the watermark is embedded in the image which was contained in the chosen ticket. The user had no sooner paid for the chosen ticket than he received the ticket at home or in his office by using his printer. At this point, copy control technology about the digital data of the ticket is needed separately, because once a dishonest buyer of a ticket gets the ticket in digital form, he can reprint it as often as he wants. The user can enter the theater conveniently with the watermark embedded ticket which is printed in from 300dpi up.

4.3 Detecting

When a user goes to a theater or a stadium with the printed ticket, the staff of a theater or a stadium wants to know whether the ticket is genuine or not. If the ticket is forged, the staff rejects the entrance.

With our watermarking scheme, the staff can judge the genuineness of the tickets using the exclusive use terminal(the scanner) within 2 seconds before the user is authorized to enter a theater or a stadium. The time required for this procedure is reasonable for the entrance without any delay.

Detection of the watermark involves scanning the printed image and looking for the correlations of the scanned image with respect to the neighboring regions of the image.

Detection is progressed as follows:

1. The printed document is scanned in from 300dpi up using the terminal (the scanner). The necessary part of the scanned is the image part. There is a possibility that the image is rotated slightly when the document is scanned.

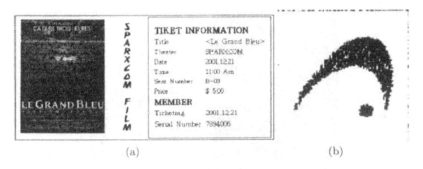

Fig. 2. (a) scanned ticket containing watermarked image with $r = 0.070$; (b) detected watermark (100×100)

2. We revise the rotation error to some degrees and cut the scanned image as the same size as the original. We denote this Iem^*. Iem^* is different with Iem because of PS distortions, geometric distortions, and cutting distortions, etc.

$$Iem^* = Iem + N,$$

where N is a random noise.

3. Divide Iem^* by the size of the embedded watermark. Since watermark is for authentication and exclusive use terminal s used by the watermark provider, it is right that the size of the embedded watermark makes public to the exclusive use terminal.

4. Look into the correlation of the division parts, and extract this correlation from each division part.

5. Extracted information from each division part lies one upon another.

6. Calculate the similarity of the embedded watermark and the extracted watermark. The embedded information was determined at a theater or a stadium, so it does not affect the security of the system that the original embedded watermark is inserted into the exclusive use terminal of the theater or the stadium.

Figure 2 shows the scanned image and the detected watermark.

5   Experimental Result

In this section, we show that our scheme satisfies the basic properties discussed in section 3. We experiment with popular devices in table 2. The results are as follows:

1. **robustness to the PS distortion** Figure 2 shows our scheme is robust to the PS distortion.

2. **robustness to slight rotation** Proposed detection algorithm corrects this rotation to some degrees and authenticates correctly.

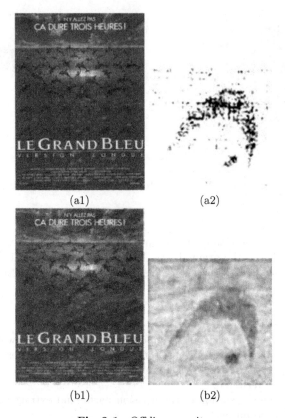

(a1)   (a2)

(b1)   (b2)

Fig. 3-1. Off-line security
(a1) scan of the rotated image after *ImageCutter.exe* (a2) detected watermark of (a1)
(b1) scan of the crumpled image   (b2) detected watermark of (b1)

3. **robustness to cropping** Our scheme, of course, is robust to cropping less than a half since the mark information is divided and inserted all over the image.

4. **robustness to folding and crumpling** We have folded a ticket(an image) into from two to eight leaves. Also, we have crumpled a ticket(an image) at random into the bargain. Experimental results show that our scheme is robust to folding and crumpling.

5. **fragility to copy** We do experiment the fragility to copy about the various brightness of the various duplicators. From these experimental results, we can conclude that our scheme is fragile to copy. We have found that the brighter the duplicator is, the better detected the watermark is . (c1) and (c2) in Figure 3 show the result of the copy using the most bright duplicator, SINDORICOH NT 4240

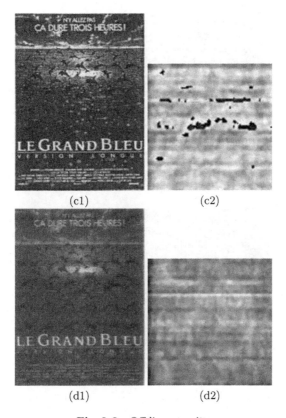

(c1)   (c2)

(d1)   (d2)

Fig. 3-2. Off-line security
(c1) scan of the copy   (c2) detected watermark of (c1)
(d1) scan of the psp distorted image   (d2) detected watermark of (d1)

Table 2. Experimental environment

Computer	Pentium III 600, Ram 128
Scanner	FUJITSU M3292DC
Printer	HP 2000c, HP-Laserjet 1100, SINDORICOH LP 3250, etc
Duplicator	SINDORICOH NT 4140, ST 4520, XEROX X-230, Cannon NP 6650, CLC-900, LC-2300
Paper	general copying papers

6. **fragility to the PSP distortion** In our experiment, we have printed and scanned in 300dpi. We can conclude that the proposed scheme is fragile to psp distortions.
7. **propriety of time requirement** In our scheme, it takes 2 or 3 seconds for the watermark to be embedded and takes 1 or 2 seconds for the document or the ticket to be authenticated. We conclude that the required time is reasonable for realization of the off-line authentication scheme.

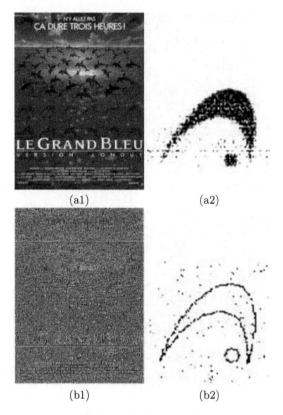

(a1)   (a2)

(b1)   (b2)

Fig. 4-1. On-line security
(a1) JPEG 95%   (a2) detected watermark of (a1)
(b1) high-pass filtering(−8/9)   (b2) detected watermark of (b1)

We also experiment the robustness to the existing various attacks on the digital information of the watermarked image without D/A and A/D transformations separately. The results show in Table 3.

Figure 3 and Figure 4 present the security of our scheme. In (c) and (d) of Figure 3, the watermarks are embedded with $r = 0.090$(high value) in order to show the powerful fragility to these attacks, but, in other experiments, the regulation factor is 0.040.

6   Conclusion and Further Research

In this paper we have presented the off-line authentication scheme using watermarks. This scheme is applicable to on-line issuing of documents, on-line ticketing etc. Off-line authentication is a new application of the watermarking scheme. As we have mentioned above, our scheme is robust to the essential and various at-

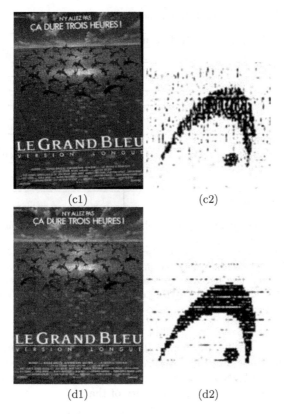

Fig. 4-2. On-line security
(c1) rotation(2 degree)   (c2) detected watermark of (c1)
(d1) Jitter attack   (d2) detected watermark of (d1)

tacks. These observations promise a significant performance improvements, but there is still much room for future work. As for the present study, the problems that it raises are as follows:

1. ***analyzing more accurate models of distortions*** Our analyzing of the embedding and the detecting processes are based on the augmentation of the regularity of the watermark and the irregularity of the various noises. A more accurate model must be considered in order to improve overall performance.
2. ***modeling the attacks more minutely*** This paper has presented the basic attack models for off-line authentication scheme. We must model the attacks more minutely and analyze the security against each attack.
3. ***analyzing potential performance improvement given side information at the detector*** Our watermarking algorithm assumes knowledge of the original and the size of the watermark at the detector. It increases the speed of the detection but is not ideal for the security. Without any ne-

Table 3. Robustness of various attacks

Attacks	Robustness	Remark
JPEG	○	20, 40, 60, 90, 95 %
truncation	○	
histogram equalization	○	
gamma correction	○	$\gamma = 0.5, 1, 1.5$
dithering	○	
filtering	○	high-pass filter, low-pass filter
adding noise	○	
rotation	△	nearest, bilinear, bicubic
Stirmark 3.1	△	
UnZign 1.2	○	
Jitter attack	○	

1. *adding noise*: gaussian(0.04, 0.09), salt and pepper(0.05, 0.09), and multiplicative noise.
2. *rotation*: The mark is detected in case that the rotation degree is less than 2.
3. *Stirmark 3.1*: The mark is detected in case that the rotation degree is less than 2. The mark is not detected after general linear geometric transformation.
4. *Jitter attack*: Copy odd rows to even rows after deletion of even rows.

cessities of these side information(full blind extraction), more sophisticated algorithms can be possibly developed. Therefore, we analyze the importance between the performance improvement given side information at the detector and the security without any necessity of that.

References

1. R.B. Wolfgang, C.I. Podilchuk, and E.J. Delp, *Perceptual Watermarks for Digital Images and Video*, Proceedings of the IEEE, vol. 87, no. 7, pp. 1108-1126, July 1999.
2. R. Ulichney, *Digital halftoning*, MIT Press, Cambridge, Massachusetts, 1987.
3. C. Cox, J. Killian, T. Leighton, and T. Shamoon, *Secure spread spectrum communication for multimedia*, Technical report, N.E.C Research Institute, 1995.
4. G. Caronni, *Assuring ownership rights for digital images*, Reliable IT Systems, VIS'95, pp. 251-265, Viweg, Germany, 1995.
5. G.C. Langelaar, J. van der Lubbe, and J. Biemond, *Copy protection for multimedia data based on labelling techniques*, 17th Symposium on Information Theory in the Benelux, May 1996.
6. A.Z. Tirkel, G.A. Rankin, R.G. van Schyndel, W.J. Ho, N.R.A. Mee, and C.F. Osborne, *Electronic watermark*, Dicta-93, pp. 666-672, December 1993.
7. A.Z. Tirkel, R.G. van Schyndel, and C.F. Osborne, *A two-dimensional digital watermark*, Proceedings of ACCV'95, pp. 378-383, 1995.
8. J. Smith and B. Comiskey, *Modulation and information hiding in images*, Proceedings of the First International Workshop in Information Hiding, LNCS 1173, Springer, pp. 207-226, 1996.

9. M. Yeung and F. Mintzer, *Invisible watermarking for image verification*, Journal of Electronic Imaging, vol. 7, pp. 578-591, July 1998.

10. M. Schneider and S.F. Chang, *A robust content based digital signature for image authentication*, Proceedings of ICIP, pp. 227-230, September 1996.

11. C.Y. Lin and S.F. Chang, *A robust image authentication method distingushing JPEG compression from malicious manipulation*, CU/CTR Technical Report 486-97-19, December 1997.

12. C.S. Lu, H.Y. Mark Liao, and C.J. Sze, *Combined watermarking for image authentication and protection*, IEEE International Conference on Multimedia and Expo (III), pp. 1415-1418, 2000.

13. C.S. Lu, H.Y. Mark Liao, S.K. Huang, and C.J. Sze, *Cocktail watermarking on images*, 3rd International Workshop on Information Hiding, LNCS 1768, pp. 333-347, September, 1999.

14. S.G. Wang and K.T. Knox, *Embedding digital watermarks in halftone screens*, In security and watermarking of multimedia contents II, Proceedings of SPIE vol. 3971, pp. 218-227, 2000.

15. C-Y Lin and S-F Chang, *Distortion Modeling and Invariant Extraction for Digital Image Print-and-Scan Process*, ISMIP99, Taipei, Taiwan, Dec., 1999.

Slide Attacks
with a Known-Plaintext Cryptanalysis

Soichi Furuya

Systems Development Lab., Hitachi, Ltd.
soichi@sdl.hitachi.co.jp

Abstract. Although many strong cryptanalytic tools exploit weaknesses in the data-randomizing part of a block cipher, relatively few general tools for cryptanalyzing on the other part, the key scheduling part, are known. A slide attack is an instance of attacks exploiting the key-schedule weakness. In this paper, currently proposed slide attacks can be still enhanced so that all currently published known-plaintext analytic technique can be applied to smaller part of a cipher with a weak key-scheduling part. As an example, we demonstrate applications of a slide attack to linear cryptanalysis, a DES variant case. In addition, we also show that our enhancement enables to declassify the unknown primitive used in a block cipher. We test a block cipher, GOST, and show how to de-classify the hidden 4-bit substitution tables.

1 Introduction

Many cryptanalyses of a block cipher are based on careful and elaborate observations on the data-randomizing part in a block cipher, and ignore the structure of corresponding key schedule. Since the differential cryptanalysis was presented by Biham and Shamir [1,2], the statistical aspects of a cryptosystem, typically the data-randomizing part, have been studied to check whether or not the cipher is secure against these attacks. Linear cryptanalysis [14] and other related works [9,10,12,19] adopt this approach, however with other statistical characteristics.

In those attacks, round keys, i.e. the outputs of a key-scheduling part, are concerned and very little analyses on a key schedule are used to improve attacks. A related-key analyses[11] is one of major intersections between analyses on data-randomizing part and ones on key-scheduling part. However, the basic idea of the related-key analysis is particularly specialized not in general cryptanalytic techniques, but only in differential cryptanalysis.

A slide attack [3] is an attack based on a particular key-schedule weakness; if a key schedule has inherent cyclicity of round keys, there exist two distinct known-plaintext pairs such that all intermediate values are conceptually identical but appears in different rounds. More precisely the coinciding intermediate values are *sliding* by one round. Biryukov and Wagner presented some novel improvements in [4], e.g., sliding attacks in the adoptive chosen plaintext-ciphertext environments. In those attacks, the attacker basically exploits the weaknesses substantially in the key schedule.

K. Kim (Ed.): ICICS 2001, LNCS 2288, pp. 214–225, 2002.

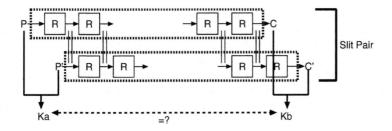

Fig. 1. Finding slit pair and deriving round key.

In this paper, we show that the slide attack can be used to enhance some types of current attacks of a cipher with a weak key-scheduling part. In fact, this enhancement is applicable to any known-plaintext cryptanalytic tools, e.g. linear and partitioning cryptanalysis. In other words, our attack exploits the combination of the weakness in the key schedule and the data-randomizing part.

More precisely, our extension allows the attack to be applicable not only to the target cipher with a *weak* round function but also a cipher with a sequence of rounds vulnerable to known-plaintext attacks. Note that the *weak* round function can be only one or two rounds of Feistel structure, whereas a sequence of round function allows four or more iterations.

We apply this technique to some block-cipher variants in order to demonstrate the effectiveness of the proposed enhancement. We treat the DES variant with four-round iterated round keys and describe its cryptanalysis that is the combination of the linear cryptanalysis and the slide attack.

We also study the GOST block cipher, which has both the simple key-scheduling algorithm and the round function with confidential S boxes. Combining investigations on both parts, we point out interesting technique to de-classify the hidden S boxes.

This paper consists of the following sections. Section two prepares the preliminaries for representations and notations. Section three describes the original ideas of slide attacks. Section four proposes an extension to the slide attack. In section five, the extension is applied to a couple of example ciphers. In section six, we concludes our works.

2   Slide Attacks

Let us assume that $K_i, (1 \leq i \leq r)$ are identical and the round function R has a structure such that the round key can be efficiently calculated out of the pair of the input and the output.

The attacker tries to find a special pair of plaintexts, hereafter we call it a *slit* pair, (P, P') such that one (for example P') is the output of the first round function of the other's (P) encryption. Once he succeeds in finding the right pair, then he can efficiently calculate the round key out of two pairs of the input and the output for a round function.

To find the concerning plaintext pair, the naive way is to collect arbitrary $2^{b/2}$ known-plaintexts, where b is the block length in bits. Thanks to the birthday paradox, there exists such a slit pair with high probability.

Given the sufficient number of known-plaintexts, he identifies whether or not each pair in collected plaintexts (P_a, P_b) is the slit pair. The validation is to check the identity of two derived round keys, namely K_P (calculated out of P_a and P_b) and K_C (calculated out of C_a and C_b), where C_a and C_b are corresponding ciphertexts of P_a and P_b. If $K_P = K_C$, the pair (P_a, P_b) can be a slit pair. Otherwise they cannot be so and he tries the next plaintext pair.

This naive way to find a slit pair can be improved more effectively, depending on the structure or the characteristics of the round function. For instance, if the round function R is a Feistel structure, then a slid pair must share the same half of data, namely upper half of C_a and lower half of C_b. In this case the number of collected plaintexts are reduced to $2^{b/4}$.

The total complexity consists of two computations: collecting the sufficient number of known-plaintexts and searching the slit pair. Let n be the number of known-plaintexts. The former takes n computations and the latter takes $T \times n(n-1)/2$, where T is the computation for the slit-pair verification.

To calculate n the number of sufficient known-plaintexts, we take the birthday paradox into account. According to the rough estimation, about 50 % of the successful attack can be achieved by using about $2^{b/2}$ known-plaintexts. To obtain higher probability, say 80%, the increase of required known plaintext is no more than a factor of 2 to 4. An example calculation of probability that a birthday collides is depicted in Fig.2.

This basic idea is easily extended to plural rounds iteration. The original paper demonstrated the extended idea applied to their DES variants, 2K-DES, where K_1 is used in odd rounds, and K_2 in even rounds. In the slide attack of 2K-DES, two-round sliding property is used. More-round sliding has not been in open publications.

Remarks on Weak Round Function: In the original slide attack, Biryukov and Wagner introduced the concept of the *weak* round function with respect to the slide attacks. In order to apply the original slide attack, a sliding gap must be so *weak* that sufficient key information is derived out of one (or two) input-output pairs of a round function. If the gap consists of more than two rounds, generally it becomes much more difficult to derive the key information.

3   Enhanced Attacks with a Known-Plaintext Attack

3.1   Our Enhancement

In this section, we enhance slide attacks so that they are applicable not only ciphers with a *weak* round function but also ones with a sequence of a round function that is vulnerable to a known-plaintext cryptanalysis.

At first we introduce the enhanced slide attack. In this context, the term "round" does not necessarily specify the exact definition of the round function

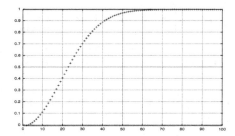

Fig. 2. Collision probability in choosing n out of 365.

of the block cipher. Instead, we intend to use the term "Round" to mean the unit of sliding, i.e., the gap. Typically the "Round" also includes the sequence of the round function.

A target cipher which we concern in this part consists of a sequence of Rounds. In addition, each Round is identically keyed, or equivalently the round-key generated in a cyclic manner. There also exist $\mathcal{D}_R(t_d, q_d)$ and $\mathcal{A}_R(t_a, q_a)$ that are the distinguisher and the key-deriving attack against the Round with q_d (or q_a) known-plaintexts and t_d (or t_a) computational time.

Because of the structure of the cipher, a slit pair exists as well as the original slide attack. At first, the attacker tries to find the slit pair by a Round. Instead of collecting a number of known-plaintexts, the attacker generates a number of (arbitrary) plaintexts and asks for $q_a - 1$ ciphertexts of multiple encryptions for each plaintext, i.e., the ciphertext, the ciphertext of the double encryption, triple-encryption and so on. If the pair of plaintexts (P_a, P_b) is a *slit* pair, then so is the corresponding pair of ciphertexts $(C_a^{(1)}, C_b^{(1)})$. Similarly the pair of the ciphertexts after n times encryption $(C_a^{(q_a-1)}, C_b^{(q_a-1)})$ also keeps the sliding property (Fig.3). Note that each pair of ciphertexts $(C_a^{(k)}, C_b^{(k)})$ is also thought of as the pair of the input and the output of the Round. Therefore the attacker can obtain the q_a pairs (including plaintext pair) of the input-output pairs of the Round function, with which the attacker can mount the known-plaintext attack $\mathcal{A}_R$.

To calculate the number of plaintexts, we apply the birthday paradox again. Consequently the naive method requires $2^{b/2}$ plaintexts to find a slit pair. An attacker use the distinguisher $\mathcal{D}_R$ to find a slit pair. For each possible pair of the collected plaintexts, the attacker invokes $\mathcal{D}_R$. If $\mathcal{D}_R$ returns yes, then the attacker treats the pair as the slit pair. Otherwise he discards the pair and try the next pair. To invoke $\mathcal{D}_R$ the attacker has to prepare q_d pairs of the input-output pairs of the Round. Hence the attacker prepares $q_d - 1$ ciphertexts for each plaintext.

We estimate the computational complexity for this attack. The attacker prepares $2^{b/2}$ plaintexts each of which is encrypted for $max(q_d, q_a)$ times. In total, it takes $max(q_d, q_a) \times 2^{b/2}$ computational time for generating data. The attacker makes $2^{b/2}(2^{b/2} - 1)/2$ times $\mathcal{D}$ invocations, and a $\mathcal{A}$ invocation. In total, the

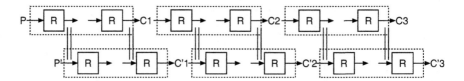

Fig. 3. A slide attack on multiple encryption.

computational complexity is estimated to be $max(q_d, q_a) \times 2^{b/2} + 2^{b/2}(2^{b/2} - 1)t_d/2 + t_a$.

In finding a slit pair, we must exploit a certain weakness of a cipher element iterated in the target cipher. Using this weakness, an attacker can distinguish a slit pair (with the weak property) and others (expectedly which holds a random property). As an example, an attacker targeting four-round iterated Feistel cipher can be interested in a probabilistic linear property, with which he test pairwise multiple ciphertexts to check if the probabilistic property of four rounds is detected. Practically, the technical way to find a slit pair is likely to relate to the way to attack the iterated cipher element. In the following, we demonstrate the typical analysis of our proposed enhancement.

3.2   Four-Round Iteration of DES

We consider a cipher holding four-round sliding property. An element of four-round iteration is not *weak* in the sense of original slide attacks by Biryukov and Wagner. Because of this reason the original slide attack cannot be applied to this cipher. Alternatively, they also proposed some advanced slide attacks where an attacker makes both encryption and decryption queries[4]. Their approach was to find a slit pair consisting of one plaintext-ciphertext pair and one ciphertext-plaintext pair.

In this section, we apply our enhancement to the DES-cipher treated in[4] which is intensionally weakened to hold four-round subkey iteration, i.e. a DES cipher with four-round sliding property. We define the model to attack. However our enhancement achieves to attack the DES variant *without* decryption query, namely adoptive chosen-plaintext attack in ECB, CBC, CFB or known-plaintext attack in OFB mode.

Let us demonstrate a simple example of a variant DES, ikDES4[1], with a simple key schedule. In ikDES4, the key schedule works as follows. The secret key K whose length is $4 \times 48 = 192$, is divided into four 48-bit strings, K_t, $(t = 1, 2, 3, 4)$. Subkeys are set as follows:

$$K_{4i-3} = K_1 \text{ for } 1 \leq i \leq 8$$
$$K_{4i-2} = K_2 \text{ for } 1 \leq i \leq 8$$
$$K_{4i-1} = K_3 \text{ for } 1 \leq i \leq 8$$
$$K_{4i} = K_4 \text{ for } 1 \leq i \leq 8$$

---

[1] ikDESn stands for Iteratively-Keyed DES of n rounds. We do not care the number of rounds in a whole cipher as long as it is multiple of n.

The data-randomizing part is identical to DES cipher[6] except for the number of rounds and lack of initial and final bitwise permutations.

First of all, a cryptanalytic tool exploited in the attack is introduced. This is the probabilistic linear relations of four rounds variant of DES data-randomizing part. This is approximated with $p_{DES4} = 1/2 - 1.95 \times 2^{-5}$, which is detectable with $C \times (p_{DES4}^{-2})$. C depends on out of how many candidates the slit pair is detected[13]. Since in any case of slide attacks (and its variants), a statistical characteristics of the correct slit pair must be identified out of a large number of incorrect pairs, C must be large enough. We, in this paper, assume that $C = 16$ gives enough to recognize out of less than 2^{64} corresponding random events.

It's easy to see that four rounds' slit pair, (P, C) and (P', C'), guarantees that each of those two pairs, (P, P') and (C, C') is a pair of input-output pair of DES-four rounds. Remember that in the original slide attack, an attacker can know no more than two input-output pairs even after he detects a slit pair. Consequently he cannot decide any information more than having about 2^{192-64} candidates, according to information theory. Although the result of this attack must be helpful to degrade the work effort of exhaustive search, this way of attacking does not use the full effect caused by iteration of round keys.

Now, an attacker tries to attack the cipher, gathering ciphertexts in the multiple encryption of the target cipher. It's easy to see that ciphers in multiple encryption keeps to be slit if the pair of plaintexts does. In this case, an attacker continues to gather double encryption ciphertext, triple encryption, and so on, until he gathers sufficient mount of pairwise data. If a pair of plaintext, (P_A, P_B), is a slit pair, the ciphers in single encryption, (C_A^1, C_B^1), the ciphers in double encryption, (C_A^2, C_B^2) and the ciphers in N times encryption, (C_A^n, C_B^n), each of which is a pair of input-output pair of four rounds. N must be the number for necessary plaintext pairs for linear cryptanalysis on four rounds DES, so that he exploits linear cryptanalysis on four round DES, stripping one or two rounds out of four applying maximum-likelihood-method on subkeys. Then the number of required input-output pair after finding a slit pair is

$$C(p'_{DES2})^{-2} = 8 \times (-20/64)^{-2} = 81.9,$$

for stripping two rounds. Note that in this analysis, we use $C = 8.0$ instead of 16, for a correct key in 2^{12} candidates in $S5$ box in round one, $S1$ box in round four. In successful attack, another parity of key bits in round three will be known to the attacker. Thirteen bits in total can be found in an attack (Fig. 4).

We omitted explanation of how to distinguish a slit pair from pools of pairs. In this case, he can use four round linear characteristic for each slit pair, (P_A, P_B), (C_A^1, C_B^1), (C_A^2, C_B^2), and so on, and check correctness of a testing pair. If all the pairs above are correctly input-output pair of DES four rounds, they must show explicit bias in statistics, whereas if they not, they will not. In this case, meeting the most bias with the correct pair requires

$$C \times (p'_{DES4})^{-2} \approx 4294.96,$$

$$p'_{DES4} = 2^2 \times (1/2 - 12/64)^2 (1/2 - 22/64).$$

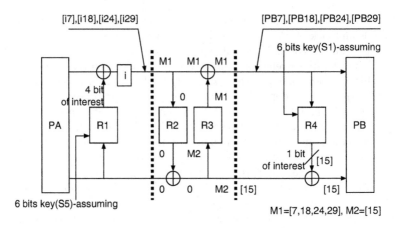

Fig. 4. Attack strategy in ikDES4.

This is the number of multiple encryptions enough to recognize the slit pair.

In terms of computational complexity in this attack, an attacker expects one slit pair in 2^{32} known-plaintexts pool. For each plaintext, he gathers 4294 ciphertext blocks, each of which is a resultant ciphertext of i-times multiple encryption, for $1 \le i \le 4294$, storing a bit information of parity masked according to linear approximation of four rounds DES. In this first stage of an attack, it takes $4294 \times 2^{32} \approx 2^{44}$ encryptions to gather data and 2^{44} times masked parity calculation. In order to save whole encryption results, about $2^{44} \times 64$-bit memory space is required. Nevertheless, 2^{44}-bit space is enough if an attacker can apply chosen plaintexts queries after finding a slit pair.

In the next stage, he tries to find a slit pair, checking bias in distribution of parity bit. For each pair out of about 2^{63} possible pairs, 4295-bit exclusive-or and bit increment for a counter. In 32-bit processor, 4295-bit exclusive-or operation will take 4295/32 clocks, since storing 4295 bits in $\lceil 4295/32 \rceil$ is possible during the first stage. $2^{63} \times \lceil 4295/32 \rceil \approx 2^{70}$ exclusive-or operations approximately correspond to 2^{62} encryption, since a DES encryption takes $45 \times 8 = 360$ cycles on a Pentium processor[18]. As for the memory space, this stage requires negligible memory space since all he needs are the maximum bias and its pair information.

In the third stage, deriving 13-bit key information takes small amount of time, like $2^{13} \times 82$ counter increments, in comparison with those on above two stages. In this stage, 2^{12} counters are required.

In total, work efforts equivalent to 2^{62} encryptions with 2^{44} chosen-plaintexts, enable to crack ikDES4, which is independent of the number of rounds as well as original slide attacks. The minimum memory requirement is about 2^{44} bit, i.e. 2000 GByte.

We summarize the results on It's dependent on the applying known-plaintext attack. It must be very light calculation so that the mount of checking is less than the work effort of key exhaustive search.

4 Key-Schedule Analyses on Block Ciphers

We describe our observations on key schedules in a couple of block ciphers, discussing applicability of our attack and effectiveness. The necessary conditions to apply our attack are very simple: a vulnerable cipher should have a sliding property (but not only ones with weak round functions); a cipher with a sliding property must be structured by a number of Round iterations and its non-negligible key space generate cyclic round keys synchronizing to iteration of round function. Most of the currently proposed block ciphers iterate identical round functions. Then our major observation begins with the structure of key schedule.

4.1 GOST

GOST is a 64-bit block cipher proposed from the former Soviet Union [7,17], keyed with 256-bit secret key and equips eight *secret* S-boxes. However, an example of S-boxes for GOST is disclosed and actually used in some applications.

We initially show a brief description of GOST cipher. A plaintext block, P, is divided into two 32-bit words, L_0 and R_0, and iterates a round function $R(L_i, R_i, K_i)$ for 32 times ($1 \leq i \leq 32$), where K_i is expanded keys generated by very simple key schedule. The ciphertext, C, is $L_{32}\|R_{32}$. The round function, R, is very simple. The input data, R, is added with the key, K_i. The result, $R_{i-1} + K_i$, is divided into eight four-bit data, each of which becomes a input of one of eight S-boxes. The results of S-boxes are concatenated to make a 32-bit word. Then eleven-bit left rotation is executed on the result and exclusive-ored with L_i, which generates the output for R_i. The other output, L_i, is R_{i-1}. In terms of key schedule of GOST, it adopts very simple one. A 256-bit key is divided into eight 32-bit words, $S_1, \ldots, S_8$. The round key, K_i is decided as follows:

$$K_i = S_{i \bmod 8} \text{ for } 1 \leq i \leq 24$$
$$K_i = S_{33-i} \quad \text{for } 25 \leq i \leq 32$$

Due to its key schedule, a slide attack chooses secret keys to hold the sliding property.

In this attack, the key to be attacked must be very particular. All the words in a key, K_i, ($1 \leq i \leq 32$), are identical. In total, 2^{32} keys are vulnerable against our attack.

Now all the 32 round functions are keyed with an identical key, so that the original slide attack is applicable to check the slit pair, that just sees identity of two halves of data of ciphertext pairs.

In this case, the original slide attack is applicable if the S boxes are known. More interestingly, our attack allows user of GOST with unknown S boxes to de-classify his cipher. The same approach has already been described in [16]. We briefly revisit the result.

Saarinen's Algorithm

1. Set K to be zero vector (namely all subkeys are zero, too),
2. Encrypt a zero-vector plaintext $(0,0)$ and find $(z,0)$ formatted sliding plaintext as the original slide attack, where (x,y) is denoted the left and right halves of plaintext or ciphertext data. Let a to be the common half data of both ciphertexts, namely a right half of zero plaintext's and a left half of sliding plaintext's.
3. The S box disclosure consists of $2^4 \times 2^4$ queries for each possible input v and output u, whether or not v is the output of the input u. Each query determines (a,b) values and the answer of queries should be given by sliding ciphertexts of $(a,0)$ and (b,a) as plaintext. Repeating these queries for all eight S boxes, 2^{11} queries are required to disclose whole contents of hidden S boxes.

As Saarinen mentions, a and b are defined by interested query of v and u. However it is very unlikely to hold *sliding property* for those two plaintexts $(a,0)$ and (b,a) even if the attacker knows z such that $z = f(0)$, where f is the zero-keyed F function of GOST. Consequently we claim that the Saarinen's algorithm, which still finds some elements of S box, lacks of flexibility to complete whole S box entries.

We introduce our enhancement to add the flexibility to the Saarinen's attack in order to fulfill the objective. Set one of 2^{32} vulnerable keys against one round sliding attack, and find a slit pair in the original way. For the next step, the attacker calculates n times multiple encryption each of plaintext of a slit pair. In each time of encryption, two ciphertexts are saved. With sufficient number of input-output pairs, all the elements in each S box are easily calculated since he knows the round key.

We consider the sufficient number of t in order to know all sixteen elements. We get the equation of the probability of choosing all sixteen elements after t times picking:

$$ p_t = \sum_{i=0}^{16} (-1)^i \binom{i}{16} \left(1 - \frac{i}{16}\right)^t . $$

In order to know all the elements in all eight S boxes, the probability to know all the elements is p_t^8. We show the graphical image of both two curves, p_t and p_t^8 in Fig.5. According to the probability, about 128 samples are enough to provide high probability to know all the elements in all eight S boxes.

The work effort of this attack includes computations for finding the slit pair (2^{32} encryptions and 2^{64} 32-bit data matching) and multiple encryptions of the slit pair to disclose S boxes (2×128). Since total amount is the sum of those two computations, that is approximated to 2^{32}.

4.2   MISTY

MISTY is a block cipher whose key schedule is designed relatively simple. After our detailed observation of MISTY's key schedule, we could find the very small

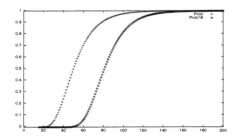

Fig. 5. Probabilities to choose all 16 elements.

Table 1. Key scheduling of MISTY.

Round keys for FO function				Round keys for FI function		
Round key	KO_{i1}	KO_{i2}	KO_{i3}	KO_{i4}	Round key	KI_{i1}	KI_{i2}	KI_{i3}
Key data	K_i	K_{i+2}	K_{i+7}	K_{i+4}	Key data	K'_{i+5}	K'_{i+1}	K'_{i+3}

Round keys for FL function	
Round key	KL_{i1}	KL_{i2}
Key data	$K_{\frac{i+1}{2}}$ (odd i)	$K'_{\frac{i+1}{2}+6}$ (odd i)
	$K'_{\frac{i}{2}+2}$ (even i)	$K_{\frac{i}{2}+4}$ (even i)

An index, $i(\le 8)$ is reduced to $i - 8$.

Round	KO_{i1}	KO_{i2}	KO_{i3}	KO_{i4}	KI_{i1}	KI_{i2}	KI_{i3}	KL_{i1}	KL_{i2}
Actual	K_i	K_{i+2}	K_{i+7}	K_{i+4}	K'_{i+5}	K'_{i+1}	K'_{i+3}	$K_{\frac{i+1}{2}}$ (odd i)	$K'_{\frac{i+1}{2}+6}$ (odd i)
								$K'_{\frac{i}{2}+2}$ (even i)	$K_{\frac{i}{2}+4}$ (even i)

An index, $j(\le 8)$ is reduced to $j - 8$.

key space of MISTY without FL functions which holds the sliding properties. However even with our enhancement of the slide attack, the key schedule of MISTY without FL functions is still resistant against slide attacks.

MISTY is based on provable security against differential and linear cryptanalyses [15]. If round keys are independent and uniformly distributed, three rounds of MISTY requires more than 2^{56} chosen(or known)-plaintext pairs for those two attacks.

In our study, we focused on the simplicity of the key-scheduling algorithm and investigated the possibility of our enhancement of slide attacks.

The key scheduling of MISTY is relatively simple. A 128-bit secret key, or eight 16-bit keys, K_i for $1 \le i \le 8$, is used to generate eight other 16-bit keys, K'_i for $1 \le i \le 8$, as follows:

$$K'_i = FI(K_i, K_{i+1}), 1 \le i \le 8,$$

where the index, 9, is reduced to 1. In the data-randomizing part, these 16 keys are used according to the following key schedule.

The first observation would agree with a specific characteristics of extended keys of a secret key that consists of identical 16-bit strings, i.e. eight K_is, for which resultant eight subkeys K_i's are identical. Therefore, for all rounds, KO_{i1}, KO_{i2}, KO_{i3}, and KO_{i4} are generated as the same 16-bit key, while KI_{i1}, KI_{i2}, and KI_{i3} are generated as other identical keys. There are 2^{16} keys, each of which serves the same keys for all rounds, apart from KL_{i1} and KL_{i2} functions. If we consider a modified model of MISTY, i.e. removing KL_{i1} and KL_{i2}, these 2^{16} keys allow for sliding properties.

In the sense of the original slide attacks, FO functions are not *weak*. However our extension of the slide attack could allow one to apply a known-plaintext attack on a round function, e.g. linear cryptanalysis to FO function. Note that the linear probability of FO function is proven to be less than 2^{-28}, whereas the input-output size is 32 bits.

The question is still open as to whether there exists a sufficiently effective known-plaintext cryptanalysis on FO round function. However, in comparison with the work effort required to find a slit pair, no less than $2^{32/2}$ to apply the birthday paradox, the effective key size for the sliding property, 2^{16}, is much smaller. From this reason, the key-scheduling algorithm of MISTY is still secure against our enhancement of slide attacks.

We also note cryptographic importance of FL functions. As a consequence of our study, the existence of FL functions and the keying rules to these functions make it very hard to find sliding property. It looks that full specification of MISTY with FL functions are most unlikely to be vulnerable against any kinds of slide attacks. In MISTY, FL functions are cheap way to make hedges against cryptanalyses exploiting particular characteristics of a cipher, such as the sliding properties.

5   Concluding Remarks

We described a novel way of combining slide attack and a known-plaintext cryptanalysis, and demonstrated some applications of the proposed enhancement. We also noted observations of key schedules properties relevant to slide attacks, discussing the applicability of our attacks.

The proposed idea enhances slide attacks from two points of view: (1) the target round function can be more generalized; and (2) the required condition for key schedule is untightened. The first point is that in our enhanced attack, the cipher does not necessarily iterate a *weak* round function. The second point is that the iterating number of identical subkeys is not limited to one or two. Theoretically, there is a possibility that a cipher with more round subkey iteration is vulnerable against our enhancement.

References

1. E. Biham, A. Shamir, "Differential Cryptanalysis of DES-like Cryptosystems," *Journal of Cryptology*, Vol.4, No.1, pp. 3-72, 1991. (The extended abstract was presented at CRYPTO'90.

2. E. Biham, A. Shamir, "Differential Cryptanalysis of the Data Encryption Standard," *Springer-Verlag*, 1993.

3. A. Biryukov, D. Wagner, "Slide attacks," *Preproceedings of FSE6, Fast Software Encryption Workshop 1999*, 1999.

4. A. Biryukov, D. Wagner, "Advanced Slide attacks," *Advances in Cryptology, - EUROCRYPT2000, LNCS Vol. 1807, Springer- Verlag*, 2000.

5. D.W. Davies, "Some Regular Properties of the 'Data Encryption Standard' algorithm," *Advances in Cryptology: Proceedings of CRYPTO82*, Plenum Press, 1983.

6. FIPS 46, "Data Encryption Standard," Federal Information Processing Standards Publication 46, U.S. Department of Commerce/National Bureau of Standards, National Technical Information Service, Springfield, Virginia, 1977 (revised as FIPS 46-1:1988, FIPS46-2:1993, FIPS46-3:1999).

7. GOST, Gosudarstvennyi Standard 28147-89, "Cryptographic Protection for Data Processing Systems," *Government Committee for the USSR for Standards*, 1989. (In Russian.)

8. L. R. Knudsen, "Cryptanalysis of LOKI91," *Advances in Cryptology, - ASIACRYPT'91, LNCS Vol. 739, Springer-Verlag*, 1991.

9. B. S. Kaliski, M. J. B. Robshaw, "Linear Cryptanalysis Using Multiple Approximations," *Advances in Cryptology, -CRYPTO'94, LNCS Vol. 839, Springer- Verlag*, 1994.

10. L. R. Knudsen, M. J. B. Robshaw, "Non-linear Approximations in Linear Cryptanalysis," *Advances in Cryptology, -EUROCRYPT'96, LNCS Vol. 1070, Springer-Verlag*, 1996.

11. J. Kelsey, B. Schneier, D. Wagner, "Key-Schedule Cryptanalysis of 3-WAY, IDEA, G-DES, RC4, SAFER, and Triple-DES," *Advances in Cryptology, -CRYPTO'96, LNCS Vol. 1109, Springer- Verlag*, 1996.

12. S. K. Langford, M. E. Hellman, "Differential- Linear Cryptanalysis," *Advances in Cryptology, -CRYPTO'94, LNCS Vol. 839, Springer-Verlag*, 1994.

13. M. Matsui, "Linear Cryptanalysis Method for DES Cipher," *Advances in Cryptology, -EUROCRYPT'93, LNCS Vol. 765, Springer-Verlag*, 1993.

14. M. Matsui, "The First Experimental Cryptanalysis of the Data Encryption Standard," *Advances in Cryptology, - CRYPTO'94, LNCS Vol.839, Springer-Verlag*, 1994.

15. M. Matsui, "New Block Encryption Algorithm MISTY," *Fast Software Encryption, 4th International Workshop, FSE'97, LNCS Vol. 1267, Springer-Verlag*, 1997.

16. M. J. Saarinen, "A chosen key attack against the secret S-boxes of GOST," unpublished, available at http://www.jyu.fi/~mjos/gost_cka.ps.

17. B. Schneier, "The GOST Encryption Algorithm," *Dr. Dobb's Journal*, Vol. 20, No. 2, 1995.

18. B. Schneier, D. Whiting "Fast Software Encryption: Designing Encryption Algorithms for Optimal Software Speed on the Intel Pentium Processor," *Fast Software Encryption, 4th International Workshop, FSE'97, LNCS Vol. 1267, Springer-Verlag*, 1997.

19. S. Vaudenay, "An experiment on DES statistical cryptanalysis," *Proc. of 3rd ACM CCCS*, 1996.

Constructions
of Cheating Immune Secret Sharing

Josef Pieprzyk[1] and Xian-Mo Zhang[2]

[1] Department of Computing
Macquarie University
Sydney, NSW 2109, Australia
josef@ics.mq.edu.au
[2] School of IT and CS, University of Wollongong
Wollongong NSW 2522, Australia
xianmo@cs.uow.edu.au

Abstract. The work addresses the problem of cheating prevention in secret sharing. Two cheating scenarios are considered. In the first one, the cheaters always submit invalid shares to the combiner. In the second one, the cheaters collectively decide which shares are to be modified so the combiner gets a mixture of valid and invalid shares from the cheaters. The secret scheme is said to be k-cheating immune if any group of k cheaters has no advantage over honest participants. The paper investigates cryptographic properties of the defining function of secret sharing so the scheme is k-cheating immune. Constructions of secret sharing immune against k cheaters are given.

1 Introduction

Secret sharing is the basic cryptographic tool that allows to define an environment in which the active entity is a group. A (t, n) threshold secret sharing scheme permits any group of t or more participants to access the secret. Any group of $t - 1$ or less participants cannot recover the secret. The group operation is normally performed by a trusted combiner who collects shares from participants, computes the result and communicates it to the members of the active group. Tompa and Woll [20] showed that a dishonest participant can cheat by providing an invalid share to the combiner. If the secret sharing in use is linear then the cheater is able to recover the valid secret from an invalid secret returned by the combiner. In effect, the cheater holds the secret while other (honest) participants are left with an invalid secret.

Cheating prevention becomes a major challenge in the distributed environment. Ideally, one would expect that a cheater should gain no advantage over honest participants. The problem can be addressed by forcing the combiner to check validity of shares before they are used to recover the secret. In the conditionally secure setting, shares can be checked using verifiable secret sharing (see [4,8,17,13]). In the unconditionally secure secret sharing, shares can be verified using a system of linear equations (see [11,2,1]). Note that share verification requires the combiner to be able to access the additional information (which also needs to be authenticated). This introduces extra complexity in the design and maintenance of secret sharing.

An alternative approach removes the main incentive for cheating. If one or more shares are invalid, then the invalid secret recovered by the combiner provides no information about the valid secret. In a sense, the cheaters position is similar to that of the honest participants except that the cheater knows that the recovered secret is invalid (in practice, the honest participants will learn about this with some delay when they try to use the invalid secret with a predictable failure).

The work in this paper covers the case where shares and the secret are binary. The non-binary case when shares are from $GF(p^t)$, was considered in [9]. Note that for the binary case, functions display some "special" characteristics not found when $p > 2$. In effect, constructions for binary case do not follow those for the case when shares are drawn from $GF(p^t)$. Moreover, design of strictly cheating immune secret sharing over $GF(p^t)$ is in general easier than over $GF(2)$.

This work uses a different concept of cheating prevention by removing the main incentive for cheating. The secret sharing is design in such a way that the group of cheaters has no advantage over honest participants. In the case of cheating, all participants (honest and dishonest) end up with an invalid secret and both honest and dishonest participants have the same probability of guessing the valid secret. This differentiate our approach from others (such as that in [3]) in which cheating prevention is done by share verification. In other words, the combiner will return secret only when all shares submitted are valid.

The work is structured as follows. Binary sequences are introduced in Section 2. An initial model of cheating is introduced in Section 3 and a lower bound on the probability of successful cheating is derived. The strengthened propagation is defined and its basic properties are investigated in Section 3.1. Secret sharing immune against k cheaters is studied in Section 3.2 and such secret sharing is constructed in Section 3.3. A generalised model of cheating where cheaters may submit a mixture of their valid and invalid shares, is considered in Section 4. Properties of secret sharing immune against the generalised cheating are examined in Section 4.1 and construction for such secret sharing is given in Section 4.2. In this paper we provide all proofs in the Appendix.

2   Binary Sequences

Let $GF(2)$ denote the binary field and V_n denote the vector space of n tuples of elements from $GF(2)$. Then each vector α can be expressed as $\alpha = (a_1, \ldots, a_n)$ where each $a_j \in GF(2)$. We consider a mapping f from V_n to $GF(2)$. f can be written as $f(x)$ or $f(x_1, \ldots, x_n)$, where $x = (x_1, \ldots, x_n)$ and each $x_j \in GF(2)$. f is also called a *function* on V_n. The *truth table* of f is a sequence defined by $(f(\alpha_0), f(\alpha_1), \ldots, f(\alpha_{2^n-1}))$, where $\alpha_0 = (0, \ldots, 0, 0)$, $\alpha_1 = (0, \ldots, 0, 1)$, $\ldots$, $\alpha_{2^n-1} = (1, \ldots, 1, 1)$. Each α_j is said to be the *binary representation* of integer j,

$j = 0, 1, \ldots, 2^n - 1$. A function f is said to be *balanced* if its truth table contains an equal number of zeros and ones.

An *affine* function f on V_n is a function that takes the form of $f(x_1, \ldots, x_n) = a_1 x_1 \oplus \cdots \oplus a_n x_n \oplus c$, where $\oplus$ denotes the addition in $GF(2)$, $a_j, c \in GF(2)$, $j = 1, 2, \ldots, n$. Furthermore f is called a *linear* function if $c = 0$. It is easy to verify that any nonzero affine function is balanced.

The *Hamming weight* of a vector $\alpha \in V_n$, denoted by $HW(\alpha)$, is the number of nonzero coordinates of α. The Hamming weight of a function f, denoted by $HW(f)$, is the number of nonzero terms in the truth table of f.

Let f be a function on V_n. We say that f satisfies the *propagation criterion with respect to* $\alpha \in V_n$ if $f(x) \oplus f(x \oplus \alpha)$ is a balanced function, where $x = (x_1, \ldots, x_n) \in V_n$ and $\alpha = (a_1, \ldots, a_n) \in V_n$. Furthermore f is said to satisfy the *propagation criterion of degree* k if it satisfies the propagation criterion with respect to every nonzero vector α whose Hamming weight is not larger than k [10]. Note that the *SAC (strict avalanche criterion)* [19] is the same as the propagation criterion of degree one.

Due to Lemma 3 of [22], we can give a k-resilient function an equivalent definition. A function f on V_n is said to be k-resilient if for every subset $\{j_1, \ldots, j_k\}$ of $\{1, \ldots, n\}$ and every $(a_1, \ldots, a_k) \in V_k$, $f(x_1, \ldots, x_n)|_{x_{j_1} = a_1, \ldots, x_{j_k} = a_k}$ is a balanced function on V_{n-k}. Additionally using Corollary 2 of [22], we can say that f is k-resilient if it is also t-resilient for any $t = 0, 1, \ldots, k$.

A vector $\alpha \in V_n$ is called a *linear structure* of f if $f(x) \oplus f(x \oplus \alpha)$ is a constant. For any function f, the zero vector on V_n is a linear structure. It is easy to verify that the set of all linear structures of a function f form a linear subspace of V_n, whose dimension is called the *linearity of* f.

Bent functions create a special class of functions. The class can be defined differently but all definitions are equivalent [12]. A function f on V_n is said to be *bent* if f satisfies the propagation criterion with respect to every nonzero vector in V_n. The sum of any bent function on V_n and any affine function on V_n is bent. Bent functions are not balanced and bent functions on V_n exist only when n is even.

3   Initial Model of Cheating

We see secret sharing as a set of distribution rules combined into a single table $\mathcal{T}$ (see [18]) with binary entries. We also assume that we are dealing with (n, n) threshold scheme where any n participants are able to determine a single entry from $\mathcal{T}$ which indicates the secret. Being more specific, the sequence of shares is $x = (x_1, \ldots, x_n)$ and the secret is $f(x)$ where $f : V_n \to \{0, 1\}$.

Our considerations are restricted to the case of (n, n) secret sharing. The general case of (n, N) secret sharing can be seen as a concatenation of (n, n) secret sharing with a system of N "consistent" linear equations. Shares are generated for N participants using the linear equations. Any n participants can get a system of linear equations with a unique solution which points out the unique row of the table $\mathcal{T}$.

Let $x = (x_1, \ldots, x_n)$ and $\delta = (\delta_1, \ldots, \delta_n)$ be two vectors in V_n. Define a vector in V_n, denoted by x_δ^+, whose j-th coordinate is x_j if $\delta_j = 1$, or 0 if $\delta_j = 0$. In addition, we denote a vector by x_δ^-, whose j-th coordinate is 0 if $\delta_j = 1$, or x_j if $\delta_j = 0$. For example, let $x = (x_1, x_2, x_3, x_4, x_5, x_6, x_7)$, and $\delta = (0, 1, 0, 1, 1, 0, 0)$ then $x_\delta^+ = (0, x_2, 0, x_4, x_5, 0, 0)$ and $x_\delta^- = (x_1, 0, x_3, 0, 0, x_6, x_7)$.

It is easy to verify the properties of operations x_δ^+ and x_δ^-: (i) $(\beta \oplus \gamma)_\delta^\pm = \beta_\delta^\pm \oplus \gamma_\delta^\pm$ holds for any two vectors β and γ in V_n, (ii) $\delta_\delta^+ = \delta$, $\delta_\delta^- = 0$, (iii) $\beta_\delta^+ \oplus \beta_\delta^- = \beta$ holds for any $\beta \in V_n$.

Given a function f on V_n. We introduce the following notations:

- Let $\alpha \in V_n$ be the sequence of shares held by the group $\mathcal{P} = \{P_1, \ldots, P_n\}$ of n participants and the secret $K = f(\alpha)$.
- The collection of cheaters is determined by the sequence $\delta = (\delta_1, \delta_2, \ldots, \delta_n)$ where P_i is a cheater if and only if $\delta_i = 1$.
- At the pooling time, the cheaters submit their shares. It is assumed that cheaters always submit invalid shares. The honest participants always submit their valid shares. We consider the vector $\alpha \oplus \delta$. From the properties of x_δ^+ and x_δ^-,

$$\alpha \oplus \delta = \alpha_\delta^- \oplus \alpha_\delta^+ \oplus \delta$$

Thus the combiner obtains $\alpha \oplus \delta$ that splits into two parts: α_δ^- – the part submitted by honest participants, and $\alpha_\delta^+ \oplus \delta$ – the part submitted by cheaters. The combiner returns an invalid secret $K^* = f(\alpha \oplus \delta)$. Note that the cheaters always change their shares.

- Let

$$\Omega_{\delta,\alpha}^* = \{x_\delta^- \mid f(x_\delta^- \oplus \alpha_\delta^+ \oplus \delta) = K^*\}$$

where α_δ^+ determines valid shares held by the cheaters. The set $\Omega_{\delta,\alpha}^*$ represents the view of the cheater after getting back K^* from the combiner. The set clearly includes also the vector α of all valid shares.

- The set

$$\Omega_{\delta,\alpha} = \{x_\delta^- \mid f(x_\delta^- \oplus \alpha_\delta^+) = K\}$$

determines a collection of rows of $\mathcal{T}$ with the correct secret K and valid shares held by the cheaters.

Example 1. Let $n = 7$, f be a function on V_7 and $\delta = (0, 1, 0, 1, 1, 0, 0)$. Furthermore let $\alpha = (a_1, a_2, a_3, a_4, a_5, a_6, a_7)$. Then $\alpha \oplus \delta = (a_1, 1 \oplus a_2, a_3, 1 \oplus a_4, 1 \oplus a_5, a_6, a_7)$. Let $K = f(\alpha)$ and $K^* = f(\alpha \oplus \delta)$. Write $x = (x_1, x_2, x_3, x_4, x_5, x_6, x_7)$. Clearly $x_\delta^- \oplus \alpha_\delta^+ = (x_1, a_2, x_3, a_4, a_5, x_6, x_7)$ and $x_\delta^- \oplus \alpha_\delta^+ \oplus \delta = (x_1, 1 \oplus a_2, x_3, 1 \oplus a_4, 1 \oplus a_5, x_6, x_7)$. Therefore $\Omega_{\delta,\alpha}^* = \{(x_1, 0, x_3, 0, 0, x_6, x_7) \mid f(x_1, 1 \oplus a_2, x_3, 1 \oplus a_4, 1 \oplus a_5, x_6, x_7) = K^*\}$ and $\Omega_{\delta,\alpha} = \{(x_1, 0, x_3, 0, 0, x_6, x_7) \mid f(x_1, a_2, x_3, a_4, a_5, x_6, x_7) = K\}$. In this example, P_2, P_4 and P_5 are cheaters and they all submit invalid shares.

The function f is called the *defining function* as it determines the secret sharing. The nonzero vector $\delta = (\delta_1, \ldots, \delta_n)$ is called the *cheating vector*, α is called the *original vector*, and $\alpha \oplus \delta$ is called the *failure vector*. The value of

$\rho_{\delta,\alpha} = \#(\Omega^*_{\delta,\alpha} \cap \Omega_{\delta,\alpha})/\#\Omega^*_{\delta,\alpha}$ expresses the probability of successful cheating with respect to δ and α. As the original vector α is always in $\Omega^*_{\delta,\alpha} \cap \Omega_{\delta,\alpha}$, the probability of successful cheating always satisfies $\rho_{\delta,\alpha} > 0$. Clearly the number of cheaters is equal to $HW(\delta)$.

Theorem 1. *Given secret and its defining function f on V_n. Then for any cheating vector $\delta \in V_n$ with $0 < HW(\delta) < n$ and any vector $\alpha \in V_n$, there exists a vector $\gamma \in V_n$ such that $\rho_{\delta,\alpha} + \rho_{\delta,\gamma} = 1$ otherwise $\rho_{\delta,\alpha} = 1$.*

3.1   Strengthened Propagation

We introduce the concept of strengthened propagation that is useful further in the paper. Let $\tau = (t_1, \ldots, t_n)$ and $\delta = (\delta_1, \ldots, \delta_n)$ be two vectors in V_n. We write $\tau \preceq \delta$ to denote the property that if $t_j = 1$ then $\delta_j = 1$. In addition, we write $\tau \prec \delta$ to denote the property that $\tau \preceq \delta$ and $\tau \neq \delta$. For example, $(0,1,0,0,1) \preceq (1,1,0,0,1)$ or precisely $(0,1,0,0,1) \prec (1,1,0,0,1)$. Clearly if $\tau \preceq \delta$ then $\tau \oplus \delta \preceq \delta$.

A function f on V_n is said to satisfy the *strengthened propagation* with respect to a nonzero vector $\delta \in V_n$ if for any vector τ with $\tau \preceq \delta$, $f(x^-_\delta \oplus \tau) \oplus f(x^-_\delta \oplus \tau \oplus \delta)$ is balanced. If f satisfies the strengthened propagation with respect to every $\delta \in V_k$ with $0 < HW(\delta) \leq k$ then f is said to satisfy the *strengthened propagation of degree k*.

We now illustrate the strengthened propagation. Let f be a function on V_4 such that $f(x_1, x_2, x_3, x_4) = x_1 x_2 \oplus x_3 x_4 \oplus x_1 x_3$. Let $\delta = (1,1,0,0)$. Choose $\tau = (0,0,0,0)$. Then $f(x^-_\delta \oplus \tau) = f(0,0,x_3,x_4) = x_3 x_4$ and $f(x^-_\delta \oplus \tau \oplus \delta) = f(1,1,x_3,x_4) = 1 \oplus x_3 x_4 \oplus x_3$. Thus $f(x^-_\delta \oplus \tau) \oplus f(x^-_\delta \oplus \tau \oplus \delta) = 1 \oplus x_3$ is balanced. Next we choose $\tau = (0,1,0,0)$. Then $f(x^-_\delta \oplus \tau) = f(0,1,x_3,x_4) = x_3 x_4$ and $f(x^-_\delta \oplus \tau \oplus \delta) = f(1,0,x_3,x_4) = x_3 x_4 \oplus x_3$. Thus $f(x^-_\delta \oplus \tau) \oplus f(x^-_\delta \oplus \tau \oplus \delta) = x_3$ is balanced. We have proved that f satisfies the strengthened propagation with respect to $\delta = (1,1,0,0)$.

Proposition 1. *Let f be a function on V_n. If f satisfies the strengthened propagation of degree k then f satisfies the propagation criterion of degree k.*

It should be noticed that the converse of Proposition 1 does not hold when $k \geq 2$. For example, $f(x_1, x_2, x_3, x_4) = x_1 x_2 \oplus x_3 x_4$ is a bent function on V_4 thus f satisfies the propagation criterion of degree 4. But f does not satisfy the strengthened propagation of degree 2. This can be seen from the following: $f(0,0,x_3,x_4) = x_3 x_4$ and $f(1,1,x_3,x_4) = 1 \oplus x_3 x_4$, and $f(0,0,x_3,x_4) \oplus f(1,1,x_3,x_4) = 1$. Therefore f does not satisfy the strengthened propagation with respect to $\delta = (1,1,0,0)$. However we can state as follows.

Proposition 2. *A function f on V_n satisfies the strengthened propagation of degree 1 if and only if f satisfies the propagation criterion of degree 1 (the SAC).*

Lemma 1. *If a function f on V_n satisfies the strengthened propagation of degree k then $f \oplus \psi$ also satisfies the strengthened propagation of degree k where ψ is any affine function on V_n.*

Lemma 2. *Let f_1 and f_2 be two functions on V_p and V_q respectively. Set $f(x) = f_1(y) \oplus f_2(z)$ where $x = (y, z)$, $y \in V_p$ and $z \in V_q$. Then (i) f is balanced if f_1 or f_2 is balanced, (ii) f satisfies the strengthened propagation of degree k if both f_1 and f_2 satisfy the strengthened propagation of degree k.*

3.2 k-Cheating Immune Secret Sharing Scheme

Clearly it is desirable that $\max\{\rho_{\delta,\alpha} | \delta \in V_n,\ \delta \neq 0,\ \alpha \in V_n\}$ is as small as possible. However if $\rho_{\delta,\alpha} < \frac{1}{2}$ for a nonzero vector δ and a vector $\alpha \in V_n$, from Theorem 1, there exists a vector $\gamma \in V_n$ such that $\rho_{\delta,\alpha} + \rho_{\delta,\gamma} = 1$ and then $\rho_{\delta,\gamma} > \frac{1}{2}$. This indicates that the case of $\min\{\rho_{\delta,\alpha} | \delta \in V_n,\ \delta \neq 0,\ \alpha \in V_n\} < \frac{1}{2}$ is not desirable. For this reason we introduce the concept of k-cheating immune secret sharing scheme.

Given secret sharing with its defining function f on V_n. Let k be an integer with $1 \leq k \leq n - 1$. The secret sharing is said to be k-*cheating immune* if $\rho_{\delta,\alpha} = \frac{1}{2}$ holds for every $\delta \in V_n$ with $1 \leq HW(\delta) \leq k$ and every $\alpha \in V_n$. The integer k is called the *order of cheating immunity* of the secret sharing.

1-cheating immune secret sharing is also called cheating immune secret sharing in [21]. The following is a characterisation of 1-cheating immune secret sharing [21]:

Theorem 2. *Given secret sharing with its defining function f on V_n. Then this secret sharing is 1-cheating immune if and only if f is 1-resilient and satisfies the SAC.*

The following result provides a relationship between k-cheating immune secret sharing and $(k - 1)$-cheating immune secret sharing:

Lemma 3. *Given secret sharing with its defining function f on V_n. Let this secret sharing be $(k-1)$-cheating immune. Then it is k-cheating immune if and only if the following two conditions are satisfied simultaneously: (i) f satisfies the strengthened propagation with respect to every vector in V_n with Hamming weight k, (ii) for any vector $\alpha \in V_n$ with $HW(\delta) = k$ and any vector $\tau \in V_n$ with $\tau \preceq \delta$, $f(x_{\bar{\delta}} \oplus \tau)$ is balanced.*

Theorem 3. *Given secret sharing with its defining function f on V_n. Then the secret sharing is k-cheating immune if and only if f is k-resilient and satisfies the strengthened propagation of degree k.*

3.3   Constructions of k-Cheating Immune Secret Sharing Scheme

Due to Theorem 3, to construct a k-cheating immune secret sharing, we need k-resilient functions satisfying the strengthened propagation of degree k. In particular we consider quadratic functions with such properties.

Proposition 3. *Let $f(x_1, \ldots, x_n)$ be a quadratic function on V_n. Let $\delta = (\delta_1, \ldots, \delta_n)$ be a nonzero vector in V_n. Set $J_\delta = \{j \mid \delta_j \neq 0, \ 1 \leq j \leq n\}$. For each integer i with $1 \leq i \leq n$ and $i \notin J_\delta$, define $D_\delta(i) = \{j \mid j \in J_\delta$ and $x_i x_j$ is a term of $f\}$. Then f satisfies the strengthened propagation with respect to δ if and only if there exists some i_0 with $1 \leq i_0 \leq n$ and $i_0 \notin J_\delta$ such that $\#D_\delta(i_0)$ is odd.*

The following will be used in constructions of desirable functions.

Corollary 1. *Let $f(x_1, \ldots, x_n)$ be a quadratic function on V_n. Then*

(i) *f satisfies the strengthened propagation with respect to $\delta = (0, \ldots, 0, 1, 0, \ldots, 0)$ where only the j-th coordinate is nonzero, if and only if there exists some s with $1 \leq s \leq n$ and $s \neq j$ such that $x_s x_j$ is a term of f,*

(ii) *f satisfies the strengthened propagation with respect to $\delta = (0, \ldots, 0, 1, 0, \ldots, 0, 1, 0, \ldots, 0)$ where only the j-th and i-th coordinates are nonzero, if and only if there exists some s with $1 \leq s \leq n$ and $s \neq j, i$ such that $x_s x_j$ is a term of f and $x_s x_i$ does not appear in f.*

The following is a restatement of a lemma in [7]:

Lemma 4. *Let a quadratic function f on V_n do not have a nonzero constant term, in other words, $f(0, \ldots, 0) = 0$. Then f is balanced if and only if there exists a nonzero linear structure $\alpha \in V_n$ such that $f(\alpha) \neq 0$.*

The following Lemma can be found in [22]:

Lemma 5. *Let f_j be a t_j-resilient function on V_{n_j}, $j = 1, \ldots, s$. Then $f_1(y) \oplus \cdots \oplus f_s(z)$ is an $(s - 1 + t_1 + \cdots + t_s)$-resilient function on $V_{n_1 + \cdots + n_s}$, where f_i and f_j have disjoint variables if $i \neq j$.*

Lemma 6. *Define two functions as follows*

$$\chi_{2k+1}(x_1, \ldots, x_{2k+1}) = x_1 x_2 \oplus x_2 x_3 \oplus \cdots \oplus x_{2k} x_{2k+1} \oplus x_{2k+1} x_1 \qquad (1)$$

$$\chi_{2k}(x_1, \ldots, x_{2k}) = x_1 \oplus x_1 x_2 \oplus x_2 x_3 \oplus \cdots \oplus x_{2k-1} x_{2k} \oplus x_{2k} x_1 \qquad (2)$$

Then

(i) *χ_{2k+1} is balanced, satisfies the strengthened propagation of degree k,*
(ii) *χ_{2k} is balanced, satisfies the strengthened propagation of degree $(k-1)$.*

Due to Theorem 3, the following constructions enable us to construct k-cheating immune secret sharing scheme.

Theorem 4. *Let k and s be positive integers with $s \geq k + 1$. Let $n_1, \ldots, n_s = 2k + 1$ or $2k + 2$, and $n = n_1 + \cdots + n_s$. Define a function on V_n such as $f(x) = \chi_{n_1}(y) \oplus \cdots \oplus \chi_{n_s}(z)$ where $x = (y, \ldots, z)$, $y \in V_{n_1}, \ldots, z \in V_{n_s}$, each χ_{n_j} has been defined in (1) or (2), and $\chi_{n_1}, \ldots, \chi_{n_s}$ have disjoint variables mutually. Then the secret sharing with the defining function f is k-cheating immune.*

Note that $n = n_1 + \cdots + n_s$, defined in Theorem 4, can be expressed as $n = (2k+1)r + (2k+2)q$ where $r \geq 0$ and $q \geq 0$ are integers. Since $2k+1$ and $2k+2$ are relatively prime, any integer can also be written as $(2k+1)r + (2k+2)q$ where r and q are integers. Furthermore it is easy to verify that any integer n with $n \geq (2k+1)^2$ can be expressed as $n = (2k+1)r + (2k+2)q$ where $r, q \geq 0$. Since $n \geq (2k+1)^2$, it is easy to verify that $s = r + q > k + 1$ where s was mentioned in Theorem 4. Using Theorem 4, we can construct k-cheating immune secret sharing with n participants where $n \geq (2k+1)^2$.

4   Generalised Model of Cheating

As before secret sharing is considered to be a set of distribution rules combined into a single table $\mathcal{T}$ (see [18]) with binary entries. We also assume that we are dealing with (n, n) threshold scheme where any n participants are able to determine a single entry from $\mathcal{T}$ which indicates the secret.

Given a function f on V_n. We introduce the following notations:

- Let $\alpha \in V_n$ be the sequence of shares held by the group $\mathcal{P} = \{P_1, \ldots, P_n\}$ of n participants and the secret $K = f(\alpha)$.
- The collection of cheaters is determined by the sequence $\delta = (\delta_1, \delta_2, \ldots, \delta_n)$ where P_i is a cheater if and only if $\delta_i = 1$.
- At the pooling time, the cheaters submit their shares. This time it is assumed that cheaters may submit a mixture of valid and invalid shares. The honest participants always submit their valid shares. Define $\tau = (\tau_1, \ldots, \tau_n)$ such that

$$\tau_j = \begin{cases} 0, \text{ if } P_j \text{ is honest or } P_j \text{ is a cheater who submits a valid share} \\ 1, \text{ if } P_j \text{ a cheater who submits an invalid share} \end{cases}$$

Clearly $\tau \preceq \delta$. We assume that there exists at least one cheater who submits invalid share, in other words, we only consider the case that τ is nonzero or $HW(\tau) > 0$.

We consider the vector $\alpha \oplus \tau$. Due to the properties of operations x_δ^+ and x_δ^-,

$$\alpha \oplus \tau = \alpha_\delta^- \oplus \alpha_\delta^+ \oplus \tau$$

The combiner obtains $\alpha \oplus \tau$ that splits into two parts: α_δ^- – the part submitted by honest participants and $\alpha_\delta^+ \oplus \tau$ the part submitted by cheaters. The combiner returns an invalid secret $K^* = f(\alpha \oplus \tau)$.

- Let

$$\Omega_{\delta,\tau,\alpha}^* = \{x_\delta^- \mid f(x_\delta^- \oplus \alpha_\delta^+ \oplus \tau) = K^*\}$$

where α_δ^+ determines valid shares held by the cheaters. The set $\Omega_{\delta,\tau,\alpha}^*$ represents the view of the cheater after getting back K^* from the combiner. The set clearly includes also the vector α of all valid shares.

– The set

$$\Omega_{\delta,\tau,\alpha} = \{x_{\delta}^- \mid f(x_{\delta}^- \oplus \alpha_{\delta}^+) = K\}$$

determines a collection of rows of $\mathcal{T}$ with the correct secret K and valid shares held by the cheaters.

Example 2. Let $n = 7$, f be a function on V_7 and $\delta = (0,1,0,1,1,0,0)$ and $\tau = (0,0,0,1,1,0,0)$. Furthermore let $\alpha = (a_1,a_2,a_3,a_4,a_5,a_6,a_7)$. Then $\alpha \oplus \tau = (a_1,a_2,a_3, 1 \oplus a_4, 1 \oplus a_5, a_6, a_7)$. Let $K = f(\alpha)$ and $K^* = f(\alpha \oplus \tau)$. Write $x = (x_1,x_2,x_3,x_4,x_5,x_6,x_7)$. Clearly $x_{\delta}^- \oplus \alpha_{\delta}^+ = (x_1,a_2,x_3,a_4,a_5,x_6,x_7)$ and $x_{\delta}^- \oplus \alpha_{\delta}^+ \oplus \tau = (x_1, a_2, x_3, 1 \oplus a_4, 1 \oplus a_5, x_6, x_7)$. Therefore $\Omega_{\delta,\tau,\alpha}^* = \{(x_1,0,x_3, 0,0, x_6,x_7) \mid f(x_1,a_2,x_3, 1 \oplus a_4, 1 \oplus a_5, x_6, x_7) = K^*\}$ and $\Omega_{\delta,\tau,\alpha} = \{(x_1,0,x_3, 0,0, x_6,x_7) \mid f(x_1,a_2,x_3, a_4,a_5,x_6,x_7) = K\}$. In this example P_2, P_4 and P_5 are cheaters but P_2 submits valid share.

From Examples 1 and 2, we can find a main difference between initial and generalised models of cheating. Clearly P_2, P_4 and P_5 are cheaters in both examples. However P_2, P_4 and P_5 all submit invalid shares in Example 1 while P_4, P_5 submit invalid shares and P_2 submits valid share in Example 2.

The function f is called the *defining function* as it determines the secret sharing. The nonzero vector $\delta = (\delta_1, \ldots, \delta_n)$ is called the *cheating vector*, the nonzero vector $\tau \preceq \delta$ is called *active cheating vector*, α is called the *original vector*, and $\alpha \oplus \tau$ is called the *failure vector*. The value of $\rho_{\delta,\tau,\alpha} = \#(\Omega_{\delta,\tau,\alpha}^* \cap \Omega_{\delta,\tau,\alpha})/\#\Omega_{\delta,\tau,\alpha}^*$ expresses the probability of successful cheating with respect to δ, τ and α. As the original vector α is always in $\Omega_{\delta,\tau,\alpha}^* \cap \Omega_{\delta,\tau,\alpha}$, the probability of successful cheating always satisfies $\rho_{\delta,\tau,\alpha} > 0$. Clearly the number of cheaters is equal to $HW(\delta)$ and the number of active cheaters is equal to $HW(\tau)$. In particular, if $\tau = \delta$, we regain the initial scheme. Therefore the initial model of cheating is a special case of the generalised model of cheating. From now, we consider secret sharing in the generalised model.

4.1   Strictly k-Cheating Immune Secret Sharing Scheme

By using the same arguments as in the proof of Theorem 1, we can prove

Theorem 5. *Given secret sharing with its defining function f on V_n. Then for any cheating vector $\delta \in V_n$, any active cheating vector $\tau \preceq \delta$ with $1 \leq HW(\tau) \leq HW(\delta) < n$, and any vector $\alpha \in V_n$, there exists a vector $\gamma \in V_n$ such that $\rho_{\delta,\tau,\alpha} + \rho_{\delta,\tau,\gamma} = 1$ otherwise $\rho_{\delta,\tau,\alpha} = 1$.*

For the same reason mentioned in Section 3.2, we introduce the concept of k-cheating immune secret sharing scheme.

Given secret sharing with its defining function f on V_n. Let k be an integer with $1 \leq k \leq n-1$. The secret sharing is said to be *strictly k-cheating immune* if the probability of successful cheating satisfies $\rho_{\delta,\tau,\alpha} = \frac{1}{2}$ for every $\delta \in V_n$ and any $\tau \preceq \delta$ with $1 \leq HW(\tau) \leq HW(\delta) \leq k$ and every $\alpha \in V_n$. The integer k is called the *order of strict cheating immunity* of the secret sharing.

Lemma 7. *Given secret sharing with its defining function f on V_n. Then the secret sharing is strictly k-cheating immune if and only if for any integer t with $0 \le t \le k - 1$, any subset $\{j_1, \ldots, j_t\}$ of $\{1, \ldots, n\}$ and any $a_1, \ldots, a_t \in GF(2)$, $f(x_1, \ldots, x_n)|_{x_{j_1} = a_1, \ldots, x_{j_t} = a_t}$, as a function on V_{n-t} with the variables $x_{i_1}, \ldots,$ $x_{i_{n-t}}$, where $\{i_1, \ldots, i_{n-t}\} \cup \{j_1, \ldots, j_t\} = \{1, \ldots, n\}$, is $(k - t)$-resilient and satisfies the strengthened propagation of degree $(k - t)$.*

Theorem 6. *Given secret sharing with its defining function f on V_n. Then the secret sharing is strictly k-cheating immune if and only if the following conditions are satisfied simultaneously: (i) f is k-resilient, (ii) for any integer t with $0 \le t \le k - 1$, any subset $\{j_1, \ldots, j_t\}$ of $\{1, \ldots, n\}$ and any $a_1, \ldots, a_t \in GF(2)$, $f(x_1, \ldots, x_n)|_{x_{j_1} = a_1, \ldots, x_{j_t} = a_t}$, as a function on V_{n-t} with the variables $x_{i_1}, \ldots, x_{i_{n-t}}$, where $\{i_1, \ldots, i_{n-t}\} \cup \{j_1, \ldots, j_t\} = \{1, \ldots, n\}$, satisfies the strengthened propagation of degree $(k - t)$.*

We indicate that the condition (ii) in Theorem 6 is more restrictive than the strengthened propagation of degree k. For example, due to Lemma 6, χ_{13} satisfies the strengthened propagation of degree 6. However χ_{13} does not satisfy the condition (ii) in Theorem 6 with $k = 3$ and $t = 2$. This can be seen from the following: $\chi_{13}(0, x_2, 0, x_4, \ldots x_{13}) = x_4 x_5 \oplus x_5 x_6 \oplus \cdots \oplus x_{12} x_{13}$, as a function on V_{11} with the variables $x_2, x_4, x_5, \ldots, x_{13}$, does not contain x_2. From (i) of Corollary 1, $\chi_{13}(0, x_2, 0, x_4, \ldots x_{13})$, as a function on V_{11} with the variables $x_2, x_4, x_5, \ldots, x_{13}$, does not satisfy the strengthened propagation of degree 1. Therefore χ_{13} does not satisfy the condition (ii) in Theorem 6 with $k = 3$ and $t = 2$.

4.2   Construction of Strictly k-Cheating Immune Secret Sharing Scheme

If f in Theorem 6 is quadratic, Theorem 6 can be simplified as follows:

Theorem 7. *Given secret sharing with its defining function f on V_n. Let f be quadratic. Then the secret sharing is strictly k-cheating immune if and only if the following two conditions are satisfied simultaneously: (i) f is k-resilient, (ii) for any integer t with $0 \le t \le k - 1$ and any subset $\{j_1, \ldots, j_t\}$ of $\{1, \ldots, n\}$, $f(x_1, \ldots, x_n)|_{x_{j_1} = 0, \ldots, x_{j_t} = 0}$, as a function on V_{n-t} with the variables $x_{i_1}, \ldots,$ $x_{i_{n-t}}$, where $\{i_1, \ldots, i_{n-t}\} \cup \{j_1, \ldots, j_t\} = \{1, \ldots, n\}$, satisfies the strengthened propagation of degree $(k - t)$.*

Construction of Strictly 2-Cheating Immune Secret Sharing Scheme

Let $s \ge 3$ be an integer, $n_1, \ldots, n_s = 5, 6$ and $n = n_1 + \cdots + n_s$. Define a function on V_n such as $f(x) = \chi_{n_1}(y) \oplus \cdots \oplus \chi_{n_s}(z)$ where $x = (y, \ldots, z)$, each χ_{n_j} has been defined in (1) or (2), and χ_{n_i}, χ_{n_j} have disjoint variables if $i \ne j$. Due to Theorem 4, the secret sharing with the defining function f is $(s - 1)$-cheating immune, where $s - 1 \ge 2$, and thus from Theorem 3, f is $(s - 1)$-resilient and satisfies the strengthened propagation of degree $s - 1$. Therefore f satisfies the condition (i) of Theorem 7 with $k = 2$.

Next we verify that f satisfies the condition (ii) of Theorem 7 with $k = 2$. Let $t = 0$ in the condition (ii) of Theorem 7 with $k = 2$. Due to Lemma 6 and Lemma 2, we know that f satisfies the strengthened propagation of degree 2. Let $t = 1$ in the condition (ii) of Theorem 7 with $k = 2$. Fix any j_0 with $1 \leq j_0 \leq n$. Note that each $1 \leq i \leq n$, x_i appears in two quadratic terms in f thus for any i with $1 \leq i \leq n$ and $i \neq j_0$, x_i appears in at least one quadratic term in $f(x)|_{x_{j_0}=0}$. From (i) of Corollary 1, we know that $f(x)|_{x_{j_0}=0}$ satisfies the strengthened propagation of degree 1.

We have proved that f satisfies the condition (ii) of Theorem 7 with $k = 2$. Therefore if we place f as the defining function of a secret sharing and then by Theorem 7 we conclude that this secret sharing is 2-cheating immune.

Note that 5 and 6 are relatively prime thus any integer can also be written as $5p + 6q$ where p and q are integers. Furthermore it is easy to verify that any integer $n \geq 20$ can be expressed as $n = 5p + 6q$ where $p \geq 0$ and $q \geq 0$ are integers. As for $n \leq 19$, n can be expressed as $n = 5p + 6q$ where $p > 0$ and $q \geq 0$ are integers when $n = 5, 6, 11, 12, 15, 16, 17$ and 18.

Therefore we can construct 2-cheating immune secret sharing with n participants where $n \geq 20$ or $n = 5, 6, 11, 12, 15, 16, 17$ and 18.

Construction of Strictly 3-Cheating Immune Secret Sharing Scheme

Set

$$h_9(x_1, x_2, x_3, x_4, x_5, x_6, x_7, x_8, x_9)$$
$$= x_1 \oplus \chi_9(x_1, x_2, x_3, x_4, x_5, x_6, x_7, x_8, x_9)$$
$$\oplus \chi_3(x_1, x_4, x_7) \oplus \chi_3(x_2, x_5, x_8) \oplus \chi_3(x_3, x_6, x_9)$$

where each χ_j has been defined in (1) or (2).

Since $(1, 1, 1, 1, 1, 1, 1, 1, 1) \in V_9$ is a nonzero linear structure of h_9 and $h_9(1, 1, 1, 1, 1, 1, 1, 1, 1) \neq 0$, from Lemma 4, we know that h_9 is balanced.

Set

$$h_{10}(x_1, x_2, x_3, x_4, x_5, x_6, x_7, x_8, x_9, x_{10})$$
$$= x_1 \oplus \chi_{10}(x_1, x_2, x_3, x_4, x_5, x_6, x_7, x_8, x_9, x_{10})$$
$$\oplus \chi_{10}(x_1, x_4, x_7, x_{10}, x_3, x_6, x_9, x_2, x_5, x_8)$$

Since $(1, 1, 1, 1, 1, 1, 1, 1, 1, 1) \in V_{10}$ is a nonzero linear structure of h_{10} and $h_{10}(1, 1, 1, 1, 1, 1, 1, 1, 1, 1) \neq 0$, from Lemma 4, we know that h_{10} is balanced.

Let $s \geq 4$ be an integer, $n_1, \ldots, n_s = 9, 10$ and $n = n_1 + \cdots + n_s$. Define a function on V_n such as $f(x) = h_{n_1}(y) \oplus \cdots \oplus h_{n_s}(z)$ where h_{n_i} and h_{n_j} have disjoint variables if $i \neq j$.

Since both h_9 and h_{10} are balanced, from Lemma 5, f is $(s - 1)$-resilient. Therefore f satisfies the condition (i) of Theorem 7 with $k = 3$.

Next we verify that f satisfies the condition (ii) of Theorem 7 with $k = 3$. Let $t = 0$ in the condition (ii) of Theorem 7 with $k = 3$. Using a straightforward verification, we know that h_9 (h_{10}) satisfies the condition mentioned in Proposition 3 for every $\delta \in V_9$ ($\delta \in V_{10}$) with $HW(\delta) = 1, 2, 3$. Thus both h_9 and h_{10} satisfy the strengthened propagation of degree 3. Due to Lemma 2, f satisfies the strengthened propagation of degree 3. Let $t = 1$ in the condition (ii) of Theorem 7 with $k = 3$. Fix any j_0 with $1 \leq j_0 \leq n$, it is easy to verify that $f(x)|_{x_{j_0}=0}$ satisfies the condition mentioned in (ii) of Corollary 1, and thus $f(x)|_{x_{j_0}=0}$ satisfies the strengthened propagation of degree 2. Let $t = 2$ in the condition (ii) of Theorem 7 with $k = 3$. Fix any j_0 and i_0 with $1 \leq j_0 < i_0 \leq n$. Note that each $1 \leq i \leq n$, x_i appears in four quadratic terms in f thus for any i with $1 \leq i \leq n$ and $i \neq j_0, i_0$, x_i appears in at least two quadratic terms in $f(x)|_{x_{j_0}=0,x_{i_0}=0}$. From (i) of Corollary 1, we know that $f(x)|_{x_{j_0}=0,x_{i_0}=0}$ satisfies the strengthened propagation of degree 1.

We have proved that f satisfies Theorem 7 with $k = 3$. Therefore if we place f as the defining function of a secret sharing and then by Theorem 7, we conclude that this secret sharing is 3-cheating immune.

Note that 9 and 10 are relatively prime thus any integer can also be written as $9p + 10q$ where p and q are integers. Furthermore it is easy to verify that any integer $n \geq 72$ can be expressed as $n = 9p + 10q$ where $p \geq 0$ and $q \geq 0$ are integers. As for $n \leq 71$, n can also be expressed as $n = 9p + 10q$ where $p \geq 0$ and $q \geq 0$ are integers when $n = 9, 10, 18, 19, 20, 27, 28, 29, 30, 36, 37, 38, 39, 40, 45, 46, 47, 48, 49, 50, 54, 55, 56, 57, 58, 59, 60, 61, 62, 63, 64, 65, 66, 67, 68, 69, 70$. Therefore we can construct 3-cheating immune secret sharing with n participants where $n \geq 72$ or $n = 9, 10, 18, 19, 20, 27, 28, 29, 30, 36, 37, 38, 39, 40, 45, 46, 47, 48, 49, 50, 54, 55, 56, 57, 58, 59, 60, 61, 62, 63, 64, 65, 66, 67, 68, 69, 70$.

Based on Theorem 7 and Proposition 3, we can continue to construct strictly k-cheating immune secret sharing scheme, $k = 4, 5, \ldots$. Due to the page limitation, this will be completed in the full paper.

5   Conclusions and Remarks

We have considered secret sharing and its resistance against cheating by a group of k dishonest participants. We have proved that the probability of successful cheating is always higher than $\frac{1}{2}$ if the participants hold binary shares. The secret scheme is said to be k-cheating immune if the probability of successful cheating is $\frac{1}{2}$ for any group of k or less participants. We have characterised k-cheating immune secret sharing scheme by examining its defining function. This characterisation enables us to construct k-cheating immune secret sharing scheme. Being more precise, we have studied two cases. In the first case, the group of cheaters always submit invalid shares. While in the second case, the group is more flexible as they collectively decide which of their shares should be modified and which should be submitted in their original form.

Acknowledgements

The first author was supported by the Large ARC Grant A00103078. The second author was supported by a Queen Elizabeth II Fellowship (227 23 1002).

References

1. M. Carpentieri. A perfect threshold secret sharing scheme to identify cheaters. *Designs, Codes and Cryptography*, 5(3):183–187, 1995.
2. M. Carpentieri, A. De Santis, and U. Vaccaro. Size of shares and probability of cheating in threshold schemes. In T. Helleseth, editor, *Advances in Cryptology - EUROCRYPT'93*, LNCS No 765, pages 118–125. Springer-Verlag, 1993.
3. C. Ding, D. Pei and A Salomaa. Chinese remainder theorem: applications in computing, coding and cryptography. World Scientific, Singapore, 1996
4. P. Feldman. A practical scheme for non-interactive verifiable secret sharing. In *Proceedings of the 28th IEEE Symposium on Foundations of Computer Science*, pages 427–437. IEEE, 1987.
5. F.J. MacWilliams and N.J.A. Sloane. *The theory of error-correcting codes*. North-Holland, Amsterdam, 1977.
6. K. Nyberg. On the construction of highly nonlinear permutations. In *Advances in Cryptology - EUROCRYPT'92*, LNCS No 658, pages 92–98. Springer-Verlag, 1993.
7. K. Nyberg and L. R. Knudsen. Provable security against differential cryptanalysis. In *Advances in Cryptology - CRYPTO'92*, LNCS No 740, pages 566–574. Springer-Verlag, 1993.
8. T.P. Pedersen. Non-interactive and information-theoretic secure verifiable secret sharing. In J. Feigenbaum, editor, *Advances in Cryptology - CRYPTO'91*, LNCS No 576, pages 129–140. Springer-Verlag, 1992.
9. J. Pieprzyk and X. M. Zhang. Cheating prevention in secret sharing over $GF(p^t)$. to appear in Indocrypt 2001.
10. B. Preneel, W. V. Leekwijck, L. V. Linden, R. Govaerts, and J. Vandewalle. Propagation characteristics of boolean functions. In *Advances in Cryptology - EURO-CRYPT'90*, LNCS No 437, pages 155–165. Springer-Verlag, 1991.
11. T. Rabin and M. Ben-Or. Verifiable secret sharing and multiparty protocols with honest majority. In *Proceedings of 21st ACM Symposium on Theory of Computing*, pages 73–85, 1989.
12. O. S. Rothaus. On "bent" functions. *Journal of Combinatorial Theory (A)*, 20:300–305, 1976.
13. B. Schoenmakers. A simple publicly verifiable secret sharing scheme and its application to electronic voting. In M. Wiener, editor, *Advances in Cryptology - CRYPTO'99*, LNCS No 1666, pages 148–164. Springer-Verlag , 1999.
14. J. Seberry, X. M. Zhang, and Y. Zheng. On constructions and nonlinearity of correlation immune functions. In *Advances in Cryptology - EUROCRYPT'93*, LNCS No 765, pages 181–199. Springer-Verlag, 1994.
15. J. Seberry, X. M. Zhang, and Y. Zheng. Nonlinearity and propagation characteristics of balanced boolean functions. *Information and Computation*, 119(1):1–13, 1995.
16. T. Siegenthaler. Correlation-immunity of nonlinear combining functions for cryptographic applications. *IEEE Transactions on Information Theory*, IT-30 No. 5:776–779, 1984.

17. M. Stadler. Publicly verifiable secret sharing. In U. Maurer, editor, *Advances in Cryptology - EUROCRYPT'96*, LNCS No 1070, pages 190–199. Springer-Verlag, 1996.
18. D.R. Stinson. *Cryptography: Theory and Practice*. CRC Press, 1995.
19. A. F. Webster and S. E. Tavares. On the design of S-boxes. In *Advances in Cryptology - CRYPTO'85*, LNCS No 219, pages 523–534. Springer-Verlag, 1986.
20. Martin Tompa and Heather Woll. How to share a secret with cheaters. In A.M. Odlyzko, editor, *Advances in Cryptology - CRYPTO'86*, LNCS No 263, pages 261–265. Springer-Verlag, 1987.
21. X. M. Zhang and J. Pieprzyk, Cheating immune secret sharing. to appear in *The Third International Conference on Information and Communication Security (ICICS)* 2001.
22. X. M. Zhang and Y. Zheng. Cryptographically resilient functions. *IEEE Transactions on Information Theory*, 43(5):1740–1747, 1997.

Appendix: Proofs of Propositions, Lemmas and Theorems

<u>The Proof of Theorem 1</u> Let $f(\alpha) = K$ and $f(\alpha \oplus \delta) = K^*$. Set $\Omega^*_{\delta,\alpha} = \{x^-_\delta \mid f(x^-_\delta \oplus \alpha^+_\delta \oplus \delta) = K^*\}$ and $\Omega_{\delta,\alpha} = \{x^-_\delta \mid f(x^-_\delta \oplus \alpha^+_\delta) = K\}$.

We partition $\Omega^*_{\delta,\alpha}$ into two parts: $\Omega^*_{\delta,\alpha} = \Omega^*_1 \cup \Omega^*_2$ where

$$\Omega^*_1 = \{x^-_\delta \mid f(x^-_\delta \oplus \alpha^+_\delta \oplus \delta) = K^*, \ f(x^-_\delta \oplus \alpha^+_\delta) = K\}$$

and

$$\Omega^*_2 = \{x^-_\delta \mid f(x^-_\delta \oplus \alpha^+_\delta \oplus \delta) = K^*, \ f(x^-_\delta \oplus \alpha^+_\delta) = K \oplus 1\}$$

Note that $\Omega^*_{\delta,\alpha} \cap \Omega_{\delta,\alpha} = \Omega^*_1$. Therefore

$$\rho_{\delta,\alpha} = \#(\Omega^*_{\delta,\alpha} \cap \Omega_{\delta,\alpha})/\#\Omega^*_{\delta,\alpha} = \#\Omega^*_1/\#\Omega^*_{\delta,\alpha} \tag{3}$$

There exist two cases to be considered: $\Omega^*_2 \neq \emptyset$, where $\emptyset$ denotes the empty set, and $\Omega^*_2 = \emptyset$.

Case 1: $\Omega^*_2 \neq \emptyset$. Then there exists a vector $\beta \in \Omega^*_2$. Thus

$$f(\beta^-_\delta \oplus \alpha^+_\delta \oplus \delta) = K^*, \ f(\beta^-_\delta \oplus \alpha^+_\delta) = K \oplus 1 \tag{4}$$

Set $\gamma = \beta^-_\delta \oplus \alpha^+_\delta$. Therefore (4) can be rewritten as $f(\gamma \oplus \delta) = K^*$, $f(\gamma) = K \oplus 1$. Clearly $\gamma^+_\delta = \alpha^+_\delta$ and $\gamma^-_\delta = \beta^-_\delta$.

Next we choose γ as the original vector. Therefore $\Omega^*_{\delta,\gamma} = \{x^-_\delta \mid f(x^-_\delta \oplus \gamma^+_\delta \oplus \delta) = K^*\}$ and $\Omega_{\delta,\gamma} = \{x^-_\delta \mid f(x^-_\delta \oplus \gamma^+_\delta) = K \oplus 1\}$. Since $\gamma^+_\delta = \alpha^+_\delta$, we have $\Omega^*_{\delta,\gamma} = \{x^-_\delta \mid f(x^-_\delta \oplus \alpha^+_\delta \oplus \delta) = K^*\}$ and $\Omega_{\delta,\gamma} = \{x^-_\delta \mid f(x^-_\delta \oplus \alpha^+_\delta) = K \oplus 1\}$. Clearly $\Omega^*_{\delta,\gamma} \cap \Omega_{\delta,\gamma} = \Omega^*_2$. Note that $\Omega^*_{\delta,\gamma}$ is identified with $\Omega^*_{\delta,\alpha}$. Therefore

$$\rho_{\delta,\gamma} = \#(\Omega^*_{\delta,\gamma} \cap \Omega_{\delta,\gamma})/\#\Omega^*_{\delta,\gamma} = \#\Omega^*_2/\#\Omega^*_{\delta,\gamma} = \#\Omega^*_2/\#\Omega^*_{\delta,\alpha} \tag{5}$$

Combining (3) and (5), and noticing $\#\Omega^*_{\delta,\alpha} = \#\Omega^*_1 + \#\Omega^*_2$, we have $\rho_{\delta,\alpha} + \rho_{\delta,\gamma} = 1$.

Case 2: $\Omega^*_2 = \emptyset$. Then $\Omega^*_{\delta,\alpha} = \Omega^*_1$. From (3), we have $\rho_{\delta,\alpha} = 1$. We have proved the theorem.

The Proof of Proposition 1 Fix any $\delta \in V_n$ with $0 < HW(\delta) \le k$. Let $\tau \preceq \delta$. Since f satisfies the strengthened propagation of degree k, $f(x_\delta^- \oplus \tau) \oplus f(x_\delta^- \oplus \tau \oplus \delta)$ is balanced. Since τ is an arbitrary vector with $\tau \preceq \delta$, $x_\delta^- \oplus \tau$ runs through every vector in V_n while τ and x_δ^- are as changed as possible. Therefore $f(x) \oplus f(x \oplus \delta)$ is balanced, i.e., f satisfies the propagation criterion with respect to δ. Since δ is an arbitrary vector in V_n with $0 < HW(\delta) \le k$, we have proved that f satisfies the propagation criterion of degree k.

The Proof of Proposition 2 The necessity is true due to Proposition 1. We now prove the sufficiency. Assume that f satisfies the SAC. Fix j with $1 \le j \le n$ and set

$$g(x_1, \ldots, x_{j-1}, x_j, x_{j+1}, \ldots, x_n)$$
$$= f(x_1, \ldots, x_{j-1}, x_j, x_{j+1}, \ldots, x_n) \oplus f(x_1, \ldots, x_{j-1}, x_j \oplus 1, x_{j+1}, \ldots, x_n)$$

Obviously $g(x_1, \ldots, x_{j-1}, 0, x_{j+1}, \ldots, x_n) = g(x_1, \ldots, x_{j-1}, 1, x_{j+1}, \ldots, x_n)$. Since f satisfies the SAC, $g(x_1, \ldots, x_{j-1}, x_j, x_{j+1}, \ldots, x_n)$ is balanced. Thus both $g(x_1, \ldots, x_{j-1}, 0, x_{j+1}, \ldots, x_n)$ and $g(x_1, \ldots, x_{j-1}, 1, x_{j+1}, \ldots, x_n)$ are balanced. This proves that f satisfies the strengthened propagation of degree 1.

The Proof of Lemma 1 In fact, since ψ is affine, $\psi(x_\delta^- \oplus \tau) \oplus \psi(x_\delta^- \oplus \tau \oplus \delta)$ is constant, where δ is any nonzero vector in V_n and τ is any vector in V_n with $\tau \preceq \delta$. Thus the lemma holds.

The Proof of Lemma 2 The part (i) can be found from Lemma 12 of [15]. By using (i) of the Lemma, we can verify the part (ii) of the lemma.

The Proof of Lemma 3 Let $\delta \in V_n$ be the cheating vector with $HW(\delta) = k$. Let τ be a vector in V_n with $\tau \preceq \delta$. Clearly $\delta_\delta^+ = \delta$, $\tau_\delta^+ = \tau$ and $\tau \oplus \delta_\delta^+ = \tau \oplus \delta$.
    Write $x = (x_1, \ldots, x_n)$. Set

$$R_0 = \{x_\delta^- \mid f(x_\delta^- \oplus \tau) = 0, f(x_\delta^- \oplus \tau \oplus \delta) = 0\}$$
$$R_1 = \{x_\delta^- \mid f(x_\delta^- \oplus \tau) = 0, f(x_\delta^- \oplus \tau \oplus \delta) = 1\}$$
$$R_2 = \{x_\delta^- \mid f(x_\delta^- \oplus \tau) = 1, f(x_\delta^- \oplus \tau \oplus \delta) = 0\}$$
$$R_3 = \{x_\delta^- \mid f(x_\delta^- \oplus \tau) = 1, f(x_\delta^- \oplus \tau \oplus \delta) = 1\} \tag{6}$$

Write $\#R_i = r_i$, $i = 0, 1, 2, 3$. Since $HW(\delta) = k$, it is easy to see that $r_1 + r_2 + r_3 + r_4 = 2^{n-k}$. By definition, it is easy to verify that

$$\rho_{\tau,\delta} = \begin{cases} \frac{r_0}{r_0+r_2} & \text{if } f(\tau \oplus \delta) = 0, f(\tau) = 0 \\ \frac{r_2}{r_0+r_2} & \text{if } f(\tau \oplus \delta) = 0, f(\tau) = 1 \\ \frac{r_1}{r_1+r_3} & \text{if } f(\tau \oplus \delta) = 1, f(\tau) = 0 \\ \frac{r_3}{r_1+r_3} & \text{if } f(\tau \oplus \delta) = 1, f(\tau) = 1 \end{cases} \tag{7}$$

Similarly,

$$\rho_{\tau \oplus \delta, \delta} = \begin{cases} \frac{r_0}{r_0+r_1} & \text{if } f(\tau) = 0, f(\tau \oplus \delta) = 0 \\ \frac{r_1}{r_0+r_1} & \text{if } f(\tau) = 0, f(\tau \oplus \delta) = 1 \\ \frac{r_2}{r_2+r_3} & \text{if } f(\tau) = 1, f(\tau \oplus \delta) = 0 \\ \frac{r_3}{r_2+r_3} & \text{if } f(\tau) = 1, f(\tau \oplus \delta) = 1 \end{cases} \tag{8}$$

Assume that the secret sharing is k-cheating immune. Let δ be the cheating vector and τ be the original vector. Since the scheme is k-cheating immune and $HW(\delta) = k$, we have $\rho_{\tau,\delta} = \frac{1}{2}$. Due to (7), $\frac{r_0}{r_0+r_2} = \frac{r_2}{r_0+r_2}$ and $\frac{r_1}{r_1+r_3} = \frac{r_3}{r_1+r_3}$. It follows that $r_0 = r_2$ and $r_1 = r_3$. On the other hand, since $\rho_{\tau\oplus\delta,\delta} = \frac{1}{2}$, due to (8), $\frac{r_0}{r_0+r_1} = \frac{r_1}{r_0+r_1}$ and $\frac{r_2}{r_2+r_3} = \frac{r_3}{r_2+r_3}$. It follows that $r_0 = r_1$ and $r_2 = r_3$. Therefore we have proved that $r_0 = r_1 = r_2 = r_3$.

Note that $\#\{x_\delta^- \mid f(x_\delta^- \oplus \tau) \oplus f(x_\delta^- \oplus \tau \oplus \delta) = 0\} = r_0 + r_3$ and $\#\{x_\delta^- \mid f(x_\delta^- \oplus \tau) \oplus f(x_\delta^- \oplus \tau \oplus \delta) = 1\} = r_1 + r_2$. Thus

$$\#\{x_\delta^- \mid f(x_\delta^- \oplus \tau) \oplus f(x_\delta^- \oplus \tau \oplus \delta) = 0\}$$
$$= \#\{x_\delta^- \mid f(x_\delta^- \oplus \tau) \oplus f(x_\delta^- \oplus \tau \oplus \delta) = 1\} \qquad (9)$$

From (9), $f(x_\delta^- \oplus \tau) \oplus f(x_\delta^- \oplus \tau \oplus \delta)$ is balanced. Since τ is an arbitrary vector in V_n with $\tau \preceq \delta$, f satisfies the strengthened propagation with respect to δ, where δ is an arbitrary vector in V_n with $HW(\delta) = k$. This proves that the condition (i) is satisfied.

We now consider the condition (ii). Note that $\#\{x_\delta^- \mid f(x_\delta^- \oplus \tau) = 0\} = r_0 + r_1$ and $\#\{x_\delta^- \mid f(x_\delta^- \oplus \tau) = 1\} = r_2 + r_3$. Since $r_0 = r_1 = r_2 = r_3$, we have $\#\{x_\delta^- \mid f(x_\delta^- \oplus \tau) = 0\} = \#\{x_\delta^- \mid f(x_\delta^- \oplus \tau) = 1\}$. This proves that $f(x_\delta^- \oplus \tau)$ is balanced, where δ is an arbitrary vector in V_n and τ is any vector in V_n with $\tau \preceq \delta$. Therefore the condition (ii) is satisfied.

Conversely assume that f satisfies (i) and (ii).

From the condition (i), for any vector $\delta \in V_n$ with $HW(\delta) = k$ and any vector $\tau \in V_n$ with $\tau \preceq \delta$, $f(x_\delta^- \oplus \tau)$ is balanced, thus we have $r_0 + r_3 = r_1 + r_2$.

From the condition (ii), $f(x_\delta^- \oplus \tau)$ is balanced, thus $r_0 + r_1 = r_2 + r_3$. By the same reasoning, $f(x_\delta^- \oplus \tau \oplus \delta)$ is also balanced, thus $r_0 + r_2 = r_1 + r_3$.

Therefore we conclude that $r_0 = r_1 = r_2 = r_3$. Due to (7), it follows that

$$\rho_{\tau,\delta} = \frac{1}{2} \qquad (10)$$

Next we prove that $\rho_{\delta,\alpha} = \frac{1}{2}$ for every $\alpha \in V_n$. Clearly $\alpha_\delta^+ \preceq \delta$, $\alpha_\delta^+ \oplus \delta \preceq \delta$. Replacing τ and $\tau \oplus \delta$ by α_δ^+ and $\alpha_\delta^+ \oplus \delta$ in (6) respectively, and using the same arguments for (10), we can prove that $\rho_{\delta,\alpha} = \frac{1}{2}$.

The Proof of Theorem 3 We prove the theorem by induction on k. Due to Theorem 2, the theorem is true when $k = 1$. Assume that the theorem is true when $1 \leq k \leq s - 1$. Consider the case of $k = s$.

We now prove the necessity. Assume that the secret sharing is s-cheating immune. Then it is also $(s - 1)$-cheating immune. Due to the assumption that the theorem is true when $1 \leq k \leq s - 1$, f is $(s - 1)$-resilient and satisfies the strengthened propagation of degree $(s - 1)$. Since the secret sharing is s-cheating immune, from the condition (i) of Lemma 3, f satisfies the strengthened propagation with respect to any vector in V_n with Hamming weight s. Therefore f satisfies the strengthened propagation of degree s. On the other hand, due to the condition (ii) of Lemma 3, for any vector $\alpha \in V_n$ with $HW(\delta) = k$ and any vector $\tau \in V_n$ with $\tau \preceq \delta$, $f(x_\delta^- \oplus \tau)$ is balanced. Combing this property and the fact that f is $(s - 1)$-resilient, we conclude that f is s-resilient.

Conversely assume that f is s-resilient and satisfies the strengthened propagation of degree s. Due to the assumption that the theorem is true when $1 \leq k \leq s - 1$, the secret sharing is $(s - 1)$-cheating immune. Since f satisfies the conditions (i) and (ii), due to Lemma 3, the secret sharing is s-cheating immune. We have proved the theorem when $k = s$. The proof is completed.

The Proof of Proposition 3 We generalise the notations J_δ and $D_\delta(i)$. For any $\tau = (\tau_1, \ldots, \tau_n) \preceq \delta$, set $J_\tau = \{j \mid \tau_j \neq 0, 1 \leq j \leq n\}$. For any i with $1 \leq i \leq n$ and $i \notin J_\delta$, define $D_\tau(i) = \{j \mid j \in J_\tau$ and $x_i x_j$ is a term of $f\}$.

It is easy to see that $x_j x_i$ is a quadratic term of $f(x_\delta^- \oplus \tau)$ if and only if $x_j x_i$ is a quadratic term of f with $j, i \notin J_\delta$. Similarly $x_j x_i$ is a quadratic term of $f(x_\delta^- \oplus \tau \oplus \delta)$ if and only if $x_j x_i$ is a quadratic term of f with $j, i \notin J_\delta$. Therefore $f(x_\delta^- \oplus \tau)$ and $f(x_\delta^- \oplus \tau \oplus \delta)$ have the same quadratic terms. Thus $f(x_\delta^- \oplus \tau) \oplus f(x_\delta^- \oplus \tau \oplus \delta)$ does not contain any quadratic term and thus we only need to consider affine terms in $f(x_\delta^- \oplus \tau)$ and $f(x_\delta^- \oplus \tau \oplus \delta)$.

First we assume that there exists some i_0 with $1 \leq i_0 \leq n$ and $i_0 \notin J_\delta$ such that $\#D_\delta(i_0)$ is odd. Since $i_0 \notin J_\delta$, we know that $i_0 \notin J_\tau$ and $i_0 \notin J_{\tau \oplus \delta}$. Note that x_{i_0} appears linearly in $f(x_\delta^- \oplus \tau)$ if and only if $\#D_\tau(i_0)$ is odd. Similarly x_{i_0} appears linearly in $f(x_\delta^- \oplus \tau \oplus \delta)$ if and only if $\#D_{\tau \oplus \delta}(i_0)$ is odd. Note that for $i_0 \notin J_\delta$, we have $\#D_\tau(i_0) + \#D_{\tau \oplus \delta}(i_0) = \#D_\delta(i_0)$. Since $\#D_\delta(i_0)$ is odd, x_{i_0} must appear linearly in $f(x_\delta^- \oplus \tau) \oplus f(x_\delta^- \oplus \tau \oplus \delta)$. This proves that $f(x_\delta^- \oplus \tau) \oplus f(x_\delta^- \oplus \tau \oplus \delta)$ is non-constant affine and then balanced. We have proved the sufficiency.

Conversely assume that f satisfies the strengthened propagation with respect to δ. We now prove the necessity by contradiction. Assume that $\#D_\delta(i)$ is even for each $i \notin J_\delta$. From the proof of the sufficiency, for each $i \notin J_\delta$, x_j cannot appear in $f(x_\delta^- \oplus \tau) \oplus f(x_\delta^- \oplus \tau \oplus \delta)$. This implies that $f(x_\delta^- \oplus \tau) \oplus f(x_\delta^- \oplus \tau \oplus \delta)$ is constant and then unbalanced. This contradicts the assumption that assume that f satisfies the strengthened propagation with respect to δ. The contradiction proves the necessity.

The Proof of Lemma 6 (i) Since $(1, \ldots, 1) \in V_{2k+1}$ is a nonzero linear structure of χ_{2k+1} and $\chi_{2k+1}(1, \ldots, 1) \neq 0$, from Lemma 4, we know that χ_{2k+1} is balanced. Let δ be a nonzero vector in V_{2k+1} with $0 < HW(\delta) \leq k$. Since $1 \leq \#J_\delta = HW(\delta) \leq k$, where J_δ has been defined in Proposition 3, there must exist an integer s with $1 \leq s \leq 2k + 1$ such that $s \in J_\delta$ and $s + 1, s + 2 \notin J_\delta$ (if $s = 2k$ then $s + 2 = 2k + 2$ is regarded as 1, and if $s = 2k + 1$ then $s + 1 = 2k + 2$ and $s + 2 = 2k + 3$ are regarded as 1 and 2 respectively). Clearly $\#D_\delta(s+1) = 1$. From Proposition 3, we know that χ_{2k+1} satisfies the strengthened propagation respect to δ. Since δ is an arbitrary nonzero vector in V_{2k+1} with $0 < HW(\delta) \leq k$. We have proved the part (i) of the lemma.

(ii) Since $(1, \ldots, 1) \in V_{2k}$ is a nonzero linear structure of χ_{2k} and $\chi_{2k}(1, \ldots, 1) \neq 0$, from Lemma 4, we know that χ_{2k} is balanced. Using the same arguments in the proof of the part (i), we complete the proof of the part (ii).

The Proof of Theorem 4 Due to Lemma 6, each χ_{n_j} is balanced. From Lemma 2, f is $(s - 1)$-resilient, where $s - 1 \geq k$. Using Lemma 6 and Lemma 2, we know that f satisfies the strengthened propagation of degree k. Using Theorem 3, we have proved that the secret sharing is k-immune.

The Proof of Lemma 7 Assume that the secret sharing is strictly k-cheating immune. Let g be a function on V_{n-t} given by $g = f(x_1, \ldots, x_n)|_{x_{j_1}=a_1,\ldots,x_{j_t}=a_t}$. Since f is the defining function on V_n of a strictly k-cheating immune secret sharing in generalised model of cheating, we know that g is the defining function on V_{n-t} of a $(k-t)$-cheating immune secret sharing in initial model of cheating. Applying Theorem 3 to g, we conclude that g is $(k-t)$-resilient and satisfies the strengthened propagation of degree $(k-t)$. We have proved the necessity. Comparing generalised model of cheating with initial model of cheating, we can invert the above reasoning and then prove the sufficiency.

The Proof of Theorem 6 Comparing Theorem 6 with Lemma 7, due to a definition of k-resilient functions mentioned in Section 2, it is easy to see the equivalence between Theorem 6 and Lemma 7.

The Proof of Theorem 7 Due to Theorem 6, we only need to prove the following lemma called Lemma (C): "let f be a quadratic function on V_n, t be an integer with $0 \leq t < n$ and $\{j_1, \ldots, j_t\}$ be a subset of $\{1, \ldots, n\}$. Then for any $a_1, \ldots, a_t \in GF(2)$, $f(x_1, \ldots, x_n)|_{x_{j_1}=a_1,\ldots,x_{j_t}=a_t}$, as a function on V_{n-t} with the variables $x_{i_1}, \ldots, x_{i_{n-t}}$, where $\{i_1, \ldots, i_{n-t}\} \cup \{j_1, \ldots, j_t\} = \{1, \ldots, n\}$, satisfies the strengthened propagation of degree s if and only if $f(x_1, \ldots, x_n)|_{x_{j_1}=0,\ldots,x_{j_t}=0}$ satisfies the strengthened propagation of degree s".

Since the necessity is obvious, we only need to prove the sufficiency. It is easy to verify that $x_j x_i$ is a quadratic term of $f(x_1, \ldots, x_n)|_{x_{j_1}=a_1,\ldots,x_{j_t}=a_t}$ if and only if $x_j x_i$ is a quadratic term of f with $j, i \notin \{j_1, \ldots, j_t\}$. Similarly $x_j x_i$ is a quadratic term of $f(x_1, \ldots, x_n)|_{x_{j_1}=0,\ldots,x_{j_t}=0}$ if and only if $x_j x_i$ is a quadratic term of f with $j, i \notin \{j_1, \ldots, j_t\}$. Then $f(x_1, \ldots, x_n)|_{x_{j_1}=a_1,\ldots,x_{j_t}=a_t}$ and $f(x_1, \ldots, x_n)|_{x_{j_1}=0,\ldots,x_{j_t}=0}$ have the same quadratic terms. Therefore $f(x_1, \ldots, x_n)|_{x_{j_1}=a_1,\ldots,x_{j_t}=a_t}$ can be expressed as

$$f(x_1, \ldots, x_n)|_{x_{j_1}=a_1,\ldots,x_{j_t}=a_t} = f(x_1, \ldots, x_n)|_{x_{j_1}=0,\ldots,x_{j_t}=0} \oplus \psi(x_{i_1}, \ldots, x_{i_{n-t}})$$

where ψ is an affine function on V_{n-t}.

Assume that $f(x_1, \ldots, x_n)|_{x_{j_1}=0,\ldots,x_{j_t}=0}$, as a function on V_{n-t} with the variables $x_{i_1}, \ldots, x_{i_{n-t}}$, where $\{i_1, \ldots, i_{n-t}\} \cup \{j_1, \ldots, j_t\} = \{1, \ldots, n\}$, satisfies the strengthened propagation of degree s. By using Lemma 1, we conclude that $f(x_1, \ldots, x_n)|_{x_{j_1}=a_1,\ldots,x_{j_t}=a_t}$, satisfies the strengthened propagation of degree s. We have proved Lemma (C) and thus the theorem is true.

Private Computation with Shared Randomness over Broadcast Channel[*]

Clemente Galdi and Pino Persiano

Dipartimento di Informatica ed Applicazioni
Università di Salerno, 84081 Baronissi (SA), Italy
{clegal,giuper}@dia.unisa.it

Abstract. In this paper we introduce a new model for private computation, the Shared Randomness over Broadcast Channel model (SRoBC for short). Following the classical model for private computation [2,4], we consider a set of n computationally unbounded honest but curious players $P_1, \ldots, P_n$ with private inputs $x_1, x_2, \ldots, x_n$. The players wish to compute a function f of their inputs in a private way. Unlike in the classical model, no private channel is available to the players but all the communication takes place using a broadcast channel. Moreover, the only available source of randomness is a shared random string.

We show that even in this minimal setting private computation is possible: we present a protocol for computing the sum modulo 2 in a t-private way in the SRoBC model. The protocol uses $n(t+1)/2$ random bits. We show that this is the optimal randomness complexity in the case each random bit is shared between two players (low-contention protocols).

We further show that, in the case $t = 1$, this protocol is optimal with respect to the randomness complexity regardless of the contention of the protocol.

Keywords: Secure Distributed Protocols; Private Function Evaluation; Broadcast Channel.

1 Introduction

The study and the design of protocols for private computation has been a fecund line of research in distributed computation and cryptography. Typically, we have a set of n players $\mathcal{P} = \{P_1, P_2, \ldots, P_n\}$ with unbounded computational resources each holding a private input that wish to compute a function f of their inputs in a *private* way. This means that after the protocol has been carried out every player knows the value of the function but has no information on the inputs of the other players except what can be deduced from his own input and the value of the function. This notion of private computation can be extended to the notion of t-private computation where the requirement that no information is obtained is required to hold for arbitrary coalitions of t players.

Two fundamental ingredients for private computations are randomness and private channels. A private channel is point-to-point communication channel

---

[*] Partially supported by a Research Grant of the Università di Salerno.

K. Kim (Ed.): ICICS 2001, LNCS 2288, pp. 244–257, 2002.
© Springer-Verlag Berlin Heidelberg 2002

between two parties P_i and P_j that can be read or written only by P_i and P_j. In fact, it is well know that no non-trivial function can be computed if the players are deterministic or if all communications have to take place using broadcast communication channels. A broadcast channel is a communication channel that can be read or written by all parties. This is unfortunate as randomness is a scarce resource and the privacy of a communication channel is most often hard to assess.

In this paper we introduce the Shared Randomness over Broadcast Channel model (SRoBC for short) for private computation. As we shall see, the SRoBC is a very restricted model for private computation but, nonetheless, we show that it is possible to perform some *non-trivial* private computations. In the SRoBC model parties can communicate only by means of a broadcast channel and have access to a shared random string. The protocol specifies for each party which random bits of the shared random string to read. Following the classical model for private computation, we assume the parties to be honest; that is they follow the protocol specification and thus read only the random bits they have been assigned.

The SRoBC model is similar to the model presented in [5,12], but we point out the following differences. In [5,12] the main objective was to obtain anonymous communication, whereas our main goal is multiparty computation. Moreover, our research aims at a quantitative study of the randomness complexity of private multiparty computation in the SRoBC model.

Another related model for private computation has been proposed by Feige *et. al.* in [8]. There, we have 2 players, each connected through a private channel to a distinguished player C whose only task is to allow the communication among the players. In this model they show, quite surprisingly, that all functions have secure protocol and that all functions in NL have efficient secure protocols. The extension of this model to n players has been considered in [10].

We stress here that our model is much more restrictive. Indeed all protocols in our model can be simulated in the standard model and thus all impossibility result apply in our model as well. Moreover in the model of [8,10] it only makes sense to consider 1-private protocols. Our model instead allows a meaningful definition of t-privacy also for $t > 1$.

Our first result is a t-private protocol for computing the modular sum of n inputs in the SRoBC using $n(t+1)/2$ random bits. We also show that this is the minimum number of bits necessary, if each random bit is shared by 2 players. The proof uses graph theoretic tools: the relation between players and random bits is encoded by a graph, with each player being a vertex and each random bit an edge, and then we show that the graph has to satisfy certain connectivity requirements.

We further show a lower bound on the number of random bits for the case of 1-private protocols with no assumption on the number of players that share each random bit. We prove that, in this case, at least n bits are required to execute the protocols and, thus, that our protocol is optimal with respect to the randomness complexity.

1.1 Notations and definitions

A protocol is said to be r-random if the total number of random bits used by the players in every execution of the protocol does not exceed r.

We consider the protocols divided in rounds of interactions. Each round is identified from the following steps: receiving messages, performing local computation, sending messages. A protocol is said to be m-round if for every possible input and for any possible choice of the random string, the protocol terminates after at most m rounds.

A protocol is said to be d-message if, for every possible input vector and for every choice of the random string, the total number of bits exchanged during the protocol execution does not exceeds d.

In the following, boldface capital letters, say $\mathbf{X}$, denote a random variable taking value on a set denoted by the corresponding capital letter, $X = \{x_1, \ldots, x_{|X|}\}$, according to some probability distribution $\{Pr_\mathbf{X}(x)\}_{x \in X}$.

Let $\mathbf{X}$ be the random variable describing the players' inputs during the execution of the algorithm, let $\mathbf{R}$ be the random variable describing the random tape the players use and let $\mathbf{M}$ be the random variable representing the messages sent by the players. Let $Y = \{i_1, \ldots, i_k\} \subset [n] = \{1, \ldots, n\}$ and $i \in [n]$ We denote by $\mathbf{X}_i$ (resp., $\mathbf{R}_i$, $\mathbf{M}_i$) the inputs (resp., the random tape, the messages) of player P_i and by $\mathbf{X}_Y$ (resp., $\mathbf{R}_Y$, $\mathbf{M}_Y$) the inputs (resp., the random tape, the messages) of the players in Y. In other words, it is possible to write $\mathbf{X}_Y = \mathbf{X}_{i_1}, \ldots, \mathbf{X}_{i_k}$, $\mathbf{X} = \mathbf{X}_1 \mathbf{X}_2 \ldots, \mathbf{X}_n$, $\mathbf{M} = \mathbf{M}_1 \mathbf{M}_2 \ldots, \mathbf{M}_n$, and $\mathbf{R} = \mathbf{R}_1 \mathbf{R}_2 \ldots, \mathbf{R}_n$. Analogously, it is possible to write $x_Y = x_{i_1}, \ldots, x_{i_k}$, $x = x_1 x_2 \ldots, x_n$, and $r = r_1 r_2 \ldots, r_n$.

Definition 1 (Privacy). *A n-player protocol for computing a function f is t-private if for any subset $T \subseteq \mathcal{P}$ of at most t players, for any input vectors x, y such that $f(x) = f(y)$ and such that $x_i = y_i$ for each $i \in T$, for any communication string m and for any random string r_T,*

$$Pr\left[\mathbf{M} = m | \mathbf{R}_T = r_T, \mathbf{X} = x\right] = Pr\left[\mathbf{M} = m | \mathbf{R}_T = r_T, \mathbf{X} = y\right]$$

where the distribution probability is taken on the random string of the other players.

2 The SRoBC Model for t-Private Computation

We consider the following model for private computation.

There is a set $\mathcal{P} = \{P_1, P_2, \ldots, P_n\}$ of n players, each holding a private input x_i taken from $\{0, 1\}$. The players have unbounded computational resources and are supposed to be honest, that means that each player follows the protocol, but curious, that is, after the execution of the protocol some players can try to infer information about the private inputs of the remaining players by collecting the messages they have received during the protocol execution. The players wish to distributively compute a function f of their private inputs

We assume that there exists a string R of random bits and the protocol specifies which bits from the string each player reads. Each player reads his "assigned" random bits in a private way and, as players are honest, each random bit will be read exclusively by those players that are supposed to read it. We encode the assignment of bits to players through a "distribution matrix" $\mathcal{M}$. A distribution matrix is a 0-1 matrix which has one row for each player and one column for each random bit of R and player i reads the j-th bit in $\mathcal{R}$ iff $\mathcal{M}(i,j) = 1$. We denote by S_i the set of indices corresponding to the bits read by player i; i.e., $j \in S_i$ iff $\mathcal{M}(i,j) = 1$.

Moreover the players are connected by means of a broadcast channel that is, each message sent by player P_i is read by all other players. Thus each player can read random bits privately, but can not send private messages to any other player.

2.1   Equivalence between Two Models

In this section we give evidence that the model presented in this paper is a non-stronger model then the standard one in the sense that all the protocols that can be executed in the SRoBC model can be simulated in the standard one. On the other hand we give simulation in the SRoBC model only for a small class of protocols that can be executed in the standard model.

Lemma 2. *Let $\mathcal{A}$ be an m-round, d-message, r-random t-private n-player protocol that computes a function f in the SRoBC model. There exists an $(m+1)$-round, $(d(n-1) + r(k-1))$-message, r-random, t-private n-player protocol $\mathcal{B}$ that computes the same function in the standard model, where k is the maximum number of players sharing the same random bit.*

Proof. To run this simulation, the players executing the protocol $\mathcal{B}$ need the following two primitive: (a) privately reading a bit from the shared random string and (b) broadcasting a message.

In the first round, the players in the standard model simulate the random bits distribution as follows: Assume the players $P_{i_1}, \dots, P_{i_k}$ share a random bit and, w.l.o.g., assume that $i_1 = min\{i_1, \dots, i_k\}$. The players P_{i_1} in $\mathcal{B}$ generates a random bit b and sends it to any other player in the set. As the communication channel is private, only players $P_{i_1}, \dots, P_{i_j}$ know the value of b. The number of messages exchanged for this simulation is, thus, $r(k-1)$. To simulate the broadcast of a message m, the sender simply sends the same message to all the players. Thus the cost for broadcasting a message is $n-1$. The correctness and the privacy of the protocol $\mathcal{B}$ follow immediately from the correctness of $\mathcal{A}$.

Let us consider the private protocols for which the behaviour of the players is independent on the inputs and on the random string. The following lemma shows that it is possible to simulate these protocols in the SRoBC model.

Lemma 3. *Let $\mathcal{A}$ be an m-round, d-message, r-random t-private n-player protocol that computes a function f in the standard model. There exists an m-round, d-message, $(r+d)$-random, t-private n-player protocol $\mathcal{B}$ that computes the same function in the SRoBC model.*

Proof. The players in the standard model needs the following two primitive: (a) privately generating a random bit and (b) sending a message on a point-to-point private channel.

Let us denote by r_i the number of random bits read by player P_i during the execution of the protocol $\mathcal{A}$. Recall that, since the protocol is independent on the inputs and on the random string, these r_is are known *a priori*. Player P_1 reads the first r_1 random bits from the shared random string, player P_2 reads r_2 new random bits from the random string and so on. In this way, each player holds the necessary private random source for his execution.

The total number of bits read in this phase is, by hypothesis, r. The shared random string has d random bits left.

We use these random bits to simulate the private channels. In particular, we associate to each (one-bit) message a random bit in the shared random string. As the protocol is independent on the inputs and on the random coins tosses, the sender and the receiver of each message sent during the execution of protocol $\mathcal{A}$ are also known *a priori*. Thus it is possible for the players to read the random bits they will need to send and receive private message in a first phase.

3 t-Private Distribution Matrices

Security and correctness of the protocols in the SRoBC model, critically depends on certain properties of the distribution matrix used to encode the association between players and random bits. We begin by giving the definition of t-private distribution matrices. As we shall see in Section 6, a t-private distribution matrix with n rows can be used to construct a t-private n-player protocol for computing the modular sum.

Let A and B be two sets. We denote by $A \oplus B$ the set defined as $(A \setminus B) \cup (B \setminus A)$. Let M be a $0 - 1$ matrix. In the following we denote by S_i the set $S_i = \{j | M(i,j) \neq 0\}$.

Definition 4 (t-Private Distribution Matrix). *A distribution matrix $M = (m_{i,j})_{i \in [n]}^{j \in [\ell]}$ is a t-private distribution matrix if the following conditions are satisfied.*

1. **(Correctness)** $\bigoplus_{i=1}^{n} S_i = \emptyset$
2. **(Privacy)** *For all $T, P \subseteq [n]$ such that $0 \leq |T| \leq t$, $P \cap T = \emptyset$ and $T \cup P \neq [n]$:*

$$\bigoplus_{i \in P} S_i \not\subseteq \bigcup_{i \in T} S_i.$$

Remark: The Correctness condition of Definition 4 is equivalent to the requirement that each element of $[\ell]$ belongs to an even number of S_is. We say that a distribution matrix M has *contention* k if each column of M has at most k ones, corresponding to the fact that each bit is shared by at most k players. The contention of a distribution matrix is an important efficiency measure and we shall consider primarily distribution matrices with contention 2. The reason for this choice is that, in a distributed environment, the larger is the number of players accessing a resource (a random bit in our case), the larger is the latency in the resource allocation (the actual reading).

Another important measure of efficiency we consider in this paper is the number of column of the distribution matrix. Indeed, this is equal to the length of the random string $\mathcal{R}$ and thus to the total number of random bits used. Our goal is to distribute to each player P_i, a subset $\mathcal{R}_i = \{r_j \text{ s.t. } j \in S_i \subseteq [\ell]\}$ in such a way that the protocol described is t-private, correct and the value of ℓ is minimum.

It should be expected that, by allowing contention larger than 2, should lead to a better randomness complexity. Surprisingly, this is not true. Indeed, in Section 7 we show a lower bound on the number of random bits for computing any boolean defining a commutative operation over the group $\{0, 1\}$. This minimal randomness complexity is achieved by our protocol using distribution matrices with contention 2.

4   Constructing a $(n - 2)$-Private Distribution Matrix

In this section we give a simple method for constructing $(n - 2)$-private distribution matrices.

We denote by I_n the n by n identity matrix and by 1^n the vector $(1, 1, \ldots, 1)$ consisting of n 1's. We define recursively $M(n)$ as follow:

$$M(2) = \begin{bmatrix} 1 \\ 1 \end{bmatrix}, \quad M(n) = \left[\begin{array}{c|c} 1^{n-1} & 00 \cdots 0 \\ \hline I_{n-1} & M(n-1) \end{array} \right] \qquad \text{for } n > 2.$$

Notice that $M(n)$ has n rows and $n(n-1)/2$ columns. In the following we denote by $S_{i,n}$ the i-th row in the matrix $M(n)$.

Lemma 5. For any k, $|\{i | k \in S_i\}| = 2$.

Proof. We proceed by induction on n. For $n = 2$, the Lemma holds obviously. Suppose the proposition holds for all $l < n$. We can identify two cases: If $k \in [n-1]$, the proposition holds since $S_{k+1,n} \cap S_{1,n} = S_{k+1,n} \cap [n-1] = k$ and $S_{i,n} \cap S_{j,n} = \emptyset$ for all $i \neq j$ with $i, j \neq 1$. If $k \in \{n, \ldots, \ell(n)\}$ the proposition holds by inductive hypothesis.

Lemma 6. For all $i \neq j$ we have that $|S_i \cap S_j| = 1$.

Proof. We proceed by induction on n with the case $n = 2$ omitted.

We can consider two cases: $i = 1$, and $i > 1$. For the first case we observe that $S_{1,n} = [n-1]$ and, for each j, $|[n-1] \cap S_{j,n}| = 1$. Thus $|S_1 \cap S_j| = 1$.

For the second case recall that, for any $i \geq 1$, $S_{i,n} \cap [n-1] = i - 1$. Thus, for any $i, j > 1$, with $i \neq j$, $S_{i,n} \cap S_{j,n} \cap [n-1] = \emptyset$. It is possible to write:

$$|S_i \cap S_j| = |S_{i,n} \cap S_{j,n} \cap [n-1]| + |S_{i,n} \cap S_{j,n} \cap \{n, \ldots, \ell(n)\}|$$
$$= |S_{i-1,n-1} \cap S_{j-1,n-1}|$$
$$= 1 \qquad \qquad \text{by inductive hypothesis.}$$

Theorem 7. *The Matrix $M(n)$ is a $(n-2)$-private distribution matrix.*

Proof. We have to prove that $M(n)$ meets the two properties of Definition 4. The Correctness condition follows from Lemma 5 since each column has exactly two 1s.

Let us consider the Privacy condition. Let T and P be two disjoint subsets of $[n]$ with $P \cup T \neq [n]$ and $0 \leq |T| \leq t$. Consider $j \in [n]$ such that $j \notin P \cup T$ and fix any $i \in P$. By Lemma 6 there exists k such that $S_i \cap S_j = \{k\}$. By Lemma 5, for all $h \neq i, j$, we have that $k \notin S_h$, since column k has exactly two 1s. Therefore we have that $k \in \bigoplus_{i \in P} S_i$ but $k \notin \bigcup_{i \in T} S_i$.

5   Graph Theoretic View
of t-Private Distribution Matrices

In this section we give a graph theoretic interpretation of distribution matrices. In particular we show that t-private distribution matrices with contention equal to 2 correspond to $(t+1)$-connected undirected graphs.

An *r-regular graph* $G = (V, E)$ is a graph in which each vertex has degree r. An undirected graph G is *connected* iff for each pair of nodes u, v in V, there exist a path from u to v in G. The graph G is *d-connected* if, after the removal of at most $d-1$ vertices (and of all the incident edges), the remaining graph is still connected. Obviously if a graph G is d connected, the degree of the each vertex in the graph is *at least d*. G is maximally connected if it is d-connected and d is the minimum degree of vertices in G.

Definition 8. *The edge matrix of a graph $G = ([n], E)$ is a boolean $n \times |E|$ matrix M such that for each edge $(u, v) \in E$ exists exactly one column j for which:*

$$M(i, j) = 1 \text{ iff } i = u \text{ or } i = v \qquad M(i, j) = 0 \ i \neq u, v.$$

The above definition, naturally gives rise to a mapping between graphs with vertex set $[n]$ and distribution matrices with n rows and contention equal to 2, by identifying the vertices of the graph with rows of distribution matrices and the edges with the columns. The mapping is 1-to-1 up to isomorphism. In particular with S_i we can identify the set of edges incident on the vertex associated to player P_i. Notice that, if we consider $\bigoplus_{i \in W} S_i$ for some $W \subseteq [n]$, we obtain the set of edges with one end-point in W and the other end-point outside W. Notice that the matrix we have constructed in the previous section is the edge matrix of a clique of size n.

Theorem 9. *A matrix M is a t-private distribution matrix with contention 2 iff. M is the edge matrix of a $(t+1)$-connected graph.*

Proof. ($\Longrightarrow$) Let M be a t-private distribution matrix with contention 2. By definition of contention, M is an edge matrix and let G be the graph with edge matrix M. By way of contradiction, suppose that, by removing a set T of most t vertices from the graph, we disconnect the graph. Let C be one of the maximal connected component obtained by removing T. Thus all the edges with an endpoint in C, have the other endpoint either in C or T. The set $\bigoplus_{i \in C} S_i$ (where $S_i = \{j | M(i,j) \neq 0\}$) contains all the edges with an end-point in C and the other end-point in T thus $\bigoplus_{i \in C} S_i \subseteq \bigcup_{i \in T} S_i$. But this means that the Privacy Condition does not hold. Contradiction.

($\Longleftarrow$) We have to prove that the edge matrix M of a $(t+1)$-connected graph G meets the properties in Definition 4.

For the Correctness condition, we observe that, by Definition 8, for each column in M, exist exactly two 1s (one for each endpoint of the edge). Thus this property holds.

Let us consider the Privacy Condition. Let $T, P \subseteq [n]$ be two disjoint sets of vertices such that $0 \leq |T| \leq t$, $P \cup T \neq [n]$. Since the graph is $(t+1)$-connected, after the removal of T from G, the graph is still connected. Thus there must exist at least one edge from P to an element that is not in P nor in T and thus $\bigoplus_{i \in P} S_i \nsubseteq \bigcup_{i \in T} S_i$ holds

Corollary 10 (Number of Random Bits). *Each t-private distribution matrix with n rows has at least $\frac{n(t+1)}{2}$ columns.*

Proof. Theorem 9 shows that exists a 1-to-1 mapping between $(t+1)$-connected graphs and t private distribution matrices. Moreover in $(t+1)$-connected graph each vertex has degree at least $t+1$. To minimize the number of edges, we can consider the class of $(t+1)$-regular maximally connected graphs. But each $(t+1)$-regular graph with n nodes has $\frac{n(t+1)}{2}$ edges.

6   A 1-Round t-Private Protocol for Computing the Sum

In this section we describe a protocol that computes the sum[1] of n bit t-privately. The protocol presented is a 1-round protocol and uses only broadcast communication channels. That is each player P_i privately reads the random bit from the string $\mathcal{R}$ according to the distribution matrix $\mathcal{M}$; computes its message y_i as function of its input and the random bits read and it broadcasts y_i. The sum of the inputs is then computed as function of all the messages sent by the players.

Recall that, in the graph-theoretic interpretation of the distribution matrices, S_i represents the set of edges incident to the vertex associated to player P_i.

In the following we say that an edge is *shared* between two players in P if both the end points of the edge belong to the set P. An edge is *known* to the coalition T if exactly one end-point of the edge is in T and the other one is in P. Finally an edge is *outgoing* P if one end-point of the edge is in P and the other end-point is in $[n] \setminus (P \cup T)$.

---

[1] All the summations in this work are intended to be modulo 2.

Theorem 11. *If there exists a t-private distribution matrix $\mathcal{M}$ with n rows than there exists s t-private n-player protocol for computing the function XOR.*

Proof. The protocol for computing the sum is the following:

Protocol XOR
Program for player P_i

Input: $x_i \in \{0, 1\}$
  t-private distribution matrix: $M = \{m_{i,j}\}$
  Shared random string $\mathcal{R} = r_1 r_2 \ldots r_\ell$

1 Let $S_i = \{j \text{ such that } \mathcal{M}(i, j) = 1\}$.
2 Compute $m_i = x_i + \sum_{j \in S_i} r_j$.
3 Publish m_i.
4 Compute $XOR = \sum_{i=1}^n m_i$.

Correctness. For the correctness of the protocol, we want the value XOR, computed by each player, to be equal to the sum of the private inputs. Indeed we have:

$$XOR = \sum_{i=1}^n m_i = \sum_{i=1}^n \left(x_i + \sum_{j \in S_i} r_j \right) = \sum_{i=1}^n x_i + \sum_{i=1}^n \sum_{j \in S_i} r_j = \sum_{i=1}^n x_i$$

The last equation holds since, by Correctness condition of Definition 4, each random bit belongs to an even number of S_i's.

Privacy. Let $T, P \subseteq [n]$ such that $0 \leq |T| \leq t$, $P \cap T = \emptyset$, $T \cup P \neq [n]$ and, w.l.o.g., assume that $P = \{1, \ldots, k\}$. Let x, y be two input vectors satisfying the requirement of Definition 1. We need to show that

$$Pr\left[\mathbf{M} = m | \mathbf{R}_T = r_T, \mathbf{X} = x\right] = Pr\left[\mathbf{M} = m | \mathbf{R}_T = r_T, \mathbf{X} = y\right].$$

Write $m = m_1, \ldots, m_n$, where m_j is the message broadcast by player P_j. For any $i \in P$, let us denote by H_i (resp., O_i, A_i) the set of random bits shared (resp., outgoing or known to the players in T) held by P_i and let us denote by $\chi(W, j)$ the characteristic function of the set W, i.e., $\chi(W, j) = 1$ if $j \in W$ or $\chi(W, j) = 0$ if $j \notin W$. By construction,

$$m_i = x_i + \sum_{j \in S_i} r_j = x_i + \sum_{j=1}^n \chi(H_i, j)r_j + \sum_{j=1}^n \chi(A_i, j)r_j + \sum_{j=1}^n \chi(O_i, j)r_j$$

Consider the following system of linear equations in the unknown r_js:

$$\begin{bmatrix} \overline{m_1} = x_1 + \sum_{j=1}^n \chi(H_1, j)r_j + \sum_{j=1}^n \chi(O_1, j)r_j \\ \vdots \\ \overline{m_k} = x_k + \sum_{j=1}^n \chi(H_k, j)r_j + \sum_{j=1}^n \chi(O_k, j)r_j \end{bmatrix} \tag{1}$$

Where $\overline{m_i} = m_i + \sum_{j=1}^{n} \chi(A_i, j) r_j$. All we need to show is that for any fixed m, for any fixed random tape r_T, i.e., the view of the coalition T, and for any possible value of the input vector x_P, there exists the same number of possible random tapes r_P, *consistent* with r_T, that are a solution for (1). By simple linear algebra, it is sufficient to show that the number of unknown (to the players in T) variables in the system is at least k. Let us denote by n_s be the number of edges shared by players in P and by n_o the number of edges outgoing P.

Informally, the idea of the proof is the following: Consider a $(k+t+1)$-vertex $(t+1)$-connected graph G. Moreover, consider a set P of k vertices and a set T of t in this graph. The number of edges in G can be computed as the sum of the number n_s of shared edges in P, the shared edges in T, the number n_a of the edges of P known to T and the edges incident on the vertex $v = [k+t+1] \setminus \{P \cup T\}$. Notice that the number of edges incident on v is given by the sum of the n_o outgoing edges and the number of edges of v known to T. We can consider the following two cases:

- $k > t$: In this case, it holds that $n_a \leq t(t+1)$ since each player in T can share each of his edges with one player in P, i.e., players in T do not share edges among them and do not share edges with the vertex v. Thus it is possible to write:

$$n_s + n_o \geq \frac{(k+t+1)(t+1)}{2} - t(t+1)$$
$$= \frac{(k-t+1)(t+1)}{2}.$$

Simple algebra shows that $n_s + n_o \geq k$ holds for every $k > t$.
- $k \leq t$: In this case, it holds that $n_a \leq kt$ since each player in P can share at most one edge with each other player in T. Also in this case, the number of edges incident on v is exactly n_o. Thus it is possible to write:

$$n_s + n_o \geq \frac{(k+t+1)(t+1)}{2} - kt.$$

As before, simple algebra shows that $n_s + n_o \geq k$ for every $k \leq t$.

7   Lower Bound on the Number of Random Bits

The protocol presented in this paper does not use private channels to run private computation. This means that the privacy of the protocols relies exclusively on the randomization of the messages.

In this section we show a lower bound on the number of random bits needed to compute any function f, 1-privately, in the broadcast model. We briefly review some definition we use in this section.

We say that a function $f : \mathcal{X}^n \to \mathcal{X}$ is *sensitive* to its i-th variable on an assignment $\mathbf{x} = (x_1, \ldots, x_n)$, if it results that $|\{f(x_1, \ldots, x_{i-1}, z, x_{x+1}, \ldots, x_n) :$

$z \in \mathcal{X}\}| = |\mathcal{X}|$. We say that f is *i-sensitive* if it is sensitive to its i-th variable for any assignment $\mathbf{x}$. The *sensitivity* of a function f is the number of indices i to which the function f is i-sensitive. For instance, the sensitivity of the function $\sum_{i=1}^{n} x_i \bmod q$ is n. More generally, if $(G, \otimes)$ is a group then the function $f : G^n \to G$ defined by $f(x_1, x_2, \ldots, x_n) = x_1 \otimes \ldots \otimes x_n$ has sensitivity n. In our proofs we use some concepts of Information Theory. We refer the reader to [7] for a complete treatment of this subject.

Lemma 12 ([3]). *In any protocol computing an i-sensitive function f, it holds that:*

$$H(\mathbf{X}_i | \mathbf{M}_i, \mathbf{R}_i) = 0,$$

where $H(\cdot|\cdot)$ is Shannon conditioned entropy function.

Corollary 13. *In any protocol computing a function f with sensitivity n, it results that:*

$$H(\mathbf{X}_Y | \mathbf{M}, \mathbf{R}_Y) = 0$$

for any $Y = \{i_1, i_2, \ldots, i_{|Y|}\} \subseteq [n]$.

Proof.

$$0 \le H(\mathbf{X}_Y | \mathbf{M}, \mathbf{R}_Y) = \sum_{k=1}^{|Y|} H(\mathbf{X}_{i_k} | \mathbf{M}, \mathbf{R}_Y, \mathbf{X}_{i_1}, \mathbf{X}_{i_2}, \ldots, \mathbf{X}_{i_{k-1}})$$

$$\le \sum_{k=1}^{|Y|} H(\mathbf{X}_{i_k} | \mathbf{M}, \mathbf{R}_Y)$$

$$\le \sum_{k=1}^{|Y|} H(\mathbf{X}_{i_k} | \mathbf{M}_{i_k}, \mathbf{R}_{i_k})$$

$$= 0$$

Lemma 14 ([3]). *Let $\mathbf{X}_1, \ldots, \mathbf{X}_n$ be independent and uniformly distributed random variables and let $\mathcal{F}$ be a function with sensitivity n. For any $W, Y \subseteq \{1, 2, \ldots, n\}$ such that $W \cap Y = \emptyset$ and $|W \cup Y| < n$ it results that*

$$H(\mathbf{X}_Y | \mathbf{X}_W, \mathbf{F}) = H(\mathbf{X}_Y).$$

The privacy requirement of the protocol can be written as follows (see [3] for a more detailed description):

Definition 15 (Privacy). *For any $Y \subset [n]$ such that $|Y| \le n - 2$ and for any $j \in [n] \setminus Y$ it holds that*

$$H(\mathbf{X}_Y | \mathbf{M}, \mathbf{R}_j, \mathbf{X}_j, \mathbf{F}) = H(\mathbf{X}_Y | \mathbf{F}). \tag{2}$$

Notice that, by Lemma 14, Equation 2 can be rewritten as follows:

$$H(\mathbf{X}_Y|\mathbf{M}, \mathbf{R}_j, \mathbf{X}_j, \mathbf{F}) = H(\mathbf{X}_Y) \tag{3}$$

Lemma 16. *For any $Y \subset [n]$ such that $|Y| \leq n - 2$ and for any $j \in [n] \setminus Y$ it holds that*

$$H(\mathbf{R}_Y|\mathbf{R}_j) \geq H(\mathbf{X}_Y).$$

Proof. By Corollary 13 and by Definition 15:

$$
\begin{aligned}
I(\mathbf{X}_Y; \mathbf{R}_Y|\mathbf{M}, \mathbf{R}_j, \mathbf{X}_j, \mathbf{F}) &= H(\mathbf{X}_Y|\mathbf{M}, \mathbf{R}_j, \mathbf{X}_j, \mathbf{F}) - H(\mathbf{X}_Y|\mathbf{M}, \mathbf{R}_j, \mathbf{X}_j, \mathbf{F}, \mathbf{R}_Y) \\
&= H(\mathbf{X}_Y) \\
&= H(\mathbf{R}_Y|\mathbf{M}, \mathbf{R}_j, \mathbf{X}_j, \mathbf{F}) - H(\mathbf{R}_Y|\mathbf{M}, \mathbf{R}_j, \mathbf{X}_j, \mathbf{F}, \mathbf{X}_Y)
\end{aligned}
$$

Thus it follows that:

$$
\begin{aligned}
H(\mathbf{R}_Y|\mathbf{R}_j) &\geq H(\mathbf{R}_Y|\mathbf{M}, \mathbf{R}_j, \mathbf{X}_j, \mathbf{F}) \\
&\geq H(\mathbf{R}_Y|\mathbf{M}, \mathbf{R}_j, \mathbf{X}_j, \mathbf{F}) - H(\mathbf{R}_Y|\mathbf{M}, \mathbf{R}_j, \mathbf{X}_j, \mathbf{X}_Y, \mathbf{F}) \\
&= H(\mathbf{X}_Y)
\end{aligned}
$$

Corollary 17. *For any $i \in [n]$ it holds that $H(\mathbf{R}_i) \geq H(\mathbf{X}_i)$.*

We are now ready to prove the following bound:

Theorem 18. *For any protocol computing an n-sensitive function in the broadcast model, if the X_i's are independent and $H(\overline{\mathbf{X}}) = H(\mathbf{X}_1) = \ldots = H(\mathbf{X}_n)$, it holds that:*

$$H(\mathbf{R}) > (n - 1)H(\overline{\mathbf{X}}).$$

Proof. If $\mathbf{R}_i$ are independent, the for each $i \in [n]$, it holds that $H(\mathbf{R}_i|\mathbf{R}_1 \ldots \mathbf{R}_{i-1}) = H(\mathbf{R}_i)$ thus, by Corollary 17 it holds that:

$$
\begin{aligned}
H(\mathbf{R}) = H(\mathbf{R}_1 \ldots \mathbf{R}_n) &= \sum_{i=1}^{n} H(\mathbf{R}_i|\mathbf{R}_1 \ldots \mathbf{R}_{i-1}) \\
&= \sum_{i=1}^{n} H(\mathbf{R}_i) \\
&\geq \sum_{i=1}^{n} H(\mathbf{X}_i) \\
&= nH(\mathbf{X})
\end{aligned}
$$

If the $\mathbf{R}_i$'s are not independent then there exists i, j such that $H(\mathbf{R}_i) > H(\mathbf{R}_i|\mathbf{R}_j) \geq H(\mathbf{X}_i)$. Let $V \subset [n]$ such that $|V| = n - 2$ and $i \notin V$.

$$
\begin{aligned}
H(\mathbf{R}_1 \ldots \mathbf{R}_n) &= H(\mathbf{R}_i) + H(\mathbf{R}_V|\mathbf{R}_i) + H(\mathbf{R}_k|\mathbf{R}_V \mathbf{R}_i) \\
&\geq H(\mathbf{R}_i) + H(\mathbf{R}_V|\mathbf{R}_i) \\
&> H(\mathbf{X}_i) + H(\mathbf{X}_V) \\
&= (n - 1)H(\overline{\mathbf{X}})
\end{aligned}
$$

Corollary 19. *For any protocol computing an n-sensitive* $f : \{0,1\}^n \to \{0,1\}$ *in the broadcast model, if the* X_i*'s are independent and* $H(\overline{\mathbf{X}}) = H(\mathbf{X}_1) = \ldots = H(\mathbf{X}_n)$, *it holds that:*

$$H(\mathbf{R}) \geq n.$$

Proof. In the case of boolean functions, $|\overline{X}| = |X_1| = \ldots = |X_n| = |F| = 2$. Thus:

$$\log(|\mathbf{R}|) \geq (n-1)\log|\overline{X}| + 1 = n$$

This corollary states that computing any boolean function in the broadcast model, 1-privately requires at least n random bits. It also implies that the protocol presented in this paper is optimal since, for the 1-private case, it uses exactly this minimal randomness.

8   Conclusions and Open Problems

In this paper we have introduced a new model for private computation, the SRoBC model. We have show that, in this model, it is possible to compute any function over a finite commutative group in a t-private way, for any possible value of t using constant number of rounds of interaction.

For the whole paper we have restricted the attention to the case in which each random bit is shared between two players and we modelled the association between players and random bits by a graph. The natural question could be "What happens if one random bit is shared among more than two players ?" In this case hypergraphs could be used to model the distribution of random bits to use. Unfortunately, it is not hard to see that the hypergraph connectivity requirement is still necessary but, not sufficient to guarantee the privacy of the protocol.

However, we have shown that in the case of boolean functions at least n bits are required to run a 1-private protocol in this model. Since this bound matches the number of random bits needed by our protocol, it is useless for 1-private protocols to allow more than 2 players to share a single random bit. It remains an open problem the study of the lower bound for number of random bits for t-privately computing a function f for any $t > 1$.

References

1. D. Beaver. Perfect privacy for two-party protocols. Technical Report TR-11-89, Harvard University, 1989.
2. M. Ben-Or, S. Goldwasser, and A. Wigderson. Completeness theorems for non-cryptographic fault-tolerant distributed computation. In *Proceedings of 20th Symposium on Theory of Computation*, pages 1–10, 1988.
3. C. Blundo, A. De Santis, G. Persiano, and U. Vaccaro. Randomness complexity of private multiparty protocols. *Computational Complexity*, 8(2):145–168, 1999.
4. D. Chaum, C. Crepeau, and I. Damgård. Multiparty unconditionally secure protocols. In *Proceedings of 20th Symposium on Theory of Computation*, pages 11–19, 1988.

5. D. Chaum. The Dining Cryptographers Problem: Unconditional sender and recipient untraceability. *Journal of Cryptology*, 1(1):65–75, 1988.
6. B. Chor and E. Kushilevitz. A communication-privacy tradeoff for modular addition. *Information Processing Letters*, 45:205–210, 1991.
7. T.M. Cover and J.A. Thomas. *Elements of Information Theory*. John Wiley & Sons, Singapore, 1991.
8. U. Feige, J. Kilian, and M. Naor. A minimal model for secure computation. In *Proceedings of 26th ACM Symposium on Theory of Computation*, pages 554–563, 1994.
9. M. Franklin and M. Yung. Secure hypergraphs: Privacy from partial broadcast. In *Proceedings of 27th ACM Symposium on Theory of Computing*, pages 36–44, 1995.
10. Y. Ishai and E. Kushilevitz. Private simultaneous messages protocols and applications. In *Proceedings of ISCTS 97*.
11. E. Kushilevitz. Privacy and communication complexity. *SIAM Journal of Disc. Mat.*, 5(2):273–284, 1992.
12. Michael Waidner. Unconditional sender and recipient untraceability in spite of active attacks. In *Advances in cryptology — EUROCRYPT '89*, volume 434 of *Lecture Notes in Computer Science*, 1989. Springer-Verlag.

An Optimistic Multi-party Fair Exchange Protocol with Reduced Trust Requirements

Nicolás González-Deleito and Olivier Markowitch

Université Libre de Bruxelles
Bd. du Triomphe – CP212
1050 Bruxelles, Belgium
{ngonzale,omarkow}@ulb.ac.be

Abstract. In 1999, Bao et al. proposed [6] a multi-party fair exchange protocol of electronic items with an offline trusted third party. In this protocol, a coalition including the initiator of the exchange can succeed in excluding a group of parties without the consent of the remaining entities. We show that every participant must trust the initiator of the protocol for not becoming a passive conspirator. We propose a new protocol in which the participants only need to trust the trusted third party. Moreover, under certain circumstances, if there are participants excluded from the exchange, they can prove that a problem occurred to an external adjudicator.

1   Introduction

During the last decade, the important growth of open networks such as the Internet has lead to the study of related security problems. Fair exchange of electronic information (contract signing, certified mail, ...) is one of these security challenges. An exchange protocol is said to be *fair* if it allows two or more parties to exchange electronic information in such a way that, at the end of the protocol, no honest party has sent anything valuable unless he has received everything he expected. Those protocols often use a trusted third party (TTP) helping the participants to successfully realize the exchange.

Depending on his level of involvement in a protocol, a TTP can be said *inline*, *online* or *offline*. Inline and online trusted third parties are both involved in each instance of a protocol, but the first one acts as a mandatory intermediary between the participants. An offline TTP is used when the participants in a protocol are supposed to be honest enough to not need external help in order to achieve fairness; the TTP will only be involved if some problem emerges. Protocols with such a TTP are called *optimistic*.

Fair exchange between two parties has been extensively studied and several solutions have been proposed in the online [7,12] as in the offline case [3,4,5,11]. The interest of [11], where an item is exchanged against a digital signature, is that the TTP produces the same signatures as those that would have been produced by the participants in a faultless scenario.

Fair exchange of electronic information between more than two parties may have applications in electronic commerce. For example, we can consider a com-

K. Kim (Ed.): ICICS 2001, LNCS 2288, pp. 258–267, 2002.
© Springer-Verlag Berlin Heidelberg 2002

mon and generic scenario with four parties describing a ring. Let one of these parties be a customer who wants to purchase an electronic item offered by a provider; the payment is realized through the customer's and the provider's banks. This is how each participant views the exchange:

- the provider provides the expected electronic item to the customer in exchange of having his bank crediting his account;
- the customer sends a payment authorization to his bank in exchange of the desired electronic information offered by the provider;
- the customer's bank carries out the payment to the provider's bank in exchange of the payment authorization sent by the customer;
- finally, the provider's bank credits the provider's account in exchange of the payment carried out by the customer's bank.

More general topologies than the ring described above are also possible. For example, we could consider an exchange in which each participant offers items to a set of parties in exchange of items offered by another set of participants. Franklin and Tsudik gave [8] a classification of multi-party exchanges, based on the two following properties: the number of items that a participant can exchange (one or several), and the disposition of the participants (describing a ring or a more general topology).

In the multi-party case, Asokan et al. described [2] a generic optimistic protocol with a general topology. This protocol, during which a participant may receive an affidavit from the TTP instead of the expected item, achieves *weak fairness*. Franklin and Tsudik presented [8] two protocols with an online TTP, and later Bao et al. proposed [6] a protocol where the TTP is offline. Both works supposed the exchange topology as being a ring, where each participant P_i offers to participant P_{i+1} message m_i in exchange of message m_{i-1} offered by participant P_{i-1}. Of course, all subscripts are mod n, where n is the number of participants in the exchange. We will omit this hereafter.

With regard to optimistic multi-party fair exchange, it could be unrealistic to think that all the participants in a protocol execution will be honest. Designing optimistic protocols for this kind of exchanges might not have, at a first glance, much sense. However, we think that even in a scenario with a dishonest participant, a protocol with an offline TTP remains more efficient than a protocol with an online TTP.

In this paper we focus on optimistic multi-party fair exchange with a ring topology. Section 3 describes the protocol proposed by Bao et al. in [6], and its trust requirements are discussed. In section 4 we propose a variant of that protocol, where trust needs are reduced.

The next section describes the concept of verifiable encryption schemes. This technique, used by Bao et al. in their multi-party fair exchange protocol [6], will also be used in section 4.

2   Verifiable Encryption Schemes

Suppose a scenario with two participants, Alice and Bob, and a trusted third party. Let E and D be the encryption and the corresponding decryption algorithms of a public-key cryptosystem. The TTP owns a public encryption key e and a secret decryption key d of this cryptosystem. Moreover, let h be a homomorphic one-way function. Alice knows a secret message m, with $h(m)$ being public.

Alice enciphers m to $c = E_e(m)$, generates a certificate $cert_A = certify(m, c, e)$ using a public algorithm $certify()$, and sends c and $cert_A$ to Bob.

Bob checks that $cert_A$ is a correct certificate by using a public algorithm $verify()$ such that $verify(c, cert_A, h(m), e) = yes$ if and only if $h(D_d(c)) = h(m)$. Bob is then convinced that c is indeed the cipher of m under key e. Later, Bob will be able to obtain m by asking the TTP to decipher c.

A verifiable encryption scheme must satisfy [6] these two properties:

- it is computationally unfeasible for Alice to generate a certificate $cert_A$ such that $verify(c, cert_A, h(m), e) = yes$, while $h(D_d(c)) \neq h(m)$;
- and it is computationally unfeasible for Bob to get m from c without knowing d.

Asokan et al. gave [3] some examples of verifiable encryption schemes implementations. Bao et al. used [6] an implementation of a non-interactive scheme, corresponding to the description above.

3   A Multi-party Fair Exchange Protocol

In this section we briefly describe the optimistic multi-party fair exchange protocol proposed by Bao et al. [6]. (We use notations as close as possible to the ones used in the original paper.) The exchange topology is a ring.

Let P_0 be the participant that initiates the protocol. The status of P_0 is known by the TTP, who also knows all $h(m_i)$, for $0 \leq i \leq n - 1$.

3.1   Main Protocol

P_0 begins the main protocol by sending to P_1 the cipher c_0 of the message m_0 along with a certificate $cert_0$ proving that c_0 is indeed the cipher of m_0, as described in the previous section.

After receiving $(c_0, cert_0)$, P_1 checks $cert_0$; if this certificate is valid, he enciphers m_1 and sends $(c_1, cert_1)$ to P_2. For $i = 2, 3, ..., n - 1$, participant P_i does similarly.

When P_0 receives $(c_{n-1}, cert_{n-1})$, he checks the certificate $cert_{n-1}$ and if it is valid, he sends the message m_0 to P_1. For $i = 1, 2, ..., n - 1$, after receiving m_{i-1} from participant P_{i-1}, P_i sends m_i to P_{i+1}.

These are the two rounds of the main protocol:

1. $P_i \rightarrow P_{i+1} : c_i, cert_i$ for $i = 0, ..., n - 1$.
2. $P_i \rightarrow P_{i+1} : m_i$ for $i = 0, ..., n - 1$.

3.2   Recovery Protocol

If all parties behave correctly, after the main protocol execution each participant should obtain his expected message. However, if some P_i does not receive m_{i-1}, he has to run the recovery protocol.

P_i begins this protocol by sending $(c_{i-1}, cert_{i-1})$ to the TTP. The latter checks the certificate $cert_{i-1}$, and, if it is valid, waits for the time[1] it takes to P_0 to obtain m_{n-1} if no problem occurs, before asking P_0 if he has received a valid $(c_{n-1}, cert_{n-1})$ during the first round of the main protocol. If P_0 answers *yes*, the TTP will accept to decipher the corresponding c_{i-1}.

The TTP does not contact P_0 each time that some other participant runs the recovery protocol: if P_0 answered *yes* in the first recovery execution then the TTP will accept further recovery requests; otherwise the TTP will reply with an *abort* message.

This call to P_0 prevents of having party P_i getting c_{i-1} deciphered without having sent $(c_i, cert_i)$.

Here are the four steps of the recovery protocol, initiated by some P_i having not received m_{i-1}. Steps 2 and 3 are only executed the first time that this protocol is invoked.

1. $P_i \rightarrow TTP : c_{i-1}, cert_{i-1}$.

2. $TTP \rightarrow P_0 : call$.

3. $P_0 \rightarrow TTP : yes$ or *abort*.

4. $TTP \rightarrow P_i : m_{i-1}$ or *abort*.

3.3   Analysis

Bao et al. gave [6] the following definition of fairness: *an exchange protocol is called a* fair exchange protocol *if after the protocol execution no participant P_i following properly the protocol is in a state where his c_i has been deciphered without having him received m_{i-1}.*

A participant P_j $(1 \leq j \leq n-1)$ sends m_j to P_{j+1} only if he has received m_{j-1}. P_{j+1} can however ask the TTP to decipher c_j; he obtains the expected m_j only if P_0 received $(c_{n-1}, cert_{n-1})$ in the first round of the main protocol. If so, P_j can also ask the TTP to decipher c_{j-1} in order to get m_{j-1}.

Since P_0 sends m_0 to P_1 after receiving a valid $(c_{n-1}, cert_{n-1})$ from P_{n-1}, it is then also possible for P_0 to recover m_{n-1} if he ever does not receive it later.

Assuming that the TTP behaves as described, after the protocol execution (both, the main and, possibly, the recovery protocol) each honest P_i either has obtained m_{i-1}, either his c_i has not been deciphered.

About Trust and Passive Conspiracies. As pointed out by Bao et al., in that protocol two or more parties can collude in order to exclude some other participants from the exchange. This happens only if P_0 belongs to this coalition.

---

[1] This time is estimated by the TTP.

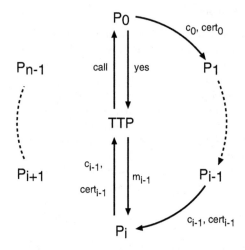

Fig. 1. An exclusion scenario

Consider that P_0 colludes with P_i (figure 1). When P_i receives $(c_{i-1}, cert_{i-1})$ from P_{i-1} in the first round of the main protocol, instead of sending $(c_i, cert_i)$ to P_{i+1}, P_i could run directly the recovery protocol, asking the TTP to decipher c_{i-1}. The TTP asks then P_0 if he has received $(c_{n-1}, cert_{n-1})$, P_0 answers with a false *yes*, and the TTP deciphers c_{i-1}.

P_i will obtain m_{i-1} without having sent $(c_i, cert_i)$. This causes no harm as long as the TTP follows the protocol properly: P_1 to P_{i-1} will be able to run the recovery protocol and obtain respectively m_0 to m_{i-2}, and P_{i+1} to P_{n-1} will not receive nor send anything: they will be simply excluded from the exchange.

By the above definition, fairness is still guaranteed for any party following properly the protocol, including those who are excluded, even if there is a participant P_i having received m_{i-1} without having sent c_i. There is here a difference between this definition of fairness and the one given by Asokan et al. [2], where *all* the participants must be in the same state at the end of the protocol. A similar approach of the former definition of fairness can be found in [9,10].

We define a passive conspirator who takes part in a coalition excluding certain participants from the exchange as someone who cannot prevent this coalition from being done and who by his idleness contributes to keep the excluded participants in ignorance of the exchange which takes place.

Even if the fairness property is respected, in the coalition described above, P_1 to P_{i-1} become passive conspirators. They must trust P_0 for not answering with a false *yes* to the TTP during the recovery protocol.

Otherwise, even if P_0 decides to send a false *abort* to the TTP during the recovery protocol, fairness is still preserved. Once the first round of the main protocol has been successfully completed, P_0 sends m_0 to P_1 and participants P_1 to P_{i-1} do similarly. If P_i decides to not send m_i, after a certain amount of time P_{i+1} will realize that he must run the recovery protocol in order to obtain m_i. However, if P_0 sends a false *abort* to the TTP, the protocol will be terminated without P_{i+1} receiving m_i. If the TTP behaves properly, P_0 will be the only participant in a non-fair state at the end of the protocol.

It must be pointed out that every participant has to not only trust the TTP for behaving properly, but must also trust P_0 for not sending a false *yes* to the TTP during the recovery protocol.

4   A Protocol with Reduced Trust Requirements

We now present a variant of the multi-party fair exchange protocol described in the previous section. In this protocol an offline trusted third party is also used. The exchange topology is a ring and the communication channels between the participants and the TTP are supposed to be resilient (data is delivered after a finite, but unknown, amount of time).

Through this section we will use the following notations:

- $\mathcal{P}$ is the set $\{P_0, P_1, ..., P_{n-1}\}$ of all the participants in the exchange.
- $A \Rightarrow \beta$: denotes participant A multicasting a message to the set of participants β.
- f_x is a flag indicating the purpose of a message in a given protocol; x is composed of a letter and a number corresponding respectively to the protocol and the message number in this protocol.
- *label* is an information identifying a protocol run, that depends, among others, on $\mathcal{P}$.
- $S_A(m)$ denotes the digital signature of participant A over the message m.
- in a protocol message, $S_A(\star)$ denotes the digital signature of A over all information preceding this signature.
- m'_i denotes the concatenation of P_i's identity and the message m_i expected by P_{i+1}. $h(m'_i)$ is supposed to be public. m'_i and $c_i = E_e(m'_i)$ are used to generate the certificate $cert_i$.

Let P_0 be the participant that initiates the protocol. We suppose that this is known by all the participants and the TTP. Moreover, the set $\mathcal{P}$ is supposed to be known by all the participants in the exchange.

4.1   Main Protocol

1. $P_i \rightarrow P_{i+1} : f_{m1}, P_{i+1}, label, c_i, cert_i, S_{P_i}(\star)$ for $i = 0, ..., n-1$.

P_0 begins the main protocol by sending to P_1 the cipher c_0 of the message m'_0, along with a certificate $cert_0$ proving that c_0 is indeed the cipher of m'_0.

After receiving $(c_0, cert_0)$ from P_0, if $cert_0$ is valid, P_1 enciphers m'_1 and sends $(c_1, cert_1)$ to P_2. For $i = 2, 3, ..., n-1$, every P_i does similarly.

2. $P_0 \Rightarrow \mathcal{P} \setminus P_0 : S_{P_0}(label)$.
   $P_0 \rightarrow P_1 : \quad f_{m2}, P_1, label, m_0, S_{P_0}(\star)$.

Upon P_0 receiving $(c_{n-1}, cert_{n-1})$, if the certificate sent by P_{n-1} is valid, P_0 multicasts his signature over the *label*, $S_{P_0}(label)$, to the set $\mathcal{P} \backslash P_0$ of participants, and sends the message m_0 to P_1.

3. $P_i \Rightarrow \mathcal{P} \backslash P_i : S_{P_0}(label).$
   $P_i \rightarrow P_{i+1} : \quad f_{m2}, P_{i+1}, label, m_i, S_{P_i}(\star).$ $\Big\}$ for $i = 1, ..., n - 1$.

When P_i $(1 \leq i \leq n-1)$ receives a valid $S_{P_0}(label)$ for the first time, he multicasts this signature to $\mathcal{P} \backslash P_i$. Upon receiving such a signature and m_{i-1} from P_{i-1}, P_i sends m_i to P_{i+1}.

4.2   Recovery Protocol

If some P_i does not receive m_{i-1} during the main protocol, he has to run the recovery protocol.

1. $P_i \rightarrow TTP : \quad f_{r1}, TTP, label, h(m'_{i-1}), S_{P_0}(label), c_{i-1}, cert_{i-1}, S_{P_i}(\star).$
2. $TTP \rightarrow P_{i-1} : S_{P_0}(label).$
3. $TTP \rightarrow P_i : \quad f_{r2}, P_i, label, m_{i-1}, S_{TTP}(\star).$

In order to get c_{i-1} deciphered by the TTP, P_i sends to the latter $(S_{P_0}(label),$ $c_{i-1}, cert_{i-1})$. If $S_{P_0}(label)$ is a valid signature of P_0 over the *label* and if the certificate $cert_{i-1}$ is correct, the TTP deciphers c_{i-1} and obtains m'_{i-1}, the concatenation of P_{i-1}'s identity and m_{i-1}. He forwards $S_{P_0}(label)$ to P_{i-1} and sends m_{i-1} to P_i.

4.3   Analysis

Before that P_0 multicasts $S_{P_0}(label)$ to $\mathcal{P} \backslash P_0$, no participant belonging to this set is able to run the recovery protocol. If the TTP behaves as described, none of them will get c_{i-1} deciphered.

P_0 could do a recovery instead of multicasting his signature over the *label*. In this case, the TTP will forward that signature to P_{n-1} and the exchange will be able to continue.

If during the main protocol some P_i does not receive the expected m_{i-1}, he can run the recovery protocol only if he provides $S_{P_0}(label)$ to the TTP. As no particular assumption has been made over the communication channels between participants, P_i could not receive the signature of P_0 over the *label*. Even in this case, P_i would remain in a fair state: if P_{i+1} contacts the TTP, this one will send $S_{P_0}(label)$ to P_i in the second step of the recovery protocol and P_i will also be able to run this protocol.

Following the definition given by Bao et al. [6], the protocol presented above is fair: at the end there will be no honest participant in a state where his c_i has been deciphered without having him received m_{i-1}.

About Trust and Passive Conspiracies. In the main protocol, m_j $(1 \leq j \leq n-1)$ is sent as soon as a valid $S_{P_0}(label)$ has been received. Otherwise, a coalition between, for example, P_0 and a participant P_i, excluding participants P_{i+1} to P_{n-1} from the exchange without the consent of participants P_1 to P_{i-1}, could exist: P_0 decides to not multicast his signature and P_i stops the exchange after receiving m_{i-1}, while P_1 to P_{i-1} are unable to inform the excluded participants that the last round of the main protocol has begun. P_1 to P_{i-1} would become passive conspirators; this situation is avoided by sending m_j after having received a valid $S_{P_0}(label)$.

During the main protocol $S_{P_0}(label)$ is multicasted to all the participants in the exchange. Suppose that some P_i refuses to realize this step of the protocol. If there exists a coalition (not including P_i) willing to exclude some participants from the exchange, P_i will not send $S_{P_0}(label)$ to the excluded participants and will become a passive conspirator. Multicasting $S_{P_0}(label)$ to *all* the others participants in the exchange prevents such a situation.

Passive conspiracies in order to exclude participants can be avoided. Therefore, the participants in the exchange do not longer need to trust P_0.

Complaint Protocol. As described above, if a participant having received the cipher and the corresponding certificate during the first round of the main protocol receives $S_{P_0}(label)$, it will be possible for him to run the recovery protocol. Otherwise, if he receives $S_{P_0}(label)$ without having received the cipher and the corresponding certificate, he will be able to prove to an external party, by executing the following *complaint protocol* with the TTP, that something went wrong during the exchange.

1. $P_i \to TTP$: $\quad f_{c1}, TTP, label, \mathcal{P}, S_{P_0}(label), S_{P_i}(\star)$.
2. $TTP \Rightarrow \mathcal{P} \setminus P_i$: $f_{c2}, \mathcal{P}, label, S_{P_0}(label), S_{TTP}(\star)$.

A participant P_i $(1 \leq j \leq n-1)$ begins that protocol by sending $(label, \mathcal{P}, S_{P_0}(label))$ to the TTP. The latter checks if $\mathcal{P}$ is consistent with the *label*, if P_i belongs to the set $\mathcal{P}$ and if $S_{P_0}(label)$ is a valid signature. If so, the TTP multicasts the signature of P_0 over the *label* to the remaining participants and asks them for the signatures obtained during the first round of the main protocol.

At this moment all the participants have received $S_{P_0}(label)$, and other excluded participants (having possibly not received this signature before) can also invoke the complaint protocol.

3. For $j = 0, ..., i-1, i+1, ..., n-1$:
$P_j \to TTP : f_{c3}, TTP, f_{m1}, label, c_{j-1}, cert_{j-1},$
$\qquad\qquad S_{P_{j-1}}(f_{m1}, P_j, label, c_{j-1}, cert_{j-1}), S_{P_j}(\star).$

4. $TTP \to P_i : f_{c4}, P_i, label, \mathcal{P}, S_{TTP}(\star).$

If after a deadline chosen by the TTP none of the remaining participants is able to present P_i's first round signature, the TTP will issue an affidavit attesting that something wrong happened during the protocol. Two cases are possible: either an honest entity was excluded from the exchange or a dishonest entity ran the complaint protocol with the help of the next participant in the ring.

$\mathcal{P}$ has been defined as a set of participants. If all the participants in $\mathcal{P} \setminus P_i$ reply to the second message of the complaint protocol, the TTP will know their disposition in the ring. However, the TTP will not be able to determine where the excluded participant P_i should be.

Otherwise, if $\mathcal{P}$ is defined as an ordered set according to the agreed topology, it will be possible for the TTP to determine the identity of the nearest participant, actively involved in the exchange, who follows P_i. At least this participant has contributed to the coalition. The TTP may not be able to identify the dishonest participant who precedes P_i because this dishonest entity may also realize a complaint protocol.

5   Conclusion

We have shown that every participant in the fair exchange protocol proposed by Bao et al. [6] must not only trust the TTP, but has to also trust P_0, the participant that initiates the protocol, for not sending a false *yes* to the TTP during the recovery protocol. Such a behavior from P_0 can lead to have a set of entities to participate, without their consent, in a coalition excluding the remaining participants. (Bao et al. also proposed [6] two modified versions of their protocol. Participants in these modified protocols have to also trust the initiator of the exchange.)

We have presented a new protocol in which participants must only trust the TTP, and where passive conspiracies in order to exclude a set of participants can be avoided. Trust requirements are reduced by increasing communication needs. It is not easy to compare this reduction of trust requirements with the resulting increase of communications. However, this communication increase is measurable, unlike trust aspects.

Moreover, the proposed protocol allows excluded honest participants having received $S_{P_0}(label)$ to prove to an external adjudicator that something went wrong during the protocol.

References

1. Martín Abadi and Roger Needham. Prudent engineering practice for cryptographic protocols. *IEEE Transactions on Software Engineering*, 22(1):6–15, January 1996.
2. N. Asokan, Matthias Schunter, and Michael Waidner. Optimistic protocols for multi-party fair exchange. Research Report RZ 2892 (# 90840), IBM Research, December 1996.
3. N. Asokan, Victor Shoup, and Michael Waidner. Optimistic fair exchange of digital signatures. Research Report RZ 2973 (#93019), IBM Research, November 1997.
4. N. Asokan and Victor Shoup. Asynchronous protocols for optimistic fair exchange. In *Proceedings of the 1998 Security and Privacy Symposium*. IEEE, 1998.
5. Feng Bao, Robert H. Deng, and Wenbo Mao. Efficient and practical fair exchange protocols with off-line TTP. In *RSP: 19th IEEE Computer Society Symposium on Research in Security and Privacy*, Washington - Brussels - Tokyo, May 1998.

6. Feng Bao, Robert Deng, Kanh Quoc Nguyen, and Vijay Vardharajan. Multi-party fair exchange with an off-line trusted neutral party. In *DEXA'99 Workshop on Electronic Commerce and Security*, Firenze, Italy, September 1999.
7. Matthew K. Franklin and Michael K. Reiter. Fair exchange with a semi-trusted third party. In *4th ACM Conference on Computer and Communications Security*, Zurich, Switzerland, April 1997.
8. Matt Franklin and Gene Tsudik. Secure group barter: Multi-party fair exchange with semi-trusted neutral parties. *Lecture Notes in Computer Science*, 1465, 1998.
9. Steve Kremer and Olivier Markowitch. A multi-party non-repudiation protocol. In *Proceedings of SEC2000 conference*, Beijing, China, August 2000.
10. Olivier Markowitch and Steve Kremer. A multi-party optimistic non-repudiation protocol. Technical Report 443, ULB, 2001. Published in the proceedings of The 3rd International Conference on Information Security and Cryptology (ICISC 2000).
11. Olivier Markowitch and Shahrokh Saeednia. Optimistic fair-exchange with transparent signature recovery. Technical Report 452, ULB, 2001. Published in the proceedings of the 5th International Conference: Financial Cryptography 2001 (FC01).
12. Jianying Zhou and Dieter Gollmann. An efficient non-repudiation protocol. In *PCSFW: Proceedings of The 10th Computer Security Foundations Workshop*. IEEE Computer Society Press, 1997.

Practical Reasoning about Accountability in Electronic Commerce Protocols

Supakorn Kungpisdan and Yongyuth Permpoontanalarp

Logic and Security Laboratory
Department of Computer Engineering
Faculty of Engineering
King Mongkut's University of Technology Thonburi
91 Pracha-utit Rd. Bangmod, Thoongkru, Bangkok, 10140, Thailand
hotkeng@hotmail.com, yongyuth@cpe.eng.kmutt.ac.th

Abstract. Accountability in electronic commerce (e-commerce) protocols is concerned with the ability to show that particular parties who engage in the protocols are responsible for some transactions. Traditionally, it is used only for resolving disputes amongst parties. Many logics were proposed for reasoning about the accountability. However, these logics lack of the application to real-world e-commerce protocols. In this paper, we show that these logics are inadequate to analyze the accountability property of such real-world protocols. We then propose a modification of the existing logics to deal with such real-world protocols. Furthermore, we propose a novel use of the accountability for specifying and analyzing the goals of e-commerce protocols, in particular client privacy property.

Keywords: Formal methods for security protocols, analysis of electronic commerce protocols

1   Introduction

In electronic commerce (e-commerce), every party requires a guarantee that any transaction that occurs is carried out in a secure fashion. In fact, it is unavoidable to prevent malicious behaviors of some parties. These malicious behaviors cause "*Disputes*". Although the system cannot prevent the dishonest party from modifying transactions, it should allow honest parties to prove their rightful behavior when a dispute amongst parties occurs. This property should be included in well-designed protocols since it gives assurances to users in that each user can prove what has actually happened in a transaction to a verifier, who may be a court or someone acting as a dispute solver. This property is called accountability.

Accountability in e-commerce protocols [1] is concerned with the ability to show that particular parties who engage in such protocols are responsible for some transactions. In particular, the accountability [1] involves the ability of a party, called a prover to convince to another party, called a verifier, that a statement is true *without revealing any secret information to the verifier*. Traditionally, the accountability is used only to resolve disputes amongst parties. In such cases, a judge would act as the verifier, and a defendant would act as a prover.

K. Kim (Ed.): ICICS 2001, LNCS 2288, pp. 268–284, 2002.

On one hand, many logics [1,3] were proposed for reasoning about the accountability in e-commerce protocols. However, these logics lack of the application to real-world e-commerce protocols. On the other hand, Herreweghen in [2] proposed an analysis of two real world protocols: SET [8] and iKP [7] on the accountability property. The analysis shows that SET lacks of the accountability whereas iKP has the property. The analysis is carried not informally since it does not use any formal logic.

In this paper, we argue that the existing logics for analyzing the accountability are *inadequate* to analyze real-world e-commerce protocols. In particular, we argue that the existing logics are not able to reason about *multiply* encrypted and/or hashed messages which are signed. Such messages are obtained from applying encryption, digital signature, hash function, or a combination of them to a plain message. Indeed, this kind of messages is that which is employed in most, if not all, e-commerce protocols.

Moreover, in order to reason about the accountability in general, it requires reasoning about verifiers whether a verifier would be convinced of some statement, after a prover sends some *non-secret* or *non-private* information to the verifier.

We argue that Kailar's logic [1] does not provide reasoning about verifiers at all. Even though Kessler and Neumann's (KN) logic [3] offers reasoning about verifier's belief as a result of the receipt of some information, it does not address the issue about proving without revealing secret or private information to verifiers. Note that this issue is a part of the definition of the accountability [1].

In this paper, we propose a modification of KN's logic to reason about the accountability of real-world e-commerce protocols. Our logic can reason not only about the accountability of complex cryptographic messages which are signed, but also about verifiers' beliefs as a result of the receipt of some information. Moreover, our logic captures the provability without revealing any secret or private information to verifiers naturally by using provers' beliefs about verifiers' possession of information. We demonstrate the practicality of our logic by showing that the result obtained from the informal analysis [2] can also be obtained from our logic.

The practical aspect of the accountability that was discussed so far in the literature [2,5] is about payment or money *only*. In particular, the money accountability is about the authorized transfer of money from customer's account to merchant's account. In this paper, we introduce another kind of accountability, called *goods* accountability, which is about the authorized order of goods by a client. The goods accountability can be used to resolve disputes on the mismatch between the goods which is ordered and that which is delivered. By using our logic, we show that both SET and iKP lack of the goods accountability.

Furthermore, we propose a novel use of the accountability for specifying and analyzing the goals of e-commerce protocols, in particular *client privacy* property. The main goal of e-commerce protocols is to ensure that after the completion of the protocols, all parties who engage in the protocols are convinced that they have authorized messages concerning transactions relevant to them. We argue that this goal can be understood as *a special case* of the accountability where the prover is the originator of a message, the verifier is the intended recipient and the statement to be proved is about some transactions.

The *client privacy* is concerned with clients' provability of their authorized payment for a transaction without revealing goods description and payment information (e.g. credit card numbers) in the transaction to banks and merchants, respectively. Thus, the client privacy can be understood as the accountability where a client is a prover, both a bank and a merchant are verifiers, and the proving statement is about the payment authorization. Moreover, both goods description and payment information are considered as client's private information, which must not be revealed to the bank and the merchant, respectively.

This paper is organized as follows. Section 2 provides the background for the accountability and the existing logics. Section 3 presents our logic. In section 4, we applied our logic to analyze SET and iKP on the two kinds of the accountability, and to specify and analyze SET on its goals. Section 5 presents the conclusion of our work.

2   Background

2.1   Formal Approach to Accountability

Kailar's Logic.
Kailar [1] is probably the first who propose a modal logic to reason about accountability. Kailar's definition of accountability is concerned with the ability to prove the association of an originator (of a message) with some action to a third party without revealing any private information to the third party. The party who can prove such a statement is called a prover whereas the third party who is convinced of the proof is called a verifier. Kailar employs the modal operator 'CanProve' to formalize the concept of accountability, i.e. P CanProve ϕ to V where P and V stand for prover and verifier, respectively, and ϕ stands for a general statement about some action.

However, Kailar's logic [1] is not suitable for analyze the real-world e-commerce protocols because of the following two reasons: firstly, Kailar's logic can analyze the signed plain message only. However, messages in real-world e-commerce protocols are not just signed plain messages, but they often are multiply encrypted and/or hashed messages which are signed, i.e. $\{h(X)\}_K$ or $\{\{X\}_{K'}\}_{K^{-1}}$, where K' is public key of a party and K^{-1} is private key of another party. Secondly, Kailar's logic does not reason about verifiers at all. Recently, it was pointed out in [2] that reasoning about verifiers is essential for analyzing real-world e-commerce protocols. We shall discuss this issue in detail in a later section.

It should be noted also that Kailar's definition for accountability is *general* in that the actions which are associated with an originator can be of any kinds.

Kessler and Neumann's Logic
Following Kailar, Kessler and Neumann (KN) [3] employs a modal logic to reason about the accountability. However, KN provides an alternative definition of the modal operator 'CanProve' by means of sending messages. KN's goal to show the accountability is to show 'P believes P CanProve ϕ to V' where P and V stands for a prover and a verifier, respectively. One way to show that P believes P CanProve ϕ to V holds is for P to believe that P can convince V to believe ϕ by sending some messages that P has to V. Thus, this logic offers reasoning about both prover's beliefs and verifier's beliefs, and in particular, prover's beliefs about verifier's beliefs.

Even though KN's logic offers reasoning about the accountability of hashed messages which are signed, its definition for dealing with such messages is *too strong* in that verifiers are able to infer the input m of any hash $h(m)$, which may be *private information*. Consider the following rule for dealing with such kind of message in KN's logic:

> *P CanProve (Q said h(X)) to V $\wedge$ V believes $\neg Q$ sees h(X)*
> *$\rightarrow$ P CanProve (Q said X) to V*

This rule states that if P can prove that Q said the hash value of X, i.e. $h(X)$, and V believes that Q does not receive $h(X)$ from anyone, then P can prove to V that Q said X. We argue that this is *not intuitive*.

Generally, to prove that Q said the input X of a hash function, both prover and verifier need to know:

a) The hashed message $h(X)$ and the input (X), and
b) That the hashed message is really obtained from applying hash function h to X.

Obviously, in KN's approach, the verifier does not need to know *a)* and *b)* but simply that the verifier must believe that the initiator does not receive the hash of the message from anyone in order to be convinced that the initiator originates the hashed message and thus its input. The verifier is convinced even though to show *b)* requires the knowledge of some private information, which should not be revealed to any third party (e.g. goods description).

Moreover, their logic does not deal with the accountability of encrypted messages which are signed, i.e. $\{\{X\}_{K'}\}_{K^{-1}}$ where K' is public key of a party and K^{-1} is private key of another party.

2.2   Herreweghen's Approach

Herreweghen [2,4] proposed the analysis of the accountability of the real-world e-commerce protocols: SET [8] and iKP [7]. The analysis shows that SET lacks of the accountability whereas iKP does not. The analysis is not formal since it is done without using any formal logic. However, the analysis is presented partly in rule-based styles.

Unlike the accountability in Kailar's approach and KN's approach, the kind of the accountability that is analyzed in [2] is *specific* in that it focuses on primitive transactions [5], which are the core of e-commerce protocols. It was proposed in [5] that a payment transaction in any e-commerce protocol consists of three types of primitive transactions:

- *Payment:* client makes the payment to merchant.
- *Value subtraction:* client allows acquirer to deduct money from his account.
- *Value claim:* merchant requests acquirer to deposit money to merchant's account.

Thus, in order to show that an e-commerce protocol has the accountability for resolving disputes, it suffices to show that at the completion of the protocol, each party in the protocol must be able to prove the authorizations of primitive transactions concerning the party to an external verifier. The following shows such kind of statements:

- *M can prove "C **authorized** payment(C, M, Amount, Date, Ref)"*
  Merchant can prove that client has sent authorized message on making the payment to him.
- *C can prove "M **authorized** payment(C, M, Amount, Date, Ref)"*
  Client can prove that merchant has sent authorized message on acknowledgement of payment to him.
- *A can prove "C **authorized** value subtraction(A, C, Amount, Date, Ref)"*
  Acquirer can prove that client has sent authorized message on requesting him to deduct money from client's account.
- *C can prove "A **authorized** value subtraction(A, C, Amount, Date, Ref)"*
  Client can prove that acquirer has sent authorized message on acknowledgement of requesting to deduct money from client's account.
- *A can prove "M **authorized** value claim(A, M, Amount, Date, Ref)"*
  Acquirer can prove that merchant has sent authorized message on requesting him to transfer money to merchant's account.
- *M can prove "A **authorized** value claim(A, M, Amount, Date, Ref)"*
  Merchant can prove that acquirer has sent authorized message on acknowledgement of transferring money to merchant's account.

The analysis in [2] shows that SET lacks of the accountability because of the following two problems:

Firstly, prover does not have key for decrypting the encrypted message, which contains the necessary information. For example, prover wants to derive PI from $\{PI\}_{K_A}$, but he does not have acquirer's private key.

Secondly, prover has to reveal the private information to verifier in order to prove the authorization. The source of this problem is that the required information is hashed together with the information which is considered to be private information to verifier e.g. $h(Price, OD)$ where $Price$ is the required information and OD is private information. To derive $Price$, prover is required to send both $Price$ and OD to verifier to convince him. Thus the verifier knows the private information.

This problem does not occur in iKP. The message format in iKP is $h(Price, h(OD))$. Thus, the prover can convince the verifier about $Price$ without revealing OD because it is hashed.

We argue that Herreweghen's analysis [2] is sensible because in order to solve disputes, prover wants to send only the necessary evidence to judge. The unnecessary private information is not what he wants to reveal for proving some statements. It is intuitive in proving something to other parties.

The disadvantages of this approach are as the following:

1) It is unsystematic since the analysis is not formal.
2) It is specific to only SET and iKP. Those rules cannot be applied to other protocols.

3   Our Logic

Our logic is based on KN's logic [3]. In particular, our logic can be seen as an extension and a simplification of KN's logic. It employs the concept of *provable authorization in the present of private information.* In order to solve disputes, a prover wants to send only the necessary information to prove some statements to a judge who acts as a verifier without revealing the unnecessary private information. With this concept, prover can prove the statement without revealing any private information to verifier. We extend KN's logic in two main aspects. Firstly, we provide axioms for the accountability of multiply encrypted and/or hashed messages which are signed in order to resolve disputes. Secondly, we propose axioms for dealing with the use accountability to specify and analyze the goals of e-commerce protocols.

Syntax

Terms

- $\{P, Q, V\}$: A set of principles that communicate with each other in the protocol.
- $\{X, Y\}$: Messages or message components sent in the protocol.
- $\{\phi, \psi\}$: The statements derived from messages.
- $\{K_P, K_Q\}$: A set of the public key of the principal P and Q, respectively.
- $\{K_P^{-1}, K_Q^{-1}\}$: A set of private key of the principal P and Q, respectively.
- $\{X\}_K$: The message X encrypted with key K.
- $h(X)$: The hash function of a message X.
- $\xrightarrow{K} Q$: Key K is the public key of Q.
- $P \xleftrightarrow{K} Q$: K is the shared key between P and Q.
- X-is-fingerprint-of-Y : X can be used as the representative (fingerprint) of Y (in other words, X may be the hashed form of Y).
- K-is-decrypting-key-for-$\{X\}_K$: Key K can be used to decrypt the encrypted message $\{X\}_K$.

Formulae

- *P believes* ϕ : A principle P believes that the statement ϕ is true by doing some actions.
- *P sees X* : Someone has sent a message X to P and P is able to read X.
- *P has X* : A principle P possesses a message X. He can send X to other principles or use it for further computations.
- *P says X* : A principle P has sent a message X.
- *P CanProve* ϕ *to Q* : A principal P can prove to Q that the statement ϕ is true by sending a message X to Q. After Q receives X, he believes that the statement ϕ is true.
- *P authorized payment(P, Q, Price, Date)* : A principle P has authorization on making the payment amount *Price* to Q on date *Date*.
- *P authorized goods-order(P, Q, OD, Date)* : A principle P has authorization on ordering goods as specified in order description *OD* to Q on date *Date*.

Axioms

Modalities: KD45-logic
K: *P believes ϕ $\wedge$ P believes ($\phi \rightarrow \psi$) $\rightarrow$ P believes ψ*
D: *P believes ϕ $\rightarrow$ $\neg$P believes $\neg\phi$*
4: *P believes ϕ $\rightarrow$ P believes P believes ϕ*
5: *$\neg$P believes ϕ $\rightarrow$ P believes $\neg$P believes ϕ*

Possession
H1: *P sees X $\rightarrow$ P has X*
H2: *(P has X_1 $\wedge$ ... $\wedge$ P has X_n) $\leftrightarrow$ P has (X_1, ... , X_n)*
    Where *(X_1, ... , X_n)* stands for a list of message X_1, X_2, ..., X_n, respectively.
H3: *P has X $\rightarrow$ P has h(X)*

The following axioms (H4, H5, and H6) are based on BAN logic [6] in order to deal with cryptographic messages.

H4: If *P* has an encrypted message $\{X\}_K$, *P* believes that *K* is shared between *P'* and *Q*, and *P* has key *K*, then *P* has *X*.

   (P has $\{X\}_K$ $\wedge$ P believes P' $\leftrightarrow$ Q $\wedge$ P has K) $\rightarrow$ P has X

H5: If *P* has an encrypted message $\{X\}_K$, *P* believes that *K* is the public key of *P'*, and *P* has its private key K^1, then *P* has *X*.

   (P has $\{X\}_K$ $\wedge$ P believes $\rightarrow$ P' $\wedge$ P has K^1) $\rightarrow$ P has X

H6: If *P* has a signed message $\{X\}_{K^{-1}}$, *P* believes that *K* is the public key of *P'*, and *P* has its verification key *K*, then *P* has *X*.

   (P has $\{X\}_{K^{-1}}$ $\wedge$ P believes $\rightarrow$ P' $\wedge$ P has K) $\rightarrow$ P has X

Comprehension
C1: If *P* has received *X*, *P* then believes that he receives *X*.

   P sees X $\rightarrow$ P believes P sees X

C2: If *P* has sent *X*, he then believes that he has sent *X*.

   P says X $\rightarrow$ P believes P says X

Seeing
SE1: *P sees (X_1,..., X_n) $\leftrightarrow$ (P sees X_1 $\wedge$ P sees X_2 $\wedge$ ... $\wedge$ P sees X_n)*

Saying
SA1: *P says (X_1, ..., X_n) $\leftrightarrow$ (P says X_1 $\wedge$ P says X_2 $\wedge$ ... $\wedge$ P says X_n)*
SA2: *P says X $\rightarrow$ P has X*

Provability

P1: *P CanProve ($\phi \rightarrow \psi$) to V*

    $\rightarrow$ (P CanProve ϕ to V $\rightarrow$ P CanProve ψ to V)

P2: *(V-is-external-party) $\wedge$ (P has X) $\wedge$ (V sees X $\rightarrow$ V believes ϕ)*

    $\rightarrow$ (P CanProve ϕ to V)

It can be seen that axiom P2 can be used to deal with the provability to an external verifier. Thus, the axiom states the accountability for resolving disputes.

P2': If V is an internal party, P has sent a message X', V receives X, and if V receives X, then he believes that ϕ is true and he has X', then P can prove to V that ϕ is true.

    (V-is-internal-party) $\wedge$ (P says X') $\wedge$ (V sees X) $\wedge$
    [V sees X $\rightarrow$ (V believes $\phi \wedge$ V has X')]
    $\rightarrow$ (P CanProve ϕ to V)

It should be noted that the verifier V in this axiom is an internal party who engages in the protocol. This axiom is intended to deal with the accountability for specifying and analyzing the goals of e-commerce protocols. We shall discuss this issue and the intuition of axiom **P2'** in section 4.4.

P3: *P has {X}$_{K^{-1}}$ $\wedge$ P has (K, X) $\wedge$ P CanProve ($\rightarrow$ Q) to V*

    $\rightarrow$ P CanProve (Q says X) to V

The following shows our axiom to deal with the provability about hashed messages. Intuitively, it states that in order to prove a statement about a hashed message, it requires both prover and verifier to know the input of the hash in order to compare the input with such hashed message.

P4: If P can prove to V that Q has sent $h(X)$ and he can prove to V that $h(X)$ is represented as fingerprint of X, P then can prove to V that Q has sent X.

    P CanProve (Q says h(X)) to V $\wedge$ P CanProve (h(X)-is-fingerprint-of-X) to V
    $\rightarrow$ P CanProve (Q says X) to V

The following shows our axiom to deal with the provability about encrypted messages. In order to prove a statement about an encrypted message, it requires both prover and verifier to know the decrypting key for such encrypted message.

P5: If P can prove to V that Q has sent the encrypted message $\{X\}_K$ and he can prove to V that the key K' can be used to decrypt this encrypted message, he can prove to V that Q has sent message X.

    P CanProve (Q says {X}$_K$) to V $\wedge$
    P CanProve (K'-is-decrypting-key-for-{X}$_K$) to V
    $\rightarrow$ P CanProve {Q says X} to V

P6:   *P CanProve (Q says (X$_1$, ..., X$_n$)} to V*
       ↔ [P CanProve (Q says X$_1$) to V ∧ ... ∧ P CanProve (Q says X$_n$) to V]

Inference Rules
MP: If ϕ and $\phi \rightarrow \psi$ then ψ
M: If ϕ is a theorem then *P* believes ϕ is a theorem
       Where theorem is a formula, which can be derived from axioms alone.

The money accountability property can be specified by the following formula:

M believes M CanProve [C says (M, Price, Date)
                  → C authorized payment(C, M, Price, Date)] to V

If merchant believes that he can prove that client has sent merchant's ID, price, and date of execution, then the authorization on *Payment* in [2] can be proved. Proving money accountability mainly concerns that the price can be proved to verifier. Goods description (*OD*) is considered to be client's private information. Similarly, the goods accountability property can be specified by the following formula:

M believes M CanProve [C says (M, OD, Date)
                  → C authorized goods-order(C, M, OD, Date)] to V

Proving goods accountability mainly concerns that the goods description can be proved to verifier. *Price* is considered to be merchant's private information.

We argue that if money accountability in [2] is sensible, goods accountability is sensible as well since goods accountability is focused on solving disputes on the mismatch between the goods which is ordered and that which is delivered, which can be occurred in practical transactions.

Apart from the axioms and rules of inference presented here, our logic contains a set of assumptions. Some of these assumptions are general whereas others are specific to protocols which are to be analyzed. For the ease of the presentation here, we discuss both kinds of assumptions in the following section.

4   Practical Reasoning about Accountability

We demonstrate how our logic works by analyzing SET [8] and iKP [7], which are the real-world and widely used protocols. In section 4.1, we describe the overview of SET protocol and the set of assumptions used in the analysis. Section 4.2 and 4.3 illustrate the analysis of proving money and goods accountability, respectively. Particularly, section 4.2 shows that the results obtained from Herreweghen's analysis [2] can also be obtained in our logic. Section 4.4 shows the analysis of accountability for specifying goals of SET.

4.1 SET Protocol Overview

SET is the most widely used credit-card based protocol using public key cryptography scheme. In SET, every participant has his own certificate. The basic SET protocol model is shown as follows:

PinitReq: $C{\to}M$:   *Initial Request*

PinitRes: $M{\to}C$: $\{TID\}_{K_M^{-1}}$

PReq: $C{\to}M$: $TID, OI, h(PI), \{h(OI), h(PI)\}_{K_C^{-1}}, \{h(OI), PI\}_{K_A}$

AuthReq: $M{\to}A$: $\{\{TID, Price, AuthReqDate, h(OI), OI, \{h(OI), h(PI)\}_{K_C^{-1}},$

                 $\{h(OI), PI\}_{K_A} \}_{K_M^{-1}} \}_{K_A}$

AuthRes: $A{\to}M$: $\{\{TID, Price, AuthDate\}_{K_A^{-1}} \}_{K_M}$

PRes: $M{\to}C$: $\{TID, AuthDate\}_{K_M^{-1}}$

Where

- $\{A, C, M\}$ stands for a set of acquirer (or bank), client, and merchant, respectively.
- $\{K_A, K_C, K_M\}$ stands for the set of public key of acquirer, client, and merchant, respectively.
- $\{K_A^{-1}, K_C^{-1}, K_M^{-1}\}$ stands for the set of private key of acquirer, client, and merchant, respectively.
- *Price* stands for amount and currency of goods.
- *OI* stands for order information. *OI* contains $\{TID, h(OD, Price)\}$.
- *PI* stands for payment information. *PI* contains $\{TID, h(OD, Price), MerchantID, Price, Client's Credit-card Information\}$.
- *TID* stands for transaction ID. *TID* contains purchase request date (*PReqDate*), which is the date of issuing purchase request.
- *AuthReqDate* stands for the date of making request by merchant for transferring money.
- *AuthDate* stands for the date of authorization made by bank.

Set of Assumptions

Notations
- *P* stands for any participant.
- $\{A, C, M\}$ stands for a set of acquirer (or bank), client, and merchant, respectively.
- *V* stands for a verifier.
- *X* stands for *A, C,* or *M*.
- *Cert$_X$* stands for the certificate of principle *X*. *Cert$_X$* contains (X, K_X), where *X* is the identity of a party *X* and K_X stands for the public key of *X*.

Assumptions
(A1) Every participant believes that he has the certificates of all participants.

     P believes P has (Cert$_A$, Cert$_C$, Cert$_M$)

(A2) P believes that if V has X's certificate, he will believe that K_X is the public key of X.

P believes $(V$ has $Cert_x \rightarrow V$ believes $(\underset{K_x}{\rightarrow} X))$

(A3) Every participant believes that he has only his private key, and nobody believes that he has the private keys of other participants.

P believes P has K_P^{-1} $\neg P$ believes P has $(K_{Q_1}^{-1}, K_{Q_2}^{-1})$

Where Q_1 and Q_2 are the parties, which are different from P.

(A4) Every participant believes that if V receives messages X and Y and the message X is $h(Y)$, V will believe that X is fingerprint of Y.

P believes $[(V$ has $(X, Y) \wedge X = h(Y)) \rightarrow V$ believes X-is-fingerprint-of-$Y]$

(A5) Every participant believes that if V receives a message encrypted with key K $(\{X\}_K)$ and the key K', and K' can be used to decrypt $\{X\}_K$, V will believe that K' is the decrypting key for $\{X\}_K$.

P believes $(V$ has $(\{X\}_K, K') \wedge X = \{\{X\}_K\}_{K'}$
$\rightarrow V$ believes K'-is-decrypting-key-for-$\{X\}_K)$

The assumptions A4 and A5 are general in that they can be used for analyzing any protocol. Indeed, together with P4 and P6, the two assumptions are used to reason about the provability of hashed messages and encrypted messages.

(A6) Every participant believes that if V has the signed message $(\{X\}_{K^{-1}})$, the key K, and the message X, and V believes that K is the public key of Q, then V will believe that Q has sent X.

P believes $[(V$ has $\{X\}_{K^{-1}} \wedge V$ has $(K, X) \wedge V$ believe $(\overset{K}{\rightarrow} Q))$
$\rightarrow V$ believes $(Q$ says $X)]$

This assumption together with axiom P2 or P2' can be used to derive similar conclusion as the axiom P3. However, this assumption would derive the conclusion, based on verifier's possession of some information whereas the axiom P3 would derive the conclusion, based on prover's possession of some information.

No Disclosure of Private Information to Verifier
(A7) Every participant does not believe that verifier has payment information and the private keys of all participants, order description in case of proving money authorization, and price in case of proving goods authorization.

$\neg P$ believes V has PI where V is not acquirer.
$\neg P$ believes V has K_P^{-1} where V is not the same party as P.

In case of Money Authorization, $\neg P$ believes V has OD
In case of Goods Authorization, $\neg P$ believes V has $Price$

By specifying the requirement "No disclosure of private information to verifier" as a set of assumptions specific to protocols, our logic formalizes intuitively the concept of the unrevealing private information to verifier when a prover proves a certain statement. This concept is essential in the accountability property for reasoning about accountability in the present of party privacy.

SET Specific Assumptions

(A8) Client and merchant believe that they have order description and price, and all participants believe that they are internal parties.

> Q believes Q has (OD, Price)   Where Q stands for C and M.
> R believes R'-is-internal-party   Where R and R' stand for A, C, and M.

(A9) Every participant believes that client has sent *PReq* and received *PRes*, merchant has sent *AuthReq* and received *AuthRes*, and acquirer has received *AuthReq* and sent *AuthRes*.

> P believes C says PReq P believes M sees PReq
> P believes M says AuthReq P believes A sees AuthReq
> P believes A says AuthRes P believes M sees AuthRes
> P believes M says PRes P believes C sees PRes

Client Privacy

(A10) Client does not believe that merchant has payment information, and client and merchant do not believe that acquirer has order description.

> $\neg C$ believes M has PI
> $\neg Q$ believes A has OD   Where Q stands for client and merchant.

General Assumptions for Proving Money Authorizations

(A11) Every participant believes that he can prove to verifier that if client has sent the message containing Merchant's ID, price, and the date of execution, then client has authorization on payment ordering.

> P believes P CanProve [C says (M, Price, Date)
> $\rightarrow C$ authorized payment(C, M, Price, Date)] to V

Assumptions for proving money authorizations for other primitive transactions are shown in Appendix.

General Assumptions for Proving Goods Authorizations

(A12) Every participant believes that he can prove to verifier that if client has sent the message containing Merchant's ID, order description, and the date of execution, then client has authorization on goods ordering.

> P believes P CanProve [C says (M, OD, Date)

$\rightarrow C$ *authorized goods-order(C, M, OD, Date)] to V*

Assumptions for proving goods authorizations for other primitive transactions are shown in Appendix.

4.2   Proving Money Accountability in SET

Proving accountability in SET, we focus on analyzing one of primitive transactions, which is *C authorized payment (C, M, Price, PReqDate)*. This means that client has authorization on making payment on the goods amount *Price* on the date of making the request *Date*.

The goal of proof:
M believes M CanProve (C authorized payment(C, M, Price, PReqDate)) to V

Where *V* stands for any external verifier.
In order to show that *M believes M CanProve (C authorized payment(C, M, Price, PReqDate)) to V*, it suffices to show that

$\quad$ *M believes M CanProve (C says (M, Price, PReqDate)) to V* $\qquad$ (x)

It is easy to see that *M believes M CanProve (C says h(PI)) to V* follows mainly by axiom P3. Since *(M, Price, PReqDate)* is in *PI*, (x) may thus be shown by using axiom P4.

However, the proof for *M believes M CanProve (h(PI)-is-fingerprint-of-PI) to V* in axiom P4 would fail since it is not the case that *M believes M has* K_A^{-1} which is required to show that *M believes M has (h(PI), PI)*. Note that *PI* is encrypted with K_A. Note also that if it were the case that *M believes M has* K_A^{-1}, the proof would still fail since it would require *M believes (V has PI)*, due to A4, which contradicts to our assumption A7.

It is also easy to see that *M believes M CanProve (C says OI) to V* follows mainly by axioms P4 and P3. Since *h(OD, Price)* is in *OI*, the proof for *M believes M CanProve (C says Price) to V* may be shown by using axiom P4. However, such proof would require *M believes V has OD* due to A4, which contradicts to our assumption A7.

4.3   Proving Goods Accountability in SET and iKP

The goal of proof:
M believes M CanProve (C authorized goods-order(C, M, OD, PReqDate)) to V

We shall discuss the goods accountability in SET first. Similarly to the proof for *M believes M CanProve (C says Price) to V* discussed in section 4.2, the proof for *M believes M CanProve (C authorized goods-order(C, M, OD, PReqDate)) to V* would fail since it would require *M believes V has Price* which contradicts to our assumption A7.

IKP [7] lacks of goods accountability due to similar reason to that in SET. The following shows a payment request from client *C* to merchant *M*.

Payment $C \rightarrow M$: $PI, \{PI, h(OI)\}_{K_C^{-1}}$

Where *PI* stands for payment information. *PI* contains *{Price, h(OI), Client's Credit-card Information}$_{K_A}$* . *OI* stands for order information. *OI* contains *(TID, Price, ClientID, MerchantID, Date, InvExpDate, h(OD))*. *InvExpDate* stands for invoice (offer) expiration date specified by merchant. *Date* stands for the date of invoice (offer) issued by merchant.

It is easy to see that *M believes M CanProve (C says h(OI)) to V* follows mainly by axiom P3. Since *(Price, h(OD)* is in *OI*, the proof for *M believes M CanProve (C says OI) to V* can be shown by using axiom P4. However, such proof would require *M believes V has Price* due to A4, which contradicts to our assumption A7.

4.4   Goals of SET and iKP as Accountability

In this section, we discuss the use of the accountability for specifying and analyzing the goals of SET and iKP protocols. For such kind of accountability, a verifier is an internal party, which involves in e-commerce protocols. Recall that the goal of e-commerce protocols is to ensure that all parties are convinced that they have authorized messages concerning primitive transactions relevant to them after the completion of the protocols.

After sending some messages to intended recipients, the originator must be able to prove the association of the originator with an intended action (or an intended message) to intended recipient(s). Such intended actions are just about primitive transactions. Such proof would ensure the originator that the intended recipients would recognize the originator's intention regarding to primitive transactions.

This intuition is formalized by axiom *P2'*. The axiom states explicitly the preconditions for the sending and receiving of messages. The rule also caters for the case where the intended recipient receives messages from an intermediate party.

Thus, the goals of e-commerce protocols can be expressed by the following:

a) C believes C CanProve (C authorized payment(C, M, Price, Date)) to M
b) M believes M CanProve (M authorized payment(C, M, Price, Date)) to C
c) C believes C CanProve (C authorized value subtraction(A, C, Price, Date)) to A
d) A believes A CanProve (A authorized value subtraction(A, C, Price, Date)) to C
e) M believes M CanProve (M authorized value claim(A, M, Price, Date)) to A
f) A believes A CanProve (A authorized value claim(A, M, Price, Date)) to M
g) C believes C CanProve (C authorized goods-order(C, M, OD, Date)) to M
h) M believes M CanProve (M authorized goods-receipt(C, M, OD, Date)) to C

It is not hard to see that in SET these goals cannot be shown due to similar reason to those for the accountability for dispute resolving discussed in sections 4.2 and 4.3. However, these goals can be shown for iKP.

With provable authorization in the present of private information, our logic can be used to specify and analyze goals of e-commerce protocols efficiently in that the originator can ensure that the recipient recognizes the intention about primitive transaction without revealing private information. This will be much benefit to protocol designers in that he can design a protocol with intended purposes.

What we demonstrate below is the analysis of client privacy using our logic.

Proving Client Privacy of SET

Client privacy can be understood as the accountability where a client is a prover, both a bank and a merchant are verifiers, and the proving statement is about the payment authorization. SET achieves client privacy if merchant cannot infer client's payment information *(PI)*, and acquirer cannot infer goods description *(OD)* from the protocol. We thus start proving client privacy by stating the goals of the proofs from *a)* and *c)*. In order to prove *a)*, it suffices to show that

$$C \text{ believes } C \text{ CanProve } (C \text{ says } (M, Price, PReqDate)) \text{ to } M$$

It is easy to see that *C believes C CanProve (C says (Price, PReqDate)) to M* follows mainly by axioms P4 and P3. Although the proof is not successful because of the lacking of merchant's ID, merchant cannot infer *PI* from the receiving message. We prove *c)* in the same way as proving *a)*. It is not hard to see that acquirer cannot infer *OD*, which is client's private information, from the receiving message.

As a result, our logic can analyze client privacy, which is an essential property of SET protocol, in that verifier can get only necessary information to prove without getting any private information.

5   Conclusion

In this paper, we show that the existing logics for reasoning about accountability are inadequate to deal with real world e-commerce protocols. We then propose an extension of the existing logics to deal with such real-world protocols. Furthermore, we demonstrate the practicality of our logic by showing that the result obtained from Herreweghen's informal analysis [2] can be also obtained formally from our logic.

Our logic can be used not only for reasoning about dispute resolution amongst parties, but also for reasoning about the goals of e-commerce protocols, in particular, client privacy property. Indeed, both kinds of reasoning can be captured *uniformly* in our logic.

For reasoning about dispute resolution, we show *formally* that SET lacks of the money accountability whereas iKP does not. Moreover, we show that both SET and iKP lack of the goods accountability. The lack of the goods accountability in our sense means that the two protocols can still provide enough evidence tokens to resolve disputes on goods description, but in order to resolve such disputes, provers must reveal their private information to verifiers. In some situations, this is undesirable.

For reasoning about the goals of e-commerce protocols, we show that such kind of reasoning can be considered as a special case of the accountability. Thus, the accountability can be seen as a fundamental property for analyzing e-commerce protocols. Indeed, other kind of properties such real-world e-commerce protocols, for example each party's requirement [9], has also been studied in [10] by using our logic presented here.

As a future work, we aim to apply our logic to analyze other kinds of e-commerce protocols, e.g. micropayment. Also, we aim to study an automated tool for our logic.

Acknowledgement

The second author would like to acknowledge supports from National Research Council of Thailand.

References

[1] R. Kailar. Accountability in Electronic Commerce Protocols. *IEEE Transaction on Software Engineering 1996*.

[2] E. V. Herreweghen. Non-Repudiation in SET: Open Issues. In the *Proceedings of the Financial Cryptography 1999*.

[3] V. Kessler and H. Neumann. A Sound Logic for Analyzing Electronic Commerce Protocols. In the *Proceedings of ESORICS'98*.

[4] E. V. Herreweghen. Using Digital signatures as Evidence of Authorizations in Electronic Credit-Card Payments. *Research report 3156, IBM Research,* June 1999.

[5] N. Asokan, E. V. Herreweghen, and M. Steiner. Towards A Framework for Handling Disputes in Payment Systems. In the *Proceedings of the 3rd USENIX workshop on Electronic Commerce*, Boston, Massachusette, August31-September3,1998.

[6] M. Burrows, M. Abadi, and R. Needham. A Logic of Authentication. *ACM Transactions in Computer Systems*, February 1990.

[7] M. Bellare, J. A. Garay, R. Hauser, A. Herzberg, H. Krawczyk, M. Steiner, G. Tsudik, E. V. Herreweghen, and M. Waidner. Design, Implementation, and Deployment of the *i*KP Secure Electronic Payment System. *IEEE Journal of Selected Areas in Communications* 2000.

[8] Mastercard and Visa. SET Protocol Specifications
    http://www.setco.org/set_specifications.html

[9] C. Meadows and P. Syverson. A Formal Specification of Requirements for Payment Transactions in the SET Protocol. In the *Proceedings of Financial Cryptography*, February 1998.

[10] S. Kungpisdan. Accountability as Fundamental Property for Electronic Commerce Protocols. Master Thesis, Department of Computer Engineering, King Mongkut's University of Technology Thonburi, Thailand, 2001, (In preparation).

Appendix

General Assumptions for Proving Money Authorizations

Every participant believes that he can prove to verifier that if merchant has sent the message containing Client's ID, price, and the date of execution, then merchant has authorization on payment receipt.

> *P believes P CanProve [M says (C, Price, Date)*
> *→ M authorized payment(C, M, Price, Date)] to V*

Every participant believes that he can prove to verifier that if acquirer has sent the message containing Client's ID, price, and the date of execution, then acquirer has authorization on value subtraction.

> *P believes P CanProve [A says (C, Price, Date)*
> *→ A authorized value subtraction(A, C, Price, Date)] to V*

Every participant believes that he can prove to verifier that if client has sent the message containing Acquirer's ID, price, and the date of execution, then client has authorization on value subtraction.

> *P believes P CanProve [C says (A, Price, Date)*
> *→ C authorized value subtraction(A, C, Price, Date)] to V*

Every participant believes that he can prove to verifier that if acquirer has sent the message containing Merchant's ID, price, and the date of execution, then acquirer has authorization on value claim.

> *P believes P CanProve [A says (M, Price, Date)*
> *→ A authorized value claim(A, M, Price, Date)] to V*

Every participant believes that he can prove to verifier that if merchant has sent the message containing Acquirer's ID, price, and the date of execution, then merchant has authorization on value claim.

> *P believes P CanProve [M says (A, Price, Date)*
> *→ M authorized value claim(A, M, Price, Date)] to V*

General Assumptions for Proving Goods Authorizations

Every participant believes that he can prove to verifier that if merchant has sent the message containing Merchant's ID, order description, and the date of execution, then merchant has authorization on goods receipt.

> *P believes P CanProve [M says (C, OD, Date)*
> *→ M authorized goods-receipt(C, M, OD, Date)] to V*

Content Extraction Signatures

Ron Steinfeld[1], Laurence Bull[1], and Yuliang Zheng[2]

[1] School of Network Computing, Monash University, Frankston 3199 Australia
ron.steinfeld,l.bull@infotech.monash.edu.au
[2] Dept. Software and Info. Systems, University of North Carolina at Charlotte,
Charlotte, NC 28223
yzheng@uncc.edu

Abstract. Motivated by emerging needs in online interactions, we define a new type of digital signature called a 'Content Extraction Signature' (CES). A CES allows the owner, Bob, of a document signed by Alice, to produce an 'extracted signature' on selected extracted portions of the original document, which can be verified (to originate from Alice) by any third party Cathy, without knowledge of the unextracted (removed) document portions. The new signature therefore achieves verifiable content extraction with minimal multi-party interaction. We specify desirable functional and security requirements from a CES (including an efficiency requirement: a CES should be more efficient in either computation or communication than the simple multiple signature solution). We propose and analyse four provably secure CES constructions which satisfy our requirements, and evaluate their performance characteristics.

Keywords: Content-extraction, fragment-extraction, content blinding, fact verification, content verification, digital signatures, provable security.

1   Introduction

During the course of a person's lifetime one has the need to use 'formal documents' such as: Birth Certificates; Marriage Certificates; Property Deeds; and Academic Transcripts etc. Such documents are issued by some recognized and trusted authority and are thereafter used by the owner in dealings with other people or organizations, to prove the truth of certain statements (based on the trust of the verifier in the issuing authority).

In an electronic world, digital signatures, together with a Public Key Infrastructure (PKI), can be used to ensure authenticity and integrity of electronic versions of formal documents. However, as digital signatures become more widely used, situations can arise where their use significantly increases the cost of desirable document processing operations, which do *not* constitute a security breach. In other words, the standard use of digital signatures sometimes places additional constraints on *legitimate* users beyond what is really required to prevent forgeries by *illegitimate* users.

In this paper we consider one particular document processing operation which is made costly by the use of signatures, namely the *extraction* of certain selected

K. Kim (Ed.): ICICS 2001, LNCS 2288, pp. 285–304, 2002.

portions of a signed document. That is, we consider the case when Bob, the owner of a document which was signed by Alice, does not wish to pass on the whole document to a third (verifying) party Cathy. Instead, Bob extracts a portion of the document and sends only that portion to Cathy. We are assuming here (and will give motivating examples later) that Alice is agreeable for Bob to perform such an extraction. Still, Cathy would like to verify that Alice is the originator of the document portion she received from Bob.

Ignoring for now our other requirements, we explain the difficulty with standard uses of digital signatures in this context. It is clear that if Alice simply signs the whole document using a standard digital signature, then Cathy will not be able to verify it unless Bob sends the whole document. The simplest solution is for Alice to divide the message into (say n) fragments and sign each fragment, giving a set of signatures to Bob, who can then forward a subset of them, corresponding to the extracted document fragments, to Cathy. Some problems with this simplistic approach include both high computation (n signatures) for Alice and Cathy as well as high communication overhead from Alice to Bob, and Bob to Cathy (up to n times the length of one signature). **We want to do better than this simple scheme in either:**

   i communication overhead savings; **OR**
   ii computation savings;

as well as:

   iii giving the signer ability to specify allowed extraction of document content; and
   iv achieving provable security.

We call schemes that perform better at document fragment extraction than the simple scheme above: **Content Extraction Signatures** (CES).

1.1   Contents of This Paper

Due to space limitations, detailed definitions and some proofs are omitted – they are included in the full version of this paper, to be available from the authors. This paper is organized as follows. In Section 2 we explain the real-life background motivating the introduction of Content Extraction Signatures, and review related work. In Section 3, we describe our desirable functional and security requirements from a CES (which differ from those of a standard digital signature – there is even a new *privacy* requirement which has no counterpart in standard digital signatures), leading to a definition of a CES as a new primitive satisfying these requirements. Section 4 presents four CES schemes meeting our definition which are provably secure under well-known cryptographic assumptions. Our CES schemes do *not* employ new cryptographic techniques. However, we believe this application for these known cryptographic techniques is novel, and in the case of the RSA schemes exploits additional properties which are useful in this application (namely to reduce communication bandwidth) but are not used to advantage in most of the original uses of these techniques. In Section 5 we close with some concluding comments.

2   Background

2.1   The Emerging Need

To motivate the rest of the paper, consider the commonplace although pertinent example illustrated in Fig. 1 below. In this example, a university (A) issues a student (B) with a formal document: an Academic Transcript (Original document). The student is required to include the formal document with a job application document sent to a prospective employer (C). Note that the Academic Transcript document is likely to include the student's personal details, in particular the student's date of birth (DOB). To avoid age-based discrimination, the student B may not wish to reveal his DOB to the employer C (indeed, in some countries it is illegal for the prospective employer to seek the applicant's DOB). The university (A) understands this and is willing to allow employers to verify academic transcripts with the DOB removed (and possibly with other fields agreed to by A removed as well, but not others which A may require to be included in any extracted document). Yet if A signs the Academic Transcript using a standard digital signature the student B must send the whole document to C to allow C to verify it.

Instead, we propose in this situation the use of a *Content Extraction Signature* (CES), as described in this paper. The university (A) uses the Sign algorithm of a CES scheme to sign the original document, divided into portions (*submessages* $m_0, m_1,...$) and produce a content extraction signature, given to student B along with the full document. The student then extracts a *subdocument* A' of the original document consisting of a selected subset of the document submessages (e.g. not including m_1, the DOB of B, but including all other submessages). He then runs an Extract algorithm of the CES scheme to produce an *extracted signature* by the university A for the extracted subdocument A'. Student B then forwards the subdocument A' and the extracted signature for A'. The employer uses the Verify algorithm of the CES to verify the extracted signature on A'. This process can also be repeated for other prospective employers as illustrated in Fig. 1.

2.2   Virtues of Content Extraction Signatures

Our approach views a document as a collection of facts or statements. The use of a CES with documents enables users to handle and process these statements in a more selective manner rather than simply using the entire document. Apart from the efficiency of only dealing with the pertinent information, more importantly, it enables the user to embrace a *minimal information model*. Once the information has been sent out, there are no mechanisms for expiring or recalling it. This is vitally important as the privacy and use of the information is at significantly more risk than any time in history due to the ease with which information can be electronically relayed and stored. This is especially true with respect to the Internet.

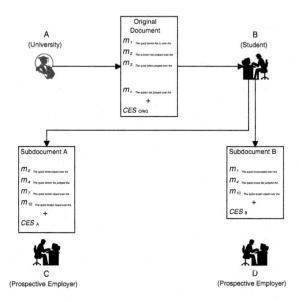

Fig. 1. A real-life application for Content Extraction Signatures.

An extraction policy is a fundamental element of our CES that enables the author to have some control over the use of the content even though the document has been forwarded to the user. A CES does not require any further interaction with the author (after the initial signing), meaning that the author is oblivious to all further uses of (and extraction from) the document by the user. It also permits the lifespan of the document to be independent to the lifespan of the author.

While there may be others, we currently envisage two types of situations that may create a need in practice to extract portions of a signed document before forwarding to the verifier:

 i *Privacy Protection:* The full document contains some sensitive/secret portions, which the document owner may not wish to reveal to some verifiers.
 ii *Bandwidth Saving:* The full document may be very long, while the verifier is only interested in a small portion of it.

2.3   Related Work

Bellare, Goldreich and Goldwasser[12] introduced 'incremental' signatures which allow a *signer* to efficiently update a signature after making an incremental operation on the signed document. Our 'Content Extraction Signatures' can be viewed as addressing the 'deletion' operation efficiently, but allowing this operation to be performed *without* the signer's secret key. While [12] use the standard signature model where this operation would constitute forgery, it is not necessarily a forgery in our 'Content Extraction Access Structure'-based CES-unforgeability definition (see section 3), which allows the signer to specify which deletions are to be regarded as forgeries.

The 'dual signature' defined for the Secure Electronic Transaction (SET) protocol [4], can be considered a special case of one of our schemes, although in that application it is again the signer who performs the 'extraction', and there are only two submessages (payment information and order information). It is therefore a different setting to ours.

The 'XML-Signature'[5] proposed recommendation of the WWW Consortium (W3C) defines a scheme which allows the signing of documents consisting of multiple online 'data objects', some of which may become unavailable after signing. However, the content extraction application and its requirements are not addressed in the recommendation. Indeed, the proposed recommendation cannot be used without modification to achieve our privacy and signer control requirements for content extraction signatures. Furthermore, our approach deals with self-contained documents.

Three of our CES schemes are modifications of batch signature generation and verification schemes [14] [6][9]. However, our schemes are a new application for these techniques. In particular, we use in a novel way the homomorphic property of the RSA schemes which is not used to save bandwidth in standard batch signing and verifying where the output (in the case of signing) is n signatures and the input (in the case of verifying) is n signatures. We are able to take advantage of this feature because in our setting all the n signatures are intended for the same user (from the point of view of batch signing) or are transmitted from a single user (but not the signer) from the point of view of batch verification. Fiat [9] mentions another application of the length-saving feature of his batching scheme for the purpose of saving bandwidth in distributed decryption.

3   Requirements and Definition of Content Extraction Digital Signatures

In this section we explain informally the requirements which a CES scheme should satisfy leading to an (informal) definition for a CES scheme and its desirable security notions. Refer to the full paper for more precise formulations of these definitions.

3.1   Terminology

First we introduce some terminology. We will assume that a document is divided into a number n of smaller *submessages*, and that such documents are represented in a form which encodes an ordering of the n submessages, their number, and allows some of them to be 'blank'. We use $[n]$ to denote the set of submessage indices $\{1, ..., n\}$. We use bold letters (e.g. $\mathbf{M} = (m_1, m_2, ..., m_n)$) to denote such a representation of documents. This representation of a document $\mathbf{M}$ allows *extraction* of a *subdocument* $\mathbf{M}'$ specified by an *extraction subset* X consisting of the indices of the submessages to be extracted and included in the subdocument. We denote by $\mathbf{M}[i]$ the i'th submessage in document $\mathbf{M}$. We use $\mathsf{Cl}(\mathbf{M})$ to denote the set of indices of 'clear' (i.e. non-blank) submessages in a document $\mathbf{M}$.

For example, if the original document is $\mathbf{M} = (m_1, m_2, m_3, m_4)$ and the extraction subset is $X = \{1, 3\}$, then the extracted subdocument is $\mathbf{M}' = (m_1, ?, m_3, ?)$, where ? denotes a blank submessage. Also, we have $\mathsf{Cl}(\mathbf{M}') = \{1, 3\}$. Note that by our definitions $\mathbf{M}'$ is distinct from the documents $(m_1, m_3, ?, ?)$ and $(m_3, ?, m_1, ?)$, which are *not* subdocuments of $\mathbf{M}$.

3.2   Functional Requirements from a CES

Here we discuss the *functional* requirements from a CES, i.e. those which are concerned with *honest* users of the scheme. The basic requirement from a CES scheme can be stated as follows.

Extraction Requirement: *A Content Extraction Signature should allow anyone, given a signed document, to extract a publicly verifiable extracted signature for a specified subdocument of a signed document, without interaction with the signer of the original document.*

As explained in the introduction, the extraction requirement can be met by a simple solution of signing each submessage seperately using a standard digital signature. In order to make this problem more interesting we have asked for an efficiency requirement as follows.

Efficiency Requirement: *A Content Extraction Signature should be more efficient, either in communication overhead (length of original or extracted signatures), or in computational requirements (signing or verifying), than the simple 'multiple signature' solution.*

An additional requirement which may be useful in some applications is the following one. It allows the extraction to continue 'downstream' as the document is passed from verifier to verifier, allowing each one to continue the extraction process. It may be considered optional.

Iterative Extraction Requirement: *A Content Extraction Signature should allow the extraction of a signature for a subdocument of a source subdocument, given the source subdocument and the extracted signature for it.*

3.3   Security Requirements from a CES

Now we discuss the *security* requirements from a CES, i.e. those which prevent undesirable *dishonest* operations.

The most basic security requirement is, as for standard digital signatures, to ensure authenticity via the *unforgeability* requirement. The unforgeability requirement for a CES must differ, however, from the standard 'existential unforgeability' one, because in the latter, any proper subdocument (i.e. a subdocument which is not equal to the original document) of the signed document is considered a 'new message' and therefore the extraction (using public knowledge only) of a valid signature for it constitutes existential forgery. As we have pointed out, however, we are considering situations where it is desirable to relax this model, and allow extraction of signatures for certain subdocuments. However, it is clear that in this model the *signer must have full control to determine which subdocuments signatures can be extracted for.* The signer may want to force some portions as manadatory for inclusion in any extracted subdocument. And it may be necessary to protect against changes in meaning of extracted portions due to certain deletions.

We will address the above observation in defining a CES scheme by allowing an additional input to the signing algorithm of the CES scheme, called the *Content Extraction Access Structure* (CEAS for short), which the signer uses to specify which subdocuments the signer is allowing signatures to be extracted for (the CEAS is therefore an encoding of the allowed extraction subsets, using the terminology introduced above). This leads to the following unforgeability requirement for a CES, assuming the strong 'chosen message' attack, where the attacker can query a CES signing oracle on documents of the attacker's choice. Note that any document **D** queried by the attacker to the signing oracle is accompanied by a corresponding CEAS (also under the attacker's choice) specifying allowed extracted subdocuments.

CES-Unforgeability Requirement: *It is infeasible for an attacker, having access to a CES signing oracle, to produce a document/signature pair* $(\mathbf{M}, \sigma)$, *such that: (i) σ passes the verification test for* **M** *and (ii)* **M** *is either (A) Not a subdocument of any document queried to the CES signing oracle, or (B) Is a subdocument of a queried document* **D**, *but not allowed to be extracted by the CEAS attached to the sign query* **D**.

We remark that the definition of CES-unforgeability above is as strong as one may hope for in our setting, i.e. it prevents forgeries for any documents which are not subdocuments of signed documents. This implies, in particular, security against 'submessage reordering attacks' (where the submessages in the forged document are the same as those in subdocument of a signed document but in different positions or order) as well as 'mixed subdocument attacks' (where the forged document contains submessages which have all appeared in signed documents but have never appeared *together* in a signed document). Note that the simple multiple signature scheme is *not* secure against both of the latter attacks and hence does not even have the CES-unforgeability property without modification.

Unlike standard signatures, where unforgeability is the only security requirement, many applications of content extraction signatures by nature may also have a *privacy* security requirement. Indeed, as mentioned in the introduction, the reason for the user to delete some submessages in the signed document prior to handing it over to the verifier may be in order to hide the sensitive content of these submessages from the verifier. But none of the above requirements on a CES exclude the possibility that the extracted signature may in general leak some information on the content of the unextracted (deleted) submessages. Indeed, this consideration has important implications on the design and efficiency of our first two proposed CES schemes (see section 4). Hence, the privacy requirement for a CES must be defined in order to allow an assessment of whether a scheme satisfies the requirement or not. We give this definition here. We note that a similar requirement has been defined in the context of 'incremental cryptography' [13].

CES-Privacy Requirement: *It is infeasible for an attacker to distinguish between two extracted signatures* σ_0 *and* σ_1, *where* σ_0 *is extracted for a subdocument* $\mathbf{D_0}$ *of the document* $(m_0, ..., m_{i-1}, M_0, m_{i+1}, ..., m_n)$ *specified by extraction subset* X *such that* $i \notin X$ *(i.e.* $\mathbf{D_0}$ *has a blank submessage in position i), and* σ_1 *is extracted for subdocument* $\mathbf{D_1}$ *of the document* $(m_0, ..., m_{i-1}, M_1, m_{i+1}, ..., m_n)$ *specified by the same extraction subset* X *(i.e.* $\mathbf{D_1} = \mathbf{D_0}$). *This holds even if the attacker can choose* (M_0, M_1), X, i, *and the remaining submessages* m_i.

The above definition is similar to the 'indistinguishability' based definition of semantic security for encryption schemes. It ensures the attacker cannot obtain information about submessages which have been blanked (i.e. those not extracted into the subdocument). Its precise formulation (included in the full paper) follows the standard two-stage 'find/guess' attack experiment [15,1]. In this formulation the attacker's find algorithm chooses (M_0, M_1), X, i, and the m_i's. A uniformly random bit b is chosen and the attacker's guess algorithm is given σ_b and tries to guess b. The requirement is that it is infeasible for guess to guess b correctly with non-negligible advantage over random guessing.

3.4   Definition of a Content Extraction Signature

Consistent with the requirements discussed above, a *Content Extraction Digital Signature* (CES) scheme $\mathcal{D}$ can be defined to consist of 4 algorithms:

1 KeyGen – Takes a security parameter k and generates a secret/public key pair (SK, PK).
2 Sign – Takes a secret key SK, a document $\mathbf{M} = (m_1, m_2, ..., m_n)$, and a content extraction access structure $CEAS$, and outputs a content extraction signature σ_{Full}.
3 Extract – Takes a document $\mathbf{M}$, a content extraction signature σ_{Full}, an extraction subset $X \subseteq [n]$ and a public key PK, and outputs an extracted signature σ_{Ext}.
4 Verify – Takes an extracted subdocument $\mathbf{M}'$, an extracted signature σ_{Ext} and a public key PK, and outputs a verification decision $d \in \{Accept, Reject\}$.

In the above definition, the main differences from a standard digital signature are the following.

The Extract algorithm allows the user to extract (from a 'full' content extraction signature σ_{Full}) a signature for the subdocument consisting of the submessages whose indexes are specified by the extraction subset X. The extracted signature σ_{Ext} can then be forwarded to the verifier along with the extracted subdocument $\mathbf{M}'$.

The 'Content Extraction Access Structure' (CEAS) is an encoding of the subsets of submessage indexes in the original document which the signer can use to specify which extracted subdocuments the user is "allowed" to extract valid signatures for. Therefore the $CEAS$ is an encoding of a collections of subsets of $[n]$.

4   Proposed Content Extraction Signature Schemes

4.1   Schemes Based on General Cryptographic Primitives

Scheme CommitVector. We build this CES scheme out of two standard cryptographic primitives:

(i) A standard digital signature scheme S with signature and verification algorithms (S, V) and key generation algorithm K. We require that S satisfies the standard unforgeability notion, i.e. it is existentially unforgeable under adaptive chosen message attacks [17].

(ii) A message commitment scheme C with committing algorithm $\mathsf{Com}(.,.)$ and common parameters generation algorithm GenPar. Given a message m, $\mathsf{Com}(m, r; cp)$ denotes the commitment to message m under random value r and common parameters cp. We require that C satisfy the following standard properties –

1 Hiding: Given $\mathsf{Com}(m, r; cp)$, the message m is semantically secure [15]
2 Binding: It is hard to find any two distinct messages $m \neq m'$ and corresponding randomness values $r \in \mathsf{Rand}(m; cp)$ and $r' \in \mathsf{Rand}(m'; cp)$ such that $\mathsf{Com}(m, r; cp) = \mathsf{Com}(m', r'; cp)$
3 Efficient Verification: Given m and r it is easy to compute $c = \mathsf{Com}(m, r; cp)$, or output 'Reject' if $r \notin \mathsf{Rand}(m; cp)$.

In (2) and (3) above, we denote by $\mathsf{Rand}(m; cp)$ the set of valid randomness values for a committed message m.

The scheme is simple. To sign a document, one commits to the submessages and signs the concatenation of the commitments. The CEAS is appended to the signature and is also appended to each submessage before committing. The randomness values used in the commitments are also appended. Extraction of a subset of submessages requires one to recompute the commitments to the removed (unextracted) submessages and append them. The randomness values for the removed submessages are removed. The result is the extracted signature. To verify the extracted signature on an extracted subdocument, one recomputes the commitments for the extracted submessages and verifies the signature on all of the commitments.

We refer the reader to Table 1 for a summary of the performance parameters of this scheme, which are discussed below. The symbols used below are defined in Section 4.3.

Computation. The main advantage of this scheme over the multiple signature one is that n signatures are replaced by 1 signature (on a message of length proportional to n), but at a cost of n commitments. This rules out the use of number-theory based commitments, which would result in little or no savings. Fortunately, an efficient provably secure statistically hiding commitment scheme can be constructed from any collision-resistant hash function [18], and there exist fast candidates for such functions, such as the well known Secure Hash Algorithm (SHA) [2]. Using this commitment scheme one can achieve a computational saving by a factor close to n over the multiple signature scheme, for sufficiently short submessages.

Signature Length. The signature length saving of this scheme over the simple one is small or non-existent. This is because one must append the commitment randomness values for extracted submessages and the commitments for the unextracted (removed) submessages. Making the reasonable assumption $l_r \geq l_c \stackrel{\text{def}}{=} (1/t)l_S$, we obtain an extracted signature length for this scheme of approximately $(1 + n/t)l_S$, compared to ml_S for the simple scheme. Note that the ratio t is typically small (say 2 to 5).

Security. The scheme meets our security requirements for CES (as defined in Section 3), and we can state the following results. The precise formulation and proof will be in the full paper.

Theorem 1. *(1) If the standard signature scheme S is existentially unforgeable under chosen message attack and the commitment scheme C is binding, then scheme* CommitVector *has the CES-Unforgeability property. (2) If the commitment scheme C is hiding, then scheme* CommitVector *has the CES-Privacy property.*

Proof. (Sketch) (1) We use a CES-Unforgeability attacker A_{CES} for scheme CommitVector to construct two attackers: A_{sig} to break the unforgeability of scheme S and A_{com} to break the binding of scheme C. We will prove our claim by showing that one of A_{sig} or A_{com} succeeds in its attack with non-negligible probability. Attacker A_{sig} works as follows. It runs A_{CES}, and answers the latter's sign queries using the signing oracle $S(.)$ for scheme S. While doing so, A_{sig} stores in a table both $\mathbf{D}_j^Q$ (the j'th document queried by A_{CES} to Sign) and $(c_j^Q[i], r_j^Q[i])$ (where $c_j^Q[i]$ denotes the commitment to the i'th submessage of $\mathbf{D}_j^Q$ under randomness $r_j^Q[i]$, used by A_{sig} to answer the j'th query to Sign). Since A_{CES} sees a perfect simulation of the real attack, it by assumption succeeds to break CES-unforgeability of CommitVector with non-negligible probability ϵ by outputting a forgery pair $(\mathbf{D}^*, \sigma^*)$. For each $i \in Cl(\mathbf{D}^*)$, denote by $c^*[i]$ the commitment to the i'th submessage of $\mathbf{D}^*$ reconstructed from randomness value $r^*[i]$ in σ^*. By definition of CES-unforgeability, $\mathbf{D}^*$ is not a subdocument of $\mathbf{D}_j^Q$ for all j. So at least one of the following three events occurs:

(i) For each j, there is an i such that $\mathbf{D}^*[i] \neq \mathbf{D}_j^Q[i]$ (or $i > length(\mathbf{D}_j^Q)$) and $c^*[i] \neq c_j^Q[i]$ or

(ii) There is a j and i such that $\mathbf{D}^*[i] \neq \mathbf{D}_j^Q[i]$ but $c^*[i] = c_j^Q[i]$ or

(iii) There exist one or more j_k such that $\mathbf{D}^*[i] = \mathbf{D^Q}_{j_k}[i]$ for all $i \in \mathsf{Cl}(\mathbf{D}^*)$ but $\mathsf{Cl}(\mathbf{D}^*) \notin CEAS_{j_k}^Q$ for all j_k.

Attacker $\mathsf{A_{sig}}$ outputs (m^*, σ^*), where $m^* = CEAS^* \parallel c^*$, where $c^* = Conc_{i=1}^n c_i^*$. The attacker $\mathsf{A_{com}}$ works as $\mathsf{A_{sig}}$, except that it produces its own signatures to answer Sign queries. It then searches for j and i as in event (ii), and if found outputs a binding collision $((\mathbf{D}^*[i], r^*[i]), (\mathbf{D_j^Q}[i], r_j^Q[i]))$.

Now note that when event (ii) occurs, attacker $\mathsf{A_{com}}$ succeeds in breaking the binding of commitment scheme $\mathcal{C}$. If event (ii) does not occur, then $\mathsf{A_{sig}}$ succeeds in breaking the unforgeability of signature scheme $\mathcal{S}$. For if event (i) occurs, the string of commitments in c^* has never been asked to S. And if event (iii) occurs, since σ^* is valid, we have $\mathsf{Cl}(\mathbf{D}^*) \in CEAS^*$ and therefore $CEAS^* \neq CEAS_{j_k}^Q$ for all j_k, meaning the $CEAS^*$ portion of m^* does not match with that corresponding to queries with indexes j_k. For other queries the commitment vector portion c^* differs since event (ii) did not occur. So here again $\mathsf{A_{sig}}$ succeeds in breaking the unforgeability of signature scheme $\mathcal{S}$. This proves that one of the attackers $\mathsf{A_{sig}}$ and $\mathsf{A_{com}}$ succeeds with probability at least $\epsilon/2$.

(2) We use a CES-Privacy attacker $\mathsf{A_{CES}}$ to construct an attacker $\mathsf{A_c}$ to break the hiding property of commitment scheme $\mathcal{C}$. Attacker $\mathsf{A_c}$ simply runs the find stage of $\mathsf{A_{CES}}$ to get a pair of submessages (m_o, m_1), and then completes the commitment to m_b given to it (for a random bit b unknown to $\mathsf{A_c}$) to form a CES signature with m_b unextracted. It then gives the CES signature to the guess stage of $\mathsf{A_{CES}}$ and uses its output to predict b. Since $\mathsf{A_{CES}}$ sees a perfect simulation, it follows that $\mathsf{A_c}$ succeeds in predicting b with $\mathsf{A_{CES}}$'s success probability, hence breaking the hiding property of commitment scheme $\mathcal{C}$, with non-negligible probability. $\qquad\square$

Scheme CommitVector

(1) KeyGen(k): (1.1) Compute $(SK_S, PK_S) \leftarrow$ K(k) (1.2) Compute $cp \leftarrow$ GenPar(k) (1.3) Output $PK = PK_S \parallel cp$ and $SK = SK_S$.

(2) Sign($\mathbf{M}, CEAS, SK$): (2.1) Parse $PK = PK_S \parallel cp$ (2.2) Compute commitments $c_i \leftarrow$ Com$(m_i, r_i; cp)$ for $i \in [n]$. (2.3) Define $c \leftarrow Conc_{i=1}^n c_i$ (2.4) Compute $\sigma_c \leftarrow$ S($CEAS \parallel c, SK$). (2.5) Output $\sigma_{Full} \leftarrow CEAS \parallel \sigma_c \parallel Conc_{i=1}^n r_i$.

(3) Extract($\mathbf{M}, \sigma_{Full}, X, PK$): (3.1) Parse $PK = PK_S \parallel cp$ (3.2) Parse $\sigma_{Full} \leftarrow CEAS \parallel \sigma_c \parallel Conc_{i=1}^n r_i$ (3.3) Compute missing commitments $c_i \leftarrow$ Com$(m_i, r_i; cp)$ for $i \in [n] - X$ (3.4) Output $\sigma_{Ext} \leftarrow CEAS \parallel \sigma_c \parallel Conc_{i \in X} r_i \parallel Conc_{i \in [n]-X} c_i$.

(4) Verify($\mathbf{M}', \sigma_{Ext}, PK$): (4.1)Parse $PK = PK_S \parallel cp$ (4.2) Parse $\sigma_{Ext} = CEAS \parallel \sigma_c \parallel Conc_{i \in Cl(\mathbf{M}')} r_i \parallel Conc_{i \in [n]-Cl(\mathbf{M}')} c_i$ (4.3) Recompute commitments $c_i \leftarrow$ Com$(m_i, r_i; cp)$ for $i \in Cl(\mathbf{M}')$ or output "Reject" if $r_i \notin$ Rand(m_i) for some i. (4.4) Compute $c \leftarrow Conc_{i=1}^n c_i$ (4.5) Verify signature $d \leftarrow$ V($CEAS \parallel c, \sigma_c, PK_S$). (4.6) Accept iff $d =$ "$Accept$" and $Cl(\mathbf{M}') \in CEAS$.

A Variant: Scheme HashTree. This scheme is a modification of the previous scheme in order to improve its main shortcoming, namely its large signature length. This length consists of two components (besides the single $\mathcal{S}$ signature):

the first is due to the randomness values for extracted submessages (this one is proportional to m) and the second is due to the commitments for the unextracted submessages (this one is proportional to $n - m$). The scheme HashTree significantly reduces the length of the second component (due to removed submessage commitments) when m is much smaller than n (i.e. when the second component dominates). It does so in a simple way: We construct a hash tree by modifying the signing algorithm to regard the set of commitments as the leaves of a binary "hash tree". The hash tree is a binary tree defined to have n leaf nodes, which are associated with the commitments $(h_1, ..., h_n)$ of the n submessages. Then, starting at the highest level of the tree below the leaves, we associate with each internal node of the tree (down to the root node) the hash value of the concatenation of its two children nodes above it. The tree has $k + 1$ levels, the lowest one consisting of a single root node, with its associated hash value h_{root}. The signer signs the root hash value h_{root}. Extraction consists of appending to the signature the hash values associated with intermediate tree nodes which are required in order to compute the root hash value from the commitments of the extracted submessages available in the subdocument. The randomness values for the extracted submessage commitments are also appended as before. Verification recomputes the root hash and verifies the signature on it.

Hash trees have been widely discussed in the literature for improving the efficiency of authentication protocols (see [16]). This scheme can be regarded a modification of the batch signature of Pavlovski and Boyd [14] except that in our application the verifier recomputes the root hash value from any number m of leaves (extracted submessage commitments) while in their application it was recomputed from only one leaf. For simplicity we have omitted some details in the above description. We refer the reader to [14] for a discussion of how to handle any number of leaves (submessages) and the requirement to use a claw-free pair of hash functions for building the tree: one for hashing the leaves and the other for hashing the intermediate tree nodes. We note also that the privacy afforded by the use of commitments in our scheme may also be useful in the batch signatures of [14], where the signer effectively performs the extraction into subdocuments of one submessage each, and may not wish one verifier from learning anything about the other submessages signed in the batch.

Computation. This scheme maintains the computational efficiency of the previous scheme, apart from an additional $2n$ hashings to build the hash tree. As mentioned before, since fast collision-resistant hash functions are available this does not reduce computational efficiency significantly except for extremely large n.

Signature Length. As for the previous scheme the extracted signature still contains the m randomness value for the extracted m submessages, so this component of signature length is still proportional to m. But in this variant, the component of signature length due to unextracted submessages (which in previous scheme was proportional to the number $n - m$ of unextracted submessages) is significantly reduced for small m. For example, in the extreme case when we extract a single submessage ($m = 1$), the previous scheme required the user to send $n - 1$ commitments for the removed submessages, while in this scheme the user only needs to send the $log_2(n)$ hash values associated with the 'sibling' nodes of the nodes on the path from the clear submessage commitment leaf node to the tree root. More generally, one can show that the worst-case overhead l for a given m is approximately $mlog_2(n/m)$ sibling hash values for $m < n/2$ and $n - m$ for $m > n/2$. It may be possible to improve somewhat on this worst-case figure by using a different tree structure. This problem is open for investigation. However, one cannot eliminate the m randomness values which need to be sent by using this approach.

Security. The security properties of the scheme can be proved as for the previous scheme. The main difference is that for this scheme, the CES-unforgeability property also relies on the collision-resistance of the hash function used to build the hash tree. Assuming this hash function was used to construct the claw-free pair of hash functions as described in [14], we have the following results, whose proof is similar to that of Theorem 1, together with the well known collision-resistance property of the Hash tree.

Theorem 2. *(1) If the standard signature scheme $\mathcal{S}$ is existentially unforgeable under chosen message attack, the commitment scheme $\mathcal{C}$ is binding, and the hash function $H(.)$ is collision-resistant then scheme* HashTree *has the CES-Unforgeability property. (2) If the commitment scheme $\mathcal{C}$ is hiding, then scheme* HashTree *has the CES-Privacy property.*

4.2   Schemes Based on RSA

The previous two CES schemes improve upon the computation of the simple multiple signature solution. But their signature length is linear in the number of submessages, due mainly to the randomness values required for the associated submessage commitments. In this section we show how to achieve the reverse tradeoff, i.e. we show RSA-based CES schemes whose **signature length is significantly shorter than that of the simple multiple signature solution**. The CES signature length for these schemes is approximately equal to that of a *single* standard RSA signature, independent of n. However, in computation these schemes do not save over that required for the (batch version) of the simple multiple-signature solution.

Scheme RSAProd. This scheme does not reduce the length of signatures communicated between the *signer* and user. However it reduces the length of *extracted* signatures communicated to the verifier.

The scheme is a modification of an RSA batch "screening" verifier, proposed by Bellare et al. [6]. This verifier is based on the homomorphic property of RSA, i.e. $h_1^d \cdot h_2^d \mod N = (h_1 \cdot h_2)^d \mod N$. The verifier multiplies the RSA signatures in a batch and checks whether the result is the signature of the product of the

---

Scheme RSAProd

(1) KeyGen(k): Pick 2 random $k/2$ bit primes p and q. Compute $N = pq$. Choose a public exponent e relatively prime to $(p-1)(q-1)$. Compute $d = e^{-1} \bmod (p-1)(q-1)$. Set $PK = (N, e)$ and $SK = (N, d)$. Output (SK, PK).

(2) Sign(M, $CEAS, SK$): Compute T: In version 1 (Non-Stateful), choose T uniformly random in $\{0,1\}^{l_{tag}}$. In version 2 (Stateful), $T \leftarrow counter$ and $counter \leftarrow counter + 1$. For each $i = 1, ..., n$ sign the submessage m_i with appended tag, sequence number and CEAS: Define $h_i \overset{\text{def}}{=} H(T \parallel i \parallel CEAS \parallel m_i)$. Compute $\sigma_i \overset{\text{def}}{=} h_i^d \bmod N$. Output $\sigma_{Full} \overset{\text{def}}{=} (CEAS \parallel T \parallel \sigma_1 \parallel ... \parallel \sigma_n)$.

(3) Extract(M, σ_{Full}, X, PK): Parse $\sigma_{Full} = CEAS \parallel T \parallel \sigma_1 \parallel ... \parallel \sigma_n$. Compute $\sigma_f \overset{\text{def}}{=} \Pi_{k \in X} \sigma_k \bmod N$. Set $\sigma_{Ext} \overset{\text{def}}{=} T \parallel CEAS \parallel \sigma_f$. Output σ_{Ext}.

(4) Verify(M', σ_{Ext}, PK): Set $X = \text{CI}(\text{M}')$. Parse $\sigma_{Ext} = T \parallel CEAS \parallel \sigma_f$. For each $k \in X$ compute $h_k \overset{\text{def}}{=} H(T \parallel k \parallel CEAS \parallel m_k)$. Test whether $\sigma_f^e = \Pi_{k \in X} h_k \bmod N$. Output "Accept" iff the test is passed, **and** $X \in CEAS$.

---

hash values. This allows "screening" of m signatures at the cost of only one RSA verification plus $2 \cdot (m-1)$ multiplications modulo N.

The crucial observation which forms the basis of this scheme is that in the content extraction signature setting, all the signatures in the 'batch' (where a batch corresponds here to the submessages contained in the extracted subdocument and their signatures) are available to the user, who can perform the signature multiplication step above prior to forwarding to the verifier. Now all the user has to send is a single product of signatures modulo N, which is the same length as a single RSA signature, independent of the number of submessages m. The verifier then only needs to check that he received a signature for the product of the submessage hash values.

Computation. The scheme needs the signer to produce n standard RSA signatures, so signer computation is identical to that of the trivial solution. The verifier's work is one signature verification plus $(m-1)$ multiplications modulo N. Our Verify algorithm is faster than the screener of [6], due to two reasons. First, we offload the signature multiplication step to the Extract algorithm. Second, we do not need to perform the 'pruning step' of eliminating duplicate messages, which is necessary in [6] in order to avoid the attack described in [7]. In our setting the sequence numbers appended to the submessages by the verifier guarantee distinctness and allow us to prove CES-unforgeability without pruning. This is an additional use for the sequence numbers besides the need to avoid submessage reordering attacks.

Signature Length. Although the full signature length from signer to user is still n times the length of a single RSA signature, once the user extracts a subdocument the resulting extracted signature has only the length of one RSA modulus (plus a small overhead for the CEAS encoding and CES tag). It is therefore much shorter than both the simple multiple signature solution and the previous two commitment-based schemes.

Security. We have the following security result for the scheme. This result holds in the *standard* model, i.e. without any random oracle assumption on the hash function $H(.)$. In this model the hash function $H(.)$ is chosen by the signer at key generation time and the algorithm for $H(.)$ is published with the signer's public key.

Theorem 3. *(1) If the Full-Domain-Hash RSA signature $m \to H(m)^d \bmod N$ (FDH − RSA) is existentially unforgeable under chosen message attack and CES-Tags are never reused by the CES Sign algorithm then scheme RSAProd has the CES-Unforgeability property (2) Scheme RSAProd has the CES-Privacy property.*

Proof. (Sketch) (1) By Lemma 1 in appendix A it suffices to prove the *'weak ordered' (WO)* CES-Unforgeability property for the simplified version of scheme RSAProd which does not append the CES tag and CEAS to submessages before hashing (i.e. only the submessage sequence number is appended). We use a 'weak ordered' CES-Unforgeability attacker A_{CES} for scheme RSAProd to construct a chosen message attacker A_{FDH} to break the existential unforgeability of the signature scheme FDH − RSA. Attacker A_{FDH} works as follows. It runs A_{CES}, and answers the latter's sign queries by simulating the Sign algorithm using the signing oracle $S_{FDH.} = (H(.))^d \bmod N$ for scheme FDH − RSA. While doing so, A_{FDH} stores in a table the documents queried by A_{CES} to Sign (we denote by D_j^Q the j'th queried document). Since A_{CES} sees a perfect simulation of the real attack, it by assumption succeeds to break WO-CES-unforgeability of RSAProd with non-negligible probability ϵ by outputting a forgery pair (D^*, σ^*). By definition of WO-CES-unforgeability, there is a submessage index i such that $D^*[i] \neq D_j^Q[i]$ for all j. Hence the message $m^* \stackrel{\text{def}}{=} i \| D^*[i]$ has never been asked by A_{FDH} to its signing oracle S_{FDH}. So A_{FDH} breaks the existential unforgeability of FDH − RSA if it can compute $\sigma_{FDH}^* \stackrel{\text{def}}{=} S_{FDH}(m^*) = H(m^*)^d \bmod N$ without querying m^* to S_{FDH}. A_{FDH} does so as follows. Since σ^* is by assumption a valid CES signature for document D^*, we have $\sigma^* = \prod_{k \in X} H(k \| D^*[k])^d \bmod N$, i.e.
$\sigma^* = \sigma_{FDH}^* \cdot R \bmod N$, where $R \stackrel{\text{def}}{=} \prod_{k \in X - \{i\}} H(k \| D^*[k])^d \bmod N$. So A_{FDH} queries the messages $k \| D^*[k]$ for all $k \in X - \{i\}$ to S_{FDH} and multiplies the answered signatures modulo N to compute R, which it then uses to compute the desired forged signature $\sigma_{FDH}^* = \sigma^*/R \bmod N$. The proof is completed by observing that the queried messages $k \| D^*[k]$ for $k \neq i$ are all distinct in their sequence prefix from the forgery message m^*.

(2) The CES-Privacy is trivial since an extracted signature is independent of the unextracted submessages. $\qquad\square$

If we use the random oracle model [11] for the hash function $H(.)$, then Theorem 3 and Lemma 2 in appendix give the following result.

Corollary 1. *If the RSA function is one-way and CES-Tags are never re-used by Sign, then then RSAProd has the CES-unforgeability property in the random oracle model.*

Remark 1: To achieve CES-unforgeability it is essential that CES tags are never re-used by the Sign algorithm. In the stateful version of Sign a tag length of

$l_T = 80$ bit suffices for up to 2^{80} CES signatures, but for the unstateful version, one needs $l_T = 160$ bit for approx. the same number of signatures to avoid birthday collisions.

Remark 2: This scheme does *not* have the optional 'Iterative Extraction' property described in section 3. Indeed, it is not hard to show that iterative extraction is as hard as inverting the RSA one-way function. However, iterative extraction would not be required in many applications. The following variant *does* allow iterative extraction.

Remark 3: It is clear from the proof of Theorem 3 that the scheme can be generalized to use any one-to-one trapdoor one-way group homomorphism in place of RSA, where the group operation and inversion are efficient.

A Variant: Scheme MERSAProd. The previous scheme achieves very low extracted signature length for the user to verifier communication. However the initial signature produced by the signer and given to the user is still n RSA signatures long. The following variant scheme MERSAProd also reduces this 'initial' signature length down to almost the length of a single RSA signature. It is a modification of a batch signature generation algorithm due to Fiat [9] and as such also gives savings in signer computation over the previous scheme. However, the scheme loses the fast verification property of the previous scheme.

Fiat [9] discovered that while standard RSA is hard to batch (see [10]), 'Multi-Exponent RSA' (MERSA) can be batched. In MERSA, the signer's public key consists of a unique RSA modulus $N = pq$ but many public exponents e_i, pairwise relatively prime and prime to $(p-1)(q-1)$ (for efficiency these n public exponents are normally chosen as the first n odd prime numbers). To each public exponent e_i there exists a corresponding secret exponent $d_i = e_i^{-1} \bmod (p-1)(q-1)$. Given a sequence $h_1, ..., h_n$ of n message hash values to be signed, Fiat's batch signing algorithm computes (more efficiently than performing n exponentiations) the n 'distinct-exponent' signatures $h_i^{d_i} \bmod N$. Fiat's algorithm is divided into 3 stages. Stage 1 outputs the product $r \overset{\text{def}}{=} \Pi_{i=1}^{n} h_i^{e/e_i} \bmod N$, where $e \overset{\text{def}}{=} \prod_{i=1}^{n} e_i$. Stage 2 signs the product and outputs $r^{1/e} = \prod_{i=1}^{n} h_i^{1/e_i} \bmod N$. Stage 3 'separates' the product into the n individual multi-exponent signatures $h_i^{1/e_i} \bmod N$. A crucial property of Fiat's algorithm for application in our scheme MERSAProd is the following one: *the separation algorithm (Stage 3) does not require the signer's secret key.*

Our scheme MERSAProd is a natural extension of scheme RSAProd to the MERSA case, modified to take advantage of the above useful property of Fiat's algorithm to save on signature length. It is described as follows. The key generation algorithm publishes in the public key an RSA modulus N and the first n_B primes e_i ($i = 1, ..., n$) as public exponents, keeping the corresponding d_i as secret exponents. The Sign algorithm runs only stages 1 and 2 of Fiat's algorithm to compute the product $r = \prod h_i^{d_i} \bmod N$, where h_i denotes the hash h_i of the i'th submessage (with appended sequence number, CES tag, and CEAS as in scheme RSAProd). Only the product r (single RSA signature length) is sent to the user (together with CEAS and tag). The Extract algorithm runs stage 3 of Fiat's algorithm to break up the product r into the individual signatures $h_i^{d_i}$ and then multiplies the ones corresponding to the extracted submessages as in scheme RSAProd, to compute $\sigma_f = \Pi_{i \in X} h_i^{d_i} \bmod N$, to achieve the same single RSA signature length of extracted signatures. The Verify algorithm verifies σ_f by checking if $\sigma_f^e = \Pi_{i \in X} h_i^{e/e_i} \bmod N$, where $e \overset{\text{def}}{=} \Pi_{i \in S} e_i$. Note that this test is equivalent to testing if $1 = \Pi_{i \in S} h_i'^{e/e_i}$ where $h_i' = h_i$ for all i except l, the smallest index in S, for which we define $h_l' = h_s / \sigma_f^{e_l} \bmod N$. Fiat's stage 1 algorithm is again used to compute the product $\Pi_{i \in S} h_i'^{e/e_i}$ efficiently.

Computation. One can show that in using Fiat's algorithm in the above scheme, the computation involved (ignoring one-off computations which are dependent only on the public exponents) for the signer (Stage 1) is bounded in the order of $n(log_2(n))^2 + log_2(N)$ multiplications mod N. Since it takes in the order of $n log_2(N)$ modular multiplications to compute n standard RSA signatures, the computational saving ratio achieved by the batching algorithm over sequential signing is about $log_2(N)/log_2(n)^2$. The extraction computation is in the same order, although we note that the user need only run Fiat's Stage 3 algorithm to extract and store the n signatures once, and then is able to combine any subset of m of them using only $m - 1$ modular multiplications. This might be useful if the user is supplying several extracted subdocuments to multiple verifiers. Finally, the verifier also performs the Stage 1 of Fiat's algorithm but using only m messages and hence has a computational load in the order of $m log_2(n) log_2(m)$ multiplications modulo N. This is equivalent to about $m log_2(m)$ low-exponent RSA verifications, so in this respect the scheme performs slightly worse (by the $log_2(m)$ factor) than the simple multiple signature solution. This is the tradeoff in this scheme compared to the previous one where verification was much faster than the simple solution.

Signature Length. Both initial signature length (as output by signer) and extracted signature length are only a little longer than than a single RSA signature length (due to appended CES tag and CEAS encoding).

Security. We can state the following. The proof is directly analogous to the proof of Theorem 3, so we do not repeat the details.

Theorem 4. *(1) If each of the n distinct-exponent FDH signature schemes* $S_i(m) \overset{\text{def}}{=} H(i\|m)^{d_i} \bmod N$ *(for $i = 1, ..., n$) are existentially unforgeable even under a chosen message attack giving acccess simultaneously to all the n oracles $S_i(,)$ and if CES-Tags are never reused by the CES Sign algorithm, then scheme* MERSAProd *has the CES-Unforgeability property (2) Scheme* MERSAProd *has the CES-Privacy property.*

Using the random oracle model for $H(.)$ we again obtain a stronger result using Lemma 3.

Table 1. Comparison of signer computation time and signer to user communication overhead. Column 'Typ. T-S' denotes typical Sign Time Saving factor. Column 'S-U Length' denotes Signer-to-User sig. length.

Scheme	Sign Time	Typ. T.S.	S-U Length
CommitVector	$T_S(nl_C) + nT_C(l_m)$	100	$l_{CEAS} + l_S + nl_r$
HashTree	$T_S(l_H) + nT_C(l_m) + 2nT_{HA}(l_H)$	100	$l_{CEAS} + l_S + nl_r$
RSAProd	$nT_S^{RSA}(l_m)$	1	$l_{CEAS} + l_T + nl_S^{RSA}$
MERSAProd	$[(log_2(n)^2/log_2(N))n + 1]T_S^{RSA}(l_m)$	1	$l_{CEAS} + l_T + l_S^{RSA}$
Trivial	$nT_S(l_m)$	1	$l_{CEAS} + l_T + nl_S$

Table 2. Comparison of user and verifier computation, user to verifier communication overhead.

Scheme	Extract Time	Extract Length	Typ. Length	Verify Time
CommitVector	negligible	$l_{CEAS} + l_S + ml_r + (n-m)l_C$	169kb	$T_V(nl_C) + mT_C(l_m)$
HashTree	negligible	$l_{CEAS} + l_S + ml_r + f(n,m)l_C$	169kb	$T_V(l_H) + mT_C(l_m) + 2nT_{HA}(l_H)$
RSAProd	mT_{MULT}	$l_{CEAS} + l_T + l_S^{RSA}$	1.28kb	$T_V^{RSA} + m(T_{MULT}^{RSA} + T_{HA}(l_m))$
MERSAProd	mT_{MULT}	$l_{CEAS} + l_T + l_S^{RSA}$	1.28kb	$mlog_2(m)T_V^{RSA}(l_m) + mT_{HA}(l_m)$
Trivial	negligible	$l_{CEAS} + l_T + ml_S$	32.3kb	$mT_V(l_m)$

Corollary 2. *If the RSA function is one-way for each of the public exponents e_i (for $i \in [n]$) and CES-Tags are never reused by* Sign, *then* MERSAProd *has the CES-unforgeability property in the random oracle model.*

4.3   Performance Summary

A summary of the performance parameters of the proposed CES schemes is given in Tables 1 and 2. We use the following symbols in the tables: n (no. of submessages in original document), m (no. submessages in extracted subdocument), T_S (sig. gen. time), T_V (sig. ver. time), T_C (commitment generation time), T_{HA} (Hashing time), N (RSA modulus), T_{MULT}^{RSA} (mult. time modulo N), l_S (signature length), l_H (hash length), l_C (commitment length), l_r (commitment randomness length), l_T (CES tag length), l_{CEAS} (CEAS encoding length), l_m (average submessage length). For algorithm running time variables the value in brackets denotes the input length. We defined $f(n,m) \stackrel{\text{def}}{=} min(mlog_2(n/m), n-m)$.

The 'Typ. Time Saving' (Approximate Sign time saving ratio over simple multiple signature scheme, or over the batched one in the case of scheme MERSAProd) and 'Typ. Length' (referring to extracted signature length) figures were estimated for the following practical example. We took $n = 100$, $m = 99$ (one unextracted submessage), $l_{CEAS} = 100$ bit (a simple CEAS specifying whether a submessage is optional or mandatory for extraction), and $l_m = 64$ bit (typ. numerical fields). For the first two schemes we assumed a DSS signature [3] with $l_S = 320$ bit, and the commitment scheme of [18], using SHA-1 [2] as the collision-resistant hash function ($k = 160$ bit), and an affine universal hash family $h_{A,b} : \{0,1\}^L \rightarrow \{0,1\}^k$ with $h(r) = Ar + b$, $A \in \{0,1\}^{L \times k}$ a Toeplitz matrix, and $b \in \{0,1\}^k$, where arithmetic is performed in the vector space $\{0,1\}^k$ over the binary field $\{0,1\}$. The commit randomness length is $l_r = 2(k + L) - 1$. To achieve a provable upper bound of $2^{-\delta}$ on distinguishing advantage of a privacy-breaking attacker, it is shown in [18] that one can take $L = 3(k + \delta) - 1$. We assumed a reasonable $\delta = 80$, giving randomness length $l_r = 2(4k + 3\delta - 1) - 1 \approx 1.7 kbit$, and commitment length $l_C = 160$ bit. For the RSA schemes, we assumed $|N| = 1024$ bit and $l_T = 160$ bit.

Notice the reversed tradeoff between the commitment-based schemes (efficient in computation but not in signature length) and the RSA schemes (efficient in signature length but not in computation), where in both cases the saving factor over the simple multiple signature scheme is in the order of n.

5 Conclusion

Motivated by emerging needs in online interactions and the need to embrace a minimal information model, we introduced a *Content Extraction Signature* (CES) as a digital signature which can be verified with knowledge of only selected extracted portions of the signed document (while hiding unextracted portions), at a communication or computation cost which is lower than the simple multiple signature solution. We defined the security and functional requirements for such signatures, and proposed four CES schemes with various performance tradeoffs. Practical application and implementation issues of our CES schemes are the subject of a forthcoming paper.

References

1. M. Bellare, A. Desai, D. Pointcheval and P. Rogaway. Relations Among Notions of Security for Public-Key Encryption Schemes. In *CRYPTO '98*, LNCS 1462.
2. NIST. Secure Hash Standard (SHS). Federal Information Processing Standards Publication 180-1. April 1995.
3. NIST. Digital Signature Standard (DSS). Federal Information Processing Standards Publication 186. November 1994.
4. MasterCard and VISA. Secure Electronic Transaction (SET) Specification Books 1-3 (Version 1.0). May 31, 1997.
5. XML Core Working Group. XML-Signature Syntax and Processing: W3C Proposed Recommendation. August 20, 2001. Available from http://www.w3.org/TR/xmldsig-core.
6. M. Bellare and J.A. Garay and T. Rabin. Fast Batch Verification for Modular Exponentiation and Digital Signatures. In *EUROCRYPT '98*, LNCS 1403. Springer-Verlag, Berlin, 1998.
7. J.S. Coron and D. Naccache. On the Security of RSA Screening. In *PKC '99*, LNCS 1560. Springer-Verlag, Berlin, 1999.
8. D. Naccache and D. M'Raihi and S. Vaudenay and D. Raphaeli. Can D.S.A be improved? In *EUROCRYPT '94*, LNCS 950. Springer-Verlag, Berlin, 1999.
9. A. Fiat. Batch RSA. In *CRYPTO '89*, LNCS 435. Springer-Verlag, Berlin, 1990.

10. H. Shacham and D. Boneh. Improving SSL Handshake Performance via Batching. In *CT-RSA 2001*, LNCS 2020. Springer-Verlag, Berlin, 2001.
11. M. Bellare and P. Rogaway. The exact security of digital signatures: How to sign with RSA and Rabin. In *EUROCRYPT '96*, LNCS 1070. Springer-Verlag, Berlin, 1996.
12. M. Bellare and O. Goldreich and S. Goldwasser. Incremental Cryptography: The Case of Hashing and Signing. In *CRYPTO '94*, LNCS 839, Springer-Verlag, Berlin, 1994.
13. M. Bellare and O. Goldreich and S. Goldwasser. Incremental Cryptography and Application to Virus Protection. In *Proc. of 27th STOC* ACM, 1995.
14. C.J. Pavlovski and C. Boyd. Efficient Batch Signature Generation Using Tree Structures. In *CrypTEC'99*. City University of Hong Kong Press, 1999.
15. S. Goldwasser and S. Micali. Probabilistic Encryption. *J. of Computer and System Sciences*, pages 270-299, vol. 28, no. 2, 1984.
16. A. Menezes and P. van Oorschot and S. Vanstone. Handbook of Applied Cryptography. CRC Press, 1997.
17. S. Goldwasser and S. Micali and R. Rivest. A Digital Signature Scheme Secure against Adaptively Chosen Message Attacks. *SIAM Journal on Computing*, pages 281-308, vol. 17, no. 2, 1988.
18. S. Halevi and S. Micali. Practical and Provably-Secure Commitment Schemes from Collision-Free Hashing. In *CRYPTO '96*, LNCS 1109. Springer-Verlag, Berlin, 1996.

A   Appendix

The following definition and Lemmas are required for security results stated in the paper. Refer to the full paper for proofs. The following definition is for a weaker notion of unforgeability than the full one defined in the paper.

Definition 1. *A CES scheme $\mathcal{D}$ is said to have the weak ordered (WO) CES-unforgeability property if the following holds: It is infeasible for an attacker, having access to a CES signing oracle, to produce a document/signature pair $(\mathbf{M}, \sigma)$, such that: (i) σ passes the verification test for $\mathbf{M}$ and (ii) $\mathbf{M}$ contains a submessage at some position i which has never appeared in position i in any document queried to the CES signing oracle.*

Lemma 1. *Any CES scheme which has the weak ordered CES-Unforgeability property can be used to construct a CES scheme which has the full CES-unforgeability property.*

The construction used in proving Lemma 1 involves choosing a unique 'CES tag' for each signed document and appending it (and a A CEAS encoding) to each submessage before applying the original CES Sign algorithm.

The following lemmas were also used in the security analysis of the RSA schemes.

Lemma 2. *(Bellare-Rogaway[11]) If $H(.)$ is a full-domain random oracle and the RSA function is one-way, then the FDH signature $S(m) \stackrel{\text{def}}{=} H(m)^d \bmod N$ is existentially unforgeable under chosen message attack.*

Lemma 3. *If $H(.)$ is a (full domain) random oracle and the RSA function is one-way for each e_i $(i = 1, ..., n)$, then the n distinct-exponent FDH signature schemes $S_i(m) \stackrel{\text{def}}{=} H(i\|m)^{d_i} \bmod N$ (for $i = 1, ..., n$) are existentially unforgeable even under a chosen message attack giving acccess simultaneously to all the n signing oracles $S_i(.)$.*

New Signcryption Schemes Based on KCDSA

Dae Hyun Yum and Pil Joong Lee

Department of Electronic and Electrical Engineering,
Pohang University of Science & Technology (POSTECH),
San 31 Hyoja-dong, Nam-gu, Pohang, Kyungbuk, 790-784, Rep. of Korea
daehyun@oberon.postech.ac.kr, pjl@postech.ac.kr
http://ist.postech.ac.kr

Abstract. A signcryption scheme is a cryptographic primitive that performs signature and encryption simultaneously, at less cost than is required by the traditional signature-then-encryption approach. We propose new signcryption schemes based on KCDSA. These are the first signcryption schemes that are based on a standardized signature scheme. We expect that these schemes will soon be applied to established KCDSA systems. We also propose a new signcryption scheme for multiple recipients which requires very small communication overhead.

1 Introduction

Signcryption, which was first proposed by Zheng[17], is a scheme that performs signature and encryption simultaneously, at a lower cost than is required by the traditional signature-then-encryption approach. Zheng's signcryption has one drawback, in that signature verification can be performed either by the recipient using his private key, or by engaging a zero-knowledge interactive protocol with a third party without disclosing the recipient's private key. Bao & Deng modified Zheng's scheme, so that the recipient's private key is no longer required for signature verification[1]. However, these two signcryption schemes are based on SDSS1 and SDSS2, which are not standardized signature schemes.

KCDSA (Korean Certificate-based Digital Signature Algorithm) is the Korean digital signature standard[8]. In this paper, we propose two signcryption schemes based on KCDSA, called "KCDSA signcryption (I), (II)" respectively. KCDSA signcryption schemes are the first signcryption schemes that are based on a standardized signature scheme. To propose these signcryption schemes, we devised a modified KCDSA that is strongly equivalent to the original KCDSA. Since a public key certificate for KCDSA can be used without modification, KCDSA signcryption (II) can be immediately applied to established KCDSA systems. We also propose KCDSA signcryption for multiple recipients with very small communication overhead.

In this paper, we use the following notations and domain parameters.

K. Kim (Ed.): ICICS 2001, LNCS 2288, pp. 305–317, 2002.
© Springer-Verlag Berlin Heidelberg 2002

Notations

$a \oplus b$: exclusive-or of two bit strings a and b.

$|A|$ denotes the bit-length of A.

(E, D) : the encryption/decryption algorithms of a symmetric key cipher.

$hash(\cdot)$: a one-way hash function.

$KDF(\cdot)$: a key derivation function.

$KH(\cdot)$: a keyed one-way hash function.

Domain Parameters

p : a large prime.

q : a large prime factor of $p - 1$.

g : a base element of order q.

2   Previous Signcryption Schemes

For most ElGamal based signature schemes, the size of the signature on a message is $2|p|$, $|p| + |q|$, or $2|q|$. To reduce the size of an ElGamal based signature, a modified "seventh generalization" method discussed in [7] can be used. With this method, two signature schemes, called SDSS1 and SDSS2, were devised[17]. These two schemes have the same signature size of $|hash(\cdot)| + |q|$, and provide provable security in the random oracle model. First, we define the following user parameters. Alice, the sender, generates a signature for a message m, and Bob, the recipient, verifies Alice's signature.

Alice's User Parameters

x_a : Alice's private key, $x_a \in Z_q^*$.

y_a : Alice's public key, $y_a = g^{x_a} \bmod p$.

Bob's User Parameters

x_b : Bob's private key, $x_b \in Z_q^*$.

y_b : Bob's public key, $y_b = g^{x_b} \bmod p$.

2.1   SDSS1 and SDSS2

Here, we briefly describe SDSS1 and SDSS2.

SDSS1

Signature Generation

Alice randomly chooses $x \in Z_q^*$, and then computes

$k = g^x \bmod p$,

$r = hash(k, m)$,

$s = x/(r + x_a) \bmod q$.

Alice sends (m, r, s) to Bob.

Signature Verification

Bob computes
$k = (y_a \cdot g^r)^s \bmod p$ and
checks whether $r = hash(k, m)$.

SDSS2

Signature Generation

Alice randomly chooses $x \in Z_q^*$, and then computes
$k = g^x \bmod p$,
$r = hash(k, m)$,
$s = x/(1 + x_a r) \bmod q$.
Alice sends (m, r, s) to Bob.

Signature Verification

Bob computes
$k = (g \cdot y_a^r)^s \bmod p$ and
checks whether $r = hash(k, m)$.

The most important characteristic of SDSS1 and SDSS2 is that $k = g^x \bmod p$ can be recovered from r, s, and other domain parameters. Note that message m is not required to obtain $k = g^x \bmod p$.

In Alice's calculation of k, if g is replaced by y_b as $k = y_b^x \bmod p$, Bob's signature verification equation should be changed into:

SDSS1 : $k = (y_a \cdot g^r)^{s \cdot x_b} \bmod p$.
SDSS2 : $k = (g \cdot y_a^r)^{s \cdot x_b} \bmod p$.

Since Bob's private key x_b is used in signature verification equation, only Bob can recover k. This is the basis of signcryption.

2.2   Zheng's Signcryption[17]

Signcryption schemes can be constructed based on SDSS1 and SDSS2. Because the case for SDSS2 is similar, we limit our description to the SDSS1 case.

Signcryption

Alice randomly chooses $x \in Z_q^*$, and then computes
$k = y_b^x \bmod p$,
$(k_1, k_2) = hash(k)$,
$c = E_{k_1}(m)$,
$r = KH_{k_2}(m)$,
$s = x/(r + x_a) \bmod q$.
Alice sends (c, r, s) to Bob.

Unsigncryption

Bob computes
$k = (y_a \cdot g^r)^{s \cdot x_b} \bmod p$,
$(k_1, k_2) = hash(k)$,
$m = D_{k_1}(c)$ to recover the plaintext message,
and then checks whether $r = KH_{k_2}(m)$ for signature verification.

First, Bob recovers $k = y_b^x \bmod p = (y_a \cdot g^r)^{s \cdot x_b} \bmod p$ from r, s and his private key x_b without knowing message m. After obtaining (k_1, k_2), he recovers the plaintext message m and verifies the signature.

The fact that unsigncryption requires Bob's private key assigns confidentiality to a message. However, this results in a drawback: only Bob, the recipient, can verify the signature. Unlike Zheng's signcryption, the validity of a traditional signature can be verified by anyone who knows the sender's public key.

Zheng's signcryption scheme requires 3 exponential computations and $|KH(\cdot)| + |q|$ communication overhead, rather than 6 exponential computations and $|p| + 2|q|$ communication overhead which occurs in a traditional DSA signature-then-ElGamal encryption.

2.3 Bao & Deng's Signcryption[1]

Bao & Deng modified Zheng's signcryption scheme so that verification of a signature no longer requires the recipient's private key. Hence, the modified scheme functions in exactly the same manner as that of the signature-then-encryption approach. The computational cost of Bao & Deng's signcryption scheme is higher than that of Zheng's scheme but lower than that of the signature-then-encryption approach.

Signcryption

Alice randomly chooses $x \in Z_q^*$, then computes
$k_1 = g^x \bmod p$, $k_2 = y_b^x \bmod p$,
$k_3 = hash(k_1)$, $k_4 = hash(k_2)$,
$c = E_{k_4}(m)$,
$r = KH_{k_3}(m)$,
$s = x/(r + x_a) \bmod q$.
Alice sends (c, r, s) to Bob.

Unsigncryption

Bob computes
$k_1 = (y_a \cdot g^r)^s \bmod p$, $k_2 = k_1^{x_b} \bmod p$,
$k_3 = hash(k_1)$, $k_4 = hash(k_2)$,
$m = D_{k_4}(c)$ to recover the plaintext message,
and then checks whether $r = KH_{k_3}(m)$ for signature verification.

If necessary, Bob may forward (m, r, s) to others, who can confirm that it originated with Alice by verifying

$k_1 = (y_a \cdot g^r)^s \bmod p,$
$k_3 = hash(k_1),$ and
$r = KH_{k_3}(m).$

Bao & Deng's signcryption scheme requires 5 exponential computations and $|KH(\cdot)| + |q|$ communication overhead rather than 6 exponential computations and $|p| + 2|q|$ communication overhead in a traditional DSA signature-then-ElGamal encryption.

3 KCDSA & Modified KCDSA

KCDSA(Korean Certificate-based Digital Signature Algorithm) is the Korean digital signature standard[8]. The use of a hash function in KCDSA allows for a proof of security in the random oracle model[15]. Moreover, the use of two hash functions in KCDSA allows for a proof of security if one of the hash functions is a random oracle and the other is collision-resistant[3]. First, we define the following user parameters.

Alice's User Parameters

x_a : Alice's private key, $x_a \in Z_q^*$, $x'_a = x_a^{-1} \bmod q$.
y_a : Alice's public key, $y_a = g^{x'_a} \bmod p$.
z_a : a hash-code of $Cert\_Data_a$, i.e., $z = hash(Cert\_Data_a)$. Here, $Cert\_Data_a$ denotes the certification data of Alice, which contains at least Alice's public key y_a, and may contain Alice's distinguished identifier or some of the domain parameters.

Bob's User Parameters

x_b : Bob's private key, $x_b \in Z_q^*$, $x'_b = x_b^{-1} \bmod q$.
y_b : Bob's public key, $y_b = g^{x'_b} \bmod p$.

3.1 KCDSA

Here, we briefly describe the KCDSA algorithm[8].

Signature Generation

Alice randomly chooses $x \in Z_q^*$, and then computes
$k = g^x \bmod p,$
$r = hash(k),$
$e = r \oplus hash(z_a, m) \bmod q,$
$s = x_a(x - e) \bmod q.$
Alice sends (m, r, s) to Bob.

Signature Verification

Bob computes
$e = r \oplus hash(z_a, m) \bmod q,$
$k = y_a^s \cdot g^e \bmod p,$
and checks whether $r = hash(k)$.

KCDSA is designed to avoid the evaluation of multiplicative inverse in normal use. It is only needed at the time of key pair generation. For comparison, in SDSS1 and SDSS2 a multiplicative inverse mod q needs to be evaluated each time a signature is generated, and in DSA[9,10] each time a signature is generated or verified. Evaluating an inverse mod q would require a small workload in the overall workload of singing/verifying on most general-purpose computers. However, it may be quite expensive in a limited computing environment, such as smart cards (See [14] for various comments on DSS, including debates on the use of inverse).

In SDSS1 and SDSS2, Bob can recover $k = g^x \bmod p$ from r, s and other domain parameters without knowing message m. However, message m is required to recover $k = g^x = y_a^s \cdot g^e \bmod p$ in KCDSA, since Bob must calculate $e = r \oplus hash(z_a, m) \bmod q$. This prevents the construction of signcryption schemes based on KCDSA. To solve this problem, we modify KCDSA so that message m is not required to calculate k.

3.2 Modified KCDSA

The basic idea of modified KCDSA is that Alice sends e instead of r, for Bob to recover k without message m. However, a side-effect of this modification is that Bob cannot obtain r from (m, e, s), because of reduction mod q in the calculation of e. Hence, we adopt the following two modifications.

First, we set $e = r \oplus hash(z_a, m)$ instead of $e = r \oplus hash(z_a, m) \bmod q$. Second, Alice sends (m, e, s) instead of (m, r, s). Note that Alice's sending e instead of r doesn't affect the security of the signature scheme because Bob could obtain e from r in KCDSA.

Signature Generation

Alice randomly chooses $x \in Z_q^*$, and then computes
$k = g^x \bmod p,$
$r = hash(k),$
$e = r \oplus hash(z_a, m),$
$s = x_a(x - e) \bmod q.$
Alice sends (m, e, s) to Bob.

Signature Verification

Bob computes
$r = e \oplus hash(z_a, m),$
$k = y_a^s \cdot g^e \bmod p,$
checks whether $r = hash(k)$.

In this modified KCDSA, Bob can recover $k = g^x = y^s \cdot g^e \bmod p$ without message m. In addition, Bob obtains (r, s), which is an original KCDSA signature. Before the construction of signcryption schemes based on modified KCDSA, we show that KCDSA and modified KCDSA are "strongly" equivalent.

Definition 1. *Two signature schemes are called strongly equivalent if the signatures of the first scheme can be transformed efficiently into signatures of the second scheme and vice versa, without knowledge of the private key[12].*

Theorem 1. *KCDSA and modified KCDSA are strongly equivalent.*

Proof Let (r, s) represent a KCDSA signature to be appended to a message m. Then $(r \oplus hash(z_a, m), s)$ is a modified KCDSA signature. Conversely, assume that (e, s) is a modified KCDSA signature. Note that e is not the reduced modulo q in modified KCDSA. Then $(e \oplus hash(z_a, m), s)$ is a KCDSA signature.
$$\text{Q.E.D. } \square$$

4   New Signcryption Schemes Based on KCDSA

4.1   KCDSA Signcryption (I) in Zheng's Model

Based on the modified KCDSA, we construct KCDSA signcryption (I) in Zheng's model. KCDSA signcryption (I) requires 3 exponential computations and $|hash(\cdot)| + |q|$ communication overhead. The advantage of KCDSA signcryption (I) over Zheng's signcryption is the removal of evaluating an inverse mod q. Here, we use the same notations as KCDSA.

Signcryption

>Alice randomly chooses $x \in Z_q^*$, and then computes
>$k = y_b^x \bmod p$,
>$r = hash(k)$, $k_1 = KDF(k)$,
>$c = E_{k_1}(m)$,
>$e = r \oplus hash(z_a, m)$,
>$s = x_a(x - e) \bmod q$.
>Alice sends (c, e, s) to Bob.

Unsigncryption

>Bob computes
>$k = (y_a^s \cdot g^e)^{x_b'} \bmod p$,
>$r = hash(k)$, $k_1 = KDF(k)$,
>$m = D_{k_1}(c)$ to recover the plaintext message,
>and then checks whether $e = r \oplus hash(z_a, m)$.

Security Arguments for KCDSA Signcryption (I). Security of KCDSA signcryption (I) is the same level as that of Zheng's signcryption. Here, we preset the key ideas of security arguments. Employed methods for security arguments are parallel with Zheng's methods.

1. Unforgeability – Malicious Bob is in the best position for forging Alice's signcryption, as he is the only person who knows x_b. Given the signcrypted text (c, e, s) of message m from Alice, Bob can use his private key x_b to decrypt c and obtain m. Thus the problem is reduced to one in which Bob is in possession of (m, e, s). This is identical to the unforgeability of the modified KCDSA.
2. Non-repudiation – A dispute between Alice and Bob can be settled by a TTP(Trusted Third Party), with a zero-knowledge proof protocol between TTP and Bob.
3. Confidentiality – We will reduce the confidentiality of another encryption scheme ES to the confidentiality of KCDSA signcryption (I). With the encryption scheme ES, the ciphertext of a message m is defined as $(u = g^x \bmod p, c = E_{k_1}(m), e - hash(k) \oplus hash(z_a, m))$ where k, k_1 and z_a are defined in the same way as in KCDSA signcryption (I). ES is a slightly modified version of a scheme whose confidentiality is relatively well-understood[2,16]. Now, assume that there is an attacker for KCDSA signcryption (I), called $A_{KCDSA(I)}$. We show how $A_{KCDSA(I)}$ can be translated into an attacker for ES, called A_{ES}. Note that for a message m, the input to $A_{KCDSA(I)}$ includes p, q, g, z_a, $y_a = g^{x'_a} \bmod p$, $y_b = g^{x'_b} \bmod p$, $u = g^x \bmod p$, $c = E_{k_1}(m)$, $e = hash(k) \oplus hash(z_a, m)$, and $s = x_a(x - e) \bmod q$. However, the input to A_{ES} includes p, q, g, z_a, $y_b = g^{x'_b} \bmod p$, $u = g^x \bmod p$, $c = E_{k_1}(m)$, and $e = hash(k) \oplus hash(z_a, m)$. One immediately identifies two numbers that correspond to y_a and s which are needed by $A_{KCDSA(I)}$ as part of its input, which are currently missing from the input to A_{ES}. Thus, in order for A_{ES} to call the attacker $A_{KCDSA(I)}$ as a sub-routine, A_{ES} must create two numbers corresponding to y_a and s in the input to $A_{KCDSA(I)}$. Call these two numbers y_a^* and s^*. y_a^* and s^* mush have the right form so that A_{ES} can fool $A_{KCDSA(I)}$. It turns out that such y_a^* and s^* can be easily created by A_{ES} as follows: (1) pick a random number s^* from $[1, \ldots, q]$. (2) let $y_a^* = u^{(s^*)^{-1}} \cdot g^{-e \cdot (s^*)^{-1}}$.

4.2   KCDSA Signcryption (II) in Bao & Deng's Model

Based on the modified KCDSA, we construct KCDSA signcryption (II) in Bao & Deng's model. The security arguments for KCDSA signcryption (II) can be developed in a similar manner as KCDSA signcryption (I).

Signcryption

Alice randomly chooses $x \in Z_q^*$, and then computes
$k_1 = g^x \bmod p$, $k_2 = y_b^x \bmod p$,

$r = hash(k_1), \; k_3 = KDF(k_2),$
$c = E_{k_3}(m),$
$e = r \oplus hash(z_a, m),$
$s = x_a(x - e) \bmod q.$
Alice sends (c, e, s) to Bob.

Unsigncryption

Bob computes
$k_1 = y_a^s \cdot g^e \bmod p, \; k_2 = k_1^{x'_b} \bmod p \;,$
$r = hash(k_1), \; k_3 = KDF(k_2) \;,$
$m = D_{k_3}(c)$ to recover the plaintext message,
and then checks whether $e = r \oplus hash(z_a, m).$

In the unsigncryption stage, Bob obtains (r, s) which is a valid KCDSA signature. If necessary, Bob may forward (m, r, s) to others, who can verify that it originated with Alice. The validity of (m, r, s) can be verified by anyone who knows Alice's public key. Hence, KCDSA signcryption (II) satisfies direct verifiability by public key.

Since (r, s) is an original KCDSA signature, a public key certificate for KCDSA can be used for KCDSA signcryption (II). Therefore, KCDSA signcryption (II) can be promptly applied to the established KCDSA systems.

KCDSA signcryption (II) requires 5 exponential computations and $|hash(\cdot)| + |q|$ communication overhead. The computational advantage of KCDSA signcryption (II) over Bao & Deng's signcryption scheme is that there is no need of inverse computation.

The comparison of signcryption schemes is summarized in Table 1.

Table 1. Comparison of signcryption schemes

	Zheng	KCDSA(I)	Bao & Deng	KCDSA(II)	Traditional
Exponential computations	3	3	5	5	6
Inverse computations	1	0	1	0	3
Communication overhead	$\|KH(\cdot)\| + \|q\|$	$\|hash(\cdot)\| + \|q\|$	$\|KH(\cdot)\| + \|q\|$	$\|hash(\cdot)\| + \|q\|$	$\|p\| + 2\|q\|$
Direct verifiability by public key	No	No	Yes	Yes	Yes
Signature standardization	No	No*	No	Yes	Yes

* While KCDSA signcryption (I) is based on KCDSA, (r, s) in KCDSA signcryption (I) is not exactly the same as that in KCDSA.

5   Signcryption Schemes for Multiple Recipients

In this section, we discuss the protocol for broadcasting a message to multiple users in a secure and authenticated manner. Signcryption must be confidential, non-forgeable, and immune to repudiation. Consistency is also important in signcryption for multiple recipients. Here, consistency refers to the case where all recipients recover an identical message from their copies of a broadcast message. The aim is to prevent a particular recipient from being excluded from the group by a dishonest sender. To satisfy these requirements, Zheng proposed the following scheme. The basic idea is to use two types of keys: a message-encryption key and a recipient-specific key.

5.1   Zheng's Signcryption for Multiple Recipients[18]

Signcryption by Alice for Multi-recipients. An input to this signcryption algorithm for multi-recipients consists of a message m to be sent to t recipients $R_1, \ldots, R_t$, Alice's private key x_a, R_i's public key y_i for all $1 \le i \le t$, p, q and g.

1. Choose a random message-encryption key v, calculate $h = KH_v(m)$, and encrypt m by $c = E_v(m, h)$.
2. Create a signcrypted text of v for each recipient $i = 1, \ldots, t$:
   (a) Choose a random number $v_i \in Z_q^*$ and calculate $k_i = y_i^{v_i} \bmod p$.
   (b) $(k_{i,1}, k_{i,2}) = hash(k_i)$.
   (c) $c_i = E_{k_{i,1}}(v)$.
   (d) $r_i = KH_{k_{i,2}}(m, h)$.
   (e) $s_i = v_i/(r_i + x_a) \bmod q$.

Alice broadcasts to all recipients $(c, c_1, r_1, s_1, \ldots, c_t, r_t, s_t)$.

Unsigncryption by Each Recipient. An input to this unsigncryption algorithm consists of a signcrypted text $(c, c_1, r_1, s_1, \ldots, c_t, r_t, s_t)$ received through a broadcast channel, along with recipient R_i's private key x_i where $1 \le i \le t$, Alice's public key y_a, p, q, and g.

1. Find out (c, c_i, r_i, s_i) in $(c, c_1, r_1, s_1, \ldots, c_t, r_t, s_t)$.
2. $k_i = (y_a \cdot g^{r_i})^{s_i \cdot x_i} \bmod p$.
3. $(k_{i,1}, k_{i,2}) = hash(k_i)$.
4. $v = D_{k_{i,1}}(c_i)$.
5. $u = D_v(c)$. Split u into m and h.
6. R_i accepts m as a valid message which originated from Alice only if both $h = KH_v(m)$ and $r_i = KH_{k_{i,2}}(u)$ hold.

5.2   KCDSA Signcryption (II) for Multiple Recipients

In contrast to Zheng's signcryption, KCDSA signcryption (II) is verifiable by using the sender's public key. Therefore, KCDSA signcryption (II) for multiple recipients can be constructed in a more efficient manner. The basic idea is the same, but we require only (r, s) instead of (r_i, s_i), where $1 \le i \le t$.

Signcryption by Alice for Multi-recipients. An input to this signcryption algorithm for multi-recipients consists of a message m to be sent to t recipients $R_1, \ldots, R_t$, Alice's private key x_a, Alice's hash-code z_a, R_i's public key y_i for all $1 \le i \le t$, p, q and g.

1. Choose a random message-encryption key v, calculate $h = hash(m, v)$, and encrypt m by $c = E_v(m, h)$.
2. Create a signcrypted text of v for each recipient $i = 1, \ldots, t$:
   (a) Choose a random number $x \in Z_q^*$ and calculate $k = g^x \bmod p$, $k_i = y_i^x \bmod p$.
   (b) $r = hash(k)$, $k_{i,1} = KDF(k_i)$.
   (c) $c_i = E_{k_{i,1}}(v)$.
   (d) $e = r \oplus hash(z_a, m, h)$.
   (e) $s = x_a(x - e) \bmod q$.

Alice broadcasts to all recipients $(c, c_1, \ldots, c_t, e, s)$.

Unsigncryption by Each Recipient. An input to this unsigncryption algorithm consists of a signcrypted text $(c, c_1, \ldots, c_t, e, s)$ received through a broadcast channel, along with recipient R_i's x_i'(inverse value of private key x_i), where $1 \le i \le t$, Alice's public key y_a, Alice's hash-code z_a, p, q, and g.

1. Find out (c, c_i, e, s) in $(c, c_1, \ldots, c_t, e, s)$.
2. $k = y_a^s \cdot g^e \bmod p$, $k_i = k^{x_i'} \bmod p$.
3. $r = hash(k)$, $k_{i,1} = KDF(k_i)$.
4. $v = D_{k_{i,1}}(c_i)$.
5. $u = D_v(c)$. Split u into m and h.
6. R_i accepts m as a valid message which originates from Alice only if both $e = r \oplus hash(z_a, m, h)$ and $h = hash(m, v)$ hold.

We examine the efficiency of the schemes.

Table 2. Comparison of signcryption schemes for multi-recipients

	Zheng	KCDSA(II)
Exponential computations	t (sender)	$t + 1$ (sender)
	2 (receiver)	3 (receiver)
Communication overhead	$t\{|v| + |KH(\cdot)| + |q|\} + |KH(\cdot)|$	$t|v| + 2|hash(\cdot)| + |q|$
Direct verifiability by public key	No	Yes
Signature standardization	No	Yes

KCDSA signcryption (II) for multiple recipients satisfies direct verifiability by a public key with 1 more exponential computation for the sender and the receiver, respectively. A significant advantage of KCDSA signcryption (II) for multiple recipients is its small communication overhead. If we assume $|v| = 128$, $|KH(\cdot)| \simeq |hash(\cdot)| \simeq |q| \simeq 160$, the required communication overhead is as follows.

Table 3. Comparison of required communication overhead

	Zheng	KCDSA(II)
Communication overhead	$t\{|v| + |KH(\cdot)| + |q|\} + |KH(\cdot)|$	$t|v| + 2|hash(\cdot)| + |q|$
$t = 5$	2400	1120
$t = 10$	4640	1760
$t = 100$	44960	13280

6   Concluding Remarks

In this paper, we proposed and analyzed KCDSA signcryption (I), (II) which are the first signcryption schemes based on a standardized signature scheme. The computational advantage of proposed schemes is the lack of a requirement for inverse computation. Since a public key certificate for KCDSA can be used without modification, we expect that KCDSA signcryption (II) will be immediately applicable to current KCDSA systems. We also proposed a new signcryption scheme for multiple recipients with small communication overhead.

We could not construct signcryption schemes based on other standardized signature schemes, such as DSA. This remains a challenging topic for future research.

Acknowledgments

The authors would like to thank the Ministry of Education of Korea for its financial support toward the Electrical and Computer Engineering Division at POSTECH through its BK21 program. This work was also supported by MSRC and Com^{2}MaC-KOSEF.

References

1. F. Bao, R. H. Deng, "A signcryption scheme with signature directly verifiable by public key," *PKC'98*, Springer-Verlag, LNCS 1431, pp. 55-59, 1998.
2. M. Bellare, P. Rogaway, "Random oracles are practical: A paradigm for designing efficient protocols," *Proceedings of the 1st ACM Conference on Computer and Communications Security*, pp. 62-73, 1993.
3. E. Brickell, D. Pointcheval, S. Vaudenay and M. Yung, "Design Validations for Discrete Logarithm Based Signature Schemes," *Proceedings of PKC 2000*, LNCS 1751, pp. 276-292, Springer-Verlag, 2000.
4. W. Diffie, M.E. Hellman, "New directions in cryptography," *IEEE Transactions on Information Theory*, 22, pp. 644-654, 1985.
5. T. ElGamal, "A public key cryptosystem and a signature scheme based on discrete logarithms," *IEEE Transactions on Information Theory*, 31(4), pp. 469-472, 1985.
6. C. Gamage, J. Leiwo, Y. Zheng, "Encrypted message authentication by firewalls," *PKC'99*, Springer-Verlag, LNCS 1560, pp. 69-81, 1999.
7. P. Horster, M. Michels, H. Petersen, "Meta-ElGamal signature schemes," *Proceedings of the 2nd ACM Conference on Computer and Communications Security*, pp. 96-107, 1994.

8. C. H. Lim, P. J. Lee, "A study on the proposed Korean digital signature algorithm," *Asiacrypt'98*, LNCS 1514, pp. 175-186, 1998.
9. NIST, "Digital signature standard (DSS)," *FIPS PUB 186*, U.S. Department of Commerce, 1994.
10. NIST, Digital signature standard, *FIPS PUB 186-2*, 2000.
11. NIST, Secure hash standard, *FIPS PUB 180-1*, 1993.
12. K. Nyberg, R. A. Rueppel, "Message recovery for signature schemes based on the discrete logarithm problem," *Eurocrypt'94*, LNCS 950, pp. 182-193.
13. D. Pointcheval, J. Stern, "Security proofs for signature schemes," *Eurocrypt'96*, Springer-Verlag, LNCS 1070, pp. 387-398, 1996.
14. R. Rivest, M. Hellman, J. Anderson, "Responses to NIST's proposal," *Comm. ACM*, 35(7), pp. 41-52, 1992.
15. D. H. Yum, P. J. Lee, "Security proof for KCDSA under the random oracle model," *CISC'99*, pp. 173-180, 1999.
16. Y. Zheng, "Improved public key cryptosystems secure against chosen ciphertext attacks," *Technical Reprt 94-1*, University of Wollongong Australia, 1994.
17. Y. Zheng, "Digital signcryption or how to achieve cost(signature & encryption) ≪ cost(signature) + cost(encryption)," *Crypto'97*, Springer-Verlag, LNCS 1294, pp. 165-179, 1997.
18. Y. Zheng, "Digital signcryption or how to achieve cost(signature & encryption) ≪ cost(signature) + cost(encryption)," full paper, http://www.pscit.monash.edu.au/~yuliang/, 1999.

An Efficient and Provably Secure Threshold Blind Signature

Jinho Kim, Kwangjo Kim, and Chulsoo Lee

IRIS (International Research center for Information Security)
ICU (Information and Communications University)
58-4 Hwaam-dong, Yusong-gu, Taejon, 305-732, S. Korea
Tel: +82-42-866-6118, Fax: +82-42-866-6154
{kman,kkj,csl100}@icu.ac.kr

Abstract. We propose an efficient and provably secure threshold blind signature scheme based on Okamoto-Schnorr blind signature. In our scheme, any t ($t < n$) out of n signers can generate a blind signature and any one can verify the signature using the same verification process used in Okamoto-Schnorr blind signature. New scheme can be used for blind signature-based voting systems with multiple administrators and secure electronic cash systems to prevent their abuse. We prove that our scheme is to be as secure as Okamoto-Schnorr's blind signature scheme and more efficient than Juang *et al.*'s scheme [12].

1 Introduction

The concept of a blind signature scheme was initially proposed by Chaum [2] in 1982. A blind signature scheme is two party protocol between a sender A and a signer B. The basic idea is as follows: A sends a piece of information m to B in a blinded form. Then B signs and returns it to A. From this signature, A can unblind and compute B's signature on message m. At the completion of the protocol, B knows neither the message m nor the signature associated with it. The purpose of a blind signature is to prevent the signer B from observing the message he signs and its resulting signature. It is widely used in many applications like secure voting protocol and electronic payment systems [1,3] when anonymity of a message is required.

In general, the voting system based on blind signature scheme [7,6,14] is managed by a single administrator, who can be empowered to authorize votes. But, a dishonest administrator can abuse this power to cast fraudulent votes. To prevent this abuse by a single administrator, we need more than one administrator using threshold signature scheme to sign a vote. By using our scheme, the power of a single administrator can be distributed to n administrators and any t ($t < n$) out of n administrators can generate a signature for a given message.

Juang and Lei initially proposed (t, n) threshold blind signature scheme in [11], and a provably secure threshold blind signature scheme [12] based on Okamoto-Schnorr blind signature. But, their work takes more time in the partial signature generation and verification processes. A user A and each signer

K. Kim (Ed.): ICICS 2001, LNCS 2288, pp. 318–327, 2002.
© Springer-Verlag Berlin Heidelberg 2002

P_i in [12] perform $O(n)$ and $O(n)$ multiplications, whereas A and P_i in our scheme perform $O(t)$ and $O(1)$ multiplications. In our scheme, we use the partial public key published by each signer P_i in key generation phase to reduce the computational complexity in the signature generation phase.

In this paper, we propose an efficient and provably secure threshold blind signature scheme and prove its security in the random oracle model. We will show the proposed scheme is as secure as Okamoto-Schnorr's blind signature scheme and is more efficient than [12]. The signature size and verification process in our scheme are the same as those of Okamoto-Schnorr's blind signature scheme.

This paper is organized as follows: In Section 2, we will review the related works such as blind signature, threshold signature, and distributed key generation. Then we propose our scheme in Section 3. In Section 4, we define some notions of security and prove the security of our scheme. The performance analysis is presented in Section 5. Finally, concluding remarks are made in Section 6.

2 Related Works

2.1 Blind Signature Scheme

Parameters
Suppose that p and q are two large primes such that $q|p-1$. g and h are elements of Z_p^* with order q. We assume that it is infeasible to compute the integer d such that $g = h^d \bmod p$, given g, h, and p.

Okamoto-Schnorr Blind Signature
This blind signature scheme designed by Okamoto [13] is an extension of the Schnorr scheme in [19]. An entity A creates a secret key $(r, s) \in Z_q \times Z_q$ and publishes the public key, $y = g^r h^s \bmod p$. Two party protocols between a sender A and a signer B are as follows:

1. B chooses $t, u \in Z_q$, and sends $a = g^t h^u \bmod p$ to A.
2. A blinds a with $\beta, \gamma, \delta \in Z_q$ into $\alpha = ag^\beta h^\gamma y^\delta \bmod p$, computes $\varepsilon = H(m, \alpha)$ and sends $e = \varepsilon - \delta \bmod q$ to B.
3. B computes $R = t - er \bmod q$ and $S = u - es \bmod q$, and sends (R, S) which satisfies $a = g^R h^S y^e \bmod p$.
4. A computes $\rho = R + \beta \bmod q$ and $\sigma = S + \gamma \bmod q$.
5. Issue a signature (α, ρ, σ) for message m.
6. We can easily verify that $\alpha = g^\rho h^\sigma y^\varepsilon \bmod p$, where $\varepsilon = H(m, \alpha)$.

Okamoto-Schnorr blind signature was proved to be provably secure under adaptively chosen message attacks in the random oracle model [17,18].

2.2 Threshold Signature Scheme

In 1979, Shamir at first defined the notion of threshold scheme [20] (often called a secret sharing scheme). The (t, n) threshold scheme proposed by Shamir is to divide a secret D into n pieces $D_1, D_2, \ldots, D_n$ in such a way that:

- Knowledge of any t $(t < n)$ or more D_i pieces makes D easily computable.
- Knowledge of any $t-1$ or fewer D_i pieces leaves D completely undetermined.

Threshold signature schemes have received considerable attention in the literature [4,8,10,16,21]. In a threshold signature scheme, digital signatures can be issued by t players rather than by one party. The secret key should be shared by n players in advance, and any set of less than t players cannot issue a digital signature for a given message.

2.3   Distributed Key Generation

Distributed key generation (DKG) defined in [9] is a secret sharing protocol without any trusted party (dealer). DKG works as a main component of threshold cryptosystems. Feldman [5] proposed a verifiable secret sharing (VSS) protocol based on discrete logarithm problem. But, his protocol has been shown to have a security flaw, since an adversary can influence the distribution of the result of his protocol to a non-uniform distribution. Moreover, Pedersen [15] proposed the first DKG protocol based on Feldman's protocol, and Gennaro et al. [9] proposed a provably secure DKG protocol. We will explain the DKG protocol in brief.

Each player P_i performs the following steps:

1. Each player chooses two random polynomials f_i, f_i' over Z_p where $f_i(x) = a_{i0} + a_{i1}x + \cdots + a_{it-1}x^{t-1}$, $f_i'(x) = b_{i0} + b_{i1}x + \cdots + b_{it-1}x^{t-1}$. Let $r_i = a_{i0} = f_i(0)$ and $r_i' = b_{i0} = f_i'(0)$. P_i broadcasts $C_{ik} = g^{a_{ik}} h^{b_{ik}} \bmod p$, for $k = 0, \ldots, t-1$. P_i computes the shares $s_{ij} = f_i(j) \bmod p$, $s_{ij}' = f_i'(j) \bmod p$, for $j = 1, \ldots, n$ and sends (s_{ij}, s_{ij}') secretly to player P_j.
2. Each player P_j verifies the shares (s_{ij}, s_{ij}'). P_j checks if

$$g^{s_{ij}} h^{s_{ij}'} = \prod_{k=0}^{t-1} (C_{ik})^{j^k} \bmod p. \tag{1}$$

If the check fails for an index i, P_j broadcasts a complaint against P_i.
   (a) Each player P_i who received a complaint defends himself by broadcasting the value (s_{ij}, s_{ij}') that satisfies Eq. (1).
   (b) Any player who either received more than $t-1$ complaints, or answered wrong value to a complaint is disqualified.
   (c) Let $H_0 := \{P_j | P_j \text{ is a not disqualified player}\}$.
   (d) The distributed secret value $r = \sum_{j \in H_0} r_i$ is not computed by any player.
3. Extracting $y = g^r \bmod p$:
   (a) Each player $P_j \in H_0$ broadcasts $A_{jk} = g^{a_{jk}} \bmod p$, for $k = 0, \ldots, t-1$
   (b) Each player P_j verifies the values broadcasted by other players in H_0, namely, for each $i \in H_0$, P_j checks if

$$g^{s_{ij}} = \prod_{k=0}^{t-1} (A_{ik})^{j^k} \bmod p. \tag{2}$$

If the check fails for an index i, P_j complains against P_i by broadcasting the value (s_{ij}, s_{ij}') that satisfies Eq. (1) but does not satisfy Eq. (2).

(c) For players P_i who received at least one valid complaint, i.e., value which satisfies Eq. (1) but does not satisfy Eq. (2), the other players run the reconstruction phase of Pedersen's VSS to compute $r_i, f_i(\cdot), A_{ik}$ for $k = 0, \ldots, t-1$

3 Threshold Blind Signature Scheme

3.1 Key Generation Protocol

We will show how to generate a random shared secret. By using the protocol presented in section 2.3, a random shared secret can be generated. But, since our scheme employs Okamoto-Schnorr blind signature scheme, we need two different secret keys. So we use a modified version of DKG protocol.

Each player P_i performs the following steps to generate a random shared secret:

1. Each player chooses three random polynomials f_i, f_i', f_i'' over Z_p where $f_i(x) = a_{i0} + a_{i1}x + \cdots + a_{it-1}x^{t-1}$, $f_i'(x) = a_{i0}' + a_{i1}'x + \cdots + a_{it-1}'x^{t-1}$, and $f_i''(x) = b_{i0} + b_{i1}x + \cdots + b_{it-1}x^{t-1}$. Let $a_{i0} = f_i(0), a_{i0}' = f_i'(0)$, and $b_{i0} = f_i''(0)$. P_i broadcasts $C_{ik} = g^{a_{ik}}h^{b_{ik}} \bmod p$ and $D_{ik} = g^{b_{ik}}h^{a_{ik}'} \bmod p$ for $k = 0, \ldots, t-1$. P_i computes the shares $s_{ij} = f_i(j) \bmod p, s_{ij}' = f_i'(j) \bmod p$, and $s_{ij}'' = f_i''(j) \bmod p$ for $j = 1, \ldots, n$ and sends $(s_{ij}, s_{ij}', s_{ij}'')$ secretly to player P_j.
2. Each player P_j verifies the shares $(s_{ij}, s_{ij}', s_{ij}'')$. P_j checks if

$$g^{s_{ij}}h^{s_{ij}''} = \prod_{k=0}^{t-1}(C_{ik})^{j^k} \bmod p \; , \; g^{s_{ij}''}h^{s_{ij}'} = \prod_{k=0}^{t-1}(D_{ik})^{j^k} \bmod p. \qquad (3)$$

If the check fails for an index i, P_j broadcasts a complaint against P_i. Then, P_i and P_j perform the same works as they do step2 (a)-(c), in section 2.3.
3. Extracting $y = g^s h^{s'} \bmod p$:
   (a) Each player $P_i \in H_0$ broadcasts $A_{ik} = g^{a_{ik}}h^{a_{ik}'} \bmod p$ for $k = 0, \ldots, t-1$.
   (b) Each player P_j checks if

$$g^{s_{ij}}h^{s_{ij}'} = \prod_{k=0}^{t-1}(A_{ik})^{j^k} \bmod p. \qquad (4)$$

If the checks fails for an index i, P_j complains against P_i by broadcasting the value $(s_{ij}, s_{ij}', s_{ij}'')$ that satisfies Eq. (3), but does not satisfy Eq. (4).
   (c) Each player P_i who received a complaint defends himself by broadcasting the value $(s_{ij}, s_{ij}', s_{ij}'')$ that satisfies Eq. (4).
   (d) Any player who either received more than $t-1$ complaints, or answered wrong value to a complaint is disqualified.

After executing this protocol, the following equations hold:

$$y = \prod_{j \in H_0} g^{a_{j0}} h^{a'_{j0}} = \prod_{j \in H_0} A_{j0}$$

$$f(x) = a_0 + a_1 x + \cdots + a_{t-1} x^{t-1}, \text{ where } a_i = \sum_{j \in H_0} a_{ji}$$

$$f'(x) = a'_0 + a'_1 x + \cdots + a'_{t-1} x^{t-1}, \text{ where } a'_i = \sum_{j \in H_0} a'_{ji}$$

$$f(i) = s_i, f'(i) = s'_i, \forall i = 1, \ldots, n$$

For convenience, let $|H_0| = n$ and the key generation protocol be denoted by

$$\{(s_1, s'_1), \ldots, (s_n, s'_n)\} \longleftrightarrow \{(a_0, a'_0) \mid y, y_j, H_0\}, \text{ for } j = 1, \ldots, n ,$$

where H_0 means an index set of disqualified players in key generation protocol, and (a_0, a'_0) is a secret key and y is a public key. For each $j \in H_0$, the value $y_j = g^{s_j} h^{s'_j}$ is the public commitment and (s_j, s'_j) is the secret key share of P_j.

3.2   Signature Generation Protocol

Let m be a message and let H be a one-way hash function. Suppose that a user A wants to get a signature from t signers, $H_1 \subseteq H_0$. For simplicity, let $H_1 = \{1, \ldots, t\}$ and let the output of key generation protocol be as follows:

$$\{(r_1, s_1), \ldots, (r_n, s_n)\} \longleftrightarrow \{(r, s) \mid y, y_j, H_0\}, \text{ for } j = 0, \ldots, n$$

The signature generation protocol starts when a user A requests each signer P_i, where $i \in H_1$, threshold blind signature by sending $\omega_i = \prod_{k=1, k \neq i}^t \frac{k}{k-i}$ for $k \in H_1$. A and P_i perform the following protocol:

1. Each P_i chooses $t_i, u_i \in Z_q$, and sends $a_i = g^{t_i} h^{u_i} \bmod p$ to A.
2. After receiving all a_i's, A computes $a = \prod_{i=1}^t a_i$ and blinds a with $\beta, \gamma, \delta \in Z_q$ into $\alpha = a g^\beta h^\gamma y^\delta \bmod p$. A computes $\varepsilon = H(m, \alpha)$ and sends $e = \varepsilon - \delta \bmod q$ to each P_i.
3. After receiving e, each P_i computes $R_i = t_i - e r_i \omega_i \bmod q$ and $S_i = u_i - e s_i \omega_i \bmod q$, and sends (R_i, S_i) which satisfies $a_i = g^{R_i} h^{S_i} y_i^{e\omega_i} \bmod p$.
4. A checks $a_i = g^{R_i} h^{S_i} y_i^{e\omega_i}$ after receiving (R_i, S_i). If the check fails for an index i, A sends e again to P_i to take a correct (R_i, S_i). Otherwise, A makes a signature (α, ρ, σ), where $\rho = \beta + \sum_{i=1}^t R_i \bmod q$, $\sigma = \gamma + \sum_{i=1}^t S_i \bmod q$.

3.3   Signature Verification

The verification process is the same as Okamoto-Schnorr blind signature scheme. A verifier accepts a signature (α, ρ, σ) on a message m if and only if

$$\alpha = g^\rho h^\sigma y^\varepsilon \bmod p, \text{ where } \varepsilon = H(m, \alpha).$$

3.4 Correctness

The validity of the signature on message m can be checked easily as follows:

$$
\begin{aligned}
\alpha &\equiv g^\rho h^\sigma y^\varepsilon \\
&\equiv g^{\beta + \sum_{i=1}^t R_i} h^{\gamma + \sum_{i=1}^t S_i} y^\varepsilon \\
&\equiv g^\beta h^\gamma (g^{\sum_{i=1}^t t_i - er} h^{\sum_{i=1}^t u_i - es}) y^\varepsilon \\
&\equiv g^\beta h^\gamma (g^{\sum_{i=1}^t t_i} h^{\sum_{i=1}^t u_i})(g^{-er} h^{-es} y^\varepsilon) \\
&\equiv a g^\beta h^\gamma y^{\varepsilon - e} \\
&\equiv a g^\beta h^\gamma y^\delta \bmod p.
\end{aligned}
$$

4 Security

In this section, we will show that our scheme is secure if and only if Okamoto-Schnorr blind signature scheme is secure, i.e., the security of the two scheme is identical.

4.1 Notions of Security

The notions of security for blind signature scheme were formally defined in [17,12] under the random oracle model.

One-More Forgery: For some polynomially bounded integer ℓ, an attacker can obtain $\ell + 1$ valid signatures after fewer than ℓ interactions with the signer.

Parallel Attack: the attacker interacts ℓ times in parallel with the signer.

We define adversaries and the notions of security for Okamoto-Schnorr blind signature scheme and (t, n) threshold blind signature scheme. Let $\tilde{A}_{OS}$ be an adversary who is allowed to execute Okamoto-Schnorr blind signature protocol. $\tilde{A}_{OS}$ can ask a signer to generate blind signatures on chosen messages. Let $\tilde{A}_D$ be an adversary who is allowed to execute the signature generation protocol. $\tilde{A}_D$ can ask any t signers to generate blind signatures on the chosen messages. $\tilde{A}_D$ also might corrupt up to $t - 1$ arbitrary signers.

Definition 1. *A blind signature scheme is secure, if no $\tilde{A}_{OS}$ can do the one-more forgery with non-negligible probability in the random oracle model under parallel attack.*

Definition 2. *A (t, n) threshold blind signature scheme is secure, if no $\tilde{A}_D$ can do the one-more forgery with non-negligible probability in the random oracle model under parallel attack.*

4.2 Proof of Security

First, we will analyze the adversary's view during the generation of a random shared secret, and build a simulator SIM, such that produces an output distribution which is polynomially indistinguishable from $\tilde{A}_D$'s view for a given public key y. Let B be the index set of corrupted players and B' be the index set of the players who publishes inconsistent values A_{im}.

Theorem 1. *For every probabilistic polynomial time adversary $\tilde{A}_D$, there exists a probabilistic polynomial time simulator SIM, such that for given y, produces an output distribution which is indistinguishable from $\tilde{A}_D$'s view.*

Proof. (sketch) In the key generation protocol, $\tilde{A}_D$ sees the following probability distribution of data produced by the uncorrupted parties:

1. The content of the random tape of $\tilde{A}_D$
2. Sharing polynomials: $f_i(\cdot), f_i'(\cdot), f_i''(\cdot)$ for $i \in B$
3. Shares: $s_j(i), s_j'(i), s_j''(i)$ for $j \in H_0, i \in B$
4. Public commitments in step 1: C_{jk}, D_{jk} for $j \in H_0, k = 1, \ldots, t-1$
5. Public commitments in step 3: A_{jk} for $j \in H_0, k = 1, \ldots, t-1$
6. Answers on a valid complaint: $s_{ij}, s_{ij}', s_{ij}''$ for $i \in B', j = 1, \ldots, n$

We use the same simulator SIM that can act as the real players in the protocol as in [9,22]. A simulator SIM can be constructed as follows:

1. Since (1) and (2) are internal values of adversary, SIM does not have to compute these values.
2. SIM performs steps 1 and 2 in subsection 3.2 on behalf of the uncorrupted players. After this step, SIM knows all shares $s_j(i), s_j'(i)$, and $s_j''(i)$ in (3) and public values in (4).
3. To compute A_{ik} for $i \in H_0, k = 1, \ldots, t-1$, SIM performs the following computations:
   - Compute $A_{ik} = g^{a_{ik}}$ for $i \in H_0 \backslash \{n\}, k = 0, \ldots, t-1$
   - Compute $A_{n0} = y \cdot \prod_{i \in H_0 \backslash \{n\}} (A_{i0})^{-1} \bmod p$
   - Compute $A_{nk} = (A_{n0})^{\lambda_{k0}} \cdot \prod_{i=1}^{t-1} (g^{s_{ni}})^{\lambda_{ki}}$ for $k = 1, \ldots, t-1$, where λ_{ki}'s are the Lagrange interpolation coefficients.
   - Broadcast A_{ik} for $i \in H_0, k = 1, \ldots, t-1$
   - $y = g^{a_{i0}} h^{a_{i0}'} = \prod_{j \in H_0} A_{j0}$
4. To handle the messages resulting from complaints, SIM acts as follows:
   - Perform for each uncorrupted player the verifications of Eq. (4) on the values A_{ik} for $i \in B$, broadcast by the players controlled by the adversary. If the verification fails for some $i \in B$ and $j \in H_0 \backslash B$, broadcast a complaint $(s_{ij}, s_{ij}', s_{ij}'')$.
   - For each valid complaint against P_i, perform the reconstruction phase of Pedersen's VSS to compute $f_i(0), f_i'(0)$ and y_i.   □

A more detailed analysis of the distribution can be found in [9,22].

Theorem 2. *For every adversary $\tilde{A}_{OS}$ who can produce one-more forgery under parallel attack, there exists an adversary $\tilde{A}_D$ who can produce one-more forgery under parallel attack.*

Proof. We can make $\tilde{A}_D$ given the adversary $\tilde{A}_{OS}$. Suppose the key generation protocol of $\tilde{A}_D$ generates a public key y. Whenever $\tilde{A}_{OS}$ asks for a signature on a message m, $\tilde{A}_D$ executes some t signers to issue a signature and returns the signature to $\tilde{A}_{OS}$. Since $\tilde{A}_{OS}$ can perform its parallel attack, $\tilde{A}_{OS}$ can produce one-more forgery. Finally, $\tilde{A}_D$ takes the output of $\tilde{A}_{OS}$ and produces one-more forgery. Hence, if $\tilde{A}_{OS}$ outputs one-more forgery under parallel attack, $\tilde{A}_D$ can output one-more forgery. □

It is known that Okamoto-Schnorr blind signature scheme is provably secure in the random oracle model [17]. Hence, we can say that there is no $\tilde{A}_{OS}$ that can produce one-more forgery under parallel attack.

Theorem 3. *If any adversary $\tilde{A}_D$ can produce one-more forgery under parallel attack, there exists an adversary $\tilde{A}_{OS}$ who can produce one-more forgery under parallel attack.*

Proof. We show how to construct $\tilde{A}_{OS}$ given the adversary $\tilde{A}_D$. For simplicity, assume $\tilde{A}_D$ corrupts players $1, \ldots, t-1 \in H_1$. $\tilde{A}_{OS}$ executes SIM described in **Theorem 1** to perform key generation protocol for a given public key y, and then runs $\tilde{A}_D$. Whenever $\tilde{A}_D$ needs a signature of m_i, $\tilde{A}_{OS}$ interacts with signer to obtain a signature of m_i. $\tilde{A}_{OS}$ has to provide $\tilde{A}_D$ the signature and the view that he sees during the signature generation protocol. $\tilde{A}_{OS}$ performs steps 1 and 2 on behalf of the players $P_1, \ldots, P_{t-1}$. After this simulation, $\tilde{A}_{OS}$ knows (a_j, R_j, S_j) for $j = 1, \ldots, t-1$. And then, $\tilde{A}_{OS}$ computes R_t, S_t, and a_t as follows.

$$\rho_i = \beta + \sum_{j=1}^{t} R_j, \quad \sigma_i = \gamma + \sum_{j=1}^{t} S_j.$$

R_t, S_t, and a_t are computed as

$$R_t = \rho_i - \beta - \sum_{i=1}^{t-1} R_i, \quad S_t = \sigma_i - \gamma - \sum_{i=1}^{t-1} S_i$$
$$a_t = g^{R_t} h^{S_t} y_t^{e\omega_t} \bmod p$$

$\tilde{A}_{OS}$ inputs (a_t, R_t, S_t) into $\tilde{A}_D$. Since the whole view of $\tilde{A}_D$ is given, $\tilde{A}_D$ can produce one-more forgery. Finally, $\tilde{A}_{OS}$ takes the output of $\tilde{A}_D$, and produces his one-more forgery. □

Corollary. *The proposed threshold blind signature scheme is secure if and only if Okamoto-Schnorr blind signature scheme is secure.*

Proof. The proof of this corollary follows immediately from **Theorems 1, 2** and **3**. □

5 Performance

In this section, we compare the computational load required to perform threshold blind signature.

Table 1 illustrates the cost of computations of blind signature scheme and threshold blind schemes. To reduce computational cost, we suppose that the

Table 1. Computational cost of the signature generation

	User A			Signer P_i		
	ADD	MUL	EXP	ADD	MUL	EXP
OS Scheme	3	3	3	2	3	2
JL Scheme	$2t+1$	$2n-t+6$	8	$2(n-t+1)$	$2n-1$	2
Our Scheme	$2t+1$	$t+5$	6	2	5	2

OS scheme: Okamoto-Schnorr blind signature scheme
JL scheme: Blind threshold scheme proposed by Lei and Juang
ADD: the number of additions
MUL: the number of multiplications
EXP: the number of exponentiations

value ω_i (for $i = 1, \ldots t$) is computed before the signature generation phase. In this case, the number of additions and multiplications done by each signer P_i in our scheme are the same as those of OS scheme. JL scheme takes more time in the partial signature generation and verification processes. In the JL scheme, the partial signature generation and verification are as follows:

- Partial signature generation
  $R_i = e(r_i + \omega_i + \sum_{j=t+1}^{n} f_j(i)) + t_i \bmod p$
  $S_i = e(s_i + \omega_i + \sum_{j=t+1}^{n} f'_j(i)) + u_i \bmod p$
- Partial signature verification
  $g^{R_i} h^{S_i} y^e \equiv a_i (\prod_{j=t+1}^{n} g^{a_{ik}})^{\omega_i e} (\prod_{j=t+1}^{n} h^{a'_{ik}})^{\omega_i e} \bmod p$

A user A and each signer P_i in our scheme perform $O(t)$ and $O(1)$ multiplications, whereas A and P_i in JL scheme perform $O(n)$ and $O(n)$ multiplications. Compared with the JL scheme, our scheme is more efficient and practical. The signature verification cost of JL scheme and our scheme are the same as that of OS scheme.

6   Conclusion

We have proposed a secure and practical threshold blind signature scheme based on Okamoto-Schnorr blind signature scheme. In our scheme, any t out of n signers can generate a blind signature and any one can verify the signature using the same verification process of Okamoto-Schnorr scheme. We have proved the proposed scheme is as secure as Okamoto-Schnorr's blind signature scheme in the random oracle model. Moreover, we claim that our scheme is more efficient and practical than [12].

References

1. D. Chaum, *Untraceable Electronic Mail, Return Addresses, and Digital Pseudonyms*, Comm. ACM 24, pp.84–88, 1981.
2. D. Chaum, *Blind Signatures for Untraceable Payments*, Advances in Cryptology-Crypto'82, pp.199–203, Springer-Verlag, 1983.

3. D. Chaum, A. Fiat, M. Naor, *Untraceable Electronic Cash*, Advances in Cryptology-Crypto'88, LNCS 403, pp.319–327, Springer-Verlag, 1988.
4. Y. Desmedt, *Threshold Cryptosystems*, Advances in Cryptology-Auscrypt'92, LNCS 718, Springer-Verlag, 1993.
5. P. Feldman, *A Practical Scheme for Non-interactive Verifiable Secret Sharing*, In Proc. 28th FOCS, pp. 427–437, 1987.
6. A. Fujioka, T. Okamoto and K. Ohta, *A Practical Secret Voting Scheme for Large Scale Election*, Advances in Cryptology-Auscrypt'92, LNCS Vol.718, pp.248–259, Springer-Verlag, 1993.
7. A. Fujioka, M. Abe, M. Ohkubo, and F. Hoshino, *An Implementation and an Experiment of a Practical and Secure Voting Scheme*, Proc. of SCIS2000, C48, Okinawa, Japan, Jan. 26–28, 2000.
8. R. Gennaro, S. Jarecki,H. Krawczyk, and T. Rabin, *Robust Threshold DSS Signatures*, Advances in Cryptology-Eurocrypt'96, LNCS Vol.1070, pp.354–371, Springer-Verlag, 1996.
9. R. Gennaro, S. Jarecki,H. Krawczyk, and T. Rabin, *Secure Distributed Key Generation for Discrete-Log Based Cryptosystems*, Advances in Cryptology-Eurocrypt'99, LNCS Vol.1592, pp.295–310, Springer-Verlag, 1999.
10. S. Jarecki and A. Lysyanskaya, *Adaptively Secure Threshold Cryptography*, Advances in Cryptology-Eurocrypt'00, LNCS Vol.1807, pp.221–242, Springer-Verlag, 2000.
11. W. S. Juang and C. L. Lei, *Blind Threshold Signatures Based on Discrete Logarithm*, Proceedings of the 2nd Asian Computing Science Conference, Lecture Notes in Computer Science 1179, Springer-Verlag, pp. 172-181, 1996.
12. W. S. Juang, C. L. Lei, and Pei-Ling Yu, *Provably Secure Blind Threshold Signatures Based on Discrete Logarithm*, Proceedings of 1999 National Computer Symposium, pp. C198-C205, 1999.
13. T. Okamoto, *Provably Secure and Practical Identification Schemes and Corresponding Signature Schemes*, Advances in Cryptology-Crypto'92, LNCS Vol.740, pp.31–53, Springer-Verlag, 1992.
14. M. Ohkubo, F. Miura, M. Abe, A. Fujioka and T. Okamoto, *An Improvement on a Practical Secret Voting Scheme*, Information Security'99, LNCS Vol.1729, pp.225–234, Springer-Verlag, 1999.
15. T.P. Pedersen, *Non-interactive and Information-theoretic Secure Verifiable Secret Sharing*, Advances in Cryptology-Crypto'91, LNCS Vol.576, pp.129–140, Springer-Verlag, 1992.
16. T.P. Pedersen, *A Threshold Cryptosystem without a Trusted Party*, Advances in Cryptology-Eurorypt'91, LNCS Vol.547, pp.522–526, Springer-Verlag, 1992.
17. D. Pointcheval and J. Stern, *Provably Secure Blind Signature Scheme*, Advances in Cryptology-Asiacrypt'96, LNCS Vol.1163, pp.252–265, Springer-Verlag, 1996.
18. D. Pointcheval and J. Stern, *Security Arguments for Digital Signatures and Blind Signatures*, Journal of Cryptology, LNCS Vol.13, Num.3, pp.361–396, Springer-Verlag, 2000.
19. C. P. Schnorr, *Efficient Identification and Signatures for Smart Cards*, Advances in Cryptology-Crypto'89, LNCS Vol.435, pp.235–251, Springer-Verlag, 1990.
20. A. Shamir, *How to Share a Secret*, Comm. ACM Vol.22, pp.612–613, 1979.
21. V. Shoup, *Practical Threshold Signatures*, Advances in Cryptology-Eurocrypt'00, LNCS Vol.1807, pp.207–220, Springer-Verlag, 1996.
22. D.R. Stinson and R. Strobl, *Provably Secure Distributed Schnorr Signatures and a (t, n) Threshold Scheme for Implicit Certificates*, Information Security and Privacy, ACISP'01, LNCS Vol.2119, pp. 417-434, Springer-Verlag, 2001.

A Multi-signature Scheme with
Signers' Intentions Secure against Active Attacks

Kei Kawauchi[1], Hiroshi Minato[2], Atsuko Miyaji[1], and Mitsuru Tada[1,*]

[1] School of Information Science,
Japan Advanced Institute of Science and Technology (JAIST),
Asahidai 1-1, Tatsunokuchi, Nomi, Ishikawa 923-1292, Japan
{kei-k,miyaji,mt}@jaist.ac.jp
[2] Department of Electrical Engineering and Computer Science,
Tufts University,
Halligan Hall, 161 College Avenue, Medford, Massachusetts 02155-5528, USA
hminato@eecs.tufts.edu

Abstract. In this paper, we propose a multi-signature scheme, in which each signer can express her *intention* associating with the message to be signed. Signers' intentions mean a kind of information which can be newly attached to a signature in signers' generating it. However, we have been introduced no multi-signature scheme dealing with intentions without loss of its efficiency.

First, we consider a multi-signature scheme realizing the concept of signers' intentions by utilizing existing schemes, and name it *primitive method*. After that, we introduce the proposed multi-signature scheme which is more efficient than the primitive method in view of the computational cost for verification and in view of the signature size. The proposed multi-signature scheme is shown to be secure even against *adaptive chosen message insider attacks*.

1 Introduction

A *multi-signature scheme*, in which plural entities (signers) jointly sign an identical message, has the advantage that it is efficient in view of the signature size and in view of the computational cost for verification. Hence we can say that a multi-signature scheme is quite useful in the following case:

- We often see a notice on a bulletin board on campus, which informs club members of an event. A notice frequently requires members to write down their names on it. It is very convenient for members to check who wants to take part in the event.

Now, we suppose that a captain of the club wants to know whether or not each member (e.g. Alice, Bob and etc.) wants to attend the event. If the name is written by him/her on the notice, it is clear that he/she wants to take part in

---

* Current affiliation of the last author: Institute of Media and Information Technology, Chiba University, Japan.

K. Kim (Ed.): ICICS 2001, LNCS 2288, pp. 328–340, 2002.

the event. But, if not, does that mean he/she does not want? The answer is No because he/she might not see the notice and also he/she does not positively express that he/she does not want to take part in the event. To make the matter sure, the captain should require members to write down their names, and also Yes or No on the notice to avoid such a problem. It is very good idea. For example, Alice may sign the notice adding the word No. On the other hand, Bob may sign it adding the word Yes. Then, we call these Yes or No *signers' intentions*. A captain may prepare a notice which has two spaces for signing. One is a space for signers who express Yes. The other is a space for signers who express No. The members put their name on one of two spaces. Unfortunately, there has been no proposal of any multi-signature schemes which efficiently handle the notice with Yes and No, namely signatures with signers' intentions. Each signer provide two secret-keys, one for expressing Yes, and the other for expressing No. It is, however, far from a good way since each entity has to manage more keys. As another countermeasure, the captain can provide two messages to be signed, one for Yes, and the other for No. Accordingly, twice verification is required for those two multi-signatures. But unlike in the first countermeasure, each entity has only to manage one key. In the example given above, signers' possible intentions are only Yes and No, and we consider that signers, in general, have choices among $\mathcal{I} := \{I_1 \ldots, I_N\}(N \geq 2)$. Each possible intention is denoted by some $I_\ell(\ell \in [1, N])$. (We can say that in the example given above, Yes and No are denoted by I_1 and I_2, respectively.) Hereafter such a multi-signature scheme in which plural message are provide and plural multi-signature are generated like in the second countermeasure, is called *a primitive method*. The details of this method are discussed in Section 3. In this paper, we introduce a *multi-signature scheme with signers' intentions* in which each signer has only to manage one key, in which one message to be signed is provided, hence in which only one multi-signature is generated, and furthermore in which only each signer can add her intention with respect to the given message. In a multi-signature scheme along the first countermeasure, each signer has to manage N keys, and in a multi-signature by the primitive method, the more the number N of signers' possible intentions gets, the more the signature size is and the more verification cost is required. On the other hand, in a multi-signature scheme with signers' intentions, the signature size is independent of N, and hence the verification cost is much smaller than that in the primitive method. Hence a multi-signature scheme with signers' intentions can be more efficient than ones constructed along the countermeasures given above. The efficiency of the proposed scheme is outstanding. Take for example distributing vacation time among office workers. Now refer to the calendar (Figure 1). Each signer establishes his/her intention by signing his name on a single day. In the proposed scheme, verification for the calendar is needed just once. Namely, the calendar can be verified by just one equation.

The security is shown with the strategy that we reduce the security of multi-signature scheme to that of multi-round identification scheme in the random oracle model [1]. To prove the security of multi-signature scheme with signers' intentions, we, for convenience' sake, consider two multi-round identification

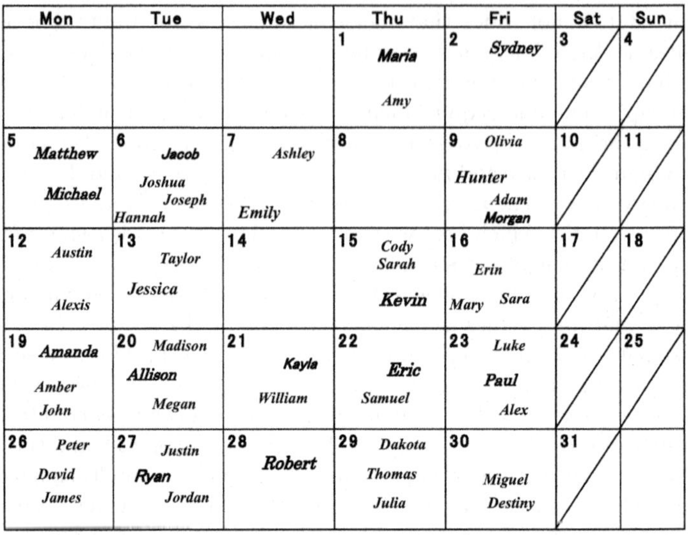

Mon	Tue	Wed	Thu	Fri	Sat	Sun
			1 *Maria* *Amy*	2 *Sydney*	3	4
5 *Matthew* *Michael*	6 *Jacob* *Joshua* *Joseph* *Hannah*	7 *Ashley* *Emily*	8	9 *Olivia* *Hunter* *Adam* *Morgan*	10	11
12 *Austin* *Alexis*	13 *Taylor* *Jessica*	14	15 *Cody* *Sarah* *Kevin*	16 *Erin* *Mary* *Sara*	17	18
19 *Amanda* *Amber* *John*	20 *Madison* *Allison* *Megan*	21 *Kayla* *William*	22 *Eric* *Samuel*	23 *Luke* *Paul* *Alex*	24	25
26 *Peter* *David* *James*	27 *Justin* *Ryan* *Jordan*	28 *Robert*	29 *Dakota* *Thomas* *Julia*	30 *Miguel* *Destiny*	31	

Fig. 1. Calendar

schemes with (prover's) intentions. We call those identification schemes ID-A and ID-B, respectively. The proof for the security of a multi-signature scheme with signers' intentions can be reduced to that for ID-A and ID-B. Concrete to say, if ID-A is secure against any polynomial-time passive adversaries, and if ID-B has zero-knowledge property, then multi-signature scheme with signers' intentions can be shown to be secure even against any polynomial-time active adversaries by using *ID-reduction technique* introduced by [7].

We can see related work as follows: In [7,10], we can see several kinds of multi-signature schemes. In [2,3,4,5], we can see a multi-signature scheme which also guarantee the signing order. The scheme given by [6] provides signing order verifiability and message flexibility.

This paper is organized as follows: In Section 2, we give the notations we use in this paper, and review the multisignature scheme given by [7]. In Sections 3, we propose the primitive method, a combination scheme of conventional multi-signatures, in which signatures with signers' intentions can be dealt with. In Section 4, we propose a new multi-signature scheme which we call a multi-signature scheme with signers' intentions. In Section 5, we give provable security for the proposed scheme. In Section 6, we evaluate the performance of the primitive method and the proposed scheme. The conclusion is given in Section 7.

2   Preliminaries

To denotes an n-tuple $(a_1, \ldots, a_n)$, we often use the bold letter $\boldsymbol{a}$. For an n-tuple $\boldsymbol{a}(= (a_1, \ldots, a_n))$ and for integer $i, j \in [1, n]$ with $(1 \leq i \leq j \leq n)$, $\boldsymbol{a}_{[i,j]}$ denotes the $(j - i + 1)$-tuple $(a_i, \ldots, a_j)$.

2.1   Multi-signature Scheme [7]

In a multi-signature scheme, plural signers (say, n signers) generate a signature for an identical message. However, we can realize such a situation by applying an ordinary (single) signature scheme n times. Then we shall extend a single signature scheme to be a multi-signature scheme so that the obtained multi-signature scheme shall satisfy the property that the signature size in the multi-signature scheme should be less than nL where L is the signature size in the single signature scheme.

In this paper, we use the multi-signature scheme, which is of the one-cycle type and is so-called a *generic* multi-signature scheme [9] obtained by translating a multi-round identification scheme.

In a multi-signature scheme, n signers $P_1, \ldots, P_n$ participate and each signer P_i publishes a public-key v_i and keeps a secret-key s_i. In the following, we describe the scheme, each P_i can query to the public random oracle function [1] $f_i : \{0,1\}^* \rightarrow \mathbf{Z}_q$. Let $\mathcal{P}$ denotes the set $\{P_1, \ldots, P_n\}$.

System parameter: System parameters p, q, g are published, and satisfy the
    following properties:
      – A trusted center publishes two large primes p and q such that $q|(p-1)$.
      – Element $g \in \mathbf{Z}_p^*$ of order q.
    System parameters are common for all schemes. Then, we omit these in latter
    schemes.

Key-generation step: Each signer $P_i \in \mathcal{P}$ provides a pair of a secret-key $s_i \in$
    $\mathbf{Z}_q$ and the corresponding public-key v_i, where $v_i := g^{s_i} \pmod{p}(i \in [1, n])$
    and n is the number of signers. When P_i registers the public-key v_i, she has
    to show that she indeed has s_i in a zero-knowledge manner. To prevent *the*
    key-generation phase attacks given by [8].

Signature generation step: Suppose that a set of signers $\mathcal{P}$ generates a multi-
    signature for a message m. The initial value y_0 is 0. For each $i \in [1, n]$, the
    following is executed.

      – P_i receives $(\boldsymbol{x}_{[1,i-1]}, y_{i-1})$, m from P_{i-1}. P_i picks up a random $r_i \in \mathbf{Z}_q$
        and computes (x_i, e_i, y_i) as follows:

$$x_i := g^{r_i} \pmod{p},$$
$$e_i := f_i(\boldsymbol{x}_{[1,i]}, m),$$
$$y_i := y_{i-1} + s_i + r_i \cdot e_i \pmod{q}.$$

    P_i sends $(\boldsymbol{x}_{[1,i]}, y_i)$, m to P_{i+1}. Also let $P_{n+1} := V$.

Verification step: Suppose that the verifier V receives a multi-signature $(\boldsymbol{x}, y_n)$
    for a message m. Then V computes $e_i := f_i(\boldsymbol{x}_{[1,i]}, m)$ for each $i \in [1, n]$. Also
    the verifier V checks the following equations:

$$g^{y_n} \stackrel{?}{\equiv} \prod_{i=1}^{n} (x_i^{e_i} \cdot v_i) \pmod{p}$$

3   Primitive Method

In Section 1, we have intuitively mentioned how we can realize a multi-signature scheme with signers' intentions. Here we present a concrete scheme of the *primitive method*. Suppose that each P_i is required her intention α_i for a message m, and that her possible intention is in a set $\mathcal{I} := \{I_1, \ldots, I_N\}$. For $\ell \in [1, N]$, let m_ℓ be the message corresponding to the intention I_ℓ for m.

Both system parameter and key-generation step are done in the same way as that of the multi-signature scheme in Section 2.

Signature generation step: Suppose that a set of signers $\mathcal{P}$ generates a multi-signature for a set of message $\{m_\ell\}$ with signers' intentions. Assume that $y_0^{(I_1)}, \ldots, y_0^{(I_N)}$ are set up to be zero. For each $i \in [1, n]$, the following is executed.

- P_i receives $(\boldsymbol{x}_{[1,i-1]}, y_{i-1}^{(I_1)}, \ldots, y_{i-1}^{(I_N)})$, $\{m_\ell\}$ and $\boldsymbol{\alpha}_{[1,i-1]}$ from P_{i-1}. P_i chooses her intention $\alpha_i \in \mathcal{I}$. Let $\alpha_i = I_\ell$. P_i picks up a random $r_i \in \mathbf{Z}_q$ and computes (x_i, e_i, y_i) as follows:

$$x_i := g^{r_i} \pmod{p},$$

$$e_i := f_i(\boldsymbol{x}_{[1,i]}^{(I_\ell)}, m),$$

$$y_i^{(I_\ell)} := y_{i-1}^{(I_\ell)} + s_i + r_i \cdot e_i \pmod{q}.$$

where $\boldsymbol{x}_{[1,i]}^{(I_\ell)}$ is defined to be $\bigcup_{j \leq i, \alpha_j = I_\ell} \{x_j\}$. For every $I_{\ell'} \in \mathcal{I} \backslash \{I_\ell\}$, let $y_i^{(I_{\ell'})} := y_{i-1}^{(I_{\ell'})}$.

P_i sends $(\boldsymbol{x}_{[1,i]}, y_i^{(I_1)}, \ldots, y_i^{(I_N)})$, $\{m_\ell\}$ and $\boldsymbol{\alpha}_{[1,i]}$ to P_{i+1}. Also let $P_{n+1} := V$.

Verification step: Suppose that the verifier V receives a multi-signature $(\boldsymbol{x}, y_n^{(I_1)}, \ldots, y_n^{(I_N)})$ for a set of message $\{m_\ell\}$ with signers' intentions $\boldsymbol{\alpha}$. Then V computes $e_i := f_i(\boldsymbol{x}_{[1,i]}^{(I_\ell)}, m_\ell)$ for each $i \in [1, n]$. Also the verifier V checks the following equations by the received $(\boldsymbol{x}, y_n^{(I_1)}, \ldots, y_n^{(I_N)})$.

$$g^{y_n^{(I_\ell)}} \stackrel{?}{\equiv} \prod_{\substack{1 \leq i \leq n \\ \alpha_i = I_\ell}}^{n} \left(x_i^{(I_\ell) e_i} \cdot v_i^{(I_\ell)} \right) \pmod{p} \quad (\forall I_\ell \in \mathcal{I})$$

The set of public-keys $\boldsymbol{v}^{(I_\ell)}$ is defined to be $\bigcup_{\alpha_i = I_\ell} \{v_i\}$, and where $\boldsymbol{x}^{(I_\ell)}$ and $e^{(I_\ell)}$ are defined as well as $\boldsymbol{v}^{(I_\ell)}$. As we can guess from the primitive method given above, the total signature size in the primitive method turns out to be $n|p| + N|q|$, by $(N-1)|q|$ which is larger than the signature size in the scheme [7].

4   Proposed Scheme

The primitive method discussed in the previous section, needs much verification cost in proportion to the number of the varieties of signers' intentions. As seen in the primitive method, as N increases, the scheme gets inefficient. Then we here propose a new multi-signature scheme with signers' intentions. In this scheme, the total signature size is independent of N, and is the same with that in the scheme [7]. The process of generating y_i, a part of signature, is very unique. And the proposed scheme is secure even against *adaptive chosen message insider attacks*.

In the following, we describe the proposed scheme, in which each P_i can query to the public random oracle function $f_i : \{0,1\}^* \to \mathbf{Z}_q$, and that anyone can access the public random oracle function $h : \{0,1\}^* \to \mathbf{Z}_q$.

Both system parameter and key-generation step are done in the same way as that of the multi-signature scheme in Section 2.

Signature generation step: Suppose that a set of signers $\mathcal{P}$ generates a multi-signature for a message m. The initial value y_0 is 0. For each $i \in [1, n]$, the following is executed.

- P_i receives $(\boldsymbol{x}_{[1,i-1]}, y_{i-1})$, m and $\boldsymbol{\alpha}_{[1,i-1]}$ from P_{i-1}. P_i chooses her intention $\alpha_i \in \mathcal{I}$, and picks up a random $r_i \in \mathbf{Z}_q$ and computes (x_i, e_i, y_i) as follows:

$$x_i := g^{r_i} \pmod{p},$$
$$e_i := f_i(\boldsymbol{x}_{[1,i]}, m, \boldsymbol{\alpha}_{[1,i]}),$$
$$y_i := y_{i-1} + s_i \cdot \theta_i + r_i \cdot e_i \pmod{q},$$

where $\theta_i := h(\alpha_i)$. P_i sends $(\boldsymbol{x}_{[1,i]}, y_i)$, m and $\boldsymbol{\alpha}_{[1,i]}$ to P_{i+1}. Also let $P_{n+1} := V$.

Verification step: Suppose that the verifier V receives a multi-signature $(\boldsymbol{x}, y_n)$ for a message m with signers' intentions $\boldsymbol{\alpha}$. Then V computes $\theta_i := h(\alpha_i)$ and $e_i := f_i(\boldsymbol{x}_{[1,i]}, m, \boldsymbol{\alpha}_{[1,i]})$ for each $i \in [1, n]$. Also the verifier V checks the following equations:

$$g^{y_n} \stackrel{?}{\equiv} \prod_{i=1}^{n} (x_i^{e_i} \cdot v_i^{\theta_i}) \pmod{p}$$

5   Security Consideration

In this section, we prove that the proposed scheme is secure against *adaptive chosen message insider attacks*.

5.1   Adversary Model

For discussion of the security of multi-signature scheme with signers' intentions, we here present the adversary model for the scheme.

MS-α Adversary. Given the system parameter (p, q, g) and the public-keys $\boldsymbol{v}$, an MS-α adversary $\mathcal{M}$ which can query to the random oracle functions $f_i(i \in [1, n])$, executes the following for each $j \in [1, Q]$ with given Q:

(S1) An MS-α adversary $\mathcal{M}$ determine a message m_j, a signer P_{i_j}, and the signer's intention $\boldsymbol{\alpha}_j \in \mathcal{I}^n$,

(S2) Generate a valid partial multi-signature for m_j in the signers' intentions $\boldsymbol{\alpha}_{j[1,i_j-1]}$ and $(\boldsymbol{x}_{[1,i_j-1]}, \boldsymbol{e}_{[1,i_j-1]}, y_{i_j-1})$ by colluding with $\mathcal{P} \backslash \{P_{i_j}\}$,

(S3) Send $(\boldsymbol{x}_{[1,i_j-1]}, \boldsymbol{e}_{[1,i_j-1]}, y_{i_j-1}, \boldsymbol{\alpha}_{j[1,i_j-1]})$ and α_{j,i_j} to P_{i_j}. To make tha adversary stronger, we assume $\mathcal{M}$ can ask P_{i_j}'s signature for P_{i_j}'s intention $\mathcal{M}$ chooses.

(S4) And get a valid partial multi-signature $(\boldsymbol{x}_{[1,i_j]}, \boldsymbol{e}_{[1,i_j]}, y_{i_j})$ and the singers' intentions $\boldsymbol{\alpha}_{[1,i_j]}$ from P_{i_j}.

After Q iterations of this step, the adversary $\mathcal{M}$ computes a multi-signature for a message m with signers' intentions $\boldsymbol{\alpha}$, where for every $j \in [1, Q]$, it must hold at least one of $m \neq m_j$ and $\boldsymbol{\alpha}_{j[i_j, i_j]} \neq \boldsymbol{\alpha}_{[i_j, i_j]}$.

Here note that in the key-generation step, each signer is required to show that she indeed has the corresponding secret-key, if Type II [7] is adopted. Hence we don't have to consider *the key generation phase attacks* given by [8].

5.2   Definition of the Security for Multi-signature Scheme with Signers' Intentions

Here we define the security of the proposed multi-signature scheme with signers' intentions

Definition 1. Suppose an MS-α adversary (probabilistic Turing machine) $\mathcal{M}$ can ask R_i-queries to f_i for each $i \in [1, n]$, and is allowed Q-time execution of the steps from (S1) to (S4). If such an MS-α adversary $\mathcal{M}$ can forge a multi-signature $(\boldsymbol{x}, \boldsymbol{e}, y_n)$ for a message m with signers' intentions $\boldsymbol{\alpha}$ in time at most t with probability at least ϵ, then we say that $\mathcal{M}$ can $(t, Q, \boldsymbol{R}, \epsilon) - break$ the *multi-signature scheme with signers' intentions*. Here, the probability is taken over the coin flips of $\mathcal{M}, f_1, \ldots, f_n$ and signing oracles $\mathcal{P}$.

Definition 2. A multi-signature scheme with signers' intentions is said to be $(t, Q, \boldsymbol{R}, \epsilon) - secure$, if there is no MS-$\alpha$ adversary which can $(t, Q, \boldsymbol{R}, \epsilon)$-break the scheme, and if for a message m, a multi-signature $(\boldsymbol{x}, \boldsymbol{e}, y_n)$ which is valid for signers' intentions $\boldsymbol{\alpha}$, is invalid for another signers' intentions $\boldsymbol{\alpha}'$ with overwhelming probability.

5.3   Identification Schemes

As we can seen in [7], the security of the multi-signature scheme given by [7] can be reduced to the security of multi-round identification scheme, from which the multi-signature scheme is derived. That means if the multi-round identification scheme is shown to be secure against polynomial-time adversaries, then it shall be shown that by *ID-reduction lemma*, in the multi-signature scheme, any adaptive chosen message insider polynomial-time adversary cannot existentially forge a signature. Also for the proposed scheme, the security of the multi-signature scheme with signers' intentions can be reduced to the security of some kinds of multi-round identification schemes. Before showing it, we first introduce two kinds of multi-round identification schemes. Those are slightly different from each other, and are necessary to prove the security of multi-signature scheme with signers' intentions.

Scheme ID-A:

The participating entities are the prover P and the verifier V, and both of them can access the public random oracle function $h : \{0,1\}^* \rightarrow \mathbf{Z}_q$.

System parameter is done in the same way as that of the multi-signature scheme in Section 2.

Key-generation step: P provides n pair of a secret-keys $s_i \in \mathbf{Z}_q$ and the corresponding public-keys v_i, where $v_i := g^{s_i} \pmod{p}(i \in [1, n])$.

Identification step: P chooses her intentions $\boldsymbol{\alpha} \in \mathcal{I}$ with $\#\boldsymbol{\alpha} = n$. First P picks up n random $r_i \in \mathbf{Z}_q$, and computes $x_i := g^{r_i} \pmod{p}(i \in [1, n])$. Then the prover P and the verifier V execute the following step for $i \in [1, n]$.

– P sends the commitment (x_i, α_i) to V, and V randomly picks up the challenge $e_i \in \mathbf{Z}_q$, and sends it to P.

After this iteration, P computes the answer

$$y := \sum_{i=1}^{n}(s_i \cdot \theta_i + r_i \cdot e_i) \pmod{q}.$$

where $\theta_i := h(\alpha_i)$. Then P sends y to V.

Receiving $(\boldsymbol{x}, y)$ and $\boldsymbol{\alpha}$, the verifier V figures out θ_i for each $i \in [1, n]$. V checks $(\boldsymbol{x}, y)$ and $\boldsymbol{\alpha}$ by following verification:

$$g^y \stackrel{?}{=} \prod_{i=1}^{n}(x_i^{e_i} \cdot v_i^{\theta_i}) \pmod{p}$$

If this equality holds, then V accepts the identification, and rejects, otherwise.

Scheme ID-B:

ID-B is different from ID-A in terms of the timing when P declares. Namely in ID-B P does before interaction between P and V.

Both system parameter and key-generation step follows that of Scheme ID-A.

Intention declaration step: The prover P publishes $\boldsymbol{\alpha} \in \mathcal{I}$ with $\#\boldsymbol{\alpha} = n$. (This distribution does not have to be uniform.)

Identification step: P picks up n random $r_i \in \mathbf{Z}_q$, and computes $x_i := g^{r_i}$ (mod p)($i \in [1, n]$). For the rest, the step is the same as the previous one.

Here we define the security for multi-round identification schemes.

Definition 3. Suppose that an ID-adversary $\mathcal{M}$ which does not have s, can pass the verification for some α in time at most t with probability at least ϵ. Then we say that ID-adversary $\mathcal{M}$ can $(t, \epsilon)-break$ the *multi-round identification schemes*.

Definition 4. We say that a multi-round identification scheme is $(t, \epsilon) - secure$, if there is no ID-adversary which can (t, ϵ)-break the scheme, $(\boldsymbol{x}, \boldsymbol{e}, y)$ which can pass the verification for intentions $\boldsymbol{\alpha} \in \mathcal{I}$, does not pass the verification for another (distinct) intentions $\boldsymbol{\alpha}'$ with overwhelming probability.

We define *the zero-knowledge property* for Scheme ID-B as follows:

Definition 5. Suppose that a polynomial-time machine $\mathcal{S}$ is given public-key $\boldsymbol{v}$ and intentions $\boldsymbol{\alpha}$. Then we say the scheme has *the perfect zero knowledge property*, if

$$\sum_{\boldsymbol{\kappa}, \boldsymbol{\lambda}, \mu} \left| \Pr[(\boldsymbol{\kappa}, \boldsymbol{\lambda}, \mu) \leftarrow [P(\boldsymbol{s}, \boldsymbol{\alpha}), V(\boldsymbol{v}, \boldsymbol{\alpha})]] - \Pr[(\boldsymbol{\kappa}, \boldsymbol{\lambda}, \mu) \leftarrow \mathcal{S}(\boldsymbol{v}, \boldsymbol{\alpha})] \right| = 0$$

Then Scheme ID-B is shown to provide the perfect zero-knowledge property by constructing a simulator $\mathcal{S}$, as follows:

- Given $\boldsymbol{v}$ and $\boldsymbol{\alpha} \in \mathcal{I}$, $\mathcal{S}$ picks up $y \in \mathbf{Z}_q$ and $\boldsymbol{e} \in \mathbf{Z}_q^n$ to compute β_i such that $y = \sum_{i=1}^n (e_i \cdot \beta_i)$ (mod q), and γ_i such that $\theta_i + e_i \cdot \gamma_i = 0$ (mod q)($i \in [1, n]$). Then $\mathcal{S}$ computes $x_i := g^{\beta_i} v^{\gamma_i}$ (mod p)($i \in [1, n]$).

Such an $(\boldsymbol{x}, \boldsymbol{e}, y)$ indeed passes the verification.

Lemma 1. Scheme ID-B has *the perfect zero-knowledge property*

Proof. We compute the following to probability of appearance of the $(2n + 1)$-tuple $(\boldsymbol{x}, \boldsymbol{e}, y)$:

- The probability of appearance of the $(2n + 1)$-tuple $(\boldsymbol{x}, \boldsymbol{e}, y)$ which can pass the verification for some $\boldsymbol{\alpha}$.
  - $\Pr\left[(\boldsymbol{\kappa}, \boldsymbol{\lambda}, \mu) \leftarrow [P(\boldsymbol{s}, \boldsymbol{\alpha}), V(\boldsymbol{v}, \boldsymbol{\alpha})]\right] = 1/q^{2n}$
  - $\Pr\left[(\boldsymbol{\kappa}, \boldsymbol{\lambda}, \mu) \leftarrow \mathcal{S}(\boldsymbol{v}, \boldsymbol{\alpha})]\right] = 1/q^{2n}$
- The probability of appearance of the $(2n + 1)$-tuple $(\boldsymbol{x}, \boldsymbol{e}, y)$ which can't pass the verification for some $\boldsymbol{\alpha}$.
  - $\Pr\left[(\boldsymbol{\kappa}, \boldsymbol{\lambda}, \mu) \leftarrow [P(\boldsymbol{s}, \boldsymbol{\alpha}), V(\boldsymbol{v}, \boldsymbol{\alpha})]\right] = 0$
  - $\Pr\left[(\boldsymbol{\kappa}, \boldsymbol{\lambda}, \mu) \leftarrow \mathcal{S}(\boldsymbol{v}, \boldsymbol{\alpha})]\right] = 0$

Thus we get that each distributions of probabilities are the same. So Scheme ID-B has *the perfect zero-knowledge property*. □

An adversary model for Scheme ID-A is given as follows.

ID-Adversary

An ID-adversary $\mathcal{M}$ is a machine, which, on input v, executes Scheme ID-A with V, and tries to pass the verification for some signers' intentions α. The ID-adversary $\mathcal{M}$ is so-called a *passive attacker*, which cannot accomplish *the attack in the middle*.

5.4   ID-Reduction Lemma

If Scheme ID-B provides the zero-knowledge property, we can obtain the following ID-reduction lemma.

Lemma 2.   (i) If there exists an MS-α adversary $\mathcal{A}_{\mathtt{a}}$ which can $(t, Q, \boldsymbol{R}, \epsilon)$-break the scheme, then there also exists an MS-α adversary $\mathcal{A}_1$ which can $(t, Q, \mathbf{1}, \epsilon_1)$-break the scheme, where $\mathbf{1}$ is the n-tuple $(1, \ldots, 1)$, and $\epsilon_1 := a_n$ with $a_0 := \epsilon$ and $a_i := \left(a_{i-1} - \frac{1}{q} \right) / R_i$.

(ii) If there exists an MS-α adversary $\mathcal{A}_1$ which can $(t, Q, \mathbf{1}, \epsilon_1)$-break the scheme, then there also exists an MS-α adversary $\mathcal{A}_{\mathtt{p}}$ which can $(t^+, 0, \mathbf{1}, \epsilon_{\mathtt{p}})$-break the scheme, where $t^+ := t + \Phi_{\mathtt{S}}$, $\Phi_{\mathtt{S}}$ is the simulation time of Q multi-signatures and $\epsilon_{\mathtt{p}} := \epsilon_1 - \frac{Q}{q}$.

(iii) If there exists an MS-α adversary $\mathcal{A}_{\mathtt{p}}$ which can $(t^+, 0, \mathbf{1}, \epsilon_{\mathtt{p}})$-break the scheme, then there also exists an ID-adversary $\mathcal{A}_{\mathtt{id}}$ which can $(t^+, \epsilon_{\mathtt{p}})$-break the scheme.

Proof. (Sketch) The proof is also the same with that of Lemma 9 in [7].   □

Lemma 3. Let $\epsilon_{\mathtt{p}} \geq \frac{2^{n+1}}{q^n}$. If there exists an ID-adversary which can $(t^+, \epsilon_{\mathtt{p}})$-break the scheme, then there exists a machine $\mathcal{M}$ which can compute a linear combination of $\boldsymbol{s}$ on input v in time t' with success probability ϵ'. Here t' and ϵ' are defined as follows:

$$ t' := \frac{t^{++}}{3\epsilon_{\mathtt{p}}} \left(2^{2n+1} + 1 \right) + \Phi_{\mathtt{L}}; \quad \epsilon' := F_1(\epsilon_{\mathtt{p}}) \prod_{i=1}^{n} \left(\frac{1}{2} F_i(\epsilon_{\mathtt{p}}) \right)^{2^{i-1}} . $$

where $t^{++} := t^+ + \Phi_{\mathtt{V}}$, $\Phi_{\mathtt{V}}$ is the time for verification, $\Phi_{\mathtt{L}}$ is the time for finding a linear combination on $\boldsymbol{s}$, and the function $F_i(\epsilon_{\mathtt{p}})$ is defined to be $1 - \left(1 - \frac{\epsilon_{\mathtt{p}}}{2^i} \right)^{2^i / \epsilon_{\mathtt{p}}}$.

Proof. (Sketch) Also for Scheme ID-A, we can obtain the Heavy row lemma like [7]. Hence we can obtain 2^n simultaneous equations with $(2^n + n - 1)$ unknowns. Among those unknowns, the n ones the secret-keys, and the rest are r components. From these equations, we can get one linear combination on only $\boldsymbol{s}$. The required time and the probability can be obtained as well as in [7].   □

By providing n linear combinations on s, we can find each s_i. Unfortunately, we cannot evaluate the probability that those equations are linear independent. In case $n = 2$, if the coefficients were uniform, then that probability would be at least $1 - \frac{2}{q}$.

Next we show one more property for security of multi-signature schemes with signers' intentions.

Lemma 4. Suppose that the tuple (x, e, y) passes the verification for signers' intentions $\alpha \in \mathcal{I}$. Then the very tuple (x, e, y) is rejected for another signers' intentions α' with overwhelming probability.

Proof. (Sketch) It comes from the following:

$$\Pr\left[(x, e, y, \alpha) \leftarrow [P(s), V(v)] : \mathtt{Ver}(v, x, e, y, \alpha') = 1 \,\middle|\, \mathtt{Ver}(v, x, e, y, \alpha) = 1\right]$$
$$\leq 1/q$$

holds for $\alpha, \alpha' \in \mathcal{I}$ with $\alpha \neq \alpha'$, where $\mathtt{Ver}$ is the verification equation.   □

Combining Lemmas 2, 3 and 4, we can obtain the following theorem.

Theorem 1. Let $\epsilon_p \geq \frac{2^{n+1}}{q^n}$. If there is no machine which can, on input v, compute a linear combination on s, in time t' with success probability ϵ', then the proposed multi-signature scheme with signers' intentions is (t, Q, R, ϵ)-*secure*.

Suppose that t and t' are bounded by a polynomial on the security parameter $|q|$. Then ϵ is non-negligible with respect to $|q|$ if and only if so is ϵ'.

6   Efficiency Consideration

We evaluate the computational amount for verification in the proposed scheme on the basis of the required number of modular-p multiplications, and also the total size of signatures. In evaluating the computational cost, more important is $\#(\bigcup_i\{\alpha_i\})$, which is the most variety of the intentions actually chosen by $\mathcal{P}$, rather than $\#\mathcal{I}$, which is the number of the intentions provided for the message.

The required number of modular-p multiplication is calculated by a simple binary method. For $(g_1^{a_1} \cdot g_2^{a_2} \cdots g_n^{a_n})$ where $(|a_1| = |a_2| = \cdots = |a_n| = |q|)$ and $(|g_1| = |g_2| = \cdots = |g_n| = |p|)$, the required number of modular-p multiplications is $\left(\frac{n}{2} + 1\right)|q| - 1$. In the computational amount for signing, there is no difference between the proposed scheme and the primitive method. It will not be discussed here. Table 1. summarizes the total size of signatures and the computational amount for verification in the primitive method and the proposed scheme. In the primitive method, the required number of modular-p multiplications is related to $\#\left(\bigcup_i\{\alpha_i\}\right)$. In other words, the primitive method loses its merit in proportion to the increase of $\#\left(\bigcup_i\{\alpha_i\}\right)$, because $\#\left(\bigcup_i\{\alpha_i\}\right)$ multi-signatures are verified in the primitive method. On the other hand, the proposed scheme is very unique. The proposed scheme has two properties simultaneously.

Table 1. Comparison of schemes

	total size of signatures	The number of the modular-p multiplications for verification
Primitive method	$n\|p\| + \#(\bigcup_i \{\alpha_i\})\|q\|$	$\left\{\dfrac{n+3\#\left(\bigcup_i\{\alpha_i\}\right)}{2}\right\}\|q\| - \#(\bigcup_i\{\alpha_i\}) + n$
Proposed scheme	$n\|p\| + \|q\|$	$\left(\dfrac{2n+3}{2}\right)\|q\| - 1$

- One is the property as a multi-signature scheme, which is suited to plural signers.
- The other is the property, which is suited to plural signers' intentions.

Roughly speaking, the former property makes the gap of the required number of modular-p multiplications between the single-signature scheme and the proposed (multi-signature) scheme. Second property, in the primitive method, the number of equations for verification (or the number of signatures) depends on the number of varieties of signers' intentions. Finally, in the proposed scheme, the number of equations for verification (or the number of signatures) do not depend on the number of signers or the number of varieties of signers' intentions.

7   Conclusion

We have proposed an idea of signers' intentions for multi-signature scheme, and have given the multi-signature scheme with signers' intentions . Then, we have shown that the proposed scheme has a computational advantage for verification, compared to the primitive method. The proposed scheme is proved to be secure against adaptive chosen message insider adversaries, by reducing it to that of two kind of multi-round identification schemes. This approach is also applicable to various multi-signature schemes such as two-cycle multi-signature schemes.

Acknowledgement

The authors would like to thank Mr.Takeshi Okamoto of JAIST for his invaluable advice and useful comments.

References

1. M. Bellare and P. Rogaway: *"Random oracles are practical: A paradigm for designing efficient protocols"*, Proceedings of the 1st Conference on Computer and Communications Security, ACM, 1993.

2. M. Burmester, Y. Desmedt, H. Doi, M. Mambo, E. Okamoto, M. Tada and Y. Yoshifuji: *"A Structured ElGamal-Type Multisignature Scheme"*, Lecture Notes in Computer Science 1751, Third International Workshop on Practice and Theory in Public Key Cryptosystems - PKC2000, Springer-Verlag, pp.466-483, 2000.

3. H. Doi, M. Mambo and E. Okamoto: *"On the Security of the RSA-Based Multisignature Scheme for Various Group Structures"*, Lecture Notes in Computer Science 1841, 5th Australasian Conference - ACISP2000, Springer-Verlag, pp.352-367, 2000.

4. H. Doi, E. Okamoto and M. Mambo: *"Multisignature Schemes for Various Group Structures"*, The 36-th Annual Allerton Conference on Communication, Control and Computing, pp.713-722, 1999.

5. H. Doi, E. Okamoto, M. Mambo and T. Uematsu: *"Multisignature Scheme with Specified Order"*, Proc. of the 1994 Symposium on Cryptography and Information security, SCIS94-2A, January 27-29, 1994.

6. S. Mitomi and A. Miyaji: *"A multisignature Scheme with Message Flexibility, Order Flexibility and Order Verifiability"*, Lecture Notes in Computer Science 1841, 5th Australasian Conference - ACISP2000, Springer-Verlag, pp.298-312, 2000.

7. K. Ohta and T. Okamoto: *"Multi-Signature Schemes Secure against Active Insider Attacks"*, IEICE transactions of fundamentals, vol. F-82-A. No.1, 1999.

8. K. Ohta and T. Okamoto: *"Generic Construction Method of Multi-Signature Schemes"*, Proc. of The 2001 Symposium on Cryptography and Information Security, SCIS01-2B, January 23-26, 2001.

9. D. Pointcheval and J. Stern: *"Security arguments for digital signatures and blind signatures"*, Journal of Cryptology, Volume 13, Number 3. pp.361-396, Springer-Verlag, 2000.

10. A. Shimbo: *"Design of a modified ElGamal Signature Scheme"*, Proc. of The 1996 Workshop on Design and Evaluation of Cryptographic Algorithms, pp.37-44, November 27, 1996.

A Distributed Light-Weight Authentication Model for Ad-hoc Networks

André Weimerskirch[2,1] and Gilles Thonet[1]

[1] Accenture Technology Labs, Sophia Antipolis, France
gilles.thonet@accenture.com
[2] Communication Security Group, Ruhr-Universität Bochum, Germany
weika@crypto.ruhr-uni-bochum.de

Abstract. In this work we present a security model for low-value trans-actions in ad-hoc networks in which we focus on authentication since this is the core requirement for commercial transactions. After a brief intro-duction to the ad-hoc networking paradigm we give a survey of various existing models and analyze them in terms of scope and applications. It is concluded that no appropriate and satisfactory model for low-value transactions in ad-hoc networks has been developed to date. Our new model is derived from previous models dealing with trust and security in ad-hoc networks, and does not require devices with strong proces-sors as public-key systems. We base the model on a recommendation and reference protocol that is inspired by human behavior and that is in accordance with the very nature of ad-hoc networks.

Keywords: Ad-hoc network, security, low-value transactions, trust, au-thentication, public-key

1 Introduction

Today microprocessors can be found almost everywhere. It is a well established trend to embed a microprocessor in nearly all electronic devices such as cellu-lar phones, televisions and video recorders. In the future this might extend to everything from coffee makers to washing machines which are then connected by a network. In combination with already existing computing devices such as desktop computers, notebooks and Personal Digital Assistants (PDAs) this could form an extremely widespread wireless network of mobile and static devices com-municating without fixed infrastructure or centralized administration. In such a self-organized network each node relies on its neighbor nodes to keep the network connected, e.g., each node routes data packets from its neighbors. Furthermore each node might take advantage of the services offered by other nodes. This type of network is usually called an *ad-hoc network* and is particularly useful when a reliable fixed or mobile infrastructure is not available – e.g. after a natural disaster – or too expensive.

The security issues for ad-hoc networks are different than the ones for fixed networks. While the security requirements are the same, namely availability, con-fidentiality, integrity, authentication, and non-repudiation, their provision must

K. Kim (Ed.): ICICS 2001, LNCS 2288, pp. 341–354, 2002.

be approached differently for ad-hoc networks. This is due to system constraints in mobile devices and frequent topology changes in the network. System constraints include low-power microprocessor, small memory and bandwidth, and limited battery power. In addition to usual denial of service attacks where a node is flooded by a vast amount of requests, availability in ad-hoc networks can be threatened by radio jamming and battery exhaustion [16]. Closely related to security is secure routing. Since each node of the network potentially acts as a repeater, a malicious node might heavily affect packet routing.

While there are many security issues in ad-hoc networks this paper focuses on *entity authentication* [14]. This is the core requirement for integrity, confidentiality and non-repudiation — if you do not know who you are talking to it does not make sense to establish a secure channel. The paper is organized as follows. Section 2 is a brief introduction to ad-hoc networking and its importance as future mobile networking model. Section 3 gives an overview of existing security models for ad-hoc networks. Then a discussion of the application domain of these models follows in Section 4. Section 5 introduces our new approach and the final section concludes the paper.

2   Ad-hoc Networks

As said previously an ad-hoc network is a wireless network made up of mobile hosts that do not require any fixed infrastructure to communicate. The basic idea behind this model is not recent: as early as the 70s, ad-hoc networks were called *packet radio networks* and were investigated for military applications almost exclusively – PRnet, developed by the American Defense Advanced Research Projects Agency (DARPA), is probably the most famous example [13]. Then, when designing the 802.11 standard for Wireless Local Area Networks (WLAN) the IEEE replaced the term packet radio network by ad-hoc network, hoping to forget the military connotation of the former one. Ad-hoc networks are frequently associated with self-organization, which means that they run solely by the operation of end-users (like the old citizen-band voice analogue network). Although pure self-organization is not required to form an ad-hoc network this feature should be understood as a basic requirement for decentralization: hosts must be capable to enter and leave the network without referring to a central authority. It is important to note that ad-hoc networks would likely neither be a replacement nor an alternative to current and future infrastructure-based networks. Their scope is more to complement the latter in cases where cost, environment or application constraints require self-organized and infrastructure-less solutions.

2.1   Main Issues

Ad-hoc networks include various architectures that range from fully mobile and decentralized radio, access, and routing technologies – the ones being investigated as part of the standardization work carried out by the Internet Engineering Task

Force (IETF) [15] – to more infrastructure-dependent standards that include an ad-hoc mode – e.g. IEEE 802.11 [11] and HiperLAN2 [10].

Here are some of the basic features one can expect to find in most ad-hoc networks:

– Communication links are wireless to guarantee mobility. Accordingly, their dependability and capacity have to be carefully scrutinized.
– Ad-hoc networks act independently from any provider. However, access points to a fixed backbone network are expected to be available if required.
– Because they do not rely on a fixed infrastructure mobile hosts have to be somehow *cooperative*. This ranges from very simple schemes for short-range networks to highly cooperative strategies in case of multi-hop wide-area networks.
– The network topology may be very dynamic, making the links and routes very unstable.
– Power management is an important system design criterion. Hosts have to be power-aware when performing such tasks as routing and mobility management.
– Finally, security is a critical issue because of the weak connectivity and of the limited physical protection of the mobile hosts. This is the focal point of this paper.

2.2   Potential Applications

Because packet radio networks have been developed initially for military purposes, potential applications are very often associated with critical situations such as battlefields and damaged areas. Other prospective applications are still at an early stage, as summarized in the following list:

– Military applications: communications between soldiers, soldiers monitoring, sensor networks for target detection and identification, ...
– Emergency situations where the existing network infrastructure is not reliable or has been damaged due to geopolitical instability, natural or man-made disaster, ...
– Provision of wireless connectivity in locations where cellular networks present an insufficient coverage, are more expensive or are wanted to be by-passed (e.g. for privacy reasons).
– Distributed networks for data collection and device monitoring: sensor networks, home networks, inter-vehicle communications, supply chain management, ...
– Creation of instant and temporary networks for ad-hoc meeting, conferences or brainstorming.

3   Previous Work

This section presents various security models for ad-hoc networks. Since a strong relationship can be identified between trust and security inherent to ad-hoc networks we start by introducing basic properties of trust and their role for security. Then, we describe a distributed trust model that we will use later in this work.

3.1   Distributed Trust Model

Before describing the Distributed Trust Model [1] it is worth mentioning some research results about trust [12]. Trust in human beings differs from trust in an IT system. We trust a human if we believe him to be benevolent, and we trust an IT system if we believe it is robust against malicious human attackers. When we trust an entity we distinguish the target of trust and its classification. The target of trust is the entity we trust, while the classification describes exactly for which aspect the entity is being trusted – while I might trust Bob in repairing a computer I might not trust him in repairing a car. Furthermore, there might be a value of trust which describes how much we trust an entity, or a ranking expressing that we trust entity A more than entity B with respect to some criteria.

Trust in distributed systems should as much as possible be based on knowledge to prevent irrational trust based on faith. There can be a hierarchy of trust relationships: explicit trust relationships are direct while implicit relationships rely on an indirect trust path. Trust can be derived by establishing a direct trust relationship using an indirect path.

The *Distributed Trust* Model [1] is a decentralized approach to trust management and uses a recommendation protocol to exchange trust-related information. The model assumes that trust relationships are unidirectional and taking place between two entities. The entities make judgments about the quality of a recommendation based on their policies, i.e., they have values for trust relationships. Also, trust is not absolute, e.g., an entity can change the trust value it received as a recommendation. The policy is not communicated, so it might not be understandable for other entities which prevents it from being misused. The recommendation protocol works by requesting a trust value in a trust target with respect to a particular classification. After getting an answer an evaluation function is used to obtain an overall trust value in the target. The protocol also allows recommendation refreshing and revocation. To do so the recommender sends the same recommendation with another recommendation value, or a neutral value to revoke. The model is suited to establishing trust relationships that are less formal, temporary, or targeting ad-hoc commercial transactions.

3.2   Password-Based Key Agreement

The work developed in [3] addresses the scenario of a group of people who want to set up a secure session in a meeting room without any support infrastructure. Desirable properties of a protocol that solves this problem are:

- *Secrecy*: only those entities that know an initial password are able to learn the session key. Perfect forward secrecy requires that an attacker who compromises one member of the group and learns all his permanent secret information is still unable to recover the session key.

- *Contributory key agreement*: the session key is formed of contributions from all entities. This ensures that if only one entity chooses its contribution key randomly all other entities will not be able to make the key space smaller.
- *Tolerance to disruption attempts*: the protocol must not be vulnerable to an attacker who is able to insert messages. It is assumed that the possibility of modifying or deleting messages in such an ad-hoc network is very unlikely.

The work describes and introduces several password-based key-exchange methods that meet these requirements. The core idea of the protocol is as follows. A weak password is sent to the group members. Each member then contributes part of the key and signs this data by using the weak password. Finally a secure session key to establish a secure channel is derived without any central trust authority or support infrastructure.

3.3   Resurrecting Duckling Security Policy

This policy was introduced in [16] and extended in [17]. It is particularly suited for devices without display, and for embedded devices which are too weak for public-key operations. The fundamental authentication problem is a solved by a secure transient association between two devices establishing a master-slave relationship. It is secure in the sense that master and slave share a common secret, and transient because the association can be terminated by the master only. Also a master can always identify its slave in a set of similar devices.

The proposed solution is called the *Resurrecting Duckling* model. The duckling is here the slave device while the mother duck is the master controller. The duckling will recognize as its mother the first entity that sends it a secret key on a secure channel, e.g., by physical contact during the device initialization. This procedure is called *imprinting*. The duckling will always obey its mother, who tells it whom to talk to through an access control list. The bond between mother and duckling is broken by death after which the duckling accepts another imprinting. Death may be caused by the mother itself, a timeout or any specific event. The whole security chain corresponds to a tree topology formed of hierarchical master-slave relationships. The root of the tree is a human being controlling all devices, and every node controls all devices in its subtree. However, if one relationship is broken the relationship to the whole subtree is also broken.

This security model can be applied to very large ad-hoc networks, e.g. networks consisting of smart dust devices [18]. A possible scenario is a battlefield of smart dust soldiers (acting as slaves or siblings) and their general (acting as the master). The master allows its slaves to communicate by uploading in each of them a highly flexible policy so that sibling entities become masters and slaves for a very short time, enough to perform one transaction. The mother duck gives the ducklings credentials that allow them to authenticate themselves.

3.4   Distributed Public-Key Management

Key management for established public-key systems requires a centralized trusted entity called *Certificate Authority* (CA). The CA issues certificates by binding a public key to a node's identity. One constraint is that the CA should always be available because certificates might be renewed or revoked. Replicating the CA improves availability. However, a central service runs contrary to the distributed structure of ad-hoc networks.

Reference [20] proposes to distribute trust to a set of nodes by letting them share the key management service, in particular the ability to sign certificates. This is done using *threshold cryptography* [6]. An $(n, t + 1)$ threshold cryptography scheme allows n parties to share the ability to perform a cryptographic operation so that any $t + 1$ parties can perform this operation jointly whereas it is infeasible for at most t parties to do so. Using this scheme the private key k of the CA is divided into n shares $(s_1, s_2, \ldots, s_n)$, each share being assigned to each special node. Using this share a set of $t + 1$ special nodes is able to generate a valid certificate. As long as t or less special nodes are compromised and do not participate in generating certificates the service can operate. Even if compromised nodes deliver incorrect data the service is able to sign certificates. Threshold cryptography can also be applied to well known signature schemes like the *Digital Signature Standard* (DSS) [7].

Another approach introduced in [9] presents a self-organized public-key infrastructure. The system replaces the centralized CA by certificate chains. Users issue certificates if they are confident about the identity, i.e. if they believe that a given public key belongs to a given user. Each user stores a list of certificates in its own repository. To obtain the certificate of another entity the requester builds a certificate chain using his repository list and implicitly trusted entity's lists until a path to an entity that has the desired certificate in its repository is found.

4   Scope of Security Models

4.1   Distributed Trust Model

The Distributed Trust Model describes how to establish trust relationships. It can be used on top of security models such as public-key systems. It is applicable to any distributed system and does not target specifically ad-hoc networks. An implementation for ad-hoc networks would be particularly vulnerable to malicious and compromised agents. Since an ad-hoc network has a very flexible topology it is also unclear if enough trustworthy entities are available to obtain a recommendation. Although it is possible to make a broadcast asking for targets offering a special service this can only be successful if the available network of explicitly and implicitly trusted entities is large enough. Finally the requester could ask for information regarding what other entities think about each other. This could be data that other entities would not want to reveal.

4.2   Password-Based Key Agreement

This model perfectly works for small groups. Authentication is done outside the IT system, e.g., the group members authenticate themselves by showing their passports or based on common knowledge[1]. The model does not suffice anymore for more complicated environments, though. Groups of people who do not know each other, or pairs of people who want to have confidential exchanges without the rest of the group be able to eavesdrop on the channel, are two examples. Another problem arises for large groups or groups at different locations. The secure channel to distribute the initial password is not available anymore. At this point it seems that existing support infrastructure is required to set up a secure channel.

4.3 Resurrecting Duckling

The Resurrecting Duckling scheme is an appropriate model for a well defined hierarchy of trust relationships. It particularly suits inexpensive devices that do not need a display or a processor to perform public-key operations. This perfectly works for a set of home devices, for instance. However, more flexible ad-hoc networks may not contain explicit trust relationships between each pair of nodes or to a centralized entity like the mother duck. Deploying a comprehensive network consisting of a hierarchy of a global mother duck and multiple subsidiary local mother ducks is very similar to a public-key infrastructure, where the mother ducks correspond to CAs, with all its advantages and drawbacks. Even the battlefield scenario raises some problems. Here the soldiers are siblings and obey to their mother, the general. If one soldier device wants to authenticate to another device it has to present its credentials [17]. The second device can then check the credentials by using its policy. But what happens if not all soldiers use the same credentials, i.e. the same secret key, to prevent it to be stolen by the enemy? If all devices use the same key the other side might invest considerable effort doing some physical attack [2] to recover the key because it would compromise all nodes. Since the devices cannot hold a list of all valid credentials it seems that a further authentication method is needed.

4.4 Distributed Public-Key Management

The Threshold Key Management system is a way to distribute a public-key system. For high-value transactions public-key systems are certainly the only way to provide a satisfactory and legal security framework. As mentioned earlier in this paper trust should as much as possible be based on knowledge. Since two entities that never met before cannot have common knowledge – a shared secret – they both have to trust a central entity, e.g., a CA to substitute this knowledge. When a user wants to prove his identity to a CA he goes there with his public key and shows his passport. The CA proves his identity and then binds his identity to his public key and signs the certificate. But how is this done when the CA is distributed? The user must prove his identity to all special nodes to prevent that a compromised node passes on faulty information. But if the CA signs certificates without proving the identity the model cannot be used for high-value transactions.

---

[1] In this case friendship or just knowing each other is considered as common knowledge.

The self-organized public-key infrastructure [9] shows similar problems. To make the system bullet-proof, the entity's identity has to be checked in the real world before users issue certificates. Furthermore, it is assumed that the certificate requester trusts each node in the recommendation chain. Finally a significant computing power and time is consumed to obtain a certificate going through the certificate chain. Each node in the chain has to perform public-key operations, first to check the received certificate for authentication (signature verification) and then to sign it before forwarding it (signature generation). This cannot be done in parallel but only one after the other until the certificate went along the entire chain.

Despite its centralized nature a central CA is preferable for applications with high-security demand. To ensure high availability the CA can be replicated. The replicated CAs are as secure as the original CA as long as the replication process is not vulnerable to attacks. The private key of the CA does not get weaker after replication. Much research has been done about efficient public-key systems – for example, a public-key system for mobile systems is presented in [8].

4.5   Different Scopes for Different Applications

No model ensures authentication in an ad-hoc network in every environment. Depending on the situation users can select the appropriate system. While for a group meeting in a small conference room the password-based key-exchange will work perfectly, for a network defined by a hierarchy of trust relationships the resurrecting duckling policy is the best alternative. To guarantee secure transactions the appropriate choice is to use a public-key system involving the hassle of getting certificates, to use a device that can perform such operations and find a reliable connection to the CA to check for revoked or renewed certificates.

Also, using no security at all is an acceptable approach for many applications. A good illustration is a huge network of cell phones where free calls can be made without using a phone service provider. Once the network has enough nodes and enough redundant connections, the no-security approach might be the best one since every user benefits from routing other users' data packets. However, none of the models presented up to now is a good solution for low-value transactions in ad-hoc networks. The next section fills this gap.

5   Distributed Light-Weight Authentication Model

5.1   Main Goals

Our new model is strongly based on human behavior when dealing with trust since the human society itself can be seen as an ad-hoc network. We use the recommendation protocol from the Distributed Trust Model to establish trust relationships and extend it by a request for references. Each entity maintains a repository of trustworthy entities such that a path between two arbitrary entities can be found by indirectly using the repositories of other entities. This idea has also been recently used for a self-organizing public-key system [9]. We also bring in the idea of threshold cryptography in the sense that compromised nodes, as long as their number is below a threshold, cannot harm the result of the operation. Since we are trying to establish a general model for low-value transactions our main goal is not to make transactions perfectly secure but rather to make the attacker's cost to get falsely authenticated higher than the value of the transaction. Thus, an attacker should not be able to make the effort once to break multiple transactions, but he should have high cost to break each single transaction.

The starting point is human behavior. To check out another person, people usually ask friends about that person. Also they will ask the target person for references. To check the identity of another person people will ask about common knowledge, e.g., a secret password or details of previous transactions. If the other person knows details which only the right person can know he is considered to be this person. If there is no explicit common knowledge a trust relationship can be derived using a trusted third party. This can be the CA in case of a public-key system or the mother duck in case of the hierarchical trust relationships. In our case trust relationships will be derived using a trust chain in the trust network of friends and friends' friends. A cooperation and feedback system introduces quality and user responsibility for recommendations to the model.

5.2   Model Description

The model works as follows. Let's assume Alice wants to check if Bob is who he pretends to be, i.e., she wants to verify Bob's identity. Note that in this case Alice and Bob are actually devices representing a user, and not human beings. Therefore users should authenticate themselves to the devices. It is worth mentioning that the model is applicable to users instead of devices but it would require some effort from the user. Alice starts asking Bob about common knowledge. This can be a secret key, but also knowledge about a recent transaction. If there is common knowledge Bob can prove his identity.

Recommendations. Otherwise, Alice starts requesting recommendations from nodes taken from her list of trustworthy entities. The classification of these requests is if the asked person believes that Bob is who he pretends to be, i.e., a request for a recommendation about Bob's identity. Note that in the simplest case possible return values can be "yes" or "no". Let's assume one of the devices Alice asks is Cathy. She can check Cathy's identity by asking her about common knowledge, e.g., a shared secret key or part of the content of their last transaction. This could be done using a challenge-response protocol [14]. Assuming that this transaction was encrypted its content cannot be eavesdropped by an attacker. Cathy then looks if Bob is on her list of trustworthy people and checks his identity, or forwards Alice's request in the same manner. Once an entity is found that knows Bob and can prove his identity, let's call him Dan, the information is sent back to Alice. She also obtains the name of all entities in the recommendation chain. To introduce anonymity the nodes in the recommendation chain could only forward the length of the chain and the result "yes" or "no". However, this would affect the feedback system that is described later on. To include some randomness Alice might ask random entities about Bob's identity. She might also ask Cathy and all other entities in the recommendation chain to do the same.

References. Asking for references is the next human behavior characteristic we use in our model. Alice can ask Bob to give his references, i.e., other devices he has done transactions with recently. Alice can then ask these devices if they know Bob. Since Bob (in this case the user behind the device) could easily set up the entities he gives as references, Alice can ask again her network of trustworthy entities if they know the references. Of course a good reference would be an entity that has a relationship with Alice and Bob. To be more convinced Alice could ask the references for references again and so on. Once a link between Alice's trust network and Bob's reference network is found (both in a recursive sense) a direct relationship between Alice and Bob can be derived.

Since Alice broadcasted her request to many nodes, and also Cathy broadcasted the request, the number of broadcasted requests is exponential. Therefore the chain length to find a relationship between Alice and Bob should be low. It has been shown that networks such as ad-hoc networks have a very small degree of separation [19]. This means that the chain between two arbitrary entities in an ad-hoc network has low length. Using both requests for recommendations and references the two trust networks are browsed starting at Alice and Bob until a connection is found. This keeps the number of involved nodes low. To further reduce the network traffic Alice and Cathy might have rankings for trustworthy entities. Alice might ask Bob for a recommendation if Bob is a cell phone but not if he is part of an on-line store. Also there can be nodes in the network particularly suited for recommendations, e.g., immobile servers that perform many transactions. These servers may be highly available through replication. Central servers, however, are not necessary for the protocol. After Alice has received the results of her request she has to evaluate the data.

Trust Evaluation. The evaluation phase allows Alice to decide if she believes in Bob's identity or not – i.e. the evaluation function maps information of all received data of the recommendation chains to a "yes" or a "no". For more sophisticated systems the function might output an upper limit for a transaction value when dealing with Bob. Before evaluation Alice can check the recommendation chains for suspicious nodes and if necessary ignore these chains for evaluation. The identification chains of Bob's references are included in the evaluation, but can be differently weighted than recommendations about Bob's identity. For the evaluation function we want to bring in the idea of threshold cryptography. If Alice receives n answers to her request, then we want that the result be not influenced by t or less malicious agents, where n is considerably larger than $2t$. How this is achieved is due to the local policy of Alice, which does not need to be public. A simple approach is to cut off t "yes" and t "no" results to the question about Bob's identity and evaluate the remaining values. A very suspicious person could configure its device such that it requires only "yes" answers while less suspicious users would allow some errors. Since an (n, t) scheme requires that exactly n nodes respond we will interpret such a scheme in the sense that $t/n \cdot 100\%$ or less compromised nodes that respond cannot affect the result. Unfortunately it is impossible to distinguish a malicious node that does not respond from a node that does not respond because of other reasons, e.g., to cover existing relationships. Management of the the trust relationships and credentials can be handled by a system as proposed in [4].

Secure Channel and Routing. Once an entity is authenticated – or both entities have authenticated mutually – a secure channel can be initiated. Since we assumed that there is no common knowledge between both entities a perfectly secure channel cannot be set up. In many cases it is sufficient not to use encryption at all. Otherwise, Alice can send Bob a secret over the trustworthy path used to derive a relationship. Bob then sends back another secret using a random path. By combining these secrets Alice and Bob can derive a shared secret key, as shown in [3]. Only trustworthy entities can obtain the secret key by eavesdropping Bob's answer. If strong processors are available a public-key method can be used. Alice generates a public/private key pair and sends Bob the public key over the trustworthy path. Bob generates a random secret, encrypts it using Alice's public key, and sends this to Alice over the trustworthy channel. Again, this method is only vulnerable to a man-in-the-middle attack mounted by a trustworthy device. The secure channel ensures confidentiality and integrity but no legal non-repudiation.

The proposed scheme can also be used for secure routing. Using a public-key method to make routing robust might be too expensive and excludes devices with weak processors. Therefore we suggest not to use any authentication method unless the loss of packets rises above a threshold. At this point the nodes might want to check out their neighbor. They can make requests about their identity, or use a similar recommendation scheme to ask about their abilities in routing data.

Cooperation. Now one might ask why any entity should participate in a request and use its battery power. We suggest a combination of incentive and punishment. Devices that give recommendations can also expect to receive answers to their requests. A device that does not want to waste battery power and therefore never answers to any request will be removed from the list of trustworthy nodes – or even put on a list of untrustworthy devices – and then receive very few or no answers to its requests. The ability and quality to authenticate other entities is therefore weakened.

Furthermore entities might receive rewards for good recommendations. For example, an on-line shop might give Cathy and Dan a prize reduction for each successful recommendation. Existing incentive schemes – for instance the ones detailed in [5] – that stimulate cooperation in ad-hoc networks can certainly be applied to our model. However, there is no solution yet for nodes that do not cooperate because they want to hide existing transaction relationships.

Feedback Information. Finally, to make the system more robust and protect it from compromised nodes, we introduce feedback messages. For example, if Alice receives from Dan via Cathy a positive recommendation about Bob's identity, but then gets cheated by Bob, she can inform Cathy and Dan about their wrong recommendation, and she might also put Bob, Cathy and/or Dan on a list of suspicious devices. Cathy and Dan can then take actions to find the security leak. The reward and feedback system introduces some quality and responsibility function to the model. The model also allows refreshing and revocation. If Dan validates Bob's identity but later on gets cheated by Bob, Dan can send a special message to Alice with negative result about Bob's identity. Of course he can also broadcast this message. Note that this does little damage to the real Bob if he can prove his identity to Alice using common knowledge.

Authentication of Users. We have not considered authentication of users by now. To solve this issue we require that users authenticate themselves to the device. This can be done using a strong one-way function, and furthermore by encrypting the file system. A problem might arise if a device is sold. Since the new user must be able to authenticate himself to the device it seems that he also takes over all trust relationships of the previous owner. To prevent this from happening the previous owner can change the device ID and delete all data on the device. The new owner can change back the ID but he cannot recover the data that include shared knowledge with the trusted devices. Therefore authentication with trusted devices will fail. This also makes clear that a simple device ID spoofing is not sufficient for a successful attack.

5.3   Security Analysis

An evaluation of the security level is very hard since the model is not based on pure mathematical foundations and also leaves the individual security policy open. So what does an attacker have to do to get falsely authenticated? Let's assume Alice wants to authenticate a device that claims to be Bob but that is actually Cathy. Cathy first had to fake references. If Alice asks these references for another reference the number of references she had to fake rises exponentially. Note that it is not enough to just fake one device and give it many IDs. Since a reference is asked about knowledge – i.e. about its history – faking a reference is much harder than just faking its ID. Furthermore Cathy had to compromise nodes from Alice's list of trustworthy entities. Finally she also had to compromise or set up enough random nodes such that the probability that Alice exactly asks these nodes is high enough. The probability is configurable by Alice by setting the threshold value (n, t) in her local security policy. Alice can also configure individually other parameters of her security policy. For example, she can reject dealing with Bob if the number of "no" requests about Bob's identity is above the threshold of t. An attacker can do a local attack by setting up a bunch of nodes at one geographical point to increase his chances for many successful attacks in this area. In this case the feedback system is able to indicate the attack before the attacker is able to regain the effort and money it took him to set up the attack. The initial cost of the attack is likely too high for the average attacker. However, we understand that our model inherits inherent problems of the Distributed Trust Model, in particular the concerns that entities need to reveal their transaction relationships. Furthermore, establishment of a secure channel is not perfectly secure. For secure channel establishment we replace a shared knowledge or trusted third party by a path in the network of trusted entities. The longer the path the higher the probability of a malicious entity among them. Therefore much effort has to be spent on finding efficient algorithms and well-sized repository lists, e.g., as done in [9]. Another challenge is the essential feedback system. It has to be efficient, detect fraud quickly, and take appropriate action to prevent repeated attacks.

6 Conclusions

In this work we have dealt with security in ad-hoc networks. We have focused on authentication since this is the core requirement to initiate a secure channel. We have described several existing models and analyzed their scope, possible scenarios but also inherent problems. Finally we have proposed a new authentication model for low-value transactions. This model is inspired by human behavior and it is consistent with the ad-hoc network paradigm. It makes use of recommendations and references to derive a trust relationship, and is scalable with respect to security. A cooperation and feedback system introduces quality and responsibility. The requirements in the device's hardware are very low. Using our model, the nodes in an ad-hoc network can establish a secure channel.

References

1. A. Abdul-Rahman and S. Hailes. A Distributed Trust Model. *New Security Paradigms Workshop 1997*, ACM, 1997.
2. R. Anderson and M. Kuhn. Tamper resistance – a cautionary note. *2nd USENIX Workshop on Electronic Commerce*, 1996.
3. N. Asokan and P. Ginzboorg. Key Agreement in Ad-hoc Networks. *Computer Communications 23*, 2000.
4. M. Blaze, J. Feigenbaum, and J. Lacy. Decentralized Trust Management. *The 17th Symposium on Security and Privacy*, IEEE Computer Society Press, 1996.
5. L. Buttyán and J.-P. Hubaux. Nuglets: a Virtual Currency to Stimulate Cooperation in Self-Organized Mobile Ad Hoc Networks. *Technical Report DSC/2001/001*, Swiss Federal Institute of Technology – Lausanne, Department of Communication Systems, 2001.

6. Y. Desmedt and Y. Frankel. Threshold cryptosystems. *Advances in Cryptology – Crypto '89*, LNCS 435, Springer-Verlag, 1990.
7. R. Gennaro, S. Jarecki, H. Krawczyk, and T. Rabin. Robust threshold DSS signatures. *Advances in Cryptology – Eurocrypt '96*, LNCS 1070, Springer-Verlag, 1996.
8. G. Horn and B. Preneel. Authentication and Payment in Future Mobile Systems. *Computer Security - ESORICS '98*, LNCS 1485, Springer-Verlag, 1998.
9. J.-P. Hubaux, L. Buttyán, and S. Čapkun. The Quest for Security in Mobile Ad Hoc Networks. *ACM Symposium on Mobile Ad Hoc Networking and Computing – MobiHOC 2001*, 2001.
10. HiperLAN2 Global Forum. *URL: http://www.hiperlan2.com.* Work in progress.
11. IEEE Computer Society LAN/MAN Standards Committee. Wireless LAN Medium Access Control (MAC) and Physical Layer (PHY) Specifications - IEEE 802.11. 1997.
12. A. Jøsang. The right type of trust for distributed systems. *New Security Paradigms Workshop 1996*, ACM, 1996.
13. J. Jubin and J.D. Tornow. The DARPRA Packet Radio Network Protocol. *Proc. IEEE, vol. 75, no. 1*, 1987.
14. A. Menezes, P.C. van Oorschot, and S.A. Vanstone. *Handbook of Applied Cryptography*, CRC Press, 1997.
15. Mobile Ad-hoc Networks (MANET). *URL: http://www.ietf.org/html.charters/manet-charter.html.* Work in progress.
16. F. Stajano and R. Anderson. The Resurrecting Duckling: Security Issues for Ad-hoc Wireless Networks. *The 7th International Workshop on Security Protocols*, LNCS 1796, Springer-Verlag, 1999.
17. F. Stajano. The Resurrecting Duckling – what next? *The 8 th International Workshop on Security Protocols*, LNCS 2133, Springler-Verlag, 2000.
18. B. Warneke, M. Last, B. Leibowitz, and K.S.J. Pister. Smart Dust: Communicating with a Cubic-Millimeter Computer. *Computer Magazine*, IEEE, Jan. 2001.
19. D. Watts and S. Strogatz. Collective dynamics of 'small-world' networks. *Nature 393*, 1998.
20. L. Zhou and Z.J. Haas. Securing Ad Hoc Networks. *IEEE Network Magazine, vol. 13, no. 6*, 1999.

Design of an Authentication Protocol for Gsm Javacards

Stelvio Cimato

Dipartimento di Informatica ed Applicazioni,
Universitá di Salerno, 84081 Baronissi (SA), Italy
cimato@dia.unisa.it

Abstract. In this work, we present an authentication protocol derived by the integration of the Kerberos and Lamport authentication schemes. The protocol is targeted to the GSM network, involving users that interact through a mobile phone, and the network provider acting as trusted authority. A prototype application which implements parts of the protocol interaction phases, has been developed in the Javacard framework and is also discussed.

Keywords: Authentication protocol, Javacard, E-commerce and M-commerce security

1 Introduction

The smartcard technology is rapidly evolving. A large number of smartcards are actually embedded in different kind of electronical devices, with very different purposes. One of the motivation of such diffusion is the rapid development of electronic engineering and of miniaturization techniques, giving smartcards a growing set of functionalities. Smartcards have not only storing capabilities, but usually embed a microprocessor with growing computational power. Currently, many producers sell on the market smartcards integrating security capabilities, such as the possibility to use the most common cipher/decipher algorithms (e.g., RSA, DES).

Indeed, one of the most diffused use of smartcards is for authentication purpose. The growing demand for secure computer systems to be used for e-commerce applications leads to the need of alternative means for authentication systems. In this context, the use of tamper-resistant hardware can enhance the security features of traditional authentication techniques. Smartcards, with their inherent security and mobility can realize secure authentication procedures, when combined with sound authentication protocols [IHA99].

In Europe, smartcards have been successfully used in the telecommunications industry. The GSM technology is winning the competition among the different proposed digital technologies, and is expanding all over the world, accounting for about the 70 percent of the world's digital mobile phones over almost 200 countries. This amounts to more than half a billion GSM mobile phones now in use worldwide, corresponding to the same number of smartcards. Every GSM phone

K. Kim (Ed.): ICICS 2001, LNCS 2288, pp. 355–368, 2002.

embeds a SIM, (Subscriber Identification Module), which is a smartcard containing the data and the necessary algorithms for accessing the mobile network. In the context of GSM technology, the research and the standardization effort lead to a number of ETSI (European Telecommunications Standard Institute) reports which describe all the standards recommended for the implementation of the protocol and for the hardware components, including the characteristics of the smartcards used as SIM modules [Ins99a,Ins99b,Ins99c]. The research is now moving to provide users with additional services, meeting the needs of the users, the needs of the network companies earning more money for the additional traffic, and finally the needs of the service companies selling those services.

As soon as new services have been added to the mobile devices, the demand for additional security mechanisms than the ones provided by the GSM network is growing. Users can be reluctant to use mobile devices for transactions and other sensible data transmission, if they are suspecting that their personal data traveling on the network (e.g., credit card number, social security number) can be revealed to an eavesdropper listening to the GSM traffic. Even if much attention has been devolved to security concerns in the design of the GSM protocols, several attacks have demonstrated some weakness in both the implementation of the algorithms and the whole infrastructure of the GSM network, compromising the security of the transmissions.

Our work presents the integration of the Kerberos protocol [SNS88] with one time password schemes [Lam81] in order to realize an authentication protocol easily usable in the context of small-value transactions. The protocol has been thought for its usage over the GSM network, involving users who interact through their mobile handsets, and the network provider acting as a trusted authority. We discuss its implementation using a GSM Javacard, the smartcard combining the two emerging technologies, the GSM standard with the Java platform. Javacards are a new generation of smartcards having the capability to execute small pieces of Java code, called *applets* [Sun99]. The Javacard proposal is one of the most promising technology, since it overcomes some of the most annoying problems for the developers of software for smartcards: standardization and lack of a friendly developing framework.

In the next Section we will overview the security issues concerning the development of m-commerce applications over the GSM network. In Section 3 we discuss briefly the two techniques which are the starting point for our work. In Section 4 we describe the protocol we propose, discussing in Section 5 the design of the applets realizing the basic functionalities. Finally in Section 6 we will draw some conclusions and plan future works.

2   Mobile Commerce and Security

According to several forecasts and study reports, in the next few years about the 25 percent of mobile subscribers worldwide and of approximately 800 million handsets in use, will be using mobile commerce. The estimated value of financial mobile transactions is expected to be about USD 66 billion by 2003. The potential of the market is more interesting in Europe, where the penetration rate of mobiles already exceeds that of PC Internet connections. Furthermore, m-commerce has the advantage of being performed in every place, without any particular additional device necessary to access the network and set up the connection. With respect to e-commerce, people do not have the need to turn on its own PC and access the network through a phone line to do banking, shopping or accessing any other mobile service. However, several surveys suggest that more than 50 percent of business leaders quote security concerns as their main reason not to embrace e-commerce [Rad].

2.1 GSM Security

The GSM network specification effort has devolved a lot of time to offer advanced security functionalities ensuring adequate protection for both the operators and the customers. In particular, authentication, confidentiality and anonymity have been crucial points considered in the design of the GSM network. However a number of weaknesses have been discovered due to both the choices and the implementation adopted.

Let us describe in more detail the authentication process. The customer and the network provider share a secret key K_i which is stored both in the SIM, embedded in the mobile phone of the customer, and in the database of the provider. Whenever the mobile phone is switched on, a challenge-response session is started to control the validity of the data contained into the SIM and to calculate a session key to cipher the transmission originated by the mobile phone over the network. In particular a random challenge $rand$ is generated and issued to the mobile, which use it as input together with the key K_i to two secret algorithms, named $A3$ and $A8$, to obtain a Signed Response, $Sres$, and a session key, K_c. The network provider will compare the response of the mobile $Sres$ with the result obtained after performing the same process, giving access to the mobile if the response is correct. Every successive transmission of data between the mobile phone and the network is ciphered using the A5 algorithm with the key K_c.

Implementation and Attacks. All the phases of authentication and data encryption are based on the secret algorithms which have been chosen by the GSM association and implemented within the embedded SIM. In practice, service providers incorporated the functionalities of the $A3$ and $A8$ algorithms in a unique algorithm named *COMP128*, receipting the suggestion of the same standard associations. The *COMP128* algorithm accepts an 128 bit input string and outputs a ciphertext 96 bit long. In this context, it is used as MAC algorithm with K_i as input parameter and $rand$ as input plaintext.

The keys K_i and K_c play a crucial role for the GSM security architecture. The knowledge of the former would allow a non-authorized user to access the network and use the resources which the real customer will pay for, with the effect of having a cloned SIM. The knowledge of the key K_c would compromise the security of the transmissions between the mobile and the network, making any malicious eavesdropper able to decrypt the information transmitted.

The $A5$ algorithm which guarantees the confidentiality of the transmissions is a stream algorithm which has been presented in two versions, $A5/1$ and $A5/2$. They differ for the security level of the cryptographic functions exploited, and were developed to respect some legal restrictions relative to the export of cryptographic software products. The $A5$ algorithm is usually implemented into mobile phones and not into the embedded SIM, to increase performance.

The choices made by the standard GSM associations revealed a number of errors and weaknesses in all over the GSM security architecture. A large part of these errors can be attributed to the fact that the GSM associations preferred to deliver part of the security to the secrecy of the used algorithms. Inevitably, all the algorithms, $A5$, $A3$, $A8$ and $COMP128$ have been made of public domain and a number of attacks have been discovered. In particular, online and offline attack to the $COMP128$ algorithm have been proposed, demonstrating that it is possible to reconstruct the key K_i after collecting 200000 authentication responses, spending at least 8 hours for an offline attack.

The $A5$ algorithm is also not much secure, since its equivalence with a 16 bit long key ciphering algorithm in its weak version, and 40 bit long key in the secure version, has been demonstrated. In [BSW00], it is showed that it is possible to break the $A5$ algorithm and reconstruct the K_c key after few seconds, listening to at least 2 minutes of transmissions.

2.2 Short Message Service and Mobile Commerce

In the context of mobile commerce, an important role is played by the Short Message Service offered by the GSM network. SMS based applications have been designed to manage commercial transactions needed for online banking and other kind of related services.

SMS are carried by the GSM network to Service Centres which coordinate the delivery of messages to and from their destinations. The centres can be functionally separate from the GSM network and more than one service centre can coexist on the same network.

To prevent the corruption or the interception of short messages, they are encrypted over the radio path between the Mobile Station and BSS using the $A5$ encryption algorithm. As showed before, the $A5$ algorithm is sufficient to control the security of messaging for normal use, however an additional end-to-end encryption layer is advisable between the sending short message subject and the receiving short message entity for critical applications such as mobile banking.

3   Background

The authentication techniques have the purpose to provide to two parties wanting to communicate, a certain degree of assurance about their reciprocal identity.

Among the main properties of identification protocols are the *completeness*, i.e., a honest party A is able to authenticate itself to the verifier B who accepts A's identity, and *soundness*, i.e., any party C different from A cannot carry out the protocol and let B to accept it as A's identity [MvOV97]. In this section we briefly review the Kerberos protocol and the one-time password scheme which we use to devise our protocol. Both are authentication techniques widely used to protect systems and data from improper access.

3.1   The Kerberos Protocol

Kerberos has been developed by MIT in 1988 [SNS88] and several organizations has adopted its protocol to provide distributed authentication services. Since then a number of modifications have been proposed as well as a large number of analysis have been conducted about its security features to recognize its limitation and weaknesses [BM90]. Currently, numerous applications (such as Unix and Windows 2000 Login, Telnet and SSH) incorporate Kerberos-based authentication services to provide secure connection mechanisms [KHDC01]. The basic Kerberos protocol involves *Alice* and *Bob*, which are the two parties wanting to communicate and an online server, the *Key Distribution Center* (KDC) which plays the role of the trusted authority. A and B share with the KDC secret keys distributed on a secure channel. The KDC provides two services: an authentication service with which each party can verify the identity of the other party; a ticket granting service which serves to establish a common shared secret between the two parties. Here is a sketch of the messages exchanged among A, B and KDC using the following notation:

- A e B are *Alice* the prover and *Bob* the verifier
- KDC is the Trusted Authority
- K_A, K_B are the secret keys which *Alice* and *Bob* share with the KDC
- K_{AB} is the session key generated by the KDC for the communication between *Alice* and *Bob*
- ID_A e ID_B are *Alice* and *Bob*' identifiers, respectively;
- E is a symmetric ciphering algorithm;
- N_A is a nonce calculated by A;
- T_A is a timestamp calculated by A;
- L is if the lifetime, indicating the validity period;

1. $A \rightarrow KDC : (ID_A, ID_B, N_A)$
2. $KDC \rightarrow A : (E_{K_A}\{ID_B, K_{AB}, N_A, L\}, ticket_B)$
3. $A \rightarrow B : (authenticator, ticket_B)$
4. $B \rightarrow A : (E_{K_{AB}}\{T_A\})$

where $ticket_B = E_{K_B}\{K_{AB}, A, L\}$ and $authenticator = E_{K_{AB}}\{T_A, ID_A\}$

The Kerberos protocol has been already integrated within smartcard hardware, moving part of the decryption and encryption operations on the ticket in a processor based smartcard [IHA99].

3.2   One Time Passwords

Authentication systems based on one time passwords have been developed firstly by Lamport [Lam81]. Such schemes eliminate any attempt of eavesdropping, since once a password is used, the same password is never used again during the successive authentication sessions. Such schemes are based on a sequence of passwords which can be generated or calculated according to a previous chosen algorithm.

In Lamport's scheme, each user A chooses an initial seed w and a one-way hash function H to calculate the password sequence $w, H(w), H(H(w)), ...,$ $H^t(w)$, where t is a fixed constant defining the number of identifications to be allowed. The verifier B is given the initial token $w^0 = H^t(w)$. For any authentication request k, $1 \leq k \leq t$, the user sends the k-th password $w^k = H^{t-k}(w)$. The scheme realizes a challenge-response protocol, where the challenge is given by the current position in the password sequence. It is worth to notice that it is impossible to calculate the next valid password w^k, knowing w^{k-1}, since H is a non invertible function. The protocol has two phases:

1. Onc time set-up: $A \rightarrow B : w^0 = H^t(w)$
2. Successive authentication : $A \rightarrow B : (ID_A, i, w^i)$

The verifier B needs to maintain a counter i_A holding the number of the visits received from A, and each time he receives the authentication request from A, he must control that $i = i_A$ and that $H(w^i) = w^{i-1}$. In this case it authenticates A, updates the value of the counter $i_A = i_A + 1$ and stores the new token w^i to be used in the next authentication.

4   The Kerberos-One-Time Protocol

The protocol we present here is the integration of the techniques illustrated above. The starting point is the Kerberos protocol and its possible adaptation to its use in the GSM network. A viable implementation of Kerberos however must take into account the particular characteristics of the parties wanting to communicate. In a typical m-commerce scenario, *Alice* and *Bob* are a client ad a server, respectively, which want to use the network provided by a particular network provider to communicate: *Alice* is using her mobile phone to set out a secure communication with *Bob* in order to finalize a transaction. In this context, the network provider plays the role of the trusted authority, with which both parties share a secret key. An efficient implementation of Kerberos must ensure a number of features:

- be as transparent as possible for the user
- limit the number of exchanged messages
- not compel users to remember long passwords to access the services

The first feature is obtained by transferring as much as possible the control of the operations to the software running on the smartcard embedded in the mobile phone. In this way we let the user remember only the Pin which authenticates himself with respect to the SIM, satisfying the third requirement too.

The second feature is obtained by using a one time password scheme to perform authentication with respect to the server *Bob*, once that a session key is shared. In this way, A and B must not interact with the KDC after the first time, and can use the shared password chain for the successive identifications.

4.1   Initialization

The first phase involves *Alice*, *Bob* and the trusted authority which plays the role of the Key Distribution Center. Both users, *Alice* and *Bob*, share a secret key with the KDC. At the end of the interaction, *Alice* and *Bob* have a shared session key, which they can use for the successive communication.

1. $A \longrightarrow KDC : (ID_A, E_{K_A}\{ID_B\})$
2. $KDC \longrightarrow A : (E_{K_A}\{ID_B, K_{AB}, w\}, ticket_B)$
3. $A \longrightarrow B : (ID_A, E_{K_{AB}}\{ID_A, w^0\}, ticket_B)$
4. $B \longrightarrow A : (ID_B, E_{K_{AB}}\{w^0 + 1\})$

where $ticket_B = (E_{K_B}\{ID_A, ID_B, K_{AB}, w^0, T\})$.
Let us analyze the protocol in more detail.

Message 1: When *Alice* needs to contact *Bob*, she sends a request to the KDC holding the identifier of *Bob*, ciphered with the key K_A she shares with the KDC.

Message 2: On the reception of the message, the KDC constructs the answer message for A, which is composed of two parts. The first one is ciphered with the key shared with *Alice* and holds the session key and the initial token w randomly generated. The second part is a ticket which is ciphered with the key shared with B and contains the identifiers of both A and B, the generated session key, the token $w^0 = H^t(w)$ and a timestamp.

Message 3: When A receives the answer from the KDC, she must decipher the first part of the message, using his shared key, to retrieve the session key, and then use it to encode the message to B. This will contain the identifier of A, the token w^0 ciphered with the new session key, and the ticket *Alice* had received from KDC. A can calculate w^0 since she knows the hash function H and the initial token w contained in the previously received message (we let t, the number of allowed identification, be a fixed parameter of our protocol).

Message 4: When B receives the message from A, he deciphers the ticket to verify that the identifiers match and retrieves the session key and the token w^0. Using the session key, he can decipher the first part of the message and control that the values contained for the identifier and w^0 match with the ones contained in the ticket. B must control also that the couple (ID_A, w^0) has not been previously used to authenticate A during the validity of the timestamp contained in the ticket. Furthermore he must control also the validity of the timestamp T, controlling that the current time T_c is contained between the starting and the expiration time values reported by the timestamp T. In the positive case, he can construct the answer to A to confirm the successful authentication, encrypting the message with the freshly generated session key.

4.2 Successive Authentications

One of the aim of the developed protocol, is to reduce as far as possible the number of messages exchanged among the involved parties. In particular we want to allow any user to authenticate with a remote server without having every time to make a request to the KDC. After the initialization phase, A and B can use the session key to communicate and authenticate exchanging the next password contained in the sequence. As usual in one time password based authentication schemes, the i-th authentication is done by sending the message containing the i-th value in the password chain:

1. $A \longrightarrow B : (ID_A, E_{K_{AB}}\{i, w^i\})$

Each time B receives an authentication request, he must decipher the message using the session key and control that $i = i_A$ and $H(w^i) = w^{i-1}$ holds.

The ticket exchanged in the initialization phase allows A and B to agree on the token w^0 representing the first valid password. Before the last authentication, our protocol provides a regeneration phase which allows the two parties to agree on a new password sequence, making it possible to continue the exchange of authenticated messages between A and B, without involving again the KDC. Within the lifetime of the session key (which can be a fixed parameter of the protocol) it is possible for A to generate a new initial token $\bar{w}$ and calculate the token $\bar{w}^0 = H^t(\bar{w})$ which will be sent to B and used for a new authentication chain:

1. $A \longrightarrow B : (ID_A, E_{K_{AB}}\{i, w^i, \bar{w}^0\},)$

Using the same mechanism, it is also possible for A to agree with B on a new session key, substituting the older one each time an authentication chain has been consumed or its lifetime has expired. In this case, however the communication sessions are not independent.

To calculate a fresh valid token, we use the J-bit Cipher Feedback procedure, which is one of the AES operation modes [DR00]. The procedure uses an *Initialization Vector IV* of 128 bit, which is divided into two parts containing j and $128 - j$ bits, respectively. After having ciphered IV (using the RC6 algorithm), it calculates the XOR between the leftmost part of the resulting bit vector with the j bits of the parameter which must be regenerated. The resulting j bits are part of the regenerated parameter, which replace also the j bits of the IV vector after shifting it j positions on the left. In this way, the feedback makes it possible to have a different input for the RC6 function at each iteration.

In our protocol, we use two input sequences *Input1* and *Input2* entered by the user during the regeneration phase. To ensure the randomness of the digits chosen by the user, the input sequences are XOR-ed with the session key shared with KDC. In this way, weak sequences, all 0s or 1s sequences, do not alter the security of the generated parameter.

4.3   Structure of Messages

The messages exchanged between *Alice* and *Bob*, have different formats according to the different phases of the protocol execution. In particular we distinguish four types of messages:

- *Authentication Request*: message sent by *Alice* the first time she requests to communicate with *Bob*;
- *Generic Authentication*: message sent each time *Alice* wants to exchange data with *Bob*,
- *Update Token*: message sent by *Alice* when she wants to renew the password sequence shared with *Bob*.
- *Update Session*: message sent by *Alice* when she wants to renew the session key shared with *Bob*.

According to the different kind of messages, the content of the message and the action that *Bob* has to take are different. To distinguish among the different type of messages, we structure the messages as follows:

$$(Code, Content, Data)$$

The *Code* field defines the type of the received message, containing different codes for the different message types. The *Content* is the part which carries the data for the authentication, while the *Data* part holds the effective content of the messages. Both the *Data* and *Content* parts are ciphered with the shared session key. For example the format of a generic authentication message is the following:

$$(C_2, (ID_A, E_{K_{AB}}\{i, w^i\}), E_{K_{AB}}\{Data\})$$

where C_2 is the code for the authentication request and i is the current index in the password chain. The format for the message of token update when *Alice* have requested $t - 1$ authentications is the following:

$$(C_3, E_{K_{AB}}\{t - 1, w^{t-1}, \bar{w}^0\}, E_{K_{AB}}\{Data\})$$

where C_3 is the code for the requested operation type, and $\bar{w}^0$ is the fresh calculated token.

4.4   Protocol Security

Being derived from the Kerberos and the One-Time protocols, our proposal combines the advantages of both authentication schemes and overcomes some of the drawbacks that the original protocols suffer with. One of the pitfall of the Kerberos protocol was in its reliance on passwords chosen by users. Password guessing attacks are successfully exploited to break the security of Kerberos-based systems when users choose poor passwords. Furthermore during the authentication phase of the Kerberos protocol, plaintext and ciphered text pairs travel on the network, making it easy for an eavesdropper to recover them and exploit cryptanalysis techniques.

In our protocol, the passwords are not chosen by the user, but are random sequences of bits stored on the smartcards. Furthermore the encryption and decryption of messages occur within the smartcards, avoiding the possibility for a malicious eavesdropper to retrieve plaintext and ciphered text associations. All the traffic between the mobile and the other party is ciphered, so that the interception of a message does not give any information to a malicious eavesdropper.

The protocol is also resistant to replay attack. In this kind of attacks, the adversary intercepts the traffic between the parties, trying to replay the message to get the identity of one of the party. The interception of the ticket request from A to the KDC or the response from the KDC to A, does not help the malicious adversary, since the messages are ciphered with the secret key held by A. The same holds also for the messages exchanged between A and B. The interception and the deletion of a message do not influence the security of the protocol, but they compel A to renew the request of credentials to the KDC, resulting in an additional message to re-initiate the communication with the other party.

The communications between A and B are ciphered using the session key which is established during the initialization phase with the help of the KDC. The use of this key should be restricted to a short time period, avoiding in this way the exposure to cryptanalytic attacks and limiting the quantity of data compromised if the key is broken [MvOV97]. In our case we allow the regeneration of the hash chain during the lifetime period of the session key. Moreover, we allow also the periodic regeneration of the session key (for example, when the hash chain is consumed), such that the amount of data that an adversary can use for a cryptoanalytic attack is reduced. The duration of the lifetime of the session key as well as the number of allowed regeneration phases must be evaluated with respect to the level of security that the protocol would like to ensure.

5   Implementation

The protocol we presented has been developed to be implemented over GSM networks. Hence we assume that the participants to the interactions are two users wanting to exchange confidential information using a mobile GSM phone, and the network operator playing the role of the trusted authority or KDC. We assume that every user connected to the mobile network of his provider, can request to the KDC the access to the additional services with an higher level of security.

Differently from Kerberos, the secret key shared between any user and the KDC is not calculated after the user login, but is stored on the smartcard which the user bought from the network operator. The key stored as a random sequence of bits is part of the personal data which are stored on the SIM at the moment of the purchase, such as the unique telephone number and the credit. The user authenticates to the SIM by using the traditional Pin insertion, while the SIM authenticates to the mobile network executing the standard procedure provided by the GSM protocol. Only when the additional authentication service is requested, the SIM manages all the phases and the interactions between the user and the KDC, and the user and the other server.

The ticket that in the Kerberos interactions could be used more than one time within its lifetime (established by the timestamp), is used just once during the initialization phase, and the timestamp does not permit its reuse. We replaced DES with RC6 in CFB mode as ciphering function, such that a 128 bit long key can be used. Furthermore, this choice is motivated by the easiness of implementation of this algorithm (w.r.t to DES) in limited machine as regards both memory and computational power, as smartcards are. The result of these implementation choices are a better security level due to the length of the used key and to the fact that all the encryption and decryption operations are performed within the smartcards. In the successive releases, we are planning to substitute the MD5 function with an implementation of SHA-1 and RC6 with AES [MvOV97,DR00]

We used GSM Javacard as target platform for the implementation of a prototype application based on our protocol. As software development framework we used the Sun's Javacards Development Toolkit [Sun99]. The prototype is still in development. For the moment we restricted our research to the development of the basic functionalities needed to implement the interactions during the execution of the protocol.

5.1   Gsm Javacards and Sim Toolkit Applications

GSM Javacard are smartcards that are compliant to the GSM API and include also a Java Virtual Machine such that incremental development of both applet and framework is possible. The main benefit of such approach, is that developers are provided with a higher level API which makes it easier programming smartcard applications, by relying on the Javacard classes. Furthermore, the Javacard framework provides the usual benefits of standardization and portability coming from the Java approach.

In Figure 1 the architecture of a Gsm Javacard is showed. The GSM Phase 2 SIMs contain the SIM Application Toolkit, which defines a set of additional standard functionalities enhancing the interaction mechanisms between the SIM and the mobile handset and the SIM and the network. GSM Javacards include the *SIM Toolkit Framework* which is composed of several components, which the most important of is the *JavaCard Runtime Environment*, responsible for the selection of the applets and the interaction with the mobile handset (through APDU commands). The *Toolkit Handler* and the *Toolkit Registry* handle the commands for the SIM Toolkit and store the data relative to the toolkit applet such that they can be accessed by the JCRE. The *File System* is the JCRE object which manages the file system of the smartcard and controls the access from the applets.

On a GSM Javacard different kinds of applets are stored. The *GSM Applet* is the first one selected after any reset of the card, while the *Loader Applet* manages the installation and disinstallation of the applets. The other two categories of applets, the *Toolkit* and the *Javacard Applet* are connected to the applications

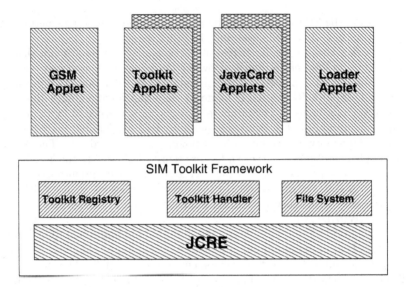

Fig. 1. The architecture of a GSM Javacard

running on the card. The Toolkit applet manages high level events, such as the interaction between the user and the SIM, while the Javacard applets elaborate the commands coming from the mobile handset. Furthermore a *Shareable Interface* is used to realize intercommunication among the applets being executed and the shared common objects.

When a Javacard is inserted into a mobile handset, the communication between the smartcard reader and the card itself occurs according to a specific protocol, called *Application Protocol Data Unit*, which encodes the different commands used to control the execution of the applications stored on the card. In a Javacard, the JCRE is responsible for the selection of the applet, and for the forwarding of the APDU commands to the applet currently in execution. Each applet can process the received APDU command, decoding the data and replying with an output containing the result of the performed operation (*status word*).

5.2   Design of the Basic Applets

For the design of the prototype application, we developed different applets which perform the basic operations needed by the proposed protocol. In particular three applets have been developed, implementing three basic functionalities requested in the different phases of the protocol:

MD5: implementing the MD5 hash function;
RC6: providing ciphering/deciphering of input messages according to the RC6 algorithm;
KOT: the main applet with the task to construct the SMS messages;

The first two applets implement basic cryptographic functions. The main applet has the task to recognize the type of the message needed in the current phase of the protocol, and to construct the needed SMS opportunely. The applet reads from a stored table *ServerTable*, the data relative to each server with which a previous authentication session has been already performed. Each entry in the table contains the server's identifier, the session key, the initial secret token, the counter for the number of authentication session already performed, and the number of utilization of the session key. The main method *Genrequest(Server)*, has in input the identifier of the server which the authentication request is directed to. The output of the method is a SMS message containing:

- the data for the successive authentication, if the server's identifier is contained in the *ServerTable*;
- a ticket request to the SMSC (playing the role of the KDC), otherwise.

In the first case, the message if ciphered using the session key shared with the server, otherwise the shared key with the SMSC is used.

The simulation environment offered by the Javacard Development Kit, simulates the execution of the JavaCard applet as running in the ROM of the smartcard. It provides two tools: the Javacard Working Desktop Environment and the APDU Tool. The first constitutes the environment where the applets are executed. It accepts in input a configuration file specifying all the applets registered and their relative data. The APDU Tool accepts in input a script file containing the APDU commands which are sent to the JCWDE for execution. The output is the status word plus possibly other additional data, which the applets return after the execution of the specified APDU command.

6 Conclusion

Integrating the features of the original Kerberos protocol, and of one-time password schemes, we presented a protocol which enhances the authentication mechanisms adopted over the GSM network. Authentication and confidentiality are two of the most important requirements of applications for mobile commerce. Our protocol can be easily used to finalize little commercial transactions among users interacting by means of mobile GSM phones. Indeed, the protocol has been implemented using the Javacard platform on a GSM Javacard.

The integration of Kerberos with smartcards has been discussed in [IHA99]. A number of advantages are coming from the use of tamper-resistant hardware, solving the problems that traditional implementations of Kerberos suffer with. The prototype we presented is relative to the user view of the system. We plan to complete the implementation realizing a whole infrastructure realizing all the interactions among the involved parties. In this way it should be possible to evaluate also the performance of the protocol with respect to other authentication techniques. Furthermore, the system could be extended to allow access also to other hardware platforms, other than mobile phones, realizing complex and interacting e-commerce infrastructure.

Acknowledgment

We would like to thank prof. A. De Santis and prof. C. Blundo for many useful conversations on the contents of this paper. Many thanks go to L. Di Maggio who worked on this project for his master thesis. We also thank the anonymous referees for their helpful comments and suggestions.

References

BM90.   S M Bellovin and M Merritt. Limitations of the kerberos authentication system. *Computer Communication Review*, 20(5):119–132, 1990.

BSW00.   A. Biryukov, A. Shamir, and D. Wagner. Real-time cryptanalysis of a5/1 gsm on a pc. In *Proceedings of the Fast Software Encryption Workshop*, New York City, USA, 2000.

DR00.   J. Daemen and V. Rijmen. The Block Cipher Rijndael. In J.J. Quisquater and B. Schneier, editors, *Smart Card Research and Applications*, volume 1820 of *Lecture Notes in Computer Science*, pages 288–296. Springer-Verlag, Berlin, 2000.

IHA99.   Naomaru Itoi, Peter Honeyman, and Ann Arbor. Smartcard integration with kerberos V5. In *Proceedings of USENIX Workshop on Smartcard Technology*, pages 51–62, Chicago, May 1999.

Ins99a.   European Telecommunications Standard Institute. Digital cellular telecommunications system (phase 2+); gsm release 1999 specifications (gsm 01.01), 1999.

Ins99b.   European Telecommunications Standard Institute. Digital cellular telecommunications system (phase 2+); specification of the sim application toolkit for the subscriber identity module - mobile equipment interface (gsm 11.14), 1999.

Ins99c.   European Telecommunications Standard Institute. Digital cellular telecommunications system (phase 2+); specification of the subscriber identity module - mobile equipment interface (gsm 11.11), 1999.

KHDC01.   O. Kornievskaia, P. Honeyman, B. Doster, and K. Coffman. Kerberized credential translation: A solution to web access control. In *Proceedings of the 10th USENIX Security Symposium*, Washington, D.C., USA, 2001.

Lam81.   Leslie Lamport. Password authentication with insecure communication. *Communications of the ACM*, 24(11):770–771, 1981.

MvOV97.   Alfred J. Menezes, Paul C. van Ooschot, and Scott A. Vanstone. *Handbook of Applied Cryptography*. Boca Raton, 1997.

Rad.   Radicchio Organization. Radicchio: The global initiative for wireless e-commerce. http://www.radicchio.org/info_center/faqs_radicchio.asp.

SNS88.   Jennifer G. Steiner, B. Clifford Neuman, and Jeffrey I. Schiller. Kerberos: An authentication service for open network systems. In *Proceedings of the USENIX Winter 1988 Technical Conference*, pages 191–202, Berkeley, CA, 1988. USENIX Association.

Sun99.   Sun Microsystems. Java Card 2.1 Application Programming Interface, 1999.

Secure Authorisation Agent for Cross-Domain Access Control in a Mobile Computing Environment

Richard Au[1], Mark Looi[1], Paul Ashley[2], and Loo Tang Seet[1]

[1] Information Security Research Centre
Queensland University of Technology,
Brisbane, Qld 4001, Australia
{w.au,m.looi,l.seet}@qut.edu.au
[2] IBM Software Group – Tivoli
11400 Burnett Road, Austin, Tx, 78758, USA
pashley@us.ibm.com

Abstract. New portable computers and wireless communication technologies have significantly enhanced mobile computing. The emergence of network technology that supports user mobility and universal network access has prompted new requirements and concerns, especially in the aspects of access control and security. In this paper, we propose a new approach using authorisation agents for cross-domain access control in a mobile computing environment. Our framework consists of three main components, namely centralised authorisation servers, authorisation tokens and authorisation agents. An infrastructure of centralised authorisation servers and application servers from different domains is proposed for supporting trust propagation to mobile hosts instantaneously. While the authorisation token is a form of static capability, the authorisation agent on the client side can be regarded as a dynamic capability to provide the functionality in client-server interactions. It works collaboratively with remote servers to provide authorisation service with finer access granularity and higher flexibility.

1 Introduction

The rapidly expanding technology of wireless communication, wireless LANs and satellite services will make information accessible at any time and anywhere around the world. In such mobile computing environment, users carrying portable computers and personal digital assistants (PDAs) equipped with wireless connections are no longer required to maintain a fixed and universally known position in the network. The general abstract view of a mobile system consists of mobile hosts interacting with the fixed network via mobile support stations. The connection between a mobile host and a mobile supporting station can be via a wireless link. As the infrastructure to support mobile computing includes not only the mobile computers themselves, but also stationary computer networks,

K. Kim (Ed.): ICICS 2001, LNCS 2288, pp. 369–381, 2002.

related issues of client/server, network control and distributed information management have to be addressed.

Since information and services will be accessible from virtually any place and time, they will be stored in a highly decentralised and distributed infrastructure. A wide variety of service providers will be accessible to mobile computers. While mobile computing lacks physical protection mechanisms as in traditional fixed networks, network access points are not necessarily under the control of the same administrative authority. Thus a new set of inter-domain access control mechanisms is needed to allow mobile users to perform security operations across different administrative domains, which may have different, or even contradictory, security policies and constraints. The approach adopted by current distributed security architecture attempts to hide details of underlying distributed system from client applications and provide transparency to users. This approach works adequately when the characteristics of the underlying system are relatively static. However, when the levels of service, which can be provided, are subject to rapid and significant fluctuations, as is the case of mobile environment, the approach starts to break down.

As a mobile user migrates throughout an internetwork, he periodically pops up in a new foreign domain. The design of mobile computing system across different security domains should have the following criteria [5]:

- Domain Separation - Interactions between users and systems in different domains result in interflow of data and information. This adds an additional dimension of security to the maintenance of the integrity, accuracy and confidentiality of internal system information. It is essential that domain-specific sensitive information should not be propagated across domains when a user is accessing multiple services in different domains.
- Transparency to Users - Authentication and authorisation in foreign domains should have minimal impact on the user interface with respect to that in home domain.
- User Information Confidentiality - All sensitive information about the user such as user's secret key, privileges and even identification data should be protected from disclosure in the foreign domain environment. Sometimes it is desirable to keep both the movements and current locations of mobile users secret as well.
- Minimal Overhead - Mobile terminals are usually less powerful than the stationary computers and supporting resources are limited. In particular, the bandwidth and reliability of wireless connections are lower than those of wired connections. Hence, the number of messages exchanged between users and various domain authorities should be kept minimal in the authorisation process.

The contribution of this paper is in the development of an authorisation framework for mobile users across different domains. While maintaining strict separation of security domains, our authorisation scheme consists of three main components:

- Infrastructure of domain-based authorisation servers for trust distribution
- Authorisation token as static capability
- Authorisation agent for dynamic access control enforcement

This paper begins with an introduction and a description of some related previous work. In section 3, the extension of our authorisation scheme is introduced. In section 4, the concept of the secure authorisation agent is explained. In section 5, the prototype using personal secure device is discussed. In section 6, some prototype interactions between the user and application server is illustrated. The paper finishes with our future work and conclusions.

2   Some Previous Work

In [2], we have proposed a new authorisation scheme with distinctive features highlighted in the following subsections.

2.1   Domain-Based Administration

There are two levels of centralisation in our authorisation scheme:

- Centralised management of users and applications using security server, which consists of authentication server and authorisation server
- Centralised access control enforcement using each application server

Centralised Authorisation Server. In our approach, with the strategy of fully centralised control for an administrator, an authorisation server is designed in each domain to facilitate the following functions :

- To manage the access control rules and application/resource information within its local domain.
- To create privilege credentials, in the form of authorisation tokens, for accessing its local applications or resources and then issue them to either local or external authenticated users in the first access request (registration).
- To administer local users by sending updated authorisation information to local application servers and foreign authorisation servers.
- To build secure communication channels with foreign authorisation servers and establish mutual trust as well as mappings of security policies across different domains.

Authorisation Managers in Application Servers. While applications and resources are distributed over multiple domains, we propose to attach an authorisation manager to each of the application servers to facilitate authorisation services. This authorisation manager follows the instructions from the central authorisation server through a trusted communication channel. When a user makes an access request by submitting an authorisation token to the application server, it is the role of the authorisation manager to handle the token and implement the access control. It makes decisions whether to grant or reject the access request according to the user's privileges submitted and access control information at the resource.

2.2   One-Shot Authorisation Token

Traditional methods of authorisation directly depend on reliable user authentication. In our approach, we separate authorisation from authentication by splitting authorisation into two stages:

- Stage of user registration - Granting rights and/or privileges (authorisation credentials) to an authenticated entity;
- Stage of access control enforcement - Using the privileges in combination with access decision rules at the resource to determine if access should be granted to the entity.

We argue that the authentication of a user is essential in the first stage of authorisation but it is no longer needed in the second one. Our scheme allow users to directly access resources securely with the pre-loaded authorisation credentials and eliminate the repeat of user authentication in every access as in the traditional mechanism.

We use the one-shot authorisation token to hold the privileges for each individual user on the network system. The token is valid for once only and it has to be renewed by the application server at the end of each access and returned to the user for next use. This technique provides a mechanism for revocation or updating user's privileges dynamically. Also security can be enhanced as forging and replaying do not work. These authorisation tokens are protected using cryptography technology and they are stored in personal portable secure devices, e.g. smart card, to provide user mobility and security.

3   Development for Mobile Computing Environments

We further develop our authorisation scheme in [2] by introducing the concepts of infrastructure of authorisation servers and client-side authorisation agents in order to cater the needs in mobile computing environments.

3.1   Infrastructure of Authorisation Servers

Referring to figure 1, the authorisation servers from different domains can be interconnected over the public Internet or WAN to form an infrastructure for providing authorisation services across multiple domains. The 'web' of authorisation servers is basically domain-centred, that is, each authorisation server can develop its own web of authorisation by interconnecting directly or indirectly with other authorisation servers in different domains. Of course, the connections require trust establishment and translation/mappings of security policies and functionality among domains involved. This infrastructure becomes a flexible and extensible backbone support for mobile users migrating across different domains. A realistic example is the virtual enterprise environment[3].

In each administrative domain, there are one or more mobile supporting stations to provide physical connecting points on the fixed network for both local and foreign mobile hosts. In general, they may be public and accessible for all users.

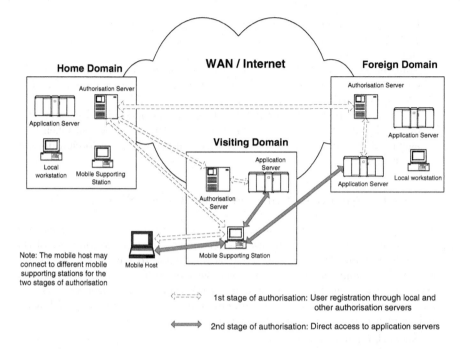

Fig. 1. Cross-Domain Authorisation in Mobile Computing Environment

3.2 Static and Dynamic Capability Approaches

In our proposed scheme, the authorisation token is a form of static capability. It is delivered to the client through the infrastructure of authorisation servers in the stage of user registration, which includes user authentication. When the client makes an access request to the application server using his authorisation token, an authorisation agent will be downloaded from the application server to the client side. It can be regarded as a form of dynamic capability. The authorisation agent executes on the client side and it works collaboratively with remote application servers to provide authorisation services with finer access granularity and higher flexibility.

Using the capability approach to provide authorisation services, we identify the following advantageous features especially in mobile computing environments:

– Mobility and Interoperability - Using the capability-based approach, authorisation tokens provide the mobility for users across multiple platforms in heterogeneous environment. A user can carry these authorisation tokens with some personal portable secure device, e.g. smart card, in a mobile host and make access request from any visiting domain.

- Direct access to application - By eliminating the traditional authentication with user identity every time, the overhead for making access to applications can be reduced.
- Multiple access activation - The authorisation tokens are tailor made by security server in individual domain. Tokens from different domains can be used simultaneously to activate multiple applications.
- Anonymity Support - Depending on the design for individual domains, it is optional to include user identity information in the authorisation tokens and agents. So it can maximize the user information privacy in foreign domain environment.

4   Secure Authorisation Agent on Client Side

Mobile computing systems allow users to access and distribute information at any time and anywhere. Mobile users are moving and thus the topology of the system is dynamically changing and it is impossible to have a static link between the user and the application server. Very often, it is difficult for both the server and the client to foresee or manage the types of heterogeneous systems that they will interact in cross domain environment. For example, the following environment contexts are subject to change:

- Network quality of service
- Capabilities of its supporting hardware

In recent years, many mobile code languages, e.g. Java and Telescript [12], have been developed. At the application level, code mobility is the capability to reconfigure dynamically the binding between the software components of the application and their physical location with a computer network. In the mobile agent paradigm, a computational component together with its state, the code and some resources required to perform the task are moved from one remote site to another.

In the design of our security architecture, we employ the concept of mobile agent in providing authorisation services for a large scale distributed system. In our scheme, after the user has submitted an authorisation token, an authorisation agent is downloaded from the application server to the client's workstation. It can be regarded as a trusted representative of the application server and it works collaboratively with the servers for the enforcement of access control. Its tasks include:

- Handling authorisation credentials
- Exchanging environmental contexts between client and server
- Providing access control and resource information to clients
- Keeping authorisation states of the workflow in the access

An authorisation agent differs from traditional client programs in several ways:

- Its life span varies, it may be valid for one access session only.

- It is tailor made by individual application server for different users or groups.
- It can be delivered on demand and installed by remote server in an automated way at the start of the access session.
- It can maintain a high degree of trust with the remote server through the trust link established in the infrastructure of security servers, provided that the security of the agent can be assured on client side.

4.1   Enhancements to Authorisation Services

In the mobile computing environment, the approach using authorisation agents enhances the functionality in existing access control systems in the following ways:

- User Customisation - A mobile agent provides better support for heterogeneous environments as it can express more the user's context and provide more interactions with the servers. In general, the functionality and robustness of access control can be enriched with an user-oriented authorisation agent on the client side. Authorisation service can be regarded as a specific function for individual user. The authorisation agent can decide dynamically what kind of facilities should be augmented for that specific user in each access session.
- Dynamic extension of application functionality - Application server can offer basic APIs and export them via the bindings of the agent environment for exploitation. The authorisation agent is tailor made specifically for that application and provides a powerful tool for the user to access the application server in the way he needs. For example, if a user wishes to browse the directory of resources available to him, he can make use of the agent to organise and set up the service for him.
- Decentralised preservation of authorisation state - In some distributed network environments, such as the Internet, the reliability would become higher if the operation can be continued even when the connection between the client and the server is temporarily cut off. For that purpose, the data of the interaction state needs to be preserved. Generally, the client takes the responsibility to preserve the state under the environment where only stateless servers are supported. The ability of the authorisation agent to carry its authorisation state around the network increases the reliability and flexibility in mobile computing environments.
- Efficient remote communication and performance optimisation - Once an authorisation agent is transported to the client side, it can raise the granularity level of the services for access control. Now a single interaction between authorisation agent and server is sufficient to specify a high number of lower level operations, which can be performed locally on the client side without involving further communication over the physical link. This self-containment results in reduction of network traffic and is actually immune from partial failure. An autonomous component encapsulates all the state involving the distributed authorisation, and can be easily traced, check-pointed, and possibly recovered locally, without any need for knowledge of the global state.

– Multi-application interactions - By using multiple authorisation tokens, a user can gain access to multiple applications, which may be in different domains, at the same time. It is possible to develop protocols for the authorisation agents to interact with one another and work cooperatively on the user's platform. As a scenario, suppose a project manager working in a team in an enterprise needs to access some applications/databases in several corporate companies. Collaborative work, e.g. the exchange of project data, among these companies can be initiated by the manager. The authorisation can be conducted in an automated way by downloading related authorisation agents which acts as representatives of their own companies.

4.2   Security Issues

Protecting the User Host. From the point of view of users, the most obvious concern is the security related to the environment in which the agent is assumed to execute. It is important that the mobile code (authorisation agent) is authenticated, verified and satisfy safety policies before execution on the client platform. There are different approaches used to protect the execution environment: sandboxing, digital "Shrink-Wrap" and Proof Carrying Code [6]. In our framework, we propose a verification code of the authorisation agent to be transferred from the application server to the client at the beginning of the access session. Then he can use it to assure the authenticity and integrity of the authorisation agent before its installation and execution. This verification codes, along with other confidential data, can be stored in secured devices, e.g. smart cards, and used whenever it is needed during the access.

Protecting the Authorisation Agent. Protecting the authorisation agent against malicious hosts and third party attackers is crucial. Some examples of attacks are denial-of-service, service overcharging, private information disclosure, and code/data modification. Protection mechanisms can be aimed at prevention or detection. Prevention mechanisms try to make it impossible to access or modify the agent data and code in a useful way. There are mainly two approaches for protection of mobile agents:

– Hardware Solution - Protection can be achieved by allowing mobile code to move to hosts that are trusted, that is, a trusted computing base is needed on the client side. Some tamper-resistant hardware, e.g. smart card or iButton [4], may be used to ensure the integrity of the runtime environments. It avoids unauthorised modification of code and state by executing the code in a physically sealed environment that is not accessible even by the owner of the system without disrupting the system itself.
– Software Solution - Protection of the mobile agent code can be achieved by setting up restricted environments, deploying cryptographic protocols between the mobile codes and the remote server [9][8]. Detection mechanisms can be developed aiming at detecting, after execution, if some illegal modification of its codes and/or state occurred in the agent lifetime. Cryptographic tracing [11] that, using execution traces and digital signatures, allows the

owner of a mobile code to detect, after termination, every possible illegal modification of the code and state.

In our framework, we propose to use a tamper-resistant hardware providing a trusted environment for the storage and execution of the authorisation agent. At the same time, cryptographic methods can be used to ensure the integrity and confidentiality of the data and authorisation states.

5   Use of Personal Secure Device

In our architecture design as shown in figure 2, we propose to use a personal secure device with computing power, e.g. Java smart card or cryptographic iButton [4], as the trusted user platform for:

- Execution of the agent manager and authorisation agents
- Storage of cryptographic keys, certificates and authorisation tokens

5.1   Central Registry

The central registry is the first program installed onto a new personal secure device before issuance to a user. It is responsible for

- handling the user private keys and certificates, and
- managing the first authorisation token and agent manager.

5.2   Agent Manager

The authorisation agent manager administers the installation and execution of all agents on the client workstation. It acts like a distributed reference monitor on the client side with the following functions:

- To verify the integrity of the authorisation agent in various authorisation stages.
- To monitor the validity of authorisation in different authorisation states.

5.3   Authorisation Agent

For one application server, there may be more than one authorisation agent downloaded to the client side. Those agents that need higher security can be transferred on the secure device and others can be installed on the client's workstation.

Authorisation Token Manager. The authorisation token manager can be regarded as one of the authorisation agents for handling the authorisation tokens on client side. This agent can be encrypted and digitally signed for online download onto the user's secure device over untrusted channel, e.g. the public Internet. Then the codes can be decrypted, verified before installation. In this way, the authorisation token manager can be installed on a new user's secure device

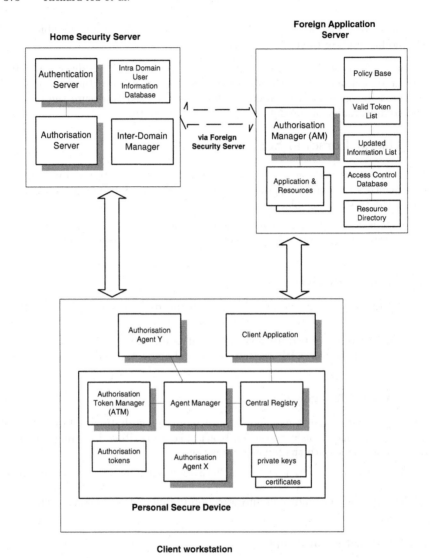

Fig. 2. Overview of Architectural Components

online in the stage of user registration. This flexibility would be an advantage in the management of a large scale distributed system.

The authorisation token manager works cooperatively with the authorisation manager on the application server side. It acts as an intelligent partner trusted by the remote application server in either on-line or off-line mode. This dynamic collaboration can manipulate the authorisation tokens with high security and provide finer granularity of access control.

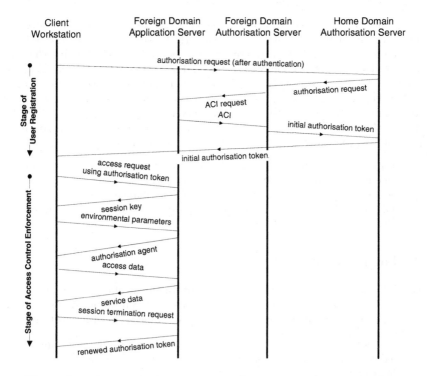

Fig. 3. Cross Domain Authorisation using Authorisation Agent and Token

Since a personal secure device is used to store personal authorisation tokens and private keys, it potentially enables the user to access the application servers from any workstation in any domain. Without lowering the degree of security and efficiency, this user mobility becomes an advantageous feature in today's dynamic world.

6   Interactions between Client and Servers

6.1   Protocols

As shown in figure 3, in order to gain access to an application in foreign domain for the first time, the user has to register through his local authorisation server. After user authentication, the local authorisation server makes an access request on behalf of the user to the authorisation server in the foreign domain. It is assumed that there are some established mappings/translations on trust, security policies and application functions through the infrastructure of authorisation servers. Then the foreign authorisation server communicates with the target application server and creates an initial authorisation token, which is sent to the user via the infrastructure of authorisation servers.

In the stage of access control enforcement, the user can make an access request to the target application server directly by submitting the authorisation token and some environment related parameters to it. After checking the privileges on the token, the application server sends a session key and an authorisation agent to the user to enable the access session. The client can verify the integrity and authenticity of the authorisation agent. Then the authorisation agent starts to operate and interact with the remote server. When the user requests to terminate the session, the application server renews the authorisation token and returns it to the client preparing for the next access. It is highlighted that

- From the viewpoint of the application server, the access process remains the same for both local and external users from other domains. It keeps track of the identity of the authorisation token rather than the user.
- Only one initial user registration is required for multiple direct accesses to the same application later on.

7 Future Work

With the new concept of authorisation agent, the focus of our research is on the security concerns and the agent's functionality. Different applications of authorisation agents can be explored and developed. Some areas of interest are:

- Collaboration of multiple authorisation agents from different domains, e.g. electronic sales agents from different servers can conduct negotiation on the client platform.
- Delegation and chaining of authorisation agents.
- Inter-domain management of authorisation tokens and agents.

8 Conclusions

In this paper, we have presented a framework for distributed systems supporting cross-domain authorisation in a mobile computing environment. It makes use of our proposed mechanism of one-shot authorisation token and authorisation agent to provide flexibility, user mobility and possibly anonymity features. Mobile users can gain access to multiple applications in different domains through the infrastructure of authorisation servers developed.

References

1. Ashley P. & Vandenwauver M. (1999). *Practical Intranet Security : An Overview of the State of the Art and Available Technologies*, Kluwer Academic Publishers.
2. Au R., Looi M. & Ashley P. (2000) *Cross-Domain One-Shot Authorisation Using Smart Cards*. Proceedings of 7th ACM Conference on Computer and Communication Security, pages 220 -227.
3. Au R., Looi M. & Ashley P. (2001) *Automated Cross-organisational Trust Es tablishment on Extranets*. Proceedings of the Workshop on Information Technology for Virtual Enterprises, pages 3-11.

4. Dallas Semiconductor Corp. (1995) *DS1954 Technical Information - Cryptographic iButton*. Available at URL: http://www.iButton.com
5. Molva R. & Samfat D. & Tsudik G. (1994) *Authentication of Mobile Users*. IEEE Network, vol 26, iss 2, pages 26-34.
6. Necula G. & Lee P. (1998) *Safe, Untrusted Agents using proof-Carrying Code*. Mobile Agents and Security, Lecture Notes in Computer Science 1419, pages 92-113.
7. Nicomette V. & Deswarte Y. (1997). *An Authorisation Scheme For Distributed Object Systems*, In Proceedings of Symposium on Security and Privacy IEEE.
8. Oppliger R. (1999). *Security Issues Related to Mobile Code and Agent-Based Systems*. Computer Communications, vol 22, pages 1165-1170.
9. Sander T. & Tschudin C. (1997) *Towards Mobile Cryptography*. International Computer Security Institute (ICSI) Technical Report 97-049.
10. Varadharajan V. (2000). *Security Enhanced Mobile Agents*. Proceedings of 7th ACM Conference on Computer and Communication Security, pages 200-209.
11. Vigna G. (1998). *Cryptographic Traces for Mobile Agents*. Mobile Agents and Security, Springer-Verlag, Lecture Notes in Computer Science 1419, pages 137 - 153.
12. White J.E. (1994). *Telescript Technology: The Foundation for the Electronic Marketplace*. White Paper by General Magic Inc, Sunnyvale CA, USA.

Protecting General Flexible Itineraries
of Mobile Agents

Joan Mir and Joan Borrell

Computer Science Dept., Universitat Autònoma de Barcelona
Edifici Q – 08193 Bellaterra, Spain
Tel: +34 93 5811470, Fax: +34 93 5813033
{jmir,jborrell}@ccd.uab.es

Abstract. Mobile agents are processes which can autonomously migrate from host to host. The migration path followed by an agent can be abstracted for programming convenience into an itinerary. This is composed of tuples of the form (server name, method specification). A flexible structure of itinerary is specified in [24] using sequence, alternative and set entries.

Despite its many practical benefits, mobile agent technology creates significant new security threats from malicious agents and hosts. In order to protect mobile agents, several itinerary protection protocols have been presented. However, they can only be applied to sequence entries.

In this paper, we introduce a protocol for each kind of entry in order to protect general flexible itineraries established in free-roaming mobile agents. Our proposal uses ElGamal cryptosystem and a public-key infrastructure. Nevertheless, any public cryptosystem may be used.

1 Introduction

Mobile agent technologies are a new paradigm to develop applications and new services on internet. These break with the traditional server-client approach and offer a general framework for information-oriented distributed applications. Thus, mobile agents are an effective choice for many applications [8].

Mobility is present with some agents in the real world, where users are represented by agents in an anonymous way, e.g. brokers or insurance agents.

A mobile agent or mobile code consists of (static) code and its current configuration, including global data structures, stack, heap and control information like the program counter, and it is expected to migrate between hosts that offer environments in which the code can be executed [4].

A mobile agent architecture consists of agents and places. A place receives agents and is an environment where an agent executes an action. So far, different architectures have been proposed for implementations. Research labs as well as companies have developed java-based systems for mobile agents, such as Odyssey [12], Voyager [20], Concordia [24], Ajanta [7] and IBM aglets [10].

Itineraries may be viewed as an specification or plan of the agent movements. A community of agents is considered to implement the movement plan correctly, if their observed behaviour corresponds to the plan.

K. Kim (Ed.): ICICS 2001, LNCS 2288, pp. 382–396, 2002.

Following an itinerary, usually fixed at the beginning, the mobile agent goes through a set of hosts to achieve different tasks. All the machines are not necessarily trustworthy. The machines might try to pull sensitive information out of the agent or change the behavior of the agent by removing, modifying or adding to its data and code.

Up to now, all previous papers devoted to itinerary protection (e.g. [2,5,21]) only consider prefixed sequence itineraries. In this paper, we propose the use of a public key cryptosystem (such as RSA, ElGamal or Paillier) in order to protect flexible itineraries from modifications. Moreover, the privacy of local data for each host is achieved. However, other environment problems such as exception treatment or status agent analysis are avoided.

The rest of the paper will proceed as follows: In Section 2, we define the flexible itinerary framework. We discuss previous works to protect itineraries against malicious attacks in Section 3. Later on, in Section 4, we present our proposal where security properties are analyzed. In Section 5, we sum up the results of this paper and introduce further research.

2   Flexible Itinerary Framework

The concept of itinerary was introduced in the Concordia system [24], and improved in [16]. It describes a set of locations to which a mobile agent is going to travel and the work to be accomplished at each location. A location is a host with agent system support. An itinerary is composed of different kinds of entries. The basic form of an entry e_i is made up off by a tuple

$$e_i = (host_i, data_i + methods_i).$$

Entries contain several other entries (recursively):

- A **Sequence** is a list $[e_1, \cdots, e_n]$ with $n \geq 1$ entries defining that the nodes specified by entry e_i $(1 \leq i < n)$ must have been visited before the nodes of entry e_{i+1} are visited.
- An **Alternative** is a list $< e_1, \cdots, e_n >$ with $n > 1$. Only one entry is visited from the whole set of entries. The entry is usually chosen by the mobile agent.
- A **Set** is a list of entries $\{e_1, \cdots, e_n\}$ with $n > 1$ specifying that the elements $e_1, \cdots, e_n$ are handled in any order as long as each element is visited exactly once.

In the following, we show an example of mobile agent itinerary [16] to clarify this definition. Let us suppose that the owner of the mobile agent is planning a romantic evening.

$$I = \{ \ (BestFlowers, buyFlowers),$$
$$< [\ (CentralTheatre, buyTicket), (KingsInn, reserveTable)],$$
$$[\ (ModernArts, buyTicket), (BeefHouse, reserveTable)] >\}$$

With this itinerary, he orders his agent to buy some flowers, a ticket to the theatre, and to reserve a table in a restaurant. The mobile agent can choose between two theatres, and depending on the theatre, it has to reserve an specific restaurant near the theatre. The sequence specifies that the mobile agent first has to go to the theatre to buy a ticket and afterwards go to the restaurant. The main set itinerary allows buying flowers either before or after the restaurant reservation.

If the complete migration path of the mobile agent is known in advance, the owner can insert appropriate entries into the agent's itinerary before dispatching it [7]. On the other hand, other solutions show that the agent is able to migrate to a yellow pages server, where it will search the next destination.

All the machines are not necessarily trustworthy. The machines might try to pull sensitive information out of the agent or change the behavior of the agent by removing, modifying or adding to its data and code. So, flexible itinerary frameworks are vulnerable to attacks. A malicious attacker could try to modify the itinerary framework, adding, deleting or changing the itinerary entries. Also, security issues like data and route privacy or source authentication must be considered. Moreover, a host could try to maliciously reuse (part of) an itinerary of a previous sent mobile agent.

We define a set of general security properties that any protocol has to satisfy in order to protect the itinerary [21,5]. The properties are the following:

P1 Non modification: It should be unfeasible for a host to modify the mobile agent itinerary (including recursive entries).

P2 Data Privacy: No host should be able to obtain/capture the particular data of other hosts.

P3 Forward Privacy: In a sequence itinerary, we should reduce the communication overhead. In such this way, each host is only informed about its predecessor and successor. Remaining route information is hidden in order to prevent traffic analysis and reveal communication relationships. This property is unfeasible with alternative and set entries.

P4 Verifiability: Every host should be able to verify that it forms part of the initial route generated by the agent's owner.

P5 Source verification: Every host should be able to verify that the mobile agent actually comes from previous coded host in the agent route.

P6 Exactly Once: Every host should be able to detect replay attacks. In other words, the host has to detect if the actual agent is a new agent or a copied agent from a previous journey.

With alternative entries, a new security property arises.

P7 Single Entry: Only one host of the list of all specified entries in the itinerary has to be visited.

3 Related Work

Mobile code approach introduces a new security issue: protection of mobile agents from malicious hosts. This new security issue is particularly challenging since it is difficult -if not impossible [3]- to protect an executing program from the interpreter in charge of its execution.

Tamper-proof-devices [23] are hardware-based devices with high security, but not suitable for open systems. There are also software-based approaches that include the computation of encrypted functions and data [14]. However, these can only be applied to polynomial functions.

In the literature, we can find methods that protect agent itineraries against attacks. The methods can be categorized into those which detect and those ones which prevent attacks.

Detection methods are protocols that detect any possible modification of the agent code, application or data. These mechanisms allow the agent's owner to check the integrity of the mobile agent, after agent migration.

- In [6] digital signatures and hash functions are used to protect the computation results established by free roaming mobile agents. The protocols enable the detection of any alteration upon its return, thus provide forward integrity and truncation resilience.
- Paper [19] presents a mechanism based on execution tracing and cryptography that allows us to detect attacks against code, state and execution flow of mobile agents.
- [16] presents a fault-tolerant protocol to ensure the execution of an agent monitoring its execution exactly once.
- A general approach to cope with the integrity of computation problem by supplementing computation results with very short proofs of correctness is presented in [1].

Prevention methods are those which protect the mobile agent from attacks.

- In [2] a sequence itinerary protection method is introduced. The protocol hides the route information of mobile agents using cryptographic tools in such a way that only the agent's homesite knows all sites to be visited.
- [21,22] present a set of approaches. Each one has a different computational complexity in order to protect sequence routes against malicious changes. Two variants with nested signatures guarantee that an attack is detected immediately while others with nested encryptions are suitable to protect the agent's route against attacks performed by the collusion of malicious contexts.
- [5] introduces the concept of enhanced hash chains improving computational complexity in sequence itinerary protection protocols.

$$e_0 \longrightarrow \boxed{host_1, m_1} \longrightarrow \boxed{host_2, m_2} \longrightarrow \cdots \longrightarrow \boxed{host_n, m_n} \longrightarrow e_{n+1}$$

Fig. 1. Sequence Itinerary

4   Our Proposal

Previous papers [2,5,21] only protect sequence itinerary entries with different security approaches. Our goal is to obtain security and flexibility for any itinerary entry type: sequence, alternative and set.

We present a different approach for each kind of entry where all aforementioned security properties are satisfied. We assume that there is a public-key infrastructure and we use ElGamal cryptosystem. Hence, the protocols are proved to be secure if Diffie-Hellman Assumption holds.

Notice that we describe all protocols based on ElGamal cryptosystem, but it is only needed for Alt 2 protocol in Section 4.2. Anywhere else, any other public key cryptosystem can be used, like RSA or Paillier cryptosystem.

Diffie-Hellman Assumption: Let $p = 2q + 1$, for primes p and q, and let m, g be generators of a subgroup of order q. Then, the quadruples (m, m^x, g, g^x) and (m, m^r, g, g^x) are indistinguishable, for random and unknown values $r, x \in Z_q$, $m, g, \in G_q$.

The mobile agent migrates between hosts that offer environments in which the code can be executed. We assume that these hosts, also known as mobile agent systems, use a non-repudiation communication protocol in order to send the agent [25,2].

4.1   Protocol to Protect Sequence Itineraries

In a sequence itinerary, there is a list of ordered entries $e_i = (h_i, m_i)$. The mobile agent has to travel from one location to another one in the specified order.

$$e_0, [e_1, e_2, \cdots, e_n], e_{n+1} \tag{1}$$

Sequence itinerary (1), noted in square brackets "[" "]", begins with e_1 entry and ends with e_n. Without loss of generality, we will always assume that there is at least one entry before and after the itinerary [see fig. 1]. If the sequence itinerary is the main itinerary, the previous entry will be $e_0 = (h_o, -)$ and the last will be $e_{n+1} = (h_o, -)$ where h_o is the mobile agent's owner (o) address with no method to execute.

Protocol Seq.:

1. Initialization Phase: $I_{n+1} = e_{n+1}$
2. For $i = n$ to 1
   (a) Sign: The owner o signs the information of the entry $e_i = (h_i, m_i)$ together with the predecessor and successor host address, and a trip marker t (see below).

$$S_o(h_{i-1}, h_i, h_{i+1}, m_i, t) \tag{2}$$

(b) Encrypt: The owner encrypts the signed information and successor itinerary with the public key of the host h_i.

$$I_i = [E_{y_i}(S_o(h_{i-1}, h_i, h_{i+1}, m_i, t), I_{i+1})] \tag{3}$$

3. Next i

Once the protocol has been employed over an itinerary, a protected itinerary I_0 is obtained.

$$I_0 = [E_{y_1}(S_o(h_o, h_1, h_2, m_1, t), [E_{y_2}(S_o(h_1, h_2, h_3, m_2, t), \cdots)]]] \tag{4}$$

Let us see an example of a small sequence itinerary. This example also allows us to discuss the security properties of our proposal protocol in a general way.

$$[(h_1, m_1), (h_2, m_2), (h_3, m_3)]$$

Last entry (h_3, m_3) is signed with the previous host address h_2, the successor host address h_o (the owner address) and the trip marker t.

$$S_o(h_2, h_3, h_o, m_3, t) \tag{5}$$

With digital signatures, active attacks can be detected. No one will be able to modify this information (Property P1) because it is signed with the private owner's key.

Communications between hosts are not anonymous and we use a non repudiation protocol. So, with the previous host address signed by the owner, every host is able to check the agent's source (P5). If the check fails, the agent should be returned to the owner. Moreover, by using the present host address, the host site will be able to detect if it is part of the initial route (P4).

The trip marker t, usually a timestamp, identifies the agent's journey and prevents replay attacks (P6). Without this value, a malicious host could change the remaining itinerary of the agent with an old mobile agent itinerary sent by the same owner. Only the owner would be able to detect the attack when the mobile agent returns. We could also employ the protocol [16] to solve this problem.

In order to protect the private data for each entry (P2), the signed information and the remaining mobile agent itinerary protected are encrypted with the public key y_3:

$$I_3 = [E_{y_3}(S_o(h_2, h_3, h_o, m_3, t)] \tag{6}$$

The final itinerary (7) is a set of nested encryptions, in the same way as the onion routing approach [17]. Each mobile agent system that receives the mobile agent has to eliminate an encryption layer to continue. Only the right host is able to decrypt the protected itinerary.

$$I_1 = [\ E_{y_1}(S_o(h_o, h_1, h_2, m_1, t), [E_{y_2}(S_o(h_1, h_2, h_3, m_2, t), \\ [\ E_{y_3}(S_o(h_2, h_3, h_o, m_3, t))])])] \tag{7}$$

The owner sends the protected itinerary to the first host, h_1. When the $host_1$ mobile agent system receives the agent, it uses its private key to decrypt the protected itinerary (7). Each site removes its decrypted information from the mobile agent's itinerary and prepares a new itinerary I_2:

$$Host_1 : S_o(h_o, h_1, h_2, m_1, t)$$
$$I_2 = [E_{y_2}(S_o(h_1, h_2, h_3, m_2, t), [E_{y_3}(S_o(h_2, h_3, h_o, m_3, t))])] \qquad (8)$$

The server uses the owner's public key to verify the message signature. Therefore, $host_1$ is able to authenticate that the message was created by the owner. The first three addresses identify the source of the mobile agent h_o, the present host h_1 and the mobile agent destination h_2. This is the minimal information needed by a host. Remaining itinerary information is encrypted to prevent traffic analysis (Property P3). Notice that we can change the onion encryption approach of the entries by a list of ciphertexts (one per host). Nevertheless this list of ciphertexts must include nested keys, in order to satisfy property P3.

Once the mobile agent has executed method m_1, it will migrate to the next host h_2. Host h_2 decrypts the itinerary and checks the security properties.

$$Host_2 : S_o(h_1, h_2, h_3, m_2, t)$$
$$I_3 = [E_{y_3}(S_o(h_2, h_3, h_o, m_3, t))] \qquad (9)$$

After method m_2 execution, the mobile agent migrates to h_3. Only the server h_3 be able to decrypt the itinerary I_3.

$$Host_3 : S_o(h_2, h_3, h_o, m_3, t)$$
$$I_4 = [\,] \qquad (10)$$

Finally, the host h_3 verifies the owner's signature and returns the mobile agent to the owner h_o when the execution method m_3 has been accomplished.

4.2   Protocol to Protect Alternative Itineraries

An alternative itinerary is made up of a list of candidate entries. Only one entry is chosen. Usually, the selected entry is chosen by the mobile agent, but we do not discuss the selection method in this paper.

$$e_0, < e_1, e_2, \cdots, e_n >, e_{n+1} \qquad (11)$$

Alternative itinerary (11), noted in angled brackets "<" ">", begins with e_1 entry and ends with e_n, but only one e_j is visited, where $1 \leq j \leq n$. Previous entry e_0 should be able to send the itinerary to any entry e_j (fig. 2), and successor entry e_{n+1} should receive the mobile agent from the selected entry e_j.

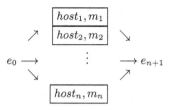

Fig. 2. Alternative Itinerary

We propose three protocols to protect alternative itineraries.

– Protocol **Alt1** encrypts each entry with its respective public key.

Protocol Alt1:

1. Initialization Phase: $I_{n+1} = e_{n+1}$
2. For $i = n$ to 1
   (a) Sign: The owner signs the entry e_i with the predecessor and successor host address, and the marker t.

$$S_o(h_o, h_i, h_{n+1}, m_i, t) \qquad (12)$$

   (b) Encrypt: The owner encrypts the signed information using the public key y_i:

$$x_i = E_{y_i}(S_o(h_o, h_i, h_{n+1}, m_i, t), I_{n+1}) \qquad (13)$$

3. Next i

4. The final protected alternative itinerary I_0 is defined by the set of tuples (h_i, x_i), where $1 \leq i \leq n$.

$$
\begin{aligned}
I_0 = < \ & (h_1, x_1), \\
& (h_2, x_2), \\
& \vdots \\
& (h_n, x_n) >
\end{aligned}
\qquad (14)
$$

Substituting the value of each x_i we obtain a larger route.

$$
\begin{aligned}
I_0 = < \ & (h_1, E_{y_1}(S_o(h_o, h_1, h_{n+1}, m_1, t), I_{n+1})), \\
& (h_2, E_{y_2}(S_o(h_o, h_2, h_{n+1}, m_2, t), I_{n+1})), \\
& \vdots \\
& (h_n, E_{y_n}(S_o(h_o, h_n, h_{n+1}, m_n, t), I_{n+1})) >
\end{aligned}
\qquad (15)
$$

When the mobile agent or the previous mobile system chooses an entry e_j, it knows the destination host address h_j and the new itinerary I_j. The new itinerary framework (16) is as a sequence protected itinerary. Therefore, all security properties (P1, $\cdots$, P6) are accomplished as in protocol **Seq**.

$$
\begin{aligned}
Host &= h_j \\
I_j &= [E_{y_j}(S_o(h_o, h_j, h_{n+1}, m_j, t), I_{n+1})]
\end{aligned}
\qquad (16)
$$

This protocol allows the source agent system to choose an entry without extra computational complexity. However, the itinerary size increases.

$$size(I_0) = \sum_{i=1}^{n}(m_i + size(I_{n+1})) \qquad (17)$$

where m_i is the method and I_{n+1} is the protected itinerary for successor entries $e_{n+1}, e_{n+2}, \cdots$

- Protocol Alt2 assumes that there is one common method m for all entries and the source agent system has to call a function f in order to obtain the itinerary. In this protocol, we use explicitly ElGamal encryption. So, encryptions $E_{y_j}(m)$ are noted as $((m)y_j^r, g^r)$.

Protocol Alt2:

1. Initialization Phase: The owner chooses random values:

$$r_1, r_2, \cdots, r_n, s$$
$$I_{n+1} = e_{n+1}$$

2. Sign: The owner signs the previous address h_0, successor address h_{n+1}, all the addresses of host entries encrypted, and the trip marker t.

$$u = S_o(h_0, h_1 g^{r_1}, h_2 g^{r_2}, \cdots, h_n g^{r_n}, h_{n+1}, m, t) \qquad (18)$$

3. For $i = n$ to 1

$$v_i = \frac{y_i^{r_i}}{g^s} \qquad (19)$$

4. Next i

5. Form of the protected itinerary I_0 as

$$I_0 =< (h_1, v_1, g^{r_1}),$$
$$(h_2, v_2, g^{r_2}),$$
$$\vdots$$
$$(h_n, v_n, g^{r_n}), ug^s > \qquad (20)$$

The mobile agent chooses one entry (h_j, v_j, g^{r_j}) and performs a function in order to arrange the itinerary for the selected host h_j.

$$f(ug^s, v_j) = ug^s \cdot v_j = ug^s \frac{uy_j^{r_j}}{ug^s} = uy_j^{r_j} \qquad (21)$$

The function $f()$ is a simple product of the arguments. In such a way we get the protected itinerary encrypted with the public key y_j.

$$(h_j, ug^{j^{r_j}}, g^{r_j}) = (h_j, (S_o(h_0, h_1 g^{r_1}, h_2 g^{r_2}, \cdots, h_n g^{r_n}, h_{n+1}, m, t))y_j^{r_j}, g^{r_j})$$

The method m and successor itinerary e_{n+1} is encrypted only once in this protocol. Hence, we resolve the size problem (17). Although we only encrypt the method m once, it is not feasible to decrypt it without the collaboration of the previous host and the selected host.

The selected host h_j decrypts the itinerary and obtains an encrypted list of candidates $h_1 g^{r_1}, h_2 g^{r_2}, \cdots, h_n g^{r_n}$. The mobile agent system knows g^{r_i}, so it is able to decrypt its own address. Thus, it is able to verify that it belongs to the initial route generated by the agent's owner (P4), and the other candidates remain anonymous.

Protocol Alt3:

1. Initialization Phase. Encrypt the remaining itinerary using a symmetric cryptosystem with a random key w: $I_{n+1} = E_w(e_{n+1})$
2. For $i = n$ to 1
   (a) Sign: The owner signs the entry e_i with the predecessor and successor host address, and the trip marker t.

$$S_o(h_o, h_i, h_{n+1}, m_i, t) \tag{22}$$

   (b) Encrypt: The owner encrypts the signed information using the public key y_i:

$$x_i = E_{y_i}(S_o(h_o, h_i, h_{n+1}, m_i, t), w) \tag{23}$$

3. Next i

4. The final protected alternative itinerary I_0 is defined by the set of tuples (h_i, x_i), where $1 \leq i \leq n$.

$$I_0 =< (\ h_1, x_1), \\ (\ h_2, x_2), \\ \vdots \\ (\ h_n, x_n), I_{n+1} > \tag{24}$$

Now, substituting the value of each x_i we obtain the route:

$$I_0 =< (\ h_1, E_{y_1}(S_o(h_o, h_1, h_{n+1}, m_1, t), w)), \\ (\ h_2, E_{y_2}(S_o(h_o, h_2, h_{n+1}, m_2, t), w)), \\ \vdots \\ (\ h_n, E_{y_n}(S_o(h_o, h_n, h_{n+1}, m_n, t), w)), I_{n+1} > \tag{25}$$

The selected entry has to decrypt remaining itinerary I_{n+1} with key w.

Protocol Alt3 solves the increasing size problem of protocol Alt1, and removes the necessity of a common method m for all entries of protocol Alt2.

The security of this proposal is equivalent to that of Alt1 and Alt2 protocols.

A malicious host h_m, previous to an alternative itinerary, is able to send mobile agent copies to different candidates. This reply attack will be detected by the successor host h_{n+1} that will receive a set of mobile agents from different sources with the same marker t (P7). If we are interested in detecting this attack as soon as possible, we must monitor the mobile agent [16].

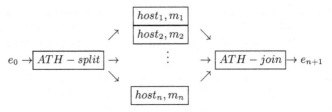

Fig. 3. Set Itinerary

4.3   Protocol to Protect Set Itineraries

A set itinerary is a list of entries $e_i = (h_i, m_i)$ which have to be visited by the mobile agent in any order.

$$e_0, \{e_1, e_2, \cdots, e_n\}, e_{n+1} \qquad (26)$$

Set itinerary (26), indicated in brackets "{" "}", begins with e_1 entry and ends with e_n. Without loss of generality, we will always assume that there is at least one entry before and after the itinerary. If the set itinerary is the main itinerary, the previous and successor entries will be $(h_o, -)$. Entries e_i, e_j where $1 \le i \le n$, $1 \le j \le n$ and $i \ne j$ are independent.

We present a preliminary version of the protocol, where an auxiliar trusted host is used in order to satisfy the security parameters. In [11] we study a different alternative without auxiliar trusted host. In figure 3 we show the communication flow, where an auxiliar trusted host (ATH) will split [7] the mobile agent in n child different agents. Once the child mobile agents are performed, they return to the auxiliar trusted host that will join them to continue the itinerary. In fact, ATH is only trusted to keep the list of hosts secret.

Protocol Set:

1. Initialization Phase: Further itineraries $e_{n+1}, e_{n+2}, \cdots$ are encrypted with a random symmetric key z.

$$I_z = E_z(e_{n+1}, e_{n+2}, \cdots)$$

2. Share: Secret random key z is shared among n entries using a (n, n) threshold secret sharing scheme [15,18].
3. For $i = n$ to 1. We prepare I_i for each entry following the structure with source host, destination host, message, shared key and trip marker encrypted.

$$I_i = (h_i, E_{y_i}(S_o(h_{ATH}, h_i, h_{ATH}, m_i, d_i, t))) \qquad (27)$$

4. Next i.

5. ATH computes

$$I_v = ((h_1, hash(d_1)), (h_2, hash(d_2)), \cdots, (h_n, hash(d_n))). \qquad (28)$$

With I_v, ATH will be able to detect cheaters when joining child agents, by checking shares returned by child agents with the hashed values of these shares. On the other hand, ATH could use a rational interpolation scheme to share the secret z [13], which allows it to detect the existence of cheaters without the use of hashed values.

6. Final form of the protected itinerary I_0.

$$I_0 = \{E_{y_{ATH}}(S_o(h_0, h_{ATH}, h_{n+1}, I_1, I_2, \cdots, I_n, I_v) I_z)\} \tag{29}$$

With a simple example we explain the protocol and also discuss the security properties. We assume a set itinerary $\{e_1, e_2, e_3\}$, with a previous entry e_0, successor entry e_4 and an auxiliar trusted host h_{ATH}. The owner uses protocol **Set** to obtain the following protected itinerary:

$$
\begin{aligned}
I_0 = \{E_{y_{ATH}} (\; & S_o(h_0, h_{ATH}, h_{n+1}, t, \\
& (h_1, E_{y_1}(S_o(h_{ATH}, h_1, h_{ATH}, m_1, d_1, t)), \\
& (h_2, E_{y_2}(S_o(h_{ATH}, h_2, h_{ATH}, m_2, d_2, t)), \\
& (h_3, E_{y_3}(S_o(h_{ATH}, h_3, h_{ATH}, m_3, d_3, t)), \\
& ((h_1, hash(d_1)), (h_2, hash(d_2)), (h_3, hash(d_3))), \\
& E_z(e_4)))\}
\end{aligned}
\tag{30}
$$

The auxiliar trusted host h_{ATH} receives the mobile agent, signed by the owner. It is able to decrypt the information of the set itinerary and it checks the security properties using the host predecessor, the host successor and the trip marker t, as in aforementioned protocols.

By using the cloning property of mobile agents [7], the auxiliar trusted host prepares a child mobile agent for each entry, in a split-phase (31). The i-child mobile agent just receives the itinerary for the entry e_i.

$$
\begin{aligned}
&1-\text{Child Mobile Agent} : E_{y_1}(S_o(h_{ATH}, h_1, h_{ATH}, m_1, d_1, t)) \\
&2-\text{Child Mobile Agent} : E_{y_2}(S_o(h_{ATH}, h_2, h_{ATH}, m_2, d_2, t)) \\
&3-\text{Child Mobile Agent} : E_{y_3}(S_o(h_{ATH}, h_3, h_{ATH}, m_3, d_3, t))
\end{aligned}
\tag{31}
$$

All child mobile agents are submitted at the same time to different host directions. So, all entries are visited at the same time without an specific order. In such this way, we optimize the itinerary execution.

Mobile agent systems receive the child mobile agent with the plain structure signed by the owner and with the private information encrypted. Once a i-child mobile agent has performed the method m_i, it comes back to the ATH with the results w_i and share d_i. This information is encrypted with the public key of ATH and signed by the mobile agent system.

$$
\begin{array}{l}
\boxed{h_1} \rightarrow E_{y_{ATH}}(S_{h_1}(w_1, d_1, t)) \rightarrow \boxed{h_{ATH}} \\
\boxed{h_2} \rightarrow E_{y_{ATH}}(S_{h_2}(w_2, d_2, t)) \rightarrow \boxed{h_{ATH}} \\
\boxed{h_3} \rightarrow E_{y_{ATH}}(S_{h_3}(w_3, d_3, t)) \rightarrow \boxed{h_{ATH}}
\end{array}
\tag{32}
$$

Only when all child mobile agents have come back to the parent agent, with both result w_i and share d_i, it is able to reconstruct the secret key z, in a joining phase.

In order to detect cheaters, ATH checks the share d_i with his hash value $(h_i, hash(d_i))$.

If all entries are visited and the child mobile agents come back, the auxiliar trusted host will recover the secret key z and will be able to decrypt the remaining itinerary $E_z(e_4)$. Thus, further itineraries remain anonymous until the set itinerary is fulfilled.

The security parameters are satisfied as above presented protocols, using the owner signature, the trip marker and encrypted private data. An auxiliar trusted host is needed to split and join phases, keeping secret the list of hosts. However, we are working in an improved version where we remove the auxiliar host.

5   Conclusions and Further Research

In this paper we focus on solutions to the problem of providing strong protection to individual mobile agents against tampering and traffic analysis attacks. Therefore, we introduce a set of new protocols to protect general flexible itineraries established in free-roaming mobile agents. We use ElGamal cryptosystem and a public key infrastructure, but any public cryptosystem could be used except in protocol Alt2. Naturally, an implementation will also require one or more certification authorities, but their role is perfectly standard and we will not discuss them here.

We have introduced one sequence, three alternative and a set itinerary protocol. Although sequence protected itineraries are not new, we improve previous versions with new security features. Alternative and set itineraries had not been protected yet. The first alternative itinerary protocol Alt1 is suitable for small routes where mobile agents only have to choose an entry. The second alternative protocol Alt2 solves the size problem using a function f. However, it has an extra computation complexity and only works if all entries have a common method m. All these issues are solved by the third alternative protocol Alt3. A preliminary version of set protected itinerary is showed using an auxiliar trusted host and a (n,n) threshold secret sharing scheme [15,18]. Set itinerary is split into several child mobile agents, one for each entry. Once a child mobile agent has performed the method, it returns to the parent mobile agent in the ATH. The parent mobile agent will continue the journey when all child mobile agents would have returned their execution results and shares. In [11] we study a different alternative without auxiliar trusted host.

We are planning a protocol implementation in order to carry out a detailed performance analysis. For implementation purposes, mobile agent systems with itinerary support, such as Ajanta [7] or Aloha [21] mobile systems, are good candidates.

Finally, we are working on enhanced hash chains in order to achieve a low computation overhead. Hash chains were introduced in [9] for password authentication and have already been used in previous sequence protection protocols [5].

Acknowledgments

This work has been partially funded by the Spanish Government Commission CICYT, through its grant TIC2000-0232-P4. Many thanks also to Camper S.A. for their financial support. Special thanks to Vanesa Daza for reading and commenting drafts of this article and the anonymous reviewers for their invaluable comments.

References

1. I. Biehl, B. Meyer, and S. Wetzel. Ensuring the Integrity of Agent-Based Computations by Short Proofs. In *Mobile Agents and Security*, Lecture Notes in Computer Science, 1419, pages 183–194. Springer-Verlag, 1998.
2. J. Borrell, S. Robles, J. Serra, and A. Riera. Securing the Itinerary of Mobile Agents through a Non-Repudiation Protocol. In *IEEE International Carnahan Conference on Security Technology*, pages 461–464, 1999.
3. D. Chess. Security Issues of Mobile Agents. In *Mobile Agents*, Lecture Notes in Computer Science, 1477, pages 1–12. Springer-Verlag, 1998.
4. D. Chess, B. Grosof, C. Harrison, D. Levine, C. Parris, and G. Tsudik. Itinerant agents for mobile computing. In *IEEE Personal Communications Magazine, 2*, pages 34–49, October 1995.
   http://www.research.ibm.com/massive.
5. J. Domingo and J Herrera. Enhanced Hash Chains for Efficient Agent Route Protection. Technical report, Submitted to Information Processing Letters, April 2000.
6. G. Karjoth, N. Asokan, and C. Gülcü. Protecting the Computation of Free-Roaming Agents. In *Mobile Agents*, Lecture Notes in Computer Science, 1477, pages 194–207. Springer-Verlag, 1998.
7. N. Karnik. *Security in Mobile Agent Systems*. PhD thesis, Department of Computer Science and Engineering, University of Minnesota, 1998.
8. D. Kotz and R. S. Gray. Mobile Agents and the Future of the Internet. In *ACM Operating System Review*, pages 7–13, 1999.
   ftp://ftp.cs.dartmouth.edu/pub/kotz/papers/kotz:future2.ps.Z.
9. L. Lamport. Password Authentication with Insecure Communications. In *Comunications of the ACM, 24(11)*, pages 770–772, 1981.
10. D. B. Lange and M. Oshima. *The Aglet cook-book*. In progress, 1997.
    http://www.trl.ibm.co.jp/aglets/aglet-book/index.html.
11. J. Mir and J. Borrell. Work in process. 2001.
12. Odyssey: Beta Release 1.0. Available as part of the Odyssey package at, 1997.
    http://www.genmagic.com/agents/.
13. J. Rifa. How to Avoid the Cheaters Succeeding in the Key Sharing Scheme. In *Designs, Codes and Cryptography*, pages 221–228. Kluwer Academic Publishers, 1993.
14. T. Sander and C. Tschudin. Protecting Mobile Agents Against Malicious Hosts. In *Mobile Agents and Security* [19], pages 137–153.
15. A. Shamir. How to share a secret. In *Communications of the ACM 22,11*, pages 612–613, November 1979.

16. M. Straßer, K. Rothermel, and C. Maifer. Providing Reliable Agents for Electronic Commerce. In *Trends in Distributed Systems for Electronic Commerce*, Lecture Notes in Computer Science, 1402, pages 241–253. Springer-Verlag, 1998.

17. P. Syverson, D. Goldchlag, and M. Reed. Anonymous Connections and Onion Routing. In *IEEE Symposium on Security and Privacy*, 1997. `www.onion-router.net/Publications/INET-97.htm`.

18. M. Tompa and H. Woll. How to Share a Secret with Cheaters. In *Journal of Cryptology*, pages 133–138. Springer-Verlag, 1988.

19. G. Vigna. Mobile Agents and Security. In *Mobile Agents and Security*, Lecture Notes in Computer Science, 1419, pages 137–153. Springer-Verlag, 1998.

20. Voyager technical overview. Technical report, ObjectSpace White Paper, 1997. `http://www.objectspace.com/voyager`.

21. D. Westhoff, M. Schneider, C. Unger, and F. Kaderali. Methods for protecting a mobile agent's route. In *Second International Information Security Workshop, ISW'99*, Lecture Notes in Computer Science, 1729, pages 51–71. Springer-Verlag, November 1999.

22. D. Westhoff, M. Schneider, C. Unger, and F. Kaderali. Protecting a Mobile Agent's Route Against Collusions. In *SAC'99*, Lecture Notes in Computer Science, 1758, pages 215–225. Springer-Verlag, 2000.

23. U. G. Wilhelm. Cryptographically Protected Objects. Technical report, Escole Polytechnique Federale de Lousanne, Switzerland, 1997.

24. D. Wong, N. Paciorek, T. Walsh, J. DiCelie, M. Young, and B. Peet. Concordia: An Infrastructure fo Collaborating Mobile Agents. In *Mobile Agents*, Lecture Notes in Computer Science, 1219, pages 86–97. Springer-Verlag, 1997. `http://www.meitca.com/HSL/Projects/Concordia`.

25. J. Zhou and D. Gollmann. Observations on Non-repudiation. In *Asiacrypt'96*, Lecture Notes in Computer Science, 1163, pages 133–144. Springer-Verlag, 1996.

RSA Speedup with Residue Number System Immune against Hardware Fault Cryptanalysis

Sung-Ming Yen[1], Seungjoo Kim[2], Seongan Lim[2], and Sangjae Moon[3]

[1] Laboratory of Cryptography and Information Security (LCIS),
Department of Computer Science and Information Engineering,
National Central University, Chung-Li, Taiwan 320, R.O.C.
yensm@csie.ncu.edu.tw
[2] Cryptographic Technology Team,
Information Security Technology Department,
Korea Information Security Agency,
78, Garak-Dong, Songpa-Gu, Seoul, Korea 138-803
{skim,seongan}@kisa.or.kr
[3] School of Electronic and Electrical Engineering,
Kyungpook National University, Taegu, Korea 702-701
sjmoon@knu.ac.kr

Abstract. This article considers the problem of how to prevent the fast RSA signature and decryption computation with residue number system (or called the CRT-based approach) speedup from a hardware fault cryptanalysis in a highly reliable and efficient approach. The CRT-based speedup for RSA signature has been widely adopted as an implementation standard ranging from large servers to very tiny smart IC cards. However, given a single erroneous computation result, a hardware fault cryptanalysis can totally break the RSA system by factoring the public modulus. Some countermeasures by using a simple verification function (e.g., raising a signature to the power of public key) or fault detection (e.g., an expanded modulus approach) have been reported in the literature, however it will be pointed out in this paper that very few of these existing solutions are both sound and efficient. Unreasonably, in these methods, they assume that a comparison instruction will always be fault free when developing countermeasures against hardware fault cryptanalysis. Researches show that the expanded modulus approach proposed by Shamir is superior to the approach of using a simple verification function when other physical cryptanalysis (e.g., timing cryptanalysis) is considered. So, we intend to improve Shamir's method. In this paper, the new concept of fault infective CRT computation and fault infective CRT recombination are proposed. Based on the new concept, two novel protocols are developed with rigorous proof of security. Two possible parameter settings are provided for the protocols. One setting is to select a small public key e and the proposed protocols can have comparable performance to Shamir's scheme. The other setting is to have better performance than Shamir's scheme (i.e., having comparable performance to conventional CRT speedup) but with a large public key. Most importantly, we wish to emphasize the importance of developing and proving the security of physically secure protocols without relying on unreliable or unreasonable assumptions, e.g., always fault free instructions.

K. Kim (Ed.): ICICS 2001, LNCS 2288, pp. 397–413, 2002.
© Springer-Verlag Berlin Heidelberg 2002

Keywords: Chinese remainder theorem (CRT), Cryptography, Factorization, Fault detection, Fault infective CRT, Fault tolerance, Hardware fault cryptanalysis, Physical cryptanalysis, Residue number system, Side channel attack.

1   Introduction

In order to provide a better support for data protection under strong cryptographic schemes (e.g., RSA [1] or ElGamal [2] systems), varieties of implementations based on tamper-proof devices (e.g., smart IC cards) are proposed. The main reason for this trend is that smart IC cards are assumed to provide high reliability and security with more memory capacity and better performance characteristics than conventional plastic cards. The CPU in a smart IC card controls its data input and output and prevents unauthorized access to a smart card. With special characteristics of computational ability, large memory capacity and security, a large variety of cryptographic applications benefit from smart IC cards. Due to this popular usage of tamper-resistance, much attention has recently been paid regarding the security issues of cryptosystems implemented on tamper-proof devices [3,4,5,6,7,8,9,10,11,12,13,14,15,16,17,18] from the view point of presence of hardware faults.

This line of research re-emerged in September 1996 when a Bellcore press release [5] reported a new kind of attack, the so-called *fault-based cryptanalysis.* In the fault-based cryptanalysis model, it is assumed that when an adversary has physical access to a tamper-proof device she may purposely induce a certain type of fault into the device. Based on a set of incorrect responses or outputs from the device, due to the presence of faults, the adversary can then extract the secrets embedded in the tamper-proof device. These attacks exploit the presence of many kinds of transient or permanent faults. These attacks are of very general nature and remain valid for a large variety of cryptosystems, e.g., the LUC public key cryptosystem [19] and elliptic curve cryptography (ECC).

In this paper, we focus our attention on public key cryptosystems in which their computation can be sped up using the Chinese remainder theorem (CRT) [20,21]. These cryptosystems may be vulnerable to the hardware fault cryptanalysis to reveal the secret key if the following three conditions are met: (1) the message to sign (or to decrypt) is known; (2) a random fault occurs during the computation of a residue number system; (3) the device outputs the faulty result. Our main objective is to emphasize the importance of a careful implementation of cryptosystems with CRT-based speedup. Suppose you are in a context involving trusted third parties (e.g., banks) where thousands of signatures being produced each day. If, for some reasons, a single signature is faulty, then the security of the whole system may be compromised. Leakage of the secret key can be avoided by making sure that either of the above three conditions will not be met.

Some existing countermeasures have been reported in the literature, however it will be pointed out in this paper that very few of these solutions are both sound and efficient. All the previous solutions employ one or more checking procedures to detect the possible faults. However, in order to develop a highly reliable CRT-based speedup technique, no checking procedure shall be assumed, because in that situation, the checking procedure itself will become most vulnerable to the hardware fault cryptanalysis and all other parts of the countermeasure will be in vain. On the other hand, some existing solutions suffer a disadvantage that the probability of producing an undetectable error is high.

The main contribution of this paper is that the new concept of *fault infective* CRT computation and *fault infective* CRT recombination are proposed. Based on the new concept, two novel protocols are developed with rigorous proof of security. The two CRT-based speedup protocols do not depend upon any checking procedure which should be absolutely error free and physical cryptanalysis immune and do not suffer the disadvantage of producing an *undetectable* error with large probability. A key point of developing a secure CRT-based computation protocol without using a checking procedure is the assurance to influence the computation of one module within the residue number system or the overall computation when an error occurred in another module within the residue number system.

Two possible parameter settings are provided for the proposed protocols. One setting is to select a small public key e and the proposed protocols can have comparable performance to Shamir's scheme. The other setting is to have better performance than Shamir's scheme (i.e., having comparable performance to conventional CRT speedup) but with a large public key.

Most importantly, in this paper, we wish to emphasize the importance of developing and proving the security of physically secure protocols without relying on unreliable or unreasonable assumptions, e.g., always fault free instructions.

2   Preliminary Background of CRT-Based Cryptanalysis

2.1   Residue Number System and Chinese Remainder Theorem

The residue number system, RNS (or often called the Chinese remainder theorem [21], CRT), tells that given a set of integers $n_1, n_2, \ldots, n_k$ that are pairwise relatively prime, then the following system of simultaneous congruences

$$s \equiv a_1 \pmod{n_1}$$
$$s \equiv a_2 \pmod{n_2}$$
$$\vdots$$
$$s \equiv a_k \pmod{n_k}$$

has a unique solution modulo $n = \prod_{i=1}^{k} n_i$.

The integer solution s to the above simultaneous congruences can be computed as

$$s = \left(\sum_{i=1}^{k} a_i \cdot N_i \cdot M_i \right) \bmod n \tag{1}$$

where $N_i = n/n_i$ and $M_i = N_i^{-1} \bmod n_i$. In this paper, the following notation

$$s = CRT(a_1, a_2, \ldots, a_k) \tag{2}$$

will be used to represent the computation of s by using the Chinese remainder theorem.

This RNS (CRT-based) structure has been widely employed in the computer engineering areas of special parallel computer architectures and some digital signal processing processors to speedup the computation.

The RNS structure was also proposed to speed up the RSA signature and decryption computation [20]. This RNS speedup for RSA computation has been widely adopted as an implementation standard with the performance of four times faster than a direct computation in terms of bit operations. The applications range from large servers to very tiny smart IC cards. For servers, there are huge amount and very frequent RSA computations to be performed. For smart IC cards, the card processors are often not powerful enough to perform real time complicated cryptographic computations, e.g., RSA signature and decryption.

2.2   The CRT-Based Cryptanalysis

Let p and q be two primes and $n = p \cdot q$. In the RSA cryptosystem, a message m is signed with a secret exponent d as $s = m^d \bmod n$. Using the residue number system approach, the value of s can be computed more efficiently from computing $s_p = m^d \bmod p$ and $s_q = m^d \bmod q$, then by using the Chinese remainder theorem (CRT) to reconstruct s.

Suppose that an error (any random error) occurs during the computation of s_p (s_p' denotes the erroneous result), but the computation of s_q is error free. Applying the CRT on both s_p' and s_q will produce a faulty signature s'. The CRT-based hardware fault cryptanalysis [13,14] enables the factorization of n by computing

$$q = \gcd((s'^e - m) \bmod n, n). \tag{3}$$

Similarly, p can be derived from a faulty signature computed by applying the CRT on a faulty s_q' and a correct s_p.

2.3   Previous Countermeasures

To counteract the above CRT-based cryptanalysis, any methodology of finding possible computation errors within $s_p = m^d \bmod p$ and $s_q = m^d \bmod q$ works for some situations. Previous research results suggest either to perform calculations twice or to apply a verification function on the computed result to detect any fault.

The first approach is very time-consuming and it cannot always provide a satisfactory solution since a permanent error (caused by a permanent hardware or software fault or implementation bug) may be undetectable, even the function is computed more than once.

The second approach suggests to verify the correctness by comparing the inversed result with the input. In the RSA signature scenario, a computed signature $s = m^d \bmod n$ can be verified by raising s to the eth power and compare whether $m \equiv s^e \pmod{n}$. Generally, this is not a satisfactory solution since the parameter e could be a large integer and this checking procedure becomes time-consuming.

Furthermore, both the above two approaches of employing a *decision procedure* to decide the correctness of computation suffer an intrinsic disadvantage which will be further studied after providing the review of Shamir's proposal.

2.4   Shamir's Countermeasure

Shamir presented a simple countermeasure in the rump session of Eurocrypt '97 [16] and applied a patent [17]. In Shamir's method, a random integer r is selected and the following two numbers are computed

$$\begin{cases} s_p = m^d \bmod pr \\ s_q = m^d \bmod qr. \end{cases} \tag{4}$$

If $s_p \equiv s_q \pmod{r}$, then it is defined to be error free and $s = CRT(s_p \bmod p, s_q \bmod q)$ is computed.

As described previously, a trivial approach of checking whether $m \equiv (CRT(s_p \bmod p, s_q \bmod q))^e \pmod{n}$ can be used to detect any possible fault occurred during the computation of s_p and s_q when the conventional CRT-based speedup is employed. However, Shamir's proposal provides an advantage that it can withstand the timing cryptanalysis on CRT-based implementation through a binary search algorithmic approach [17,22]. In the conventional CRT-based speedup, $m_p = m \bmod p$ has to be computed before the evaluation of $s_p = m^d \bmod p$ and a value of m will affect the computation time of m_p depending on whether m is larger or less than p.

2.5   Remarks on Shamir's Countermeasure

There exist at least three nontrivial drawbacks in Shamir's proposal for preventing CRT-based hardware cryptanalysis.

Theoretically, the probability of producing an undetectable error is equal to $1/r$. The setting of k-bit integer r suffers the probability of $1/2^k$ to leak the secret key. Small value of r cannot provide satisfactory protection, while larger value of r decreases the overall performance. The above probability neither depends on whether r is changed for each computation nor depends on whether r is a secret parameter.

Notice that any checking based approach, including Shamir's method and any previous results, suffers an intrinsic disadvantage that not only the RNS computation but also the *checking (decision) procedure* will be vulnerable to the hardware fault attack. How to guarantee the feasibility of designing and implementing an error free checking procedure becomes another big challenge.

In Shamir's method, expanded modulus is used in the computations of $s_p = m^d \bmod pr$ and $s_q = m^d \bmod qr$. For security reasons, the bit length of r (i.e., $|r|$) should be values of at least 20 or 30 and even larger which requires two registers (one for s_p and the other for s_q) with substantially different word length than other registers. Furthermore, for security reasons, larger values of $|r|$ may be necessary at a later time after the prototype implementation and this requires a variable length of register. These two limitations will bring nontrivial implementation (or modification) disadvantages on any existing systems, including general purpose computers or smart IC cards.

3   Why Not Using a Decision Procedure to Be against Hardware Fault Attack

Conventionally, many countermeasures against hardware fault cryptanalysis implicitly assume the usage of decision procedures. Both Shamir's method and the straightforward approach of checking whether $s^e \overset{?}{\equiv} m \pmod n$ strictly depend on the reliability of a decision procedure. From the viewpoint of low level implementation of this decision procedure, it often totally relies on the status of the *zero flag* of a processor. For example, a checking of $a \overset{?}{=} b$ is implemented by a "SUB a,b" instruction by subtracting b from a (or "CMP a,b" instruction which compares the equality of a and b), then a "JZ" (jump if zero) instruction is conditionally performed depending on the status of the zero flag.

The zero flag is a bit of the status register in a processor. So, if an attacker can induce a random fault into the status register, then the conditional jump instruction may perform falsely. This may bring a big problem in the CRT-based hardware fault cryptanalysis since even a single success of attack will totally factorize n.

The above approach of modifying a decision procedure to check the correctness of a cryptographic computation is feasible for most of the cases. This is extremely different to many assumptions made in some reported hardware fault cryptanalysis where a single bit flipping is required such that no other parts of the processor can be affected during the fault induction since those information (error free) will be necessary in the attack. On the contrary, in the above decision procedure modifying approach, the result (say $a - b$) is no longer required after the comparison. This is especially true when the instruction "CMP a,b" is used to compare the equality of a and b because the result of $a - b$ is explicitly ignored. So, an attacker can easily induce random noisy signals into the processor (especially the location of the central processing unit-CPU) in order to obtain a complementary status of the zero flag. The attacker does not have to worry about the corruption of the computation result $a - b$. Therefore, if an attacker can introduce a random error into the zero flag, he has 50% of probability to foil the correctness of the checking procedure.

Someone may argue that some techniques may be employed to protect the status register in a processor to avoid the above attack. However, as well known that any error detection or correction technique has its limitation, so a design of highly reliable CRT-based attack immune protocol cannot be guaranteed. Furthermore, it is not possible for all commercial processors to adopt such enhancement on the status register. It is also not proven that the enhancement of the status register will solve all the problems. A protocol level design with rigorous security proof without the assumption on the reliability of the underlying processor is a much better solution to counteract hardware fault cryptanalysis, especially the CRT-based attack since a single success of attack will totally break the system.

The above observation explains why the usage of a decision procedure violates the guarantee of developing a highly reliable hardware fault immune cryptographic protocol. In this paper, the idea of *fault infective* CRT computation is proposed so that security of a protocol does not rely on any decision procedure.

4   Two Secure CRT-Based Computation Protocols Based on Fault Infection

In order to develop a highly reliable CRT-based speedup, no checking procedure will be assumed hereafter, because in that situation, the procedure will become most vulnerable to the CRT-based hardware fault cryptanalysis and all other parts of the countermeasure will be in vain.

A key point of developing a secure CRT-based computation protocol without using a checking procedure is the assurance to influence the computation of s_q or the overall computation of s when an error occurred in the computation of s_p. This property should also apply when a faulty s'_q is produced.

In this paper, the above concept is called the *fault infective CRT* computation. This makes the Eq. 3 invalid and the CRT-based hardware fault cryptanalysis no longer workable.

4.1   The First Protocol – CRT-1 Protocol

Let $n = p \cdot q$ as usual RSA system and $\delta = p - q$. The smart card also prepares another set of key pair such that $d_r = d - r$ where r is a small integer (with the property of $\gcd(r, \phi(n)) = 1$) selected in order to let $e_r \equiv d_r^{-1} \pmod{\phi(n)}$ be a small integer. Notice that in this setting we don't assume that e is also a small integer. On the other hand, another setting for a small e but with a large e_r is also possible. Performance of CRT-1 protocol with these two parameter settings can be found in a latter section.

The following protocol suggests a way to speed up the RSA signature computation via the residue number system technique, while at the same time being immune against the hardware fault cryptanalysis. Furthermore, the protocol does not depend upon any checking procedure (e.g., double computation) which should be absolutely error free and physical cryptanalysis immune. Nor will the proposed protocol suffer the disadvantage of producing an *undetectable* error with large probability as in Shamir's method. In Shamir's method, it is hard to define the value of $|r|$ (bit length of r) in order to balance both performance and security. There is a trade-off between these two contradicting requirements.

Step-1 Compute both $k_p = \lfloor m/p \rfloor$ and $k_q = \lfloor m/q \rfloor$.

Step-2 Compute m^{d_r} mod n via a conventional residue number system as

$$\begin{cases} s_p = m^{d_r} \bmod p \\ s_q = \hat{m}^{d_r} \bmod q \end{cases} \tag{5}$$

where

$$\hat{m} = ((s_p^{e_r} \bmod p) + k_p \cdot \delta) \bmod q. \tag{6}$$

Step-3 A conventional CRT operation and some extra manipulation are employed to compute the required signature as

$$s = CRT(s_p, s_q) \cdot (\tilde{m}^r) \bmod n$$

where

$$\tilde{m} = (s_q^{e_r} \bmod q) + k_q \cdot q. \tag{7}$$

Lemma 1. *Given a prime p and a nonzero integer e such that $\gcd(e, \phi(p)) = 1$ where $\phi(p) = p - 1$, it assures that s_1^e mod $p \neq s_2^e$ mod p if both s_1 and s_2 are integers in $\mathbf{Z}_p$ and $s_1 \neq s_2$.*

Proof. The lemma is true for the trivial case of $e = 1$. For the case of $e > 1$, the lemma will be proven by contradiction. Suppose that $s_1 \neq s_2$ and s_1^e mod $p = s_2^e$ mod p, then $(s_1 \cdot s_2^{-1})^e$ mod $p = 1$ since $\gcd(s_2, p) = 1$ and s_2^{-1} mod p exists. This can be rewritten as s^e mod $p = 1$ (where $s = (s_1 \cdot s_2^{-1})$ mod p) and $s \neq 1$ since $s_1 \neq s_2$. Based on fundamental number theory, this requires $e | \phi(p)$ or alternatively $\gcd(e, \phi(p)) \neq 1$ if $e \neq 1$. This however contradicts the assumption of $\gcd(e, \phi(p)) = 1$. $\square$

The above Lemma 1 can be extended into the following general case which enables the famous RSA scheme to be a cryptosystem with one-to-one mapping between the message and the cipher.

Lemma 2. *Given an integer n and a nonzero integer e such that $\gcd(e, \phi(n)) = 1$, it assures that s_1^e mod $n \neq s_2^e$ mod n if both s_1 and s_2 are integers in $\mathbf{Z}_n$ and $s_1 \neq s_2$.*

Lemma 3. *Give two integers p and q such that $p > q$. Let A and B be any two distinct integers in the range of $[0, p-1]$. The number of integer pair (A, B) such that A mod $q = B$ mod q is $2\lfloor p/q \rfloor (p \bmod q) + q\lfloor p/q \rfloor (\lfloor p/q \rfloor - 1)$.*

Proof. It can be verified that for each A such that $A \equiv x$ (mod q) for x in $[0, (p \bmod q) - 1]$, there are $\lfloor p/q \rfloor$ integers B which satisfy A mod $q = B$ mod q. Totally, there are $(p \bmod q)(\lfloor p/q \rfloor + 1)$ such integers A.

On the other hand, for each A such that $A \equiv x$ (mod q) for x in $[(p \bmod q), q - 1]$, there are $\lfloor p/q \rfloor - 1$ integers B which satisfy A mod $q = B$ mod q. Totally, there are $(q - (p \bmod q))\lfloor p/q \rfloor$ such integers A.

Therefore, there are totally

$$\lfloor p/q \rfloor (p \bmod q)(\lfloor p/q \rfloor + 1) + (\lfloor p/q \rfloor - 1)(q - (p \bmod q))\lfloor p/q \rfloor$$
$$= 2\lfloor p/q \rfloor (p \bmod q) + q\lfloor p/q \rfloor (\lfloor p/q \rfloor - 1)$$

integer pairs (A, B) can be found. $\square$

In the following description, the number of possible integer pairs (A, B) in the Lemma 3 will be denoted as E.

Theorem 1. *Given a faulty s'_p in the CRT-1 protocol, the probability of generating a faulty s'_q is $1 - \frac{E}{p(p-1)}$ if $p > q$.*

Proof. Since $s_p \neq s'_p$, it follows from the Lemma 1 that $s_p^{e_r} \bmod p \neq s_p'^{e_r} \bmod p$ because $\gcd(e_r, \phi(p)) = 1$. So, there are totally $p(p-1)$ possible error patterns (s_p, s'_p) between s_p and s'_p. If $p > q$, it follows from the Lemma 3 that E out of the total $p(p-1)$ possible error patterns will lead to a same $\hat{m}$ in the Eq. 6, i.e., $((s_p^{e_r} \bmod p) + k_p \cdot \delta) \bmod q = ((s_p'^{e_r} \bmod p) + k_p \cdot \delta) \bmod q$. Therefore, the probability of generating a faulty s'_q when given a faulty s'_p is $1 - \frac{E}{p(p-1)}$. □

Theorem 2. *Given a faulty s'_p in the CRT-1 protocol, it assures that a faulty s'_q will be generated if $p < q$.*

Proof. Same as in the Theorem 1, $s_p^{e_r} \bmod p \neq s_p'^{e_r} \bmod p$ and there are totally $p(p-1)$ possible error patterns (s_p, s'_p) between s_p and s'_p. However, if $p < q$ none of these (s_p, s'_p) pairs will lead to a same $\hat{m}$ in the Eq. 6 and will produce an error free s_q. □

The Theorem 2 suggests that it is better to let $p < q$ when implementing the CRT-1 protocol. However, the Theorem 1 makes sure that the hardware fault cryptanalysis is not feasible on the CRT-1 protocol even if $p > q$ since $1 - \frac{E}{p(p-1)} \approx 1$ for very large value of $n = p \cdot q$ and $|p| \approx |q|$.

Theorem 3. *Given e_r and a faulty $s' = CRT(s_p, s'_q) \cdot (\tilde{m}'^r) \bmod n$ (where $\tilde{m}' = (s_q'^{e_r} \bmod q) + k_q \cdot q$), the complexity of finding p by computing $\gcd((CRT(s_p, s'_q))^{e_r} - m, n)$ is $O(n)$.*

Proof. Based on the Lemma 1, every faulty s'_q will produce a unique $\tilde{m}' = (s_q'^{e_r} \bmod q) + k_q \cdot q$ and there are totally $q - 1$ possible erroneous s'_q. These $q - 1$ possible erroneous $\tilde{m}'$ will also produce $q - 1$ possible different $\tilde{m}'^r \bmod n$ (let the set of these $q - 1$ integers form a set **S**) based on the Lemma 2 since $\gcd(r, \phi(n)) = 1$.

From the given faulty s' and all these different $\tilde{m}'^r \bmod n$, there are $q - 1$ possible $CRT(s_p, s'_q)$ can be obtained and only one of these values can be employed to factorize n by computing $p = \gcd((CRT(s_p, s'_q))^{e_r} - m, n)$. However, the attacker cannot tell which value in $\mathbf{Z}_n$ will be an element in the set **S**, even the parameter r is given, because of the unknown faulty $\tilde{m}'$. Therefore, all values in $\mathbf{Z}_n$ could be the random integer $\tilde{m}'^r \bmod n$ and this concludes the proof. □

Theorem 1, Theorem 2, and Theorem 3 analyze the security issues of the CRT-1 protocol. They demonstrate that a random fault occurred in one of the two RNS computation modules will not reveal the factorization of n. A brute force guessing of $O(n)$ complexity seems to be the only possible way to factorize n, even a faulty s'_q has been produced by the attacker. This does not require the integers e_r and r to be secret parameters.

4.2   The Second Protocol – CRT-2 Protocol

Let $n = p \cdot q$ as usual RSA system. The smart card also prepares another set of key pair such that $d_r = d - r$ where r is a small integer (with the property of $\gcd(r, \phi(n)) = 1$) selected in order to let $e_r \equiv d_r^{-1} \pmod{\phi(n)}$ be a small integer. Noticeably, we don't assume that e is also a small integer.

Step-1 Compute both $k_p = \lfloor m/p \rfloor$ and $k_q = \lfloor m/q \rfloor$.

Step-2 Compute $m^{d_r} \bmod n$ via a conventional residue number system as

$$\begin{cases} s_p = m^{d_r} \bmod p \\ s_q = m^{d_r} \bmod q. \end{cases} \tag{8}$$

Step-3 A conventional CRT operation and some extra manipulation are employed to compute the required signature as

$$s = CRT(s_p, s_q) \cdot (\hat{m}^r) \bmod n$$

where

$$\hat{m} = \lfloor \frac{((s_p^{e_r} \bmod p) + k_p \cdot p) + ((s_q^{e_r} \bmod q) + k_q \cdot q)}{2} \rfloor. \tag{9}$$

Notice that in the above CRT-2 protocol given any faulty s'_p or s'_q (or both) with random faults, a random faulty $\hat{m}'$ in Eq. 9 will be generated.

Theorem 4. *Given e_r and a faulty $s' = CRT(s'_p, s_q) \cdot (\hat{m}'^r) \bmod n$ (where $\hat{m}' = \lfloor \frac{((s_p'^{e_r} \bmod p) + k_p \cdot p) + ((s_q^{e_r} \bmod q) + k_q \cdot q)}{2} \rfloor$), the complexity of finding q by computing $\gcd((CRT(s'_p, s_q))^{e_r} - m, n)$ is $O(n)$.*

Proof. The proof can be obtained in a similar way as in the derivation of Theorem 3 and is ignored here. □

The complexity of finding p by computing $\gcd((CRT(s_p, s'_q))^{e_r} - m, n)$ when given e_r and a faulty $s' = CRT(s_p, s'_q) \cdot (\hat{m}'^r) \bmod n$ (where $\hat{m}' = \lfloor \frac{((s_p^{e_r} \bmod p) + k_p \cdot p) + ((s_q'^{e_r} \bmod q) + k_q \cdot q)}{2} \rfloor$) is also $O(n)$.

5   Secure Implementation of CRT Recombination

Previous designs of secure CRT based speedup focus on developing fault infective computations of both s_p and s_q. It assumes that a CRT recombination as described in Eq. 1 is fault free.

In the following, it will be pointed out that an inappropriate CRT recombination may leak the secret prime factors of n if a hardware fault was induced.

In the following analysis both s_p and s_q are assumed to be error free since the faulty conditions have been discussed extensively in the previous section. As described in Eq. 1, given s_p and s_q, one of the possible CRT recombinations computes

$$CRT(s_p, s_q)$$
$$= (s_p \cdot q \cdot q^{-1}(\bmod\ p) + s_q \cdot p \cdot p^{-1}(\bmod\ q))\ \bmod\ n$$
$$= (s_p \cdot X_p + s_q \cdot X_q)\ \bmod\ n \qquad (10)$$

where both X_p and X_q can be pre-computed and stored in advance. The above method is often called Gauss's algorithm [21, p.68]. For the Step-3 of the proposed CRT-1 protocol, it can be easily verified that if either the value of X_p or X_q is incorrect due to a hardware fault (induced by an attacker or an EEPROM error), then it follows that

$$\begin{cases} s' \bmod p \neq s_p \cdot (\tilde{m}^r)\ \bmod\ p \\ s' \bmod q \neq s_q \cdot (\tilde{m}^r)\ \bmod\ q. \end{cases} \qquad (11)$$

Therefore, the CRT-based cryptanalysis by computing $\gcd((s'^e - m)\ \bmod\ n, n)$ does not give p nor q.

There is a well known improved CRT recombination, called Garner's algorithm [21, pp.612–613], which computes

$$CRT(s_p, s_q) = s_p + p \cdot [(p^{-1}(\bmod\ q) \cdot (s_q - s_p))\ \bmod\ q]. \qquad (12)$$

Unfortunately, for the Step-3 of the proposed CRT-1 protocol, it can be verified that if the value of $p^{-1}(\bmod\ q)$ is incorrect due to a hardware fault, then it follows that

$$\begin{cases} s' \bmod p = s_p \cdot (\tilde{m}^r)\ \bmod\ p \\ s' \bmod q \neq s_q \cdot (\tilde{m}^r)\ \bmod\ q. \end{cases} \qquad (13)$$

So, based on the CRT-based cryptanalysis, the composite integer n can be factorized by computing

$$p = \gcd((s'^e - m)\ \bmod\ n, n).$$

In fact, the above hardware fault attack on the CRT recombination is generic and can be applied to the original CRT speedup, Shamir's proposal, and the CRT-2 protocol proposed in this paper. However, the original Garner's algorithm can be modified as follows in order to avoid the above mentioned hardware fault cryptanalysis

$$CRT(s_p, s_q)$$
$$= s_p + [(p \cdot p^{-1}(\bmod\ q)) \cdot (s_q - s_p)]\ \bmod\ n$$
$$= s_p + [X \cdot (s_q - s_p)]\ \bmod\ n \qquad (14)$$

where $X = p \cdot p^{-1}(\bmod\ q)$ can be pre-computed and stored in advance.

The above suggested modification requires a little more computation than its original version. However, it can be easily verified that if the value of X is incorrect due to a hardware fault, the CRT recombination is still secure against the above mentioned attack.

With a similiar idea of fault infective computations of both s_p and s_q, Gauss's algorithm (see Eq. 10) and the modified Garner's algorithm (see Eq. 14) also exhibit a fault infection property such that both

$$\begin{cases} CRT(s_p, s_q) \bmod p \neq s_p \\ CRT(s_p, s_q) \bmod q \neq s_q \end{cases} \tag{15}$$

if either the value of X_p or X_q is incorrect and the value of X is incorrect, respectively. Therefore, no factorization of n is possible by computing $\gcd((CRT(s_p, s_q)^e - m) \bmod n, n)$ if both s_p and s_q are error free (or both are incorrect). Evidently, a careful selection of a CRT recombination procedure helps against the CRT based hardware fault attack substantially.

6   Key Pairs Selection and Performance Analysis

6.1   The Process of Selecting Key Pairs

In order to have better performance, the parameter e_r should be small enough to keep the overall overhead be negligible. As described previously, we don't assume that the public key e is a small integer. The generic assumption that e can be any integer between $(3, \phi(n) - 1)$ subject to $\gcd(e, \phi(n)) = 1$ is widely accepted in the design and analysis of any general purpose cryptosystem involving the RSA system. So, a small integer e_r ($\ll p - 1$ and $\ll q - 1$) can be selected such that $\gcd(e_r, \phi(n)) = 1$, then $d_r \equiv e_r^{-1} \pmod{\phi(n)}$ is computed.

The integer $d = d_r + r$ (where r is a small random integer) is tested whether $\gcd(d, \phi(n)) = 1$ in order to guarantee the existence of $e \equiv d^{-1} \pmod{\phi(n)}$. If such d and e are found, they are considered as the RSA private and public keys. Certainly, if more than one key pairs are found, then a pair with smaller e can be selected subject to the private key d being still a secure one.

6.2   Performance of Shamir's Protocol

Notice that in the CRT-based attack given a single erroneous computation result can totally break the RSA system by factoring the public modulus. Therefore, for a practical and reliable implementation of Shamir's idea, $|r|$ (bit length of r) will not be negligibly small and this will lead to a slow down of RSA signature computation for some degree.

Let the typical time for evaluating a modular multiplication $a \cdot b \bmod p$ (or $a \cdot b \bmod q$) be t_m. Then, using the standard square-and-multiply exponentiation [21] algorithm, the time for evaluating $m_p^{d_p} \bmod p$ (or $m_q^{d_q} \bmod q$) in a conventional CRT-based speedup is roughly $1.5|p|t_m$ and it will be denoted as T. In the above derivation, $m_p = m \bmod p$, $m_q = m \bmod q$, $d_p = d \bmod \phi(p)$, and $d_q = d \bmod \phi(q)$. For the conventional CRT-based speedup approach, the total computation time will be $2T$ excluding the cost for one extra CRT recombination (i.e., Eq. 1 or in notation as Eq. 2) which is considered to be negligible.

In Shamir's fault immune CRT-based speedup scheme, the total computation time will be

$$2T \left(\frac{|pr|}{|p|} \right)^3 \approx 2T \left(1 + \frac{|r|}{|p|} \right)^3 \tag{16}$$

excluding the cost for one extra CRT recombination and two modulo r operations which are considered to be negligible. Recall that the computational complexity of a modular multiplication is $O(\ell^2)$ (where ℓ is the bit length of all operands) in terms of bit operations. The computational complexity of a square-and-multiply exponentiation algorithm is $O(\ell)$. These lead to the above performance analysis of Shamir's proposal.

As described previously, the most serious reason of the CRT-based hardware fault attack is that given a *single* erroneous computation result can totally break the RSA system by factoring the public modulus. The selection of $|r| = 32$ (as suggested by Shamir) may not be enough for a highly reliable implementation. Therefore, performance of the CRT-based approach may be slowed down substantially when a higher security level will be required, e.g., with $|r| = 96$ so that $P_{miss} = 10^{-29}$ where P_{miss} is the probability of undetectability of attack. For example, let $|p| \approx |q| = 512$ (or $|n| = 1024$) and $|r| = 96$, then the overhead is at least about 68% when compared with the conventional CRT-based speedup approach even without considering many implementation overhead or inconveniency of evaluation when modulo pr and qr but not just modulo p and q.

6.3 Implementation Details and Performance of the Two New Protocols

There are some differences between the CRT-1 and the CRT-2 protocols. In the CRT-2 protocol, we do not need to compute or to store the parameter $\delta = p - q$. Furthermore, both s_p and s_q (see Eq. 8) in the step-2 of the CRT-2 protocol can be computed in parallel if there are two independent processors. However, in the CRT-1 protocol, s_p (see Eq. 5) should be computed prior to s_q since $\hat{m}$ is dependent on the value of s_p.

However, if we consider the real implementation details of these two protocols, we conclude that the CRT-1 protocol needs fewer memory (register) space. This may sometimes be important for an implementation in a small device like a smart IC card. In the following rough analysis, a full length register and a half length register are used to store values of $|p \times q|$ bits and $|p|$ (or $|q|$) bits, respectively. In the CRT-1 protocol, five half length temporary registers are used to store δ, k_p, k_q, s_p, and s_q. We ignore the storage for e_r since it is assumed to be a small integer. Also, three full length temporary registers are used to store d_r, $CRT(s_p, s_q)$, and both $(s_p^{er} \bmod p) + k_p \cdot \delta$ and $(s_q^{er} \bmod q) + k_q \cdot q$ by storage reusing.

In the CRT-2 protocol, four half length temporary registers are used to store k_p, k_q, s_p, and s_q. However, four full length temporary registers are used to store d_r, $CRT(s_p, s_q)$, $(s_p^{er} \bmod p) + k_p \cdot p$, and $(s_q^{er} \bmod q) + k_q \cdot q$.

The two proposed protocols CRT-1 and CRT-2 can perform almost equally well but with some minor additional modular multiplications and ordinary multiplications when compared with the conventional CRT-based speedup approach if small parameters r and e_r will be used. The selection and generation of such parameters have already been suggested in this section.

It can be verified that the worse case performance of the proposed CRT-based protocols will take about twice the effort (i.e., with 100% overhead) of the original CRT-based approach. This is for the case that we choose an RSA public key e first in order to guarantee that it will be a very small integer. However, in this situation it will be difficult to guarantee (it is still an open problem) that a small e_r can be found easily, even a small r can be easily selected. Noticeably, even in this situation the proposed protocols still perform twice so much the better than a computation without any CRT-based approach. Recall that in Shamir's protocol, let $|p| \approx |q| = 512$ (or $|n| = 1024$) and $|r| = 96$, then the overhead is at least about 68%. Most importantly, the two proposed protocols do not assume any checking procedure (which may be vulnerable to a hardware fault cryptanalysis easily) and do not suffer the disadvantage of undetectable faults.

7   Conclusions

In this paper, two novel CRT-based protocols are proposed to speed up the RSA signature or decryption computation via the residue number system technique, while at the same time being immune against the hardware fault cryptanalysis. Most importantly, the two protocols do not assume the existence of any error free and physical cryptanalysis immune *checking procedure* (e.g., by double checking, by using a final verification function – say raising a RSA signature to eth power, or by some kind of redundant computation – say Shamir's expanded modulus approach) or any form of *hardware checking module*. Nor will the proposed protocols suffer the disadvantage of producing an undetectable error with large probability as in Shamir's method. Therefore, if carefully implemented, Chinese remaindering based cryptosystem implementations are not more vulnerable than usual cryptosystem implementations.

As to the problem of how to enhance the proposed protocols against timing cryptanalysis, the interested readers can refer to Appendix A.

Acknowledgments

This work was supported by the Korea Information Security Agency (KISA) under contract R&D project 2001-S-092.

References

1. R.L. Rivest, A. Shamir, and L. Adleman, "A method for obtaining digital signatures and public-key cryptosystem," *Commun. of ACM*, vol. 21, no. 2, pp. 120–126, 1978.
2. T. ElGamal, "A public key cryptosystem and a signature scheme based on discrete logarithms," *IEEE Trans. Inf. Theory*, vol. 31, no. 4, pp. 469–472, 1985.

3. R. Anderson and M. Kuhn, "Tamper resistance – a cautionary note," In *Proceedings of the 2nd USENIX Workshop on Electronic Commerce*, pp. 1–11, 1996.
4. R. Anderson and M. Kuhn, "Low cost attacks on tamper resistant devices," In *Pre-proceedings of the 1997 Security Protocols Workshop*, Paris, France, 7–9th April 1997.
5. Bellcore Press Release, "New threat model breaks crypto codes," Sept. 1996, available at URL <http://www.bellcore.com/PRESS/ADVSRY96/facts.html>.
6. D. Boneh, R.A. DeMillo, and R.J. Lipton, "On the importance of checking cryptographic protocols for faults," In *Advances in Cryptology – EUROCRYPT '97*, LNCS 1233, pp. 37–51, Springer-Verlag, 1997.
7. F. Bao, R.H. Deng, Y. Han, A. Jeng, A.D. Narasimbalu, and T. Ngair, "Breaking public key cryptosystems on tamper resistant devices in the presence of transient faults," In *Pre-proceedings of the 1997 Security Protocols Workshop*, Paris, France, 1997.
8. Y. Zheng and T. Matsumoto, "Breaking real-world implementations of cryptosystems by manipulating their random number generation," In *Pre-proceedings of the 1997 Symposium on Cryptography and Information Security*, Fukuoka, Japan, 29th January–1st February 1997. An earlier version was presented at the rump session of *ASIACRYPT '96*.
9. I. Peterson, "Chinks in digital armor – Exploiting faults to break smart-card cryptosystems," *Science News*, vol. 151, no. 5, pp. 78–79, 1997.
10. M. Joye, J.-J. Quisquater, F. Bao, and R.H. Deng, "RSA-type signatures in the presence of transient faults," In *Cryptography and Coding*, LNCS 1355, pp. 155–160, Springer-Verlag, 1997.
11. D.P. Maher, "Fault induction attacks, tamper resistance, and hostile reverse engineering in perspective," In *Financial Cryptography*, LNCS 1318, pp. 109–121, Springer-Verlag, Berlin, 1997.
12. E. Biham and A. Shamir, "Differential fault analysis of secret key cryptosystems," In *Advances in Cryptology – CRYPTO '97*, LNCS 1294, pp. 513–525, Springer-Verlag, Berlin, 1997.
13. A.K. Lenstra, "Memo on RSA signature generation in the presence of faults," September 1996.
14. M. Joye, A.K. Lenstra, and J.-J. Quisquater, "Chinese remaindering based cryptosystems in the presence of faults," *Journal of Cryptology*, vol. 12, no. 4, pp. 241-245, 1999.
15. M. Joye, F. Koeune, and J.-J. Quisquater, "Further results on Chinese remaindering," Tech. Report CG-1997/1, UCL Crypto Group, Louvain-la-Neuve, March 1997.
16. A. Shamir, "How to check modular exponentiation," presented at the rump session of *EUROCRYPT '97*, Konstanz, Germany, 11–15th May 1997.
17. A. Shamir, "Method and apparatus for protecting public key schemes from timing and fault attacks," United States Patent 5991415, November 23, 1999.
18. S.M. Yen and M. Joye, "Checking before output may not be enough against fault-based cryptanalysis," *IEEE Trans. on Computers*, vol. 49, no. 9, pp. 967–970, Sept. 2000.
19. P.J. Smith and M.J.J. Lennon, "LUC: A new public key system," In *Ninth IFIP Symposium on Computer Security*, Elsevier Science Publishers, pp. 103–117, 1993.
20. J.-J. Quisquater and C. Couvreur, "Fast decipherment algorithm for RSA public-key cryptosystem," *Electronics Letters*, vol. 18, no. 21, pp. 905–907, 1982.
21. A.J. Menezes, P.C. van Oorschot, and S.A. Vanstone. *Handbook of applied cryptography*. CRC Press, 1997.

22. P. Kocher, "Timing attacks on implementations of Diffie-Hellman, RSA, DSS, and other systems," In *Advances in Cryptology – CRYPTO '96*, LNCS 1109, pp. 104–113, Springer-Verlag, 1996.
23. B.S. Kaliski Jr. and M.J.B. Robshaw, "Comments on some new attacks on cryptographic devices," RSA Laboratories Bulletin, no. 5, July 1997.
24. W. Schindler, "A timing attack against RSA with the Chinese Remainder Theorem," In *Cryptographic Hardware and Embedded Systems – CHES 2000*, LNCS 1965, pp. 109–124, Springer-Verlag, 2000.
25. C.D. Walter, "Montgomery's exponentiation needs no final subtractions," *Electronics Letters*, vol. 35, no. 21, pp. 1831–1832, 1999.
26. C. Hachez and J.-J. Quisquater, "Montgomery exponentiation with no final subtractions: Improved results," In *Cryptographic Hardware and Embedded Systems – CHES 2000*, LNCS 1965, pp. 293–301, Springer-Verlag, 2000.

A Enhancement for Timing Attack Resistance

As described previously, Shamir's method can withstand a timing cryptanalysis of a binary search approach [17,22]. This is a very important reason that makes Shamir's method be superior to the conventional simple verification method.

In the proposed CRT-1 and CRT-2 protocols, another binary search timing cryptanalysis is possible over the $k_p = \lfloor m/p \rfloor$ and $k_q = \lfloor m/q \rfloor$ operations. Suppose an operation of $\lfloor a/q \rfloor$ will take a significant less time if $a < q$. An attacker can intentionally provide a larger m_ℓ and a smaller m_s iteratively in order to approximate the value of p by observing the timing statistics.

In the following, a novel idea is proposed to enhance the security of CRT-1 and CRT-2 in order to be against the above two timing cryptanalyses. In order to be against the binary search timing attack over modulo p (or modulo q) operation, a random integer $R = k \cdot p$ (where k is a random integer) is selected such that $R \gg p$. Then, the original process to obtain the parameter m_p (or m_q) for CRT computation is modified as

$$m_p = (m + R) \bmod p$$

where R is used as a *pre-masking* but no *post-masking* is necessary since $R \bmod p = 0$. To compute m_q, the random integer R is selected to be $k \cdot q$.

In order to be against the binary search timing attack over $\lfloor m/p \rfloor$ (or $\lfloor m/q \rfloor$) operation, a random integer of the form $R = k \cdot p$ is also applicable. Then, the original process to compute k_p (or k_q) is modified as

$$k_p = \lfloor (m + R)/p \rfloor - k$$

where R is used as a *pre-masking* and k is used as a *post-masking* since $\lfloor R/p \rfloor = k$. To compute k_q, the random integer R is selected to be $k \cdot q$.

If a random integer R ($\gg p, q$) is added with the message m prior to the modulo p or $\lfloor m/p \rfloor$ operation, the timing characteristic of whether or not modulo p or $\lfloor m/p \rfloor$ has been performed (by intentionally providing a larger m_ℓ or a smaller m_s) cannot be accessible by an attacker.

With the same property of foiling binary search timing cryptanalysis to discover p (or q), the above proposed technique suffers less overhead than Shamir's method from the viewpoint of implementation. In Shamir's method, expanded modulus is required which brings some implementation inconvenience or complexity for most of existing processors. However, no expanded modulus is necessary in the above proposal.

A remark is given in the following to consider another recently reported binary search timing cryptanalysis proposed by Schindler [24] over a RSA implementation with both CRT speedup and Montgomery exponentiation algorithm. Schindler's research showed that if a conventional Montgomery multiplication with possible final subtraction is employed, then a binary search timing attack can be efficiently performed to find p (or q). Schindler suggested one possible countermeasure by using a conventional blinding technique [22,23] (or sometimes called masking technique). Another suggestion is employment of two recently developed Montgomery exponentiations with no final subtraction [25,26]. In this approach, both the values of $R^e \bmod n$ and $R^{-1} \bmod n$ (where R is a large random integer) are precomputed in advance and are used as a pre-masking and a post-masking, respectively. In fact, the blinding technique can also be used to counteract the above two binary search timing attacks on either modulo p or $\lfloor m/p \rfloor$ operation.

As described previously, there are two possible parameter settings for the two proposed protocols. One setting will have a small public key e and the other setting will have a large e. Note especially that the blinding technique can be applicable to both parameter settings either with small e or large e since $R^e \bmod n$ is precomputed only once. Further updating does not need to raise a value to eth power. However, it should be carefully noticed that the blinding technique itself cannot withstand the CRT-based hardware fault cryptanalysis!

A Countermeasure against One Physical Cryptanalysis May Benefit Another Attack

Sung-Ming Yen[1], Seungjoo Kim[2], Seongan Lim[2], and Sangjae Moon[3]

[1] Laboratory of Cryptography and Information Security (LCIS),
Department of Computer Science and Information Engineering,
National Central University, Chung-Li, Taiwan 320, R.O.C.
yensm@csie.ncu.edu.tw
[2] Cryptographic Technology Team,
Information Security Technology Department,
Korea Information Security Agency,
78, Garak-Dong, Songpa-Gu, Seoul, Korea 138-803
{skim,seongan}@kisa.or.kr
[3] School of Electronic and Electrical Engineering,
Kyungpook National University,
Taegu, Korea 702-701
sjmoon@knu.ac.kr

Abstract. Recently, many research works have been reported about how physical cryptanalysis can be carried out on cryptographic devices by exploiting any possible leaked information through side channels. In this paper, we demonstrate a new type of safe-error based hardware fault cryptanalysis which is mounted on a recently reported countermeasure against simple power analysis attack. This safe-error based attack is developed by inducing a temporary random computational fault other than a temporary memory fault which was explicitly assumed in the first published safe-error based attack (in which more precisions on timing and fault location are assumed) proposed by Yen and Joye. Analysis shows that the new safe-error based attack proposed in this paper is powerful and feasible because the cryptanalytic complexity (especially the computational complexity) is quite small and the assumptions made are more reasonable. Existing research works considered many possible countermeasures against each kind of physical cryptanalysis. This paper and a few previous reports clearly show that a countermeasure developed against one physical attack does not necessarily thwart another kind of physical attack. However, almost no research has been done on dealing the possible mutual relationship between different kinds of physical cryptanalysis when choosing a specific countermeasure. Most importantly, in this paper we wish to emphasize that a countermeasure developed against one physical attack if not carefully examined may benefit another physical attack tremendously. This issue has never been explicitly noticed previously but its importance can not be overlooked because of the attack found in this paper. Notice that almost all the issues considered in this paper on a modular exponentiation also applies to a scalar multiplication over an elliptic curve.

K. Kim (Ed.): ICICS 2001, LNCS 2288, pp. 414–427, 2002.
© Springer-Verlag Berlin Heidelberg 2002

Keywords: Cryptography, Exponentiation, Hardware fault cryptanalysis, Physical cryptanalysis, Power analysis attack, Side channel attack, Square-multiply exponentiation, Timing attack.

1   Introduction

In order to provide a better support for data protection under strong cryptographic schemes (e.g., RSA [1] or ElGamal [2] systems), varieties of implementations based on tamper-proof devices (e.g., smart IC cards) are proposed. The main reason for this trend is that smart IC cards are assumed to provide high reliability and security with more memory capacity and better performance characteristics than conventional plastic cards. The central processing unit (CPU) in a smart IC card controls its data input and output and prevents unauthorized access to a smart card. With special characteristics of computational ability, large memory capacity and security, a large variety of cryptographic applications benefit from smart IC cards. Due to this popular usage of tamper-resistance, much attention has recently been paid regarding the security issues of cryptosystems implemented on tamper-proof devices [3,4]. Because of the tremendous differences between this branch of new cryptanalysis and the conventional pure mathematical approaches [5], these new approaches are specifically called the *physical cryptanalysis.*

Two categories are identified among the physical cryptanalysis, one approach will interfere the target device to analyze (called an active attack) and the other approach will not interrupt a normal operation (called a passive attack).

Many works have been reported on the active physical cryptanalysis [6,7,8,9,10,11,12,13,14,15,16,17,18,19] from the viewpoint of presence of hardware faults. This line of research re-emerged in September 1996 when a Bellcore press release [6] reported a new kind of attack, the so-called *fault-based cryptanalysis.* In the fault-based cryptanalysis model, it is assumed that when an adversary has physical access to a tamper-proof device she may purposely induce a certain type of fault into the device. Based on a set of incorrect outputs from a device, due to the presence of faults, an adversary may extract the secret key embedded in the tamper-proof device. These attacks exploit the presence of many kinds of transient or permanent faults. These attacks are of very general nature and remain valid for a large variety of cryptosystems, e.g., LUC public key cryptosystem [20] and elliptic curve cryptography (ECC) [21]. The above hardware fault cryptanalysis is one kind of active attack such that the normal operations of a cryptographic scheme are interfered (except by a *safe-error* proposed in [19]) and an incorrect result will be produced.

There are two well recognized and studied passive attacks have been reported in the literature, i.e., the timing cryptanalysis and the power analysis attack.

The *timing cryptanalysis* was originally proposed by Kocher [22] in 1996 and was further improved in [23,24,25,26]. In a timing cryptanalysis, it is assumed that an adversary has physical access to a device and can collect some timing information of a normal cryptographic operation. Based on the observation and simple statistical analysis on a large set of timing information, an adversary may extract the secret key employed during the normal cryptographic operation. Some considerations about timing cryptanalysis can be found in [27].

The *power analysis attack* (another kind of passive attack) was also proposed originally by Kocher [28,29] in which both simple power analysis (SPA)[1] and differential power analysis (DPA) are considered. SPA is conducted by observing a single power consumption trace of an execution process in order to extract secret key. On the other hand, DPA is conducted by the observation and simple statistical analysis on a large set of collected power consumption traces. Many further research results on power analysis attack can be found in the literature, among them, works specific for public key cryptographic schemes are [30,31,32,33,34,35] and works specific for symmetric key cryptosystems are [36,37,38,39,40,41,42,43,44,45,46,47].

For the past few years, much attentions have been paid to propose new physical attacks and to develop many countermeasures against each kind of physical cryptanalysis. However, extremely few attention has been paid on considering the mutual relationship bewteen different kinds of physical cryptanalysis when choosing a specific countermeasure except that a few research works pointed out a straightforward result that a countermeasure against one physical attack does not necessarily thwart another kind of physical attack.

The main contribution of this paper is to consider a nontrivial issue of whether a countermeasure developed against one physical attack if not carefully examined may benefit another physical attack tremendously. Based on the research of this paper, the answer of the above question is definite. Our main objective is to emphasize that the above situation will complex the design and analysis of developing countermeasures against physical cryptanalysis quite a large. This extremely important issue has never been explicitly reported previously in the open literature but its importance can not be overlooked because of the attack found in this paper.

As a case study in order to assure our above claim, a new type of *safe-error* based hardware fault cryptanalysis mounted on a recently reported countermeasure [30] against SPA is proposed. This new safe-error based attack is conducted by inducing a temporary random *computational fault* other than a temporary *memory fault* which was explicitly assumed in the first published safe-error based attack proposed by Yen and Joye [19] (in which more precisions on timing and fault location are required). Analysis shows that the new safe-error based attack proposed in this paper is powerful and feasible because the cryptanalytic complexity (especially the computational complexity) is quite small and the assumptions made are much more reasonable. Notice that the new attacks considered in this paper on a modular exponentiation also applies to a scalar multiplication over an elliptic curve.

---

[1] In 1997, this has independently been considered by S.M. Yen in a research project supported in part by the Computer & Communication Research Laboratories, Industrial Technology Research Institute, R.O.C. under contract G3-87052.

2   Preliminary Background

Since a physical cryptanalysis (an engineering approach of attack) exploits some special properties within an implementation of cryptography, thus we need to take a closer examination of some particular implementation details to understand existing attacks and to develop new types of attacks. In this section, a conventional and an enhanced version (against SPA) of exponentiation computation algorithms will be reviewed.

2.1   Binary Exponentiation

The Algorithm 1 (see Fig. 1) is a commonly used algorithm for computing RSA cryptosystem $m^d \bmod n$ where the exponent d is expressed in binary form as $d = \sum_{i=0}^{k-1} d_i 2^i$. This conventional method is usually referred to as the left-to-right binary exponentiation algorithm [48].

Input: $m, d = (d_{k-1}, \ldots, d_0)_2, n$
Output: $A = m^d \bmod n$

1.1 $A \leftarrow 1$
1.2   <u>for</u> i <u>from</u> $k - 1$ <u>downto</u> 0
1.3 $A \leftarrow A^2 \bmod n$
1.4   <u>if</u> $(d_i = 1)$ <u>then</u> $A \leftarrow A \cdot m \bmod n$
1.5   <u>endfor</u>

Fig. 1. Left-to-right binary exponentiation – Algorithm 1.

2.2   SPA Immune Left-to-Right Binary Exponentiation

It is well known that a simple power analysis is conducted by observing a single power consumption trace of an execution process which implements a cryptographic system. Because the electrical power consumed for an execution process at any given time is related to both the instruction being executed and the data (it could be a part of a secret key) being manipulated. Power consumption analysis enables one to distinguish visually one instruction from other instructions. Therefore, branch instructions of secret key dependent in an algorithm should be avoided or at least be minimized. If squaring and general purpose multiplication perform slightly different in terms of power consumption (and sometimes also slightly different in term of timing) or it performs different in terms of power consumption after the conditional branch in Line 1.4 (either to compute a modular multiplication immediately or back to a **for** loop control procedure), then the conventional left-to-right binary exponentiation algorithm in Fig. 1 may be vulnerable to SPA since an explicit secret key dependent decision and branch instruction exists (Line 1.4 in Fig. 1).

In [30], Coron proposed a modified double-and-add algorithm for elliptic curve cryptosystems computation which was announced to be resistant against SPA. A SPA immune left-to-right binary exponentiation algorithm given in Fig. 2 can be obtained directly from Coron's proposal. In this paper, the idea in Fig. 2 is called the *square-and-multiply-always* exponentiation[2].

$$
\begin{aligned}
&\texttt{Input:}\quad m, d = (d_{k-1}, \ldots, d_0)_2, n \\
&\texttt{Output:}\ A[0] = m^d \bmod n
\end{aligned}
$$

2.1 $A[0] \leftarrow 1$
2.2   <u>for</u> i <u>from</u> $k-1$ <u>downto</u> 0
2.3 $A[0] \leftarrow A[0]^2 \bmod n$
2.4 $A[1] \leftarrow A[0] \cdot m \bmod n$
2.5 $A[0] \leftarrow A[d_i]$
2.6   <u>endfor</u>

Fig. 2. SPA immune left-to-right binary exponentiation – Algorithm 2.

We note that the step 2.4 and step 2.5 can be combined into a more concise form as

$$A[\bar{d}_i] \leftarrow A[0] \cdot m \bmod n \tag{1}$$

where $\bar{d}_i$ denotes the binary complement of d_i. However, this helps not much on the performance and does not help on the robustness against the proposed hardware fault-based cryptanalysis in the following section.

2.3   DPA Countermeasures and Some Remarks

In [30, Section 5], Coron also reported some countermeasures against DPA based on some enhancements of the above SPA immune version. Two solutions in the context of modular exponentiation are briefly reviewed in the following.

The first solution suggests to randomly encode the original secret exponent d to be $d' = d + k\phi(n)$ where k is a random integer and $\phi(n)$ is the Euler totient function of n. Since $m^d \equiv m^{d'} \pmod{n}$, the result will be correct but each power trace monitored by an adversary will be independent. This idea can also be found in a patent by Shamir [18].

The second solution suggests to employ a blinding technique to randomize the base integer m to be $m' = mr \bmod n$ where r is a random integer. This makes the attacks infeasible since the blinded base integer m' is unknown to an adversary. Finally, the required result $m^d \bmod n$ is computed by $(m'^d) \cdot (r^d)^{-1} \bmod n$ where

---

[2] It was pointed out by one reviewer that the "square-and-multiply-always" idea is due to Kocher (in one patent) but not due to Coron. Coron simply applied it to elliptic curve implementations. However, no such exact patent by Kocher has been found by the authors until now.

$(r^d)^{-1}$ mod n can be precomputed. A slightly different version of the idea can be found in [22,27].

Remark: Notice that in the above two DPA countermeasures, a base integer is raised to the secret exponent (either in the form of d or d', but in fact does not make any difference in terms of attack). Therefore, any possible weakness from the viewpoint of SPA can totally break the above two DPA immune enhanced versions. This clearly shows the importance of developing a good SPA countermeasure and motivates the research of this paper.

3   New Fault-Based Attack

In [19], a potential fault-based attack was reported by Yen and Joye where secret key bits leak only through the information of whether or not the hardware device produces a faulty output after a temporary *memory fault* was induced. The attack exploits novely the interaction between an exponentiation algorithm and an interleaved modular multiplication algorithm [49,50,51]. In the original paper, this type of cryptanalysis is called a *safe-error* based physical cryptanalysis.

As the name of the attack suggested, during the execution of an algorithm, if for some reasons, some bits of a register are maliciously modified but it afterward will be overwritten by another value. Thus, the error in a register will be cleared and the final result of the algorithm could be correct. Therefore, the modified value in the register will not damage the correctness of the computation. Such kind of temporary error is called a *memory safe-error* (abbreviated as M safe-error).

In the following subsection, another kind of safe-error model is introduced where a temporary random *computational safe-error* (abbreviated as C safe-error) within the arithmetic-logic unit (ALU) will be induced and secret key bits also leak only through the information of whether or not the hardware device produces a faulty output.

3.1   Proposed Attack on SPA Immune Algorithm 2

Firstly, it is easy to observe that when a computational error is introduced within the ALU during the operation $A[0] \leftarrow A[0]^2$ mod n (see Line 2.3 in Fig. 2), then it will also force the next operation $A[1] \leftarrow A[0] \cdot m$ mod n to be incorrect. This is evident because the correct value of $A[0]$ is required during the evaluation of $A[1]$. Therefore, the final result of the algorithm will be incorrect but no useful information can be obtained by an attacker.

However, if a *computational safe error* is introduced into the ALU during the operation $A[1] \leftarrow A[0] \cdot m$ mod n (see Line 2.4 in Fig. 2), then this error will be ignored if the secret key bit d_i is zero. This is because that the faulty result $A[1]$ will not be selected in the next operation $A[0] \leftarrow A[d_i]$ (see Line 2.5 in Fig. 2) if $d_i = 0$. Eventually, the final result of the algorithm will be correct (if no other fault of any type will occur in the following operations) and the secret key bit

```
for i from k − 1 downto 0
    Introduce a computational safe-error into ALU during
                the step 2.4 and complete the exponentiation
    if (output refused or incorrect) then
        d_i ← 1
    else
        d_i ← 0
    endif
endfor
```

Fig. 3. Proposed attack (left-to-right version).

$d_i = 0$ can be extracted by exploiting the nature of this C safe-error. Otherwise, the final result will be incorrect and $d_i = 1$ can be extracted.

Some designs may have a final verification function to check the correctness of a result (e.g., by raising a RSA signature to eth power) prior to send out the computed result. However, this checking procedure does not help against the proposed attack (this is also the case in [19]) since a refusing reaction implies an incorrect result. The above attack is sketched hereafter in Fig. 3.

Notice that the above C safe-error based attack does not work upon the conventional exponentiation algorithm (see Fig. 1), however it does work upon a modified version in order to counteract SPA (see Fig. 2). The above attack illustrates our claim that a countermeasure developed against one physical attack if not carefully examined may benefit another physical attack tremendously (this will be demonstrated in the following feasibility analysis). This issue has never been noticed previously but the importance can not be overlooked because of the above attack.

One main reason for the above C safe-error attack is because of the structure of step 2.3 to step 2.5 in which one normal operation and one dummy operation are performed, then one of these two results is selected depending on the secret key bit d_i. An attacker tries to locate the dummy operation and to influence its computation with a random fault. If an attacker guesses correctly, then the final result will be error free since result of the dummy operation will be ignored. Otherwise, the final result will be incorrect. Any cryptographic implementation algorithm with the above structure may suffer an attack by exploiting C safe-error.

3.2   Feasibility of the Attack

Basically, the feasibility of a hardware fault-based cryptanalysis can be measured from the viewpoints of:

(1) controllability of a fault location;
(2) timing precision of introducing a fault;
(3) fault type assumed (e.g., random fault or flipping fault);
(4) number of bits of fault occurred.

It will be shown that the above proposed computational safe-error is extremely feasible under all the above four considerations.

For the issue of fault location, hardware fault-based cryptanalysis is usually characterized by a very precise controllability of fault location; a loose controllability of fault location; or no controllability of fault location. The first two cases ask a fault to be induced into its right position. Notice that M safe-error proposed in [19] is of the second case of fault controllability. In fact, if a fault is to be induced into a register, even in the second case, it requires more or less precise controllability of fault location since a register is often implemented as a very narrow shape of silicon area. In the proposed C safe-error attack, controllability of fault location has been minimized extensively because an ALU is often implemented as a very large and non-narrow shape of silicon area.

For the issue of fault occurrence time, a fault can be introduced into a hardware with a very precise timing controllability (e.g., when one or a few bits are accessed); a loose timing controllability (e.g., during only a percentage of the period of an operation); or minimized timing controllability (e.g., during the whole period of an operation). M safe-error proposed in [19] is of the second case of timing controllability. But, timing controllability of the proposed C safe-error is of the third case. Notice that in the case of the above proposed C safe-error attack, both step 2.3 and step 2.4 are the most time consuming operations. It implies that the time period within which a random fault should be induced is large and this will ease the attack. The only requirement is that a fault or multiple faults occurred at multiple instances (maybe caused by any possible imprecision of a fault generation equipment) are introduced within a single operation $A[0] \cdot m \bmod n$. In the SPA immune left-to-right binary exponentiation (see Fig. 2), each iteration needs two modular multiplications and the total number of modular multiplications to be performed within the algorithm are always $2k$. Therefore, a very precise estimate of the computation time for a single modular multiplication (e.g., $A[0] \cdot m \bmod n$) can be easily obtained after a few simple experiments by dividing the time to compute $m^d \bmod n$ by $2k$. These timing samples can even be obtained from some different cards.

Three types of fault are commonly referred to when analyzing a hardware fault-based cryptanalysis; they are stuck at 1 or 0 fault, bit flipping fault, and random fault. Many existing attacks assume the existence of stuck fault or flipping fault. However, these two kinds of fault are not easy to induce (at the right place and at the right time) and are considered as impractical or controversial models. On the other hand, random faults (especially those without precise specification of location and time) are generally considered as practical models and are more easy to induce. In C safe-error based attack proposed in this paper, we assume only the existence of random faults. In fact, many possible ways to introduce one or more faults within a single operation $A[0] \cdot m \bmod n$ inside an ALU can be possible. For the case of multiple faults, the faults can either be occurred at the same time over many locations inside an ALU or be occurred sequentially and distributed over many time instances within a single operation.

Input: $m, d = (d_{k-1}, \ldots, d_0)_2, n$
Output: $A = m^d \bmod n$

4.1 $A \leftarrow 1; B \leftarrow m$
4.2   <u>for</u> i <u>from</u> 0 <u>to</u> $k-1$
4.3   <u>if</u> $(d_i = 1)$ <u>then</u> $A \leftarrow A \cdot B \bmod n$
4.4 $B \leftarrow B^2 \bmod n$
4.5   <u>endfor</u>

Fig. 4. Right-to-left binary exponentiation – Algorithm 4.

Input: $m, d = (d_{k-1}, \ldots, d_0)_2, n$
Output: $A[0] = m^d \bmod n$

5.1 $A[0] \leftarrow 1; B \leftarrow m$
5.2   <u>for</u> i <u>from</u> 0 <u>to</u> $k-1$
5.3 $A[1] \leftarrow A[0] \cdot B \bmod n$
5.4 $A[0] \leftarrow A[d_i]$
5.5 $B \leftarrow B^2 \bmod n$
5.6   <u>endfor</u>

Fig. 5. SPA immune right-to-left binary exponentiation – Algorithm 5.

For the issue of number of bits of fault, many existing attacks assume that only a small number of bits (some attacks assume the extreme case of single-bit fault) will be faulty during the cryptanalysis. Generally speaking, it is more difficult to induce a single-bit fault (or limited number of faulty bits) precisely than to induce a general type of fault with any number of faulty bits. In C safe-error based attack, any fault pattern within an ALU can be possible. Notice that in M safe-error attack [19], although it does not limit how many bits of fault to be induced into a register, however it does require that the bits to be corrupted should be no longer required. Also, the fault should be introduced within a single register.

4   Extension of C Safe-Error to Other Implementations

Although we demonstrate C safe-error based attack under the left-to-right binary exponentiation technique in a previous section, it can be easily verified that the attack still works when the right-to-left exponentiation technique [48] (see Fig. 4) is employed for computing $m^d \bmod n$.

The SPA immune right-to-left binary exponentiation algorithm given in Fig. 5 is based on the similar idea from Fig. 2.

The step 5.3 and step 5.4 in Fig. 5 can be combined into a more concise form as

$$A[\bar{d}_i] \leftarrow A[0] \cdot B \bmod n.$$

```
for i from 0 to k - 1
    Introduce a computational safe-error into ALU during
                     the step 5.3 and complete the exponentiation
    if (output refused or incorrect) then
        d_i ← 1
    else
        d_i ← 0
    endif
endfor
```

Fig. 6. Proposed attack (right-to-left version).

However, this helps not much on the performance and does not help on the robustness against C safe-error attack. The procedure in Fig. 6 summarizes the attack to recover a RSA secret exponent d.

Notice that the proposed C safe-error attack also applies to any possible square-and-multiply-always algorithm variants employing k-ary exponentiation or sliding-window exponentiation [48, Section 14.6]. Furthermore, C safe-error attack also applies to Coron's original algorithm [30].

5   Related Recent Works

Some other related works (also by some authors of this paper) on the issues considered in this paper are summarized in the following. The complete results of these works will be published recently.

In a recent article [33], discussions on the security of square-and-multiply-always algorithm [30] against DPA have been given and a new countermeasure was developed that is secure against power attacks. However, a recent researche[3] shows that the algorithm (as an efficient countermeasure against DPA and timing attack) proposed in [33] is vulnerable to M safe-error attack [19]. Notice that the original algorithm in [30] can be M safe-error attack free.

In order to resist a reportd timing cryptanalysis [25], a modified implementation of the Rijndael cipher (i.e., the AES) was suggested by the authors of Rijndael and was widely adopted in many Rijndael implementations [41]. However, a recent researche [52] shows that the modified implementation is much more vulnerable to DPA than the original implementation will be.

Another related work can be found in [53] showing how an improved cryptosystem can fail by observability analysis on the response received from the cryptosystem. This work however does not consider directly the mutual relationship between different kinds of physical cryptanalysis and countermeasures.

The above recent research results show more cases that a countermeasure developed against one physical attack if not carefully examined may benefit another physical attack tremendously. The research results also emphasize again

---

[3] An attack on [33] was discovered by S.M. Yen and M. Joye and a technical report is under prepared.

that when choosing countermeasures against physical attacks, an indepth research on the possible mutual relationship between different kinds of physical cryptanalysis is necessary.

6   Conclusions

For the past few years, much attention has been paid to design efficient countermeasures for each kind of physical cryptanalysis. However, almost no research has been done on dealing the possible mutual relationship between different kinds of physical cryptanalysis when choosing a specific countermeasure.

The first contribution of this paper is that a new type of safe-error attack (called a computational safe-error attack) has been identified. This C safe-error attack is mounted upon a recently reported countermeasure against SPA. Analysis shows that C safe-error attack proposed in this paper is powerful and feasible because the cryptanalytic complexity is quite small and the cryptanalytic assumptions made (controllability of fault location and controllability of timing) are much more reasonable when compared with all previous existing hardware fault-based attacks.

The proposed C safe-error attack mounted upon a binary modular exponentiation (by square-and-multiply) also applies to more advanced exponentiation algorithms and also applies to a scalar multiplication (by double-and-add) over an elliptic curve proposed by Coron [30] which was announced to be resistant against SPA.

The second contribution of this paper is that the research result clearly shows that a countermeasure developed against one physical attack if not carefully examined may benefit another physical attack tremendously. Our main objective is to emphasize that the above observation will complex the design and analysis of developing countermeasures against physical cryptanalysis quite a large. The main idea of square-and-multiply-always algorithm is straightforward and correct in the context of SPA, but this does not apply automatically to another physical attack. The worst thing is that it opens a backdoor for safe-error based hardware fault cryptanalysis, but this can be avoided easily in the original exponentiation computation by using Algorithm 1 (see Fig. 1) if a precaution suggested in [19] is taken.

Acknowledgments

This work was supported in part by the Korea Information Security Agency (KISA) under contract R&D project 2001-S-092.

References

1. R.L. Rivest, A. Shamir, and L. Adleman, "A method for obtaining digital signatures and public-key cryptosystem," *Commun. of ACM*, vol. 21, no. 2, pp. 120–126, 1978.

2. T. ElGamal, "A public key cryptosystem and a signature scheme based on discrete logarithms," *IEEE Trans. Inf. Theory*, vol. 31, no. 4, pp. 469–472, 1985.
3. R. Anderson and M. Kuhn, "Tamper resistance – a cautionary note," In *Proceedings of the 2nd USENIX Workshop on Electronic Commerce*, pp. 1–11, 1996.
4. R. Anderson and M. Kuhn, "Low cost attacks on tamper resistant devices," In *Pre-proceedings of the 1997 Security Protocols Workshop*, Paris, France, 7–9th April 1997.
5. D. Boneh, "Twenty years of attacks on the RSA cryptosystem," *Notices of the AMS*, vol. 46, no. 2, pp. 203–213, Feb 1999.
6. Bellcore Press Release, "New threat model breaks crypto codes," Sept. 1996, available at URL <http://www.bellcore.com/PRESS/ADVSRY96/facts.html>.
7. D. Boneh, R.A. DeMillo, and R.J. Lipton, "On the importance of checking cryptographic protocols for faults," In *Advances in Cryptology – EUROCRYPT '97*, LNCS 1233, pp. 37–51, Springer-Verlag, 1997.
8. F. Bao, R.H. Deng, Y. Han, A. Jeng, A.D. Narasimbalu, and T. Ngair, "Breaking public key cryptosystems on tamper resistant devices in the presence of transient faults," In *Pre-proceedings of the 1997 Security Protocols Workshop*, Paris, France, 1997.
9. Y. Zheng and T. Matsumoto, "Breaking real-world implementations of cryptosystems by manipulating their random number generation," In *Pre-proceedings of the 1997 Symposium on Cryptography and Information Security*, Fukuoka, Japan, 29th January–1st February 1997. An earlier version was presented at the rump session of *ASIACRYPT '96*.
10. I. Peterson, "Chinks in digital armor – Exploiting faults to break smart-card cryptosystems," *Science News*, vol. 151, no. 5, pp. 78–79, 1997.
11. M. Joye, J.-J. Quisquater, F. Bao, and R.H. Deng, "RSA-type signatures in the presence of transient faults," In *Cryptography and Coding*, LNCS 1355, pp. 155–160, Springer-Verlag, 1997.
12. D.P. Maher, "Fault induction attacks, tamper resistance, and hostile reverse engineering in perspective," In *Financial Cryptography*, LNCS 1318, pp. 109–121, Springer-Verlag, Berlin, 1997.
13. E. Biham and A. Shamir, "Differential fault analysis of secret key cryptosystems," In *Advances in Cryptology – CRYPTO '97*, LNCS 1294, pp. 513–525, Springer-Verlag, Berlin, 1997.
14. A.K. Lenstra, "Memo on RSA signature generation in the presence of faults," September 1996.
15. M. Joye, A.K. Lenstra, and J.-J. Quisquater, "Chinese remaindering based cryptosystems in the presence of faults," *Journal of Cryptology*, vol. 12, no. 4, pp. 241-245, 1999.
16. M. Joye, F. Koeune, and J.-J. Quisquater, "Further results on Chinese remaindering," Tech. Report CG-1997/1, UCL Crypto Group, Louvain-la-Neuve, March 1997.
17. A. Shamir, "How to check modular exponentiation," presented at the rump session of *EUROCRYPT '97*, Konstanz, Germany, 11–15th May 1997.
18. A. Shamir, "Method and apparatus for protecting public key schemes from timing and fault attacks," United States Patent 5991415, November 23, 1999.
19. S.M. Yen and M. Joye, "Checking before output may not be enough against fault-based cryptanalysis," *IEEE Trans. on Computers*, vol. 49, no. 9, pp. 967–970, Sept. 2000.
20. P.J. Smith and M.J.J. Lennon, "LUC: A new public key system," In *Ninth IFIP Symposium on Computer Security*, Elsevier Science Publishers, pp. 103–117, 1993.

21. I.F. Blake, G. Seroussi, and N.P. Smart. *Elliptic curves in cryptography.* vol. 265 of London Mathematical Society Lecture Note Series, Cambridge University Press, 1999.
22. P. Kocher, "Timing attacks on implementations of Diffie-Hellman, RSA, DSS, and other systems," In *Advances in Cryptology – CRYPTO '96*, LNCS 1109, pp. 104–113, Springer-Verlag, 1996.
23. J.F. Dhem, F. Koeune, P.A. Leroux, P. Mestre, J.J. Quisquater, and J.L. Willems, "A practical implementation of the timing attack," *Technical Report CG-1998/1*, UCL Crypto Group, Université catholique de Louvain, June 1998.
24. J.F. Dhem, F. Koeune, P.A. Leroux, P. Mestre, J.J. Quisquater, and J.L. Willems, "A practical implementation of the timing attack," In *Proceedings of CARDIS '98 – Third Smart Card Research and Advanced Application Conference*, UCL, Louvain-la-Neuve, Belgium, Sep. 14-16, 1998.
25. F. Koeune and J.-J. Quisquater, "A timing attack against Rijndael," *Technical Report CG-1999/1*, Université catholique de Louvain, June 1999.
26. W. Schindler, "A timing attack against RSA with the Chinese Remainder Theorem," In *Cryptographic Hardware and Embedded Systems – CHES 2000*, LNCS 1965, pp. 109–124, Springer-Verlag, 2000.
27. B.S. Kaliski Jr. and M.J.B. Robshaw, "Comments on some new attacks on cryptographic devices," RSA Laboratories Bulletin, no. 5, July 1997.
28. P. Kocher, J. Jaffe and B. Jun, "Introduction to differential power analysis and related attacks," 1998, available at URL
    <http://www.cryptography.com/dpa/technical>.
29. P. Kocher, J. Jaffe and B. Jun, "Differential power analysis," In *Advances in Cryptology – CRYPTO '99*, LNCS 1666, pp. 388–397, Springer-Verlag, 1999.
30. J.-S. Coron, "Resistance against differential power analysis for elliptic curve cryptosystems," In *Cryptographic Hardware and Embedded Systems – CHES '99*, LNCS 1717, pp. 292–302, Springer-Verlag, 1999.
31. T.S. Messerges, E.A. Dabbish, and R.H. Sloan, "Power analysis attacks of modular exponentiation in smartcards," In *Cryptographic Hardware and Embedded Systems – CHES '99*, LNCS 1717, pp. 144–157, Springer-Verlag, 1999.
32. C. Clavier, J.-S. Coron, and N. Dabbous, "Differential power analysis in the presence of hardware countermeasures," In *Cryptographic Hardware and Embedded Systems – CHES 2000*, LNCS 1965, pp. 252–263, Springer-Verlag, 2000.
33. K. Okeya and K. Sakurai, "Power analysis breaks elliptic curve cryptosystems even secure against the timing attack," In *Advances in Cryptology – INDOCRYPT 2000*, LNCS 1977, pp. 178–190, Springer-Verlag, 2000.
34. C.D. Walter, "Sliding windows succumbs to big mac attack," In *Pre-proceedings of Workshop on Cryptographic Hardware and Embedded Systems – CHES 2001*, pp. 291–304, May 13-16, 2001.
35. C. Clavier and M. Joye, "Universal exponentiation algorithm: A first step towards provable SPA-resistance," In *Pre-proceedings of Workshop on Cryptographic Hardware and Embedded Systems – CHES 2001*, pp. 305–314, May 13-16, 2001.
36. T.S. Messerges, E.A. Dabbish, and R.H. Sloan, "Investigations of power analysis attacks on smartcards," In *Proceedings of USENIX Workshop on Smartcard Technology*, pp. 151–161, May 1999.
37. L. Goubin and J. Patarin, "DES and differential power analysis–The duplication method," In *Cryptographic Hardware and Embedded Systems – CHES '99*, LNCS 1717, pp. 158–172, Springer-Verlag, 1999.

38. E. Biham and A. Shamir, "Power analysis of the key scheduling of the AES candidates," In *Proceedings of the Second Advanced Encryption Standard (AES) Candidate Conference*, pp. 115–121, March 1999, available at URL <http://csrc.nist.gov/encryption/aes/round1/Conf2/aes2conf.html>.

39. S. Chari, C.S. Jutla, J.R. Rao, and P. Rohatgi, "A cautionary note regarding evaluation of AES candidates on smart-cards," In *Proceedings of the Second Advanced Encryption Standard (AES) Candidate Conference*, pp. 133–147, March 1999, available at URL
<http://csrc.nist.gov/encryption/aes/round1/Conf2/aes2conf.html>.

40. S. Chari, C.S. Jutla, J.R. Rao, and P. Rohatgi, "Towards sound approaches to counteract power-analysis attacks," In *Advances in Cryptology – CRYPTO '99*, LNCS 1666, pp. 398–412, Springer-Verlag, 1999.

41. J. Daemen and V. Rijmen, "Resistance against implementation attacks: A comparative study of the AES proposals," In *Proceedings of the Second Advanced Encryption Standard (AES) Candidate Conference*, pp. 122–132, March 1999, available at URL <http://csrc.nist.gov/encryption/aes/round1/Conf2/aes2conf.html>.

42. P.N. Fahn and P.K. Pearson, "IPA: A new class of power attacks," In *Cryptographic Hardware and Embedded Systems – CHES '99*, LNCS 1717, pp. 173–186, Springer-Verlag, 1999.

43. T.S. Messerges, "Securing the AES finalists against power analysis attacks," In *Proceedings of Fast Software Encryption Workshop – FSE 2000*, LNCS 1978, Springer-Verlag, 2000.

44. J.-S. Coron and L. Goubin, "On boolean and arithmetic masking against differential power analysis," In *Cryptographic Hardware and Embedded Systems – CHES 2000*, LNCS 1965, pp. 231–237, Springer-Verlag, 2000.

45. T.S. Messerges, "Using second-order power analysis to attack DPA resistant software," In *Cryptographic Hardware and Embedded Systems – CHES 2000*, LNCS 1965, pp. 238–251, Springer-Verlag, 2000.

46. L. Goubin, "A sound method for switching between boolean and arithmetic masking," In *Pre-proceedings of Workshop on Cryptographic Hardware and Embedded Systems – CHES 2001*, pp. 3–15, May 13-16, 2001.

47. M. Akkar and C. Giraud, "An implementation of DES and AES, secure against some attacks," In *Pre-proceedings of Workshop on Cryptographic Hardware and Embedded Systems – CHES 2001*, pp. 315–325, May 13-16, 2001.

48. A.J. Menezes, P.C. van Oorschot, and S.A. Vanstone. *Handbook of applied cryptography*. CRC Press, 1997.

49. G.R. Blakley, "A computer algorithm for the product AB modulo M," *IEEE Transactions on Computers*, vol. 32, no. 5, pp. 497–500, May 1983.

50. K.R. Sloan, Jr., Comments on "A computer algorithm for the product AB modulo M," *IEEE Transactions on Computers*, vol. 34, no. 3, pp. 290–292, March 1985.

51. Ç.K. Koç, "RSA hardware implementation," Technical Report TR 801, RSA Laboratories, Redwood City, April 1996

52. S.M. Yen and S.Y. Tseng, "Differential power cryptanalysis of a Rijndael implementation," LCIS Technical Report TR-2K1-9, Dept. of Computer Science and Information Engineering, National Central University, Taiwan, May 3, 2001.

53. M. Joye, J.-J. Quisquater, S.M. Yen, and M. Yung, "Observability analysis – detecting when improved cryptosystems fail," In *Proceedings of the CT-RSA 2002 Conference*, 2002. (to appear)

A Fast Scalar Multiplication Method with Randomized Projective Coordinates on a Montgomery-Form Elliptic Curve Secure against Side Channel Attacks

Katsuyuki Okeya[1], Kunihiko Miyazaki[1], and Kouichi Sakurai[2]

[1] Hitachi, Ltd., Systems Development Laboratory,
292, Yoshida-cho, Totsuka-ku, Yokohama, 244-0817, Japan
{ka-okeya,kunihiko}@sdl.hitachi.co.jp
[2] Kyushu University,
Graduate School of Information Science and Electrical Engineering,
6-10-1, Hakozaki, Higashi-ku, Fukuoka, 812-8581, Japan
sakurai@csce.kyushu-u.ac.jp

Abstract. In this paper, we propose a scalar multiplication method that does not incur a higher computational cost for randomized projective coordinates of the Montgomery form of elliptic curves. A randomized projective coordinates method is a countermeasure against side channel attacks on an elliptic curve cryptosystem in which an attacker cannot predict the appearance of a specific value because the coordinates have been randomized. However, because of this randomization, we cannot assume the Z-coordinate to be 1. Thus, the computational cost increases by multiplications of Z-coordinates, 10%. Our results clarify the advantages of cryptographic usage of Montgomery-form elliptic curves in constrained environments such as mobile devices and smart cards.

Keywords: Elliptic Curve Cryptosystem, Montgomery Form, Side Channel Attacks, Randomized Projective Coordinates

1 Introduction

The randomized projective coordinates method is a countermeasure against side channel attacks on an elliptic curve cryptosystem. The current method has a problem in that the randomized projective coordinates increase the computational cost of scalar multiplication. In this paper, we propose a scalar multiplication method that does not incur a higher computational cost for randomized projective coordinates of the Montgomery form of elliptic curves. This method is fast and secure against side channel attacks.

1.1 Side Channel Attacks and Randomized Projective Coordinates

Kocher et al. were first to proposed the side channel attack [Koc, Koc96, KJJ98, KJJ99] in which an attacker infers secret information using leaked data from a

K. Kim (Ed.): ICICS 2001, LNCS 2288, pp. 428–439, 2002.

cryptographic device while it executes cryptographic procedures. Later, Coron generalized the side channel attacks include to elliptic curve cryptosystems [Cor99], and proposed a countermeasure using randomized projective coordinates. Okeya and Sakurai have confirmed that the randomized projective coordinates method prevents side channel attacks [OS00].

1.2   A Montgomery-Form Elliptic Curve

Montgomery introduced a non-standard form of an elliptic curve [Mon87], which is called the Montgomery form, defined by the equation $E^M : By^2 = x^3 + Ax^2 + x$. The standard elliptic curve form in an elliptic curve cryptosystem [Kob87, Mil86] is defined by the equation $E : y^2 = x^3 + ax + b$, which is called a (short) Weierstrass form. Okeya and Sakurai proposed a fast scalar multiplication method using randomized projective coordinates that is secure against side channel attacks on a Montgomery-form elliptic curve [OS00]. Compared with other scalar multiplication methods which focus on speed up (and are vulnerable to side channel attacks), this scalar multiplication method is just as fast on a Weierstrass-form elliptic curve, but its computational cost is ten per cent higher on a Montgomery-form elliptic curve.

1.3   Our Contributions

Our scalar multiplication method avoids the above-mentioned increase in computational cost.

We compute an operation of points with randomized expression and points without randomized expression, the result of the operation is with randomized expression, and its computational cost is the same as the computational cost of an operation of points without randomized expression. In contrast, if we compute an operation of points with randomized expression, then we get a randomized result, but the computational cost increases. In the case of a Montgomery-form elliptic curve, we compute addition of two points in randomized projective coordinates using these difference point in unrandomized projective coordinates, namely, the Z-coordinate of the point to be 1, the resultant point is in randomized projective coordinates, and no extra field operations are required. The scalar multiplication method with randomized projective coordinates on a Montgomery-form elliptic curve is applicable to other types of elliptic curve, such as the Weierstrass form, and other coordinate systems, such as Jacobian coordinates. The scalar multiplication method on a Montgomery-form elliptic curve is effective against the simple power analysis (SPA) attack [OKS00, OS00] and is much faster than other scalar multiplication methods which prevent SPA attack on other types of elliptic curve [Cor99, JQ01, JT01, LS01].

The traditional scalar multiplication method of 160 bits with randomized projective coordinates on a Montgomery-form elliptic curve needs $1627.4M$, whereas our scalar multiplication method of 160 bits needs $1468.4M$. The traditional scalar multiplication method without randomized projective coordinates [Mon87] needs $1467.4M$, so our method has equivalent speed. Compared with the window method of 160 bits on a Weierstrass-form elliptic curve [CMO98], our method is six per cent faster (the window method needs $1566.4M$ and is vulnerable to side channel attacks).

Section 2 is a survey of side channel attacks and randomized projective coordinates. Section 3 descibes the original scalar multiplication method with randomized projective coordinates on a Montgomery-form elliptic curve. Section 4 describes our scalar multiplication method with randomized projective coordinates and discusses its security versus side channel attacks and its computational cost.

2   Side Channel Attacks and Randomized Projective Coordinates

2.1   Side Channel Attacks

In cryptographic devices such as smart cards, data other than input data and output data may 'leak out' during cryptographic procedures. The computation timing of cryptographic procedures is one such kind of data. So is power consumption because the smart card uses an external source. Kocher et al. developed the side channel attack in which an attacker infers stored secret information in a cryptographic device by using such leaked data [Koc, Koc96, KJJ98, KJJ99]. This type of attack, which include timing attack, SPA attack, and Differential Power Analysis (DPA) attack, render smart cards particularly vulnerable.

A timing attack [Koc, Koc96] is a side channel attack in which an attacker infers the secret information by using computation time as leaked data. Some methods of timing attack use to statistical analysis to reveal the secret information, others infer it from a one-time computation. An SPA attack [Koc, Koc96] is a side channel attack in which an attacker infers the secret information by using power consumption as leaked data. SPA attack reveals the secret information by direct observation of a device's power consumption without the need for statistical analysis. A DPA attack [KJJ98, KJJ99] is a side channel attack in which an attacker infers the secret information by using statistical analysis of power consumption. This attack is the most powerful side channel attack.

Kocher et al. envisioned mainly DES [DES] and RSA [RSA78] as targets for side channel attacks. Coron generalized DPA attack to include elliptic curve cryptosystems [Cor99].

2.2   Randomized Projective Coordinates

In addition to generalizing the DPA attack to an elliptic curve, Coron proposed some countermeasures against it. In particular, the randomized projective coordinates method uses projective coordinates of points on an elliptic curve, in which each coordinate is multiplied by a random number. That is, when we compute an elliptic operation such as a scalar multiplication, we generate a random number r, and express a point $P = (x, y)$ on the elliptic curve as a point

(rx, ry, r) in projective coordinates. The point (rx, ry, r) is same to the point P, because $(X, Y, Z) = (rX, rY, rZ)$ holds for any $r \neq 0$ in projective coordinates.

Okeya and Sakurai proposed two requirements for preventing side channel attacks [OS00]. One requirement is independency of secret information and computation procedures. The other is randomization of computing objects. The first requirement is equivalent to preventing an SPA attack. The randomized projective coordinates method fulfills the second requirement.

3   Montgomery-Form Elliptic Curve

3.1   Definition of Montgomery-Form Elliptic Curve

Let $p(\geq 3)$ be a prime, and $\mathbf{F}_{p^n}$ be a finite field with characteristic p. A Montgomery-form elliptic curve over $\mathbf{F}_{p^n}$ is defined as follows:

$$E^M : By^2 = x^3 + Ax^2 + x,$$

where $A, B \in \mathbf{F}_{p^n}$ and $B(A^2 - 4) \neq 0$.

The operations on the Montgomery-form elliptic curve are as follows. For a point P on E^M, we denote kP for k times point of P, and $mP = (X_m, Y_m, Z_m)$, $nP = (X_n, Y_n, Z_n), (m - n)P = (X_{m-n}, Y_{m-n}, Z_{m-n})$ in projective coordinates. Then $(m + n)P = mP + nP$ without Y-coordinate using the difference point $(m - n)P$ is computed by the following [Mon87]:

Addition formulae $(m \neq n)$

$$X_{m+n} = Z_{m-n}[(X_m - Z_m)(X_n + Z_n) + (X_m + Z_m)(X_n - Z_n)]^2$$
$$Z_{m+n} = X_{m-n}[(X_m - Z_m)(X_n + Z_n) - (X_m + Z_m)(X_n - Z_n)]^2$$

Doubling formulae $(m = n)$

$$4X_n Z_n = (X_n + Z_n)^2 - (X_n - Z_n)^2$$
$$X_{2n} = (X_n + Z_n)^2 (X_n - Z_n)^2$$
$$Z_{2n} = (4X_n Z_n)((X_n - Z_n)^2 + ((A + 2)/4)(4X_n Z_n))$$

In other words, X_{m+n}/Z_{m+n} and X_{2n}/Z_{2n} are equal to the x-coordinates of $(m + n)P$ and $2nP$ in affine coordinates, respectively. M and S respectively denote $\mathbf{F}_{p^n}$-operations of multiplication and squaring. Then, the addition and doubling formulae above need $4M + 2S$ and $3M + 2S$, respectively. In the addition formulae, if we can assume the Z-coordinate $Z_{m-n} = 1$, its computational cost is $3M + 2S$.

3.2   Scalar Multiplication Method

The scalar multiplication algorithm on a Montgomery-form elliptic curve [Mon87] is as follows:

Algorithm 1

INPUT A scalar value d and the x-coordinate of a point $P = (x, y)$.

OUTPUT The X and the Z coordinates of the scalar-multiplied point dP.

1. $i \leftarrow |d| - 1$
2. Compute a point $2P$ from the point P using the doubling formulae. Here, the point P is expressed by $P = (x, y, 1)$ in projective coordinates and used.
3. $m \leftarrow 1$
4. If $i = 0$, then go to Step 13. Otherwise go to Step 5.
5. $i \leftarrow i - 1$
6. If $d_i = 0$, then go to Step 7. Otherwise go to Step 10.
7. Compute a point $(2m + 1)P$ from the points $mP, (m + 1)P$ and P using the addition formulae.
8. Compute a point $2mP$ from the point mP using the doubling formulae.
9. $m \leftarrow 2m$, and go to Step 4.
10. Compute a point $(2m + 1)P$ from the points $mP, (m + 1)P$ and P using the addition formulae.
11. Compute a point $(2m + 2)P$ from the point mP using the doubling formulae.
12. $m \leftarrow 2m + 1$, and go to step 4.
13. Output the X and the Z coordinates of the point mP as the X and the Z coordinates of the scalar-multiplied point dP, respectively.

Here, $|d|$ denotes the bit length of d, and d_i the i-th bit of d. That is, $d = \sum_{i=0}^{|d|-1} d_i 2^i, d_i \in \{0, 1\}$.

Each addition in Step 7 and Step 10 needs $3M + 2S$, because the Z-coordinate of the difference point P is 1. Each doubling in Step 2, Step 8 and Step 11 needs $3M + 2S$. The number of the iterations of Step 7-9 or Step 10-12 is $|d| - 1$. Thus, the computational cost of this algorithm is $(6|d| - 3)M + (4|d| - 2)$.

This algorithm does not compute the Y-coordinate of the scalar multiplied point dP. However, we should note that we can easily compute the Y-coordinate using the y-coordinate recovery method on a Montgomery-form elliptic curve citeOS01. This y-coordinate recovery method computes the Y-coordinate of dP from X and Z coordintes of dP and $(d+1)P$, and the point P. The computational cost of the y-coordinate recovery is $12M + S$.

3.3   Scalar Multiplication Method
with Randomized Projective Coordinates

The scalar multiplication algorithm with randomized projective coordinates on a Montgomery-form elliptic curve [OS00] is as follows[1]:

---

[1] This algorithm does not compute the Y-coordinate of the scalar multiplied point dP, either. However, we should note that we can easily compute the Y-coordinate using the y-coordinate recovery method on a Montgomery-form elliptic curve [OS01].

Algorithm 2
    INPUT A scalar value d and the x-coordinate of a point $P = (x, y)$.
    OUTPUT The X and the Z coordinates of the scalar-multiplied point dP.

1. Generate a random number $r \in \mathbf{F}_{p^n} - \{0\}$.
2. The point P is expressed by $P = (rx, ry, r)$ in projective coordinates. Howevwer, we do not need to compute ry.
3. $i \leftarrow |d| - 1$
4. Compute a point $2P$ from the point P using the doubling formulae.
5. $m \leftarrow 1$
6. If $i = 0$, then go to Step 15. Otherwise go to Step 7.
7. $i \leftarrow i - 1$
8. If $d_i = 0$, then go to Step 9. Otherwise go to Step 12.
9. Compute a point $(2m + 1)P$ from the points $mP, (m + 1)P$ and P using the addition formulae.
10. Compute a point $2mP$ from the point mP using the doubling formulae.
11. $m \leftarrow 2m$, and go to Step 6.
12. Compute a point $(2m + 1)P$ from the points $mP, (m + 1)P$ and P using the addition formulae.
13. Compute a point $(2m + 2)P$ from the point mP using the doubling formulae.
14. $m \leftarrow 2m + 1$, and go to Step 6.
15. Output the X and the Z coordinates of the point mP as the X and the Z coordinates of the scalar-multiplied point dP, respectively.

Each addition in Step 9 and Step 12 needs $4M + 2S$, because the Z-coordinate of the difference point P is r, and $r \neq 1$ in general, so the Z-coordinate of the difference point P is not 1. We need to compute the multiplication $r \cdot x$ in Step 2. Thus, the computational cost of this algorithm is $(7|d| - 3)M + (4|d| - 2)S$. Compared with the original scalar multiplication, the computational cost increases by $|d|M$.

4   Proposed Method

Our scalar multiplication algorithm with randomized projective coordinates on a Montgomery-form elliptic curve is as follows[2]:

Algorithm 3
    INPUT A scalar value d and the x-coordinate of a point P.
    OUTPUT The X and the Z coordinates of the scalar-multiplied point dP.

1. Generate a random number $r \in \mathbf{F}_{p^n} - \{0\}$.
2. The point P is expressed by $P = (rx, ry, r)$ in projective coordinates. However, we do not need to compute ry.

---

[2] This algorithm does not compute the Y-coordinate of the scalar multiplied point dP, either. However, we should note that we can easily compute the Y-coordinate using the y-coordinate recovery method on a Montgomery-form elliptic curve [OS01].

3. $i \leftarrow |d| - 1$
4. Compute a point $2P$ from the point P using the doubling formulae.
5. $m \leftarrow 1$
6. If $i = 0$, then go to Step 15. Otherwise go to Step 7.
7. $i \leftarrow i - 1$
8. If $d_i = 0$, then go to Step 9. Otherwise go to Step 12.
9. Compute a point $(2m + 1)P$ from the points $mP, (m + 1)P$ and P using the addition formulae. Here, the point P without randomization, namely $(x, y, 1)$ in projective coordinates, is used.
10. Compute a point $2mP$ from the point mP using the doubling formulae.
11. $m \leftarrow 2m$, and go to Step 6.
12. Compute a point $(2m + 1)P$ from the points $mP, (m + 1)P$ and P using the addition formulae. Here, the point P without randomization, namely $(x, y, 1)$ in projective coordinates, is used.
13. Compute a point $(2m + 2)P$ from the point $(m + 1)P$ using the doubling formulae.
14. $m \leftarrow 2m + 1$, and go to Step 6.
15. Output the X and the Z coordinates of the point mP as the X and the Z coordinates of the scalar-multiplied point dP, respectively.

4.1   Validity

Firstly, we show the validity of this algorithm, that is, the algorithm outputs the scalar-multiplied point. We need to show each addition in Step 9 and Step 12 is equivalent to the results using (rx, ry, r) in projective coordinates. Let X''_m, Z''_m and X''_{m+1}, Z''_{m+1} be the X-coordinate and the Z-coordinate of the point mP and $(m+1)P$ in projective coordinates in Step 9, respectively, and X''_{2m+1} and Z''_{2m+1} be the X-coordinate and the Z-coordinate of the computed point in projective coordinates in Step 9, respectively. Let X'_m, Z'_m and X'_{m+1}, Z'_{m+1} be the X-coordinate and the Z-coordinate of the point mP and $(m + 1)P$ in projective coordinates in Step 9 of Algorithm 2, respectively, and X'_{2m+1} and Z'_{2m+1} be the X-coordinate and the Z-coordinate of the computed point in projective coordinates in Step 9 of Algorithm 2, respectively. Then, we simply need to show $X''_{2m+1}/Z''_{2m+1} = X'_{2m+1}/Z'_{2m+1}$ assuming $X''_m/Z''_m = x_m$, $X''_{m+1}/Z''_{m+1} = x_{m+1}$, $X'_m/Z'_m = x_m$ and $X'_{m+1}/Z'_{m+1} = x_{m+1}$, because $X'_{2m+1}/Z'_{2m+1} = x_{2m+1}$, and we get the result. Here, x_m, x_{m+1} and x_{2m+1} respectively denote the x-coordinate of $mP, (m + 1)P$ and $(2m + 1)P$ in affine coordinates.

Proposition 1. $X''_{2m+1}/Z''_{2m+1} = X'_{2m+1}/Z'_{2m+1}$

Proof.

$$\frac{X''_{2m+1}}{Z''_{2m+1}} = \frac{1 \cdot [(X''_{m+1} - Z''_{m+1})(X''_m + Z''_m) + (X''_{m+1} + Z''_{m+1})(X''_m - Z''_m)]^2}{x[(X''_{m+1} - Z''_{m+1})(X''_m + Z''_m) - (X''_{m+1} + Z''_{m+1})(X''_m - Z''_m)]^2}$$

$$= \frac{[(x_{m+1} - 1)(x_m + 1) + (x_{m+1} + 1)(x_m - 1)]^2}{x[(x_{m+1} - 1)(x_m + 1) - (x_{m+1} + 1)(x_m - 1)]^2}$$

$$= \frac{r[(X'_{m+1} - Z'_{m+1})(X'_m + Z'_m) + (X'_{m+1} + Z'_{m+1})(X'_m - Z'_m)]^2}{rx[(X'_{m+1} - Z'_{m+1})(X'_m + Z'_m) - (X'_{m+1} + Z'_{m+1})(X'_m - Z'_m)]^2}$$

$$= \frac{X'_{2m+1}}{Z'_{2m+1}} \qquad \qquad \qquad \square$$

By the same token, we get the case of Step 12. Therefore, we have shown the validity of our method.

4.2 Efficiency

Secondly, we estimate the computational cost of this algorithm. Each addition in Step 9 and Step 12 needs $3M + 2S$, because the Z-coordinate of the difference point P is 1. We need to compute $r \cdot x$ in Step 2. Thus, the computation cost of this algorithm is $(6|d| - 2)M + (4|d| - 2)S$. This cost is equivalent to that of the scalar multiplication method without randomized projective coordinates.

Next, we compare our method with scalar multiplication methods which focus on speed up on a Weierstrass-form elliptic curve in terms of efficiency. The window method using mixed coordinates is one of the fastest scalar multiplication methods [CMO98] on a Weierstrass-form elliptic curve. The computational cost $T^{Wei}(w, l)$ of this method with bit length l and window width w is estimated using the following equation [CMO98, LH00]:

$$T^{Wei}(w, l) = wI + \left(\frac{8l}{w + 2} + 4l + 5 \cdot 2^{w-1} - 2w - 14 \right) M$$

$$+ \left(\frac{3l}{w + 2} + 4l + 2^{w-1} - 2w - 2 \right) S.$$

Here, I denotes an $\mathbf{F}_{p^n}$-operation of inversion. Under a reasonable assumption that $I = 30M$ and $S = 0.8M$ for prime fields [LH00], we get $T^{Wei}(w, l) = 1566.4M$ for the bit length $l = 160$ and the window width $w = 4$. Thus for the given bit length and window width, the window method is 6% slower than our method.

4.3 Security

Finally, we show the immunity of our algorithm to side channel attacks. According to Okeya and Sakurai [OS00], we have to check the following two requirements:

1. Independency of secret information and computation procedures.
2. Randomization of computing objects.

First, let's examine the requirement 1. For any scalar value with a fixed bit length, we execute Step 9-11 then return Step 6 if $d_i = 0$ at Step 8, and we execute Step 12-14 then return Step 6 if $d_i = 1$ at Step 8. In any case, we compute an elliptic addition then an elliptic doubling, so the executed computations are identical. Hence, the computation procedures are identical for any scalar value with the fixed bit length. Therefore, our proposed method satisfied the requirement 1.

To show that the requirement 2 is satisfied, first, we define the following function $e(s,t)$:

$$e(s,t) = 2^{2s} + t(2^{s+2} - 2^s) \qquad \begin{cases} s = 0, 1, \cdots \\ t = 0, 1, \cdots \end{cases}$$

Then, we get the following equations by direct computation:

Lemma 1.

$$e(s+1, 0) = e(s, 2^s)$$
$$e(s+1, 2t) = 4e(s, t)$$
$$e(s+1, 2t+1) = 2(e(s,t) + e(s, t+1))$$

Next, we define the function $e(m)$ as $e(m) = e(|m| - 1, m - 2^{|m|-1})$ for $m = 1, 2, \cdots$.

On the other hand, we can denote $X''_m = r_m X_m, Z''_m = r_m Z_m$ with some integer r_m by Proposition 1 and the relation $X_m/Z_m = x_m$, where X''_m, Z''_m, X_m, Z_m respectively denote the X and the Z coordinates of the point mP of Algorithm 3 and the X and the Z coordinates of the point mP of Algorithm1. This r_m is expressed in the following by using the random number r of Step 1 and the function $e(m)$:

Proposition 2. $r_m = r^{e(m)}$

Proof. We will prove this relation by induction with regards to m. We can easily check $r_1 = r, r_2 = r^4, e(1) = 1, e(2) = 4$ by direct computation. Next, we will show $r_{2m} = r^{e(2m)}, r_{2m+1} = r^{e(2m+1)}$ if $d_i = 0$, and $r_{2m+1} = r^{e(2m+1)}, r_{2m+2} = r^{e(2m+2)}$ if $d_i = 1$ assuming $r_m = r^{e(m)}, r_{m+1} = r^{e(m+1)}$.

First, we will show the case $d_i = 0$.

$$\begin{aligned}
X''_{2m} &= (X''_m + Z''_m)^2 (X''_m - Z''_m)^2 \\
&= (r_m X_m + r_m Z_m)^2 (r_m X_m - r_m Z_m)^2 \\
&= r_m^4 (X_m + Z_m)^2 (X_m - Z_m)^2 \\
&= r^{4e(m)} X_{2m}
\end{aligned}$$

According to Lemma 1, $4e(m) = e(2m)$ holds. Thus, we have $r_{2m} = r^{e(2m)}$.

$$\begin{aligned}
X''_{2m+1} &= [(X''_{m+1} - Z''_{m+1})(X''_m - Z''_m) + (X''_{m+1} + Z''_{m+1})(X''_m + Z''_m)]^2 \\
&= [(r_{m+1}X_{m+1} - r_{m+1}Z_{m+1})(r_m X_m + r_m Z_m) \\
&\quad + (r_{m+1}X_{m+1} + r_{m+1}Z_{m+1})(r_m X_m - r_m Z_m)]^2 \\
&= r_{m+1}^2 r_m^2 [(X_{m+1} - Z_{m+1})(X_m + Z_m) + (X_{m+1} + Z_{m+1})(X_m - Z_m)]^2 \\
&= r^{2(e(m+1)+e(m))} X_{2m+1}
\end{aligned}$$

If $|m + 1| = |m|$, according to Lemma 1, $2(e(m + 1) + e(m)) = e(2m + 1)$ holds. Thus, we have $r_{2m+1} = r^{e(2m+1)}$. If $|m + 1| = |m| + 1$, that is, $m = 2^{|m|} - 1$, according to Lemma 1, $e(m + 1) = e(|m|, 0) = e(|m| - 1, 2^{|m|-1})$ holds. Thus, we have $r_{2m+1} = r^{e(2m+1)}$ by Lemma 1.

In the same way, we can show the case $d_i = 1$. Therefore, $r_m = r^{e(m)}$ for any m by induction. □

According to Proposition 2, we have $X''_m = r^{e(m)} X_m, Z''_m = r^{e(m)} Z_m$. To put it more concretely, X''_m and Z''_m are multiplication results by some power of the random number r and X_m and Z_m, respectively. Let g be the greatest common measure between $e(m)$ and $p^n - 1$ (the order of multiplication group of the finite field $\mathbf{F}_{p^n}$). Then, the number of possible values of $r^{e(m)}$ when r varies in $\mathbf{F}_{p^n} - \{0\}$ is $(p^n - 1)/g$. Let k be some factor of $p^n - 1$, and we regard the probability that k divide $e(m)$ as $1/k$. Then, $e(m)$ and $p^n - 1$ easily have a small greatest common measure, however, scarcely have a large greatest common measure. Hence, the number of possible values of $r^{e(m)}$ is sufficiently large. Thus, we can say that X''_m, Z''_m are well-randomized. Therefore, our proposed method satisfies requirement 2.

4.4 Comparison with Other Side Channel Attack Prevention Methods

Coron proposed an SPA-immune scalar multiplication method on a Weierstrass-form elliptic curve using dummy additions [Cor99]. This method always computes an addition per bit of a scalar value, then discards the result if the bit is equal to zero. The computational cost of the addition $J + A \rightarrow J^m$ of this method is $9M + 5S$, where J, A and J^m mean Jacobian coordinates, affine coordinates and modified Jacobian coordinates, respectively. The computational cost of the doubling $J^m \rightarrow J$ of this method is $3M + 4S$. The total computational cost of this method for 160 bits is $3072.0M$, assuming $S = 0.8M$.

Liardet and Smart proposed a scalar multiplication method on a Jacobian-form elliptic curve [LS01]. Addition and doubling on a Jacobian-form elliptic curve are computed using the same formulae. This property prevents SPA attacks. The computational cost of addition on a Jacobian-form elliptic curve is $16M$. This method permits a sliding window method, and the number of additions for 160 bits is about 192.2. The total computational cost of this method is $3075.2M$, assuming $S = 0.8M$.

Joye and Quisquater proposed a scalar multiplication method on a Hessian-form elliptic curve [JQ01]. Addition and doubling on a Hessian-form elliptic curve are computed using the same formulae. This property prevents SPA attacks. The computational cost of addition on a Hessian-form elliptic curve is $12M$. This method permits a sliding window method, and the number of additions for 160 bits is about 192.2. The total computational cost of this method is $2306.4M$, assuming $S = 0.8M$.

The technique of randomized (projective) coordinates is applicable to scalar multiplication methods of such types of elliptic curves. However, compared with

Table 1. Immunity to side channel attacks and computational cost

Method of computation	Immunity to side channel attacks	Computational cost (160 bits)
M-form original	Vulnerable (to DPA)	1467.4M
M-form+RPC	Immune	1627.4M
Our proposed method	Immune	1468.4M
W-form mixed coordinates	Vulnerable	1566.4M
M-form original	Immune to SPA	1467.4M
W-form SPA-immune	Immune to SPA	3072.0M
J-form	Immune to SPA	3075.2M
H-form	Immune to SPA	2306.4M

"M-form original" means the original scalar multiplication method on a Montgomery-form elliptic curve (Algorithm 1). "M-form+RPC" means the scalar multiplication method with randomized projective coordinates on a Montgomery-form elliptic curve (Algorithm 2). "Our proposed method" means our scalar multiplication method with randomized projective coordinates on a Montgomery-form elliptic curve (Algorithm 3). "W-form mixed coordinates" means the window method using mixed coordinates on a Weierstrass-form elliptic curve [CMO98]. "W-form SPA-immune" means Coron's SPA-immune scalar multiplication method on a Weierstrass-form elliptic curve [Cor99]. "J-form" means the scalar multiplication method on a Jacobian-form elliptic curve [LS01]. "H-form" means the scalar multiplication method on a Hessian-form elliptic curve [JQ01].

the scalar multiplication method on a Montgomery-form elliptic curve, these scalar multiplication methods are much slower. Therefore, it is inefficient to counter side channel attacks (DPA attack) by using the technique of randomized (projective) coordinates on Weierstrass, Jacobian or Hessian form.

Table 1 summarizes the immunities (to side channel attacks) and computational costs of the methods discussed in this paper.

Acknowledgements

The authors would like to thank the anonymous referees for their constructive comments. The first two authors also thank Dr. Kazuo Takaragi and Dr. Hiroshi Yoshiura of Hitachi Ltd. who afforded the opportunity to study this work.

References

CMO98. Cohen, H., Miyaji, A., Ono, T., *Efficient Elliptic Curve Exponentiation Using Mixed Coordinates*, Advances in Cryptology - ASIACRYPT '98, LNCS1514, (1998), 51-65.

Cor99. Coron, J.S., *Resistance against Differential Power Analysis for Elliptic Curve Cryptosystems*, Cryptographic Hardware and Embedded Systems (CHES'99), LNCS1717, (1999), 292-302.

DES.   National Bureau of Standards, *Data Encryption Standard*, Federal Information Processing Standards Publication 46 (FIPS PUB 46), (1977).

JQ01.   Joye, M., Quisquater, J.J., *Hessian elliptic curves and side-channel attacks*, Cryptographic Hardware and Embedded Systems (CHES'01), LNCS2162, (2001), 402-410.

JT01.   Joye, M., Tymen, C., *Protections against differential analysis for elliptic curve cryptography: An algebraic approach*, Cryptographic Hardware and Embedded Systems (CHES'01), LNCS2162, (2001), 377-390.

Kob87.   Koblitz, N., *Elliptic curve cryptosystems*, Math. Comp. 48, (1987), 203-209.

Koc.   Kocher, C., *Cryptanalysis of Diffie-Hellman, RSA, DSS, and Other Systems Using Timing Attacks*, Available at http://www.cryptography.com/

Koc96.   Kocher, C., *Timing Attacks on Implementations of Diffie-Hellman, RSA,DSS, and Other Systems*, Advances in Cryptology - CRYPTO '96, LNCS1109, (1996), 104-113.

KJJ98.   Kocher, C., Jaffe, J., Jun, B., *Introduction to Differential Power Analysis and Related Attacks*, Available at http://www.cryptography.com/dpa/technical/

KJJ99.   Kocher, C., Jaffe, J., Jun, B., *Differential Power Analysis*, Advances in Cryptology - CRYPTO '99, LNCS1666, (1999), 388-397.

LH00.   Lim, C.H., Hwang, H.S., *Fast implementation of Elliptic Curve Arithmetic in $GF(p^m)$*, Public Key Cryptography (PKC2000) LNCS1751, (2000), 405-421.

LS01.   Liardet, P.Y., Smart, N.P., *Preventing SPA/DPA in ECC systems using the Jacobi form*, Cryptographic Hardware and Embedded System (CHES'01), LNCS2162, (2001), 391-401.

Mil86.   Miller, V.S., *Use of elliptic curves in cryptography*, Advances in Cryptology - CRYPTO '85, LNCS218,(1986),417-426.

Mon87.   Montgomery, P.L., *Speeding the Pollard and Elliptic Curve Methods of Factorizations*, Math. Comp. 48, (1987), 243-264

OKS00.   Okeya, K., Kurumatani, H., Sakurai, K., *Elliptic Curves with the Montgomery - Form and Their Cryptographic Applications*, Public Key Cryptography (PKC2000), LNCS1751, (2000), 238-257.

OS00.   Okeya, K., Sakurai, K., *Power Analysis Breaks Elliptic Curve Cryptosystems even Secure against the Timing Attack*, Progress in Cryptology - IN-DOCRYPT 2000, LNCS1977, (2000), 178-190.

OS01.   Okeya, K., Sakurai, K., *Efficient Elliptic Curve Cryptosystems from a Scalar Multiplication Algorithm with Recovery of the y-Coordinate on a Montgomery-Form Elliptic Curve*, Cryptographic Hardware and Embedded System (CHES'01), LNCS2162, (2001), 126-141.

RSA78.   Rivest, R.L., Shamir, A., Adleman, L., *A Method for Obtaining Digital Signatures and Public-Key Cryptosystems*, Communications of the ACM, Vol.21, No.2, (1978), 120-126.

DPA Countermeasure
Based on the "Masking Method"

Kouichi Itoh, Masahiko Takenaka, and Naoya Torii

FUJITSU LABORATORIES LTD.
64 Nishiwaki, Ohkubo-cho, Akashi, 674-8555, Japan
{kito,takenaka}@flab.fujitsu.co.jp, torii.naoya@jp.fujitsu.com

Abstract. We propose a new differential power analysis (DPA) countermeasure based on the "masking method" proposed by Messerges [11]. We also evaluate the security of our method by introducing a new idea of "probabilistic DPA." Its processing speed is as fast as that of the straight-forward implementation, and it requires little RAM so it is suitable for low-cost smartcards.

Keywords: DPA, countermeasure, masking method, Rijndael

1 Introduction

The differential power analysis (DPA), proposed by Kocher et al., is a strong attack that enables extraction of a secret key stored in a cryptographic device, such as smartcard. In this attack, the attacker monitors the power consumed by the cryptographic devices, then statistically analyzes the collected data to extract the key. In their original paper [10], Kocher et al. described a DPA attack against the DES. In [3], DPA attacks against Rijndael, Twofish and other AES candidates have also been described. In [6] and [11], the DPA attacks against the public key algorithms such as RSA and elliptic curve cryptography (ECC) have been described.

Many countermeasures have been proposed for securing symmetric key algorithms against DPA attacks. Goubin and Patarin proposed a duplication method [8]. Chari et al. proposed a technique in which the data is divided into k shares [4]. Messerges proposed a masking method [12]. Coron and Goubin proposed a method for converting between Boolean and arithmetic masking [7]. Akkar et al. proposed a transformed-masking method [2]. Many countermeasures have also been proposed for protecting pubic key algorithms [5,6,9,11].

For the public key algorithms, several effective countermeasures are proposed, that is, the processing speed and required RAM size are almost the same as those of straight-forward implementation. For the symmetric key algorithms, all countermeasures have disadvantages. That is, the encryption process is slower and/or more RAM is required in comparison with a straight-forward implementation. They are thus unsuitable for implementation in low-cost smartcards, which are slow and have little RAM.

K. Kim (Ed.): ICICS 2001, LNCS 2288, pp. 440–456, 2002.

We propose a DPA countermeasure for the securing symmetric key algorithm. It is an improved countermeasure of Messeges' masking method [12]. With our method, the encryption process is faster and less RAM size is required. It is thus suitable for implementation in low-cost smartcards. In this paper, we describe our method specified to Rijndael algorithm [14]. It can be also applied to other symmetric key algorithms, such as DES.

In Messerges' method, the input data and intermediate values are "masked" by XORing a random value, and data is encrypted with the masked value. At the end of the encryption, the data is unmasked and the output has the same value that by a straight-forward implementation of the Rijndael algorithm. That is, encryption is processed by using two values, masked value T' and random mask value RM. If T is the intermediate value in the straight-forward implementation, T' and RM satisfy $T = T' \oplus RM$.

We improve the masking method in processing speed and smaller RAM size. This is achieved by randomly choosing one mask value from a fixed set of mask values previously prepared. In this paper, we call Messerges' masking method the "random-value masking method" and our method the "fixed-value masking method."

We evaluated the security of our method against DPA attacks by using the "probabilistic DPA", a new idea that is an extension of conventional DPA. We extended Kocher's original DPA [10] (an attack against a load operation for an Sbox table look-up) and Chari's binary DPA [3] (an attack against a load/store operation for key XORing) to a probabilistic DPA. The security of our method depends on the set of the fixed mask values prepared in advance. We found that our method is secure against probabilistic DPA attacks when the set satisfies certain criteria.

We describe the random-value masking method in section 2, our fixed-value masking method in section 3, and our security evaluation of the fixed-value masking method against probabilistic DPA attacks in section 4. We compare the fixed and random-value masking methods in section 5.

2   Random-Value Masking Method

As mentioned above, the random-value masking method is a DPA countermeasure proposed by Messerges [12]. Its application to the Rijndael algorithm is shown in algorithm 1, and its construction is shown in figure 1.

- *RandomSboxUpdate* is shown in algorithm 2, and *ByteSub_RM* is shown in algorithm 3.
- *MixColumn* is a matrix calculation using polynomial multiplication over $GF(2^8)$, and *ShiftRow* is a byte permutation process. Both *MixColumn* and *ShiftRow* are linear functions and are the same as those of the original Rijndael algorithm. Details of these functions have been described in [14].
- *nr* is the number of rounds: $nr = 10$ if the length of the secret key is 128-bit, $nr = 12$ if the length is 192-bit, and $nr = 14$ if the length is 256-bit.

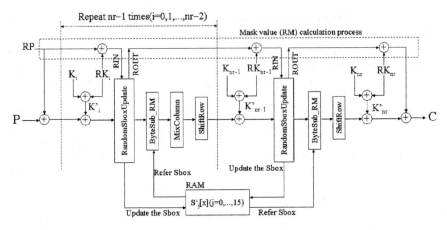

Fig. 1. Construction of random-value masking method (Rijndael algorithm)

In algorithm 1, T' is a "masked" intermediate value, and RM is a random mask value. The mask values are generated by the random-number generation function on line 2. The plaintext, P, and sub-keys, $K_i (i = 0, 1, .., nr)$, are masked using these mask values on line 3. Let T be the intermediate value calculated in the straight-forward implementation of the algorithm; note that T' and RM always satisfy $T = T' \oplus RM$ on lines 4-16. Finally, T' is unmasked using RM on line 17.

In algorithm 2, the table data for Sbox is updated according to the input random mask value, RIN, and the updated data is stored in RAM.

Algorithm 1 $RandomMasking_{RijndaelEnc}$

$RandomMasking_{RijndaelEnc}(P)$
1 /* K_i : subkey $(i = 0, 1, ..., nr)$ */
2 $(RP, RK_0, ..., RK_{nr}) = GenerateRandomMask();$ /* set initial mask */
3 $RM = RP;\ T' = P \oplus RM;$ for $(i = 0; i \le nr;\ i++)\ K'_i = K_i \oplus RK_i;$ /* initial masking */
4 for $(i=0; i < nr - 1\ ; i++)$ {
5      $RM = RM \oplus RK_i;\ T' = T' \oplus K'_i$
6      $RM = RandomSboxUpdate(S'_0[x], S'_1[x], ..., S'_{15}[x], RM);$
7   /* Update random Sbox $S'_0, S'_1, ..S'_{15}$ */
8      $T' = ByteSub\_RM(\ T', S'_0[x], ..., S'_{15}[x]);$
9      $RM = MixColumn(RM);\ T' = MixColumn(T');$
10     $RM = ShiftRow(RM);\ T' = Shift(T');$
11 }
12 $RM = RM \oplus RK_{nr-1};\ T' = T' \oplus K'_{nr-1};$
13 $RM = RandomSboxUpdate(S'_0[x], S'_1[x], ... , S'_{15}[x], RM);$
14 $T' = ByteSub\_RM(T', S'_0[x], ..., S'_{15}[x]);$
15 $RM = ShiftRow(RM);\ T' = Shift(T');$

16 $RM = RM \oplus K_{nr}; T' = T' \oplus K'_{nr};$
17 $C = T' \oplus RM;$ /* unmasking process */
18 output C;

Algorithm 2 *RandomSboxUpdate*

$RandomSboxUpdate(S'_0[x], S'_1[x], ..., S'_{15}[x], RIN)$
1 $(rin_{15}, rin_{14}, ..., rin_0) = RIN$;
2 $ROUT = GenerateRandomNumber();$ $(rout_{15}, rout_{14}, ..., rout_0) = ROUT$;
3 for $(j = 0 ; j < 16; j + +)$
4   for $(x = 0 ; x < 256; x + +)$ $S'_j[x] = S_j[x \oplus rin_j] \oplus rout_j;$ /* This data is stored in the RAM */
5 output $ROUT$;

Algorithm 3 *ByteSub_RM*

$ByteSub\_RM(X, S'_j[x])$
1 $(x_{15}, x_{14}, ..., x_0) = X$;
2 for $(j = 0 ; j < 16; j + +)$ $x_j = S'_j[x_j]$;
3 $X = (x_{15}, x_{14}, ..., x_0)$
4 output X;

Although the random-value masking method is secure against DPA attacks,
- it incurs a high processing overhead.
As shown in algorithm 1, it must calculate both the masked intermediate value T' and the random mask value RM. Moreover, the table data for Sbox must be updated in *RandomSboxUpdate* at every round. These calculations increase the required processing time and the overhead for encryption processing. The random-value masking method also requires
- a lot of RAM.
In *RandomSboxUpdate*, so much RAM is needed. If the table data for 16 Sboxes is updated one time, 4K Bytes of RAM are needed. In a practical implementation, the table data for only one Sbox is updated at a time by combining *RandomSboxUpdate* and *ByteSub_RM*. However, at least 256 Bytes of RAM are still needed, which is still much for low-cost smartcards.

3   Fixed-Value Masking Method

Our "fixed-value masking method" is faster and requires less RAM than the random-value masking method.

In our method, q sets of fixed mask values are prepared. One of the sets is randomly chosen at the start of encryption and used to mask the input data and intermediate values. Encryption is processed with those masked data. The unmasked data is output at the end of the encryption process. The table data for Sbox can also be chosen from the q sets of pre-computed data corresponding to the chosen set of fixed mask values. The table data is pre-computed and stored in ROM. As a result, the steps to calculate the mask values and update the table data for Sbox are omitted in our method.

Although our method is faster and requires less RAM than the random-value masking method, it requires more ROM. However while low-cost smartcards do not have much RAM, they do have a lot of ROM. For example, ST Microelectronics' ST16 smartcard has 128 Bytes of RAM and 6K Bytes of ROM. The fixed-value masking method is thus suitable for low-cost smartcards, as described in section 5.

3.1   Fixed-Value Masking Method (Type I)

Sbox $S'_j[x]$ in algorithm 2 depends on mask value RIN, so if it is directly replaced with a pre-computed Sbox, the number of Sboxes is $16 \times (number\ of\ rounds) \times (number\ of\ mask\ values) = 16 \times nr \times q$. This is too many to store in ROM. We thus propose the Type I fixed-value masking method. In this method, common pre-computed Sbox data is used in every round, so the number of Sboxes is reduced to $16q$. The Rijndael encryption algorithm using this method is shown in algorithm 4, and its construction is in figure 2. $ByteSub\_FM\_16$ is shown in algorithm 5. $ShiftRow$ and $MixColumn$ are the same as in algorithm 1.

Algorithm 4 $FixedMasking_{RijndaelEnc}$

$FixedMasking_{RijndaelEnc}(P)$
1 /* $K'_{i,r}$: masked subkey $(i = 0, 1, ..., nr, r = 0, 1, ..., q - 1)$ */
2 $r = GenerateRandomNumber();$ /* choose $r = 0, 1, ..., q - 1$ */
3 $T' = P;$
4 for $(i = 0; i < nr - 1; i++)$ {
5      $T' = T' \oplus K'_{i,r};$
6      $T' = ByteSub\_FM\_16(T', r);$
7      $T' = MixColumn(T');$
8      $T' = ShiftRow(T');$
9 }
10 $T' = T' \oplus K'_{nr-1,r};$
11 $T' = ByteSub\_FM\_16(T', r);$
12 $T' = ShiftRow(T');$
13 $T = T' \oplus K'_{nr,r};$ /* Unmasking process is also contained in this process */
14 output $T;$

In algorithm 4, T' is the intermediate masked value, and $K'_{i,r}$ is a masked sub-key $(i = 0, 1, .., nr, r = 0, 1, .., q - 1)$. Let K_i be the original sub-key and $FK_{i,r}$ be the fixed mask value, $K'_{i,r}$ satisfies $K'_{i,r} = K_i \oplus FK_{i,r}$ $(i = 0, 1, ..., nr, r = 0, 1, ..., q - 1)$. Let FIN_r and $FOUT_r$ $(r = 0, 1, ..., q - 1)$ be arbitrary 128-bit constant values; $FK_{i,r}$ satisfies (1) for any $i = 0, 1, ..., nr$ and $r = 0, 1, ..., q - 1$.

$$FK_{i,r} = \begin{cases} FIN_r\ (i = 0) \\ ShiftRow(MixColumn(FOUT_r)) \oplus FIN_r\ (i = 1, 2, ..., nr - 1) \\ ShiftRow(FOUT_r)\ (i = nr) \end{cases} \quad (1)$$

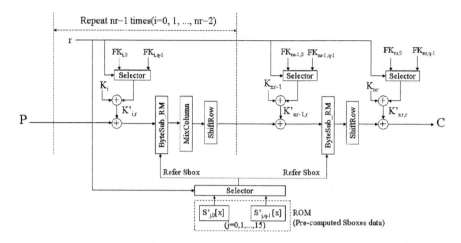

Fig. 2. Construction of fixed-value masking method (Rijndael algorithm)

To begin, one of the fixed mask values is chosen by generating random number $r = 0, 1, ..., q - 1$ on line 2. Then, lines 3-13 are calculated using this random number. If T is the intermediate value in the straight-forward implementation, T' in algorithm 4 satisfies $T = T' \oplus FIN_r$ by the processes on lines 5 and 10, $T = T' \oplus FOUT_r$ by the processes on lines 6 and 11, $T = T' \oplus MixColumn(FOUT_r)$ by the process on line 7, $T = T' \oplus ShiftRow(MixColumn(FOUT_r))$ by the process on line 8, and $T = T' \oplus ShiftRow(FOUT_r)$ by the process on line 12. (Note that $ShiftRow$ and $MixColumn$ are linear functions.) The unmasking process can be omitted because $FK_{nr,r} = ShiftRow(FOUT_r)$. The set of pre-computed Sbox data used in $ByteSub\_FM\_16$ can be commonly used in every i-th round because T' on lines 5 and 10 satisfies $T = T' \oplus FIN_r$, and the value of FIN_r is independent of the round.

Algorithm 5 $ByteSub\_FM\_16(X,r)$

1 $(x_{15}, x_{14}, .., x_0)=X$;
2 for $(j = 0; j < 16; j + +)$ $x_j = S'_{j,r}[x_j]$;
3 $X = (x_{15}, x_{14}, ..., x_0)$;
4 output X;

In algorithm 5, $S'_{j,r}[x]$ is a pre-computed Sbox. Let $S[x]$ be the Sbox used in the straight-forward implementation, $S'_{j,r}[x]$ satisfies (2), where $(fin_{15,r}, fin_{14,r}, ..., fin_{0,r}) = FIN_r$ and $(fout_{15,r}, fout_{14,r}, ..., fout_{0,r}) = FOUT_r$.

$$S'_{j,r}[x] = S[x \oplus fin_{j,r}] \oplus fout_{j,r} (0 \le j \le 15, 0 \le r \le q - 1 \text{ and } 0 \le x \le 255)(2)$$

3.2 Fixed-Value Masking Method (Type II)

The number of Sboxes used in algorithm 5 is $16 \times (number\ of\ mask\ values) = 16q$, but this is still too many to store in ROM. We propose the Type II fixed-value masking method to further reduce the number of Sboxes. In this method, $S'_{j,r}[x]$ is commonly for every $j = 0, 1, ..., 15$, so that the number of Sboxes is reduced to the number of mask values, i.e. q. This is achieved by satisfying (3) and replacing $ByteSub\_FM\_16$ with $ByteSub\_FM\_1$, as shown in algorithm 6.

$$fin_{0,r} = fin_{1,r} = ... = fin_{15,r}, fout_{0,r} = fout_{1,r} = ... = fout_{15,r} \quad (3)$$

Algorithm 6 $ByteSub\_FM\_1$

$\quad ByteSub\_FM\_1(X, r)$
1 $(x_{15}, x_{14}, ..., x_0) = X;$
2 for $(j = 0; j < 16; j + +)$ $x_j = S'_r[x_j];$
3 $X = (x_{15}, x_{14}, ..., x_0);$
4 output $X;$

Remarks:
In the Type-II method, the fixed mask values satisfy $FK_{i,r} = FIN_r \oplus FOUT_r$ $(1 \leq i \leq nr - 1)$. (Note that $FOUT_r = ShiftRow(MixColumn(FOUT_r))$ under condition (3).) Therefore, it can be seen that if $FK_{i,r}$ satisfy our criteria described in section 4, $FK_{i,r} = FIN_r \oplus FOUT_r = const$ $(1 \leq i \leq nr - 1, 0 \leq r \leq q - 1)$ for $q = 2$. (Note that $FIN_1 \oplus FOUT_1 = \overline{FIN_0} \oplus \overline{FOUT_0} = FIN_0 \oplus FOUT_0$.) There may be some attacks against the fixed-value masking method by taking account of this property.

Two ways are considered for avoiding such attacks, that is, setting q greater than 2 and selecting r at every round. In the latter, the key masking values are represented $FK_0 = FIN_{r_0}$, $FK_i = FIN_{r_i} \oplus FOUT_{r_{i-1}} (i = 1, 2, ..., nr - 1)$ and $FK_{nr} = ShiftRow(FOUT_{r_{nr-1}})$, where $r_0, r_1, ..., r_{nr-1}$ are random number satisfy $0 \leq r_0, r_1, ..., r_{nr-1} \leq q - 1$.

3.3   Advantages of Fixed-Value Masking Method

As mentioned above, the fixed-value masking method has two advantages over the random-value masking method.

- Processing overhead is lower because the mask value calculation and Sbox data update processes are omitted.
- The table data for Sbox can be stored in ROM, so less RAM is required.

4   Security Evaluation of the Fixed-Value Masking Method against DPA Attacks

In this section, we discuss our evaluation of the fixed-value masking method against DPA attacks using the new idea of "probabilistic DPA", an extension of the conventional DPA. We extend two conventional DPA attacks. One, proposed by Kocher et al. [10], is a DPA attack against a table look-up operation. The other, proposed by Chari et al. [3], is a binary DPA attack against a key

XORing operation. We call a DPA against a table look-up operation an "Sbox DPA", a DPA against key XORing operation a "key XORing DPA." We call the extension of the two DPA to probabilistic DPA a "probabilistic Sbox DPA" and a "probabilistic key XORing DPA."

For the probabilistic DPA attacks, while steps for making the DPA trace are the same as those for conventional DPA attacks, the analysis of the secret key is different. This is because the size and of the spikes appearing in the DPA trace, differ among those attacks. If the spike is large enough, it is easy to analyze the key. If it is almost zero, analysis is very hard. In the probabilistic DPA attacks, the size of the spikes depends on the prepared set of fixed mask values, so the fixed-value masking method is vulnerable to probabilistic DPA attacks under certain conditions.

We evaluated the security of the fixed-value masking method by representing the size of the spikes with a condition concerning the set of the fixed mask value, and identified the criteria for making it secure against these attacks. We found that an attacker must perform $O(2^{128})$ calculations to analyze a 128-bit secret key by using a probabilistic Sbox and key XORing DPA.

We describe the Sbox and key XORing DPA in section 4.2, the probabilistic Sbox and key XORing DPA in section 4.3, and the security criteria for the fixed-value masking method in section 4.4.

4.1 Notations

In the Rijndael algorithm, the first 128 bits of the secret key are directly used as sub-key K_0. Therefore, analyzing K_0 will reveal information critical to analyze the secret key of the Rijndael algorithm. In particular, if the length of the secret key is 128-bit, analyzing K_0 will directly reveal the secret key itself. Hereafter, we use the following notation to describe the analysis of K_0 using Sbox DPA and key XORing DPA.

$$\begin{cases} K = (k_{15}, .., k_0) = K_0 \\ K'_r = (k'_{15,r}, .., k'_{0,r}) = K'_{0,r} \\ FK_r = (fk_{15,r}, .., fk_{0,r}) = FK_{0,r} \\ FIN_r = (fin_{15,r}, .., fin_{0,r}) = FIN_{0,r} \\ FOUT_r = (fout_{15,r}, .., fout_{0,r}) = FOUT_{0,r} \\ P = (p_{15}, ..p_0), \text{where } P \text{ is input plain text} \end{cases}$$

- N : number of the monitoring.
- $V_m(time)$: power consumption trace at $time$ of m-th monitoring ($m = 0, 1, ., N-1$), where input plaintext P differs for each m.
- $bit(x, d)$: d-th bit value of $x(d = 0, 1, ..., 7)$.
- $f(P, K)$: represents arbitrary intermediate value appearing during monitoring $V_m(time)$. (e.g., $p_j, p_j \oplus k'_{j,0}$ and $S'_0[p_j \oplus k'_{j,0}]$). This function is used to divide $V_m(time)$ into two sets and to calculate the difference between their averages.

- $\delta 1_d(f(P,K))$, $\delta 0_d(f(P,K))$: set of $V_m(time)$ divided by d-th bit value of $f(P,K)$ appearing during monitoring $V_m(time)$, i.e. defined as $\delta 1_d(f(P,K)) = \{V_m(time)|bit(f(P,K),d) = 1\}$, $\delta 0_d(x) = \{V_m(time)|bit(f(P,K),d) = 0\}$.
- Δ_d : DPA trace, which is a differential of average power consumption of $V_m(time) \in \delta 1_d(f(P,K))$ and $V_m(time) \in \delta 0_d(f(P,K))$; defined as

$$\Delta_d = \frac{2}{N} \sum_{V_m \in \delta 1_d(f(P,K))} V_m(time) - \frac{2}{N} \sum_{V_m \in \delta 0_d(f(P,K))} V_m(time).$$

4.2   Sbox DPA and Key XORing DPA

Sbox DPA. In an Sbox DPA attack, the attacker monitors the power consumption when the Sbox output value is loaded into the register.

For some $j = 0, 1, ..., 15, d = 0, 1, .., 7$ the attacker guesses that the key value is k_j, then calculates Δ_d for

$$f(P,K) = S[p_j \oplus k_j]. \tag{4}$$

If k_j is correct, a spike will appear in Δ_d at $time = time_{loadS}$ when the Sbox output value is loaded. Let ϵ_{loadS} be the size of the spike; it can be represented

$$\epsilon_{loadS} = \frac{2}{N}VD_{loadS} \tag{5}$$

where

$$VD_{loadS} = \sum_{V_m \in \delta 1_d(S[p_j \oplus k_j])} V_m(time_{loadS}) - \sum_{V_m \in \delta 0_d(S[p_j \oplus k_j])} V_m(time_{loadS}).$$

By repeating this analysis for $j = 0, 1, ..., 15$, the attacker can analyze key K.

Key XORing DPA. In a key XORing attack, the attacker monitors the power consumption of a load/store instruction while key XORing process $t_j = p_j \oplus k_j$ is running. The attacker analyzes k_j by assuming that the key XORing process is implemented with following code.

$$\begin{aligned} &load\ p_j; \\ &load\ k_j; \\ &xor\ t_j, p_j, k_j; \\ &store\ t_j. \end{aligned} \tag{6}$$

This assumption is reasonable for an 8-bit micro-controller because it has few registers and data is normally stored in RAM immediately after calculation. For some $j = 0, 1, ..., 15$ and $d = 0, 1, .., 7$, the attacker calculates Δ_d in case $f(P,K) = p_j$. Then, two spikes appear in Δ_d at $time = time_{loadp_j}$ when $load\ p_j$

is executed, and at $time = time_{store t_j}$ when store t_j is executed. Let $\epsilon_{load p_j}$ and $\epsilon_{store t_j}$ be the size of these spikes; they are represented as (7), (8) and (9).

$$\epsilon_{load p_j} = \frac{2}{N} \left\{ \sum_{V_m \in \delta 1_d(p_j)} V_m(time_{load p_j}) - \sum_{V_m \in \delta 0_d(p_j)} V_m(time_{load p_j}) \right\} \quad (7)$$

$$\epsilon_{store t_j} = \frac{2}{N} \left\{ \sum_{V_m \in \delta 1_d(p_j)} V_m(time_{store t_j}) - \sum_{V_m \in \delta 0_d(p_j)} V_m(time_{store t_j}) \right\}$$

$$= \frac{2}{N} \left\{ \sum_{V_m \in \delta 1_d(p_j \oplus k_j)} V_m(time_{store t_j}) - \sum_{V_m \in \delta 0_d(p_j \oplus k_j)} V_m(time_{store t_j}) \right\} \quad (8)$$

$$(\text{if } bit(k_j, d) = 0)$$

$$= \frac{2}{N} \left\{ \sum_{V_m \in \delta 0_d(p_j \oplus k_j)} V_m(time_{store t_j}) - \sum_{V_m \in \delta 1_d(p_j \oplus k_j)} V_m(time_{store t_j}) \right\} \quad (9)$$

$$(\text{if } bit(k_j, d) = 1)$$

Equations (7), (8) and (9) show that the attacker can analyze the d-th bit of k_i by using (10) if the correlation of the power consumption to the data operand reveals that the characteristics of the *load* and *store* operations are similar.

$$\begin{array}{l} \text{Sign of } \epsilon_{load p_j} \text{ and } \epsilon_{store t_j} \text{ are the same} \Rightarrow (d\text{-th bit of } k_j) = 0 \\ \text{Sign of } \epsilon_{load p_j} \text{ and } \epsilon_{store t_j} \text{ are different} \Rightarrow (d\text{-th bit of } k_j) = 1 \end{array} \quad (10)$$

In our experiment of key XORing DPA using the code shown in (6), this attack did not work on our target processor. We think this is because the power consumption differed between the load and store operations. We thus tried the same experiment, but using the code shown in (11).

$$\begin{array}{l} load\ p_j; \\ load\ k_j; \\ xor\ t_j, p_j, k_j; \\ store\ t_j; \\ \vdots \\ load\ t_j. \end{array} \quad (11)$$

In this experiment, three sets of spikes appeared for *load* p_j, *store* t_j, and *load* t_j when Δ_d was calculated. We were able to analyze the key by comparing the sign of the spikes on the left for *load* p_j and *load* t_j. The Δ_d for $d = 0, 1, .., 7$ and k_j =0xe9 are shown in figure 3. (Note that the spike on the left in each set of spikes was used to analyze the key.)

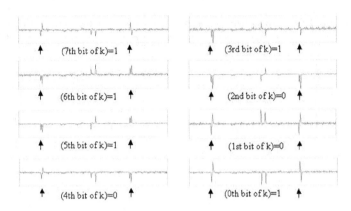

Fig. 3. Spikes that appeared in key XORing DPA attack using code *load* p_j; *load* k_j; *xor* t_j, p_j, k_j; *store* t_j; ...*load* t_j; (for k_j=0xe9).

4.3   Probabilistic DPA

We extended the Sbox and key XORing DPA attacks described in section 4.2 to a probabilistic DPA attack, and evaluated the security of the fixed-value masking method. In probabilistic DPA, Δ_d is calculated in the same way as described in section 4.2, but the size of the spikes depends on the set of fixed mask values. If spikes are large enough, the attacker can get information about the key; if they are almost 0, he cannot. For example, if all q mask values are the same, such as $FK_0 = FK_1 = ... = FK_{q-1}$ in the Type II method, the attacker can analyze $K'_r = K \oplus FK_r$ by using key XORing DPA. Finally, K can easily be revealed by a brute-force attack for FK_r, which has only 2^8 (as shown by (1) and (3)) possible values. We determined the size of the spikes by using probability function $\beta_d(x_r)$,

$$\beta_d(x_r) = prob[bit(x_r, d) = 0] \text{ for } r = 0, 1, ..., q - 1$$

and identified the conditions under which the fixed-value masking method is secure.

Probabilistic Sbox DPA. In the fixed-value mask method, there are q possible Sbox output values, $S'_{j,r}[p_j \oplus k'_{j,r}]$, for any $r = 0, 1, ..., q - 1$. And $S'_{j,r}[p_j \oplus k'_{j,r}]$ satisfies

$$S'[p_j \oplus k'_{j,r}] = S[p_j \oplus k'_{j,r} \oplus fin_{j,r}] \oplus fout_{j,r} = S[p_j \oplus k_j] \oplus fout_{j,r}. \quad (12)$$

The only difference between (4) and (12) is the constant value $fout_{j,r}$. ($S'_{j,r}[]$ and $fin_{j,r}$ disappear.) Noticing this fact, the attacker guesses that the key is k_j and calculates Δ_d in case $f(P, K) = S[p_j \oplus k_j]$ as described in **Sbox DPA** of section 4.2.

$$\epsilon_{loadS'_{j,r}} = \frac{2}{N} \left\{ \sum_{V_m \in \delta 1_d(S[p_j \oplus k_j])} V_m(time_{loadS'_{j,r}}) \right.$$

$$\left. - \sum_{V_m \in \delta 0_d(S[p_j \oplus k_j])} V_m(time_{loadS'_{j,r}}) \right\}$$

By taking into account that $bit(S[p_j \oplus k_j], d) = bit(S[p_j \oplus k_j] \oplus fout_{j,r}, d)$ holds with probability $\beta_d(fout_{j,r})$, we can present the size of spike $\epsilon_{loadS'_{j,r}}$ for load $S'_{i,r}$ as follows if the attacker guesses the correct k_j.

$$\epsilon_{loadS'_{j,r}} = \frac{2}{N} \left\{ \beta_d(fout_{j,r}) \sum_{V_m \in \delta 1_d(S[p_j \oplus k_j] \oplus fout_{j,r})} V_m(time_{loadS'_{j,r}}) \right.$$

$$+(1 - \beta_d(fout_{j,r})) \sum_{V_m \in \delta 0_d(S[p_j \oplus k_j] \oplus fout_{j,r})} V_m(time_{loadS'_{j,r}})$$

$$-\beta_d(fout_{j,r}) \sum_{V_m \in \delta 0_d(S[p_j \oplus k_j] \oplus fout_{j,r})} V_m(time_{loadS'_{j,r}})$$

$$\left. -(1 - \beta_d(fout_{j,r})) \sum_{V_m \in \delta 1_d(S[p_j \oplus k_j] \oplus fout_{j,r})} V_m(time_{loadS'_{j,r}}) \right\}$$

$$= \frac{2(2\beta_d(fout_{j,r}) - 1)}{N} V D_{loadS'_{j,r}} \tag{13}$$

where

$$V D_{loadS'_{j,r}} = \sum_{V_m \in \delta 1_d(S[p_j \oplus k_j] \oplus fout_{j,r})} V_m(time_{loadS'_{j,r}})$$

$$- \sum_{V_m \in \delta 0_d(S[p_j \oplus k_j] \oplus fout_{j,r})} V_m(time_{loadS'_{j,r}}).$$

Equation (13) indicates that if $\beta_d(fout_{j,r}) = 0$ or 1, $\epsilon_{loadS'_{j,r}}$ represents a DPA spike almost the same size as that shown in (5); however, since $\beta_d(fout_{j,r})$ is close to 1/2, $\epsilon_{loadS'_{j,r}}$ is close to 0. (Note that $V D_{loadS}$ and $V D_{loadS'_{j,r}}$ are to be the same.) Therefore, following (14) is considered to be most secure.

$$\beta_d(fout_{j,r}) = 1/2 \text{ for any } j = 0,..,15 \text{ and } d = 0,..,7 \tag{14}$$

On above proof, the attacker divides $V_m()$ into two sets by an 1-bit value of the Sbox to make DPA trace Δ_d. (single-bit Sbox DPA attack). Besides using a single-bit Sbox DPA attack, the attacker can also make DPA trace by dividing $V_m()$ into two sets by an plural bit of the Sbox (multiple-bit Sbox DPA attack). The fixed masking method is also secure against a multiple-bit Sbox DPA attack if (14) is satisfied, because the hamming weight of the Sbox output value is balanced.

Probabilistic Key XORing DPA. In this section, we discuss the case where an attacker launches a probabilistic key XORing DPA attack by using the code shown in (11).

In the fixed-value masking method, key XORing calculates $t_j = p_j \oplus k'_{j,r}$ for a random $r = 0, 1, ..., q - 1$.

The attacker calculates Δ_d in case $f(P, K) = p_j$, in the same way as described in **Key XORing DPA** of section 4.2. Three sets of spikes will appear for $load\ p_j$, $store\ t_j$ and $load\ t_j$. The size of the spike for $load\ t_j$ is represented

$$\epsilon_{loadt_j} = \frac{2}{N} \left\{ \sum_{V_m \in \delta 1_d(p_j)} V_m(time_{loadt_j}) - \sum_{V_m \in \delta 0_d(p_j)} V_m(time_{loadt_j}) \right\}.$$

By taking into account that $bit(p_j, d) = bit(p_j \oplus k'_{j,r}, d)$ holds with probability $\beta_d(k'_{j,r})$, ϵ_{loadt_j} is represented

$$
\begin{aligned}
\epsilon_{loadt_j} = \frac{2}{N} \Bigg\{ &\beta_d(k'_{j,r}) \sum_{V_m \in \delta 1_d(p_j \oplus k'_{j,r})} V_m(time_{loadt_j}) \\
&+ (1 - \beta_d(k'_{j,r})) \sum_{V_m \in \delta 0_d(p_j \oplus k'_{j,r})} V_m(time_{loadt_j}) \\
&- \beta_d(k'_{j,r}) \sum_{V_m \in \delta 0_d(p_j \oplus k'_{j,r})} V_m(time_{loadt_j}) \\
&- (1 - \beta_d(k'_{j,r})) \sum_{V_m \in \delta 1_d(p_j \oplus k'_{j,r})} V_m(time_{loadt_j}) \Bigg\} \\
= &\frac{2(2\beta_d(k'_{j,r}) - 1)}{N} VD_{loadt_j}.
\end{aligned}
\tag{15}
$$

where

$$VD_{loadt_j} = \sum_{V_m \in \delta 1_d(p_j \oplus k'_{j,r})} V_m(time_{loadt_j}) - \sum_{V_m \in \delta 0_d(p_j \oplus k'_{j,r})} V_m(time_{loadt_j}).$$

By comparing the signs of (7) and (15), we see that

- If the signs of ϵ_{loadp_j} and ϵ_{loadt_j} are the same, the attacker guesses that the d-th bit of $k'_{j,r} = 0$ for any $r = 0, 1, ..q-1$. (This happens when $\beta_d(k'_{j,r}) = 1$.)
- If the signs of ϵ_{loadp_j} and ϵ_{loadt_j} are different, the attacker guesses that the d-th bit of $k'_{j,r} = 1$ for any $r = 0, 1, ..q-1$. (This happens when $\beta_d(k'_{j,r}) = 0$.)

If $\beta_d(k'_{j,r}) = 0$ or 1, the attacker can guess the d-th bit of $k'_{j,r}$ for any $r = 0, 1, .., q-1$ by noting the sign of the spikes without knowing the value of $\beta_d(k'_{j,r})$. If $0 < \beta_d(k'_{j,r}) < 1/2$ or $1/2 < \beta_d(k'_{j,r}) < 1$, the attacker may be able to get the probabilistic value of the d-th bit of $k'_{j,r}$, but cannot analyze the information about k_j. If $\beta_d(k'_{j,r}) = 1/2$, the attacker cannot guess the d-th bit of $k'_{j,r}$ because the size of the spike is 0.

4.4   Security Criteria for Fixed-Value Masking Method against Probabilistic DPA Attacks

Based on the discussion in section 4.3, we present two lemmas as the criteria for securing the fixed-value masking method against probabilistic DPA attacks. The criteria cover both the Type I and Type II methods.

We exclude the case $q = 1$, because it is clear that $\beta_d(fout_{j,r}) = 0$ or 1 for any $j = 0, 1, .., 15$ and $d = 0, 1, .., 7$ which indicates that K is completely revealed by using Sbox DPA.

Lemma 1. In Type I and Type II methods when $q \geq 2$, the attacker can not get any information about the key by using a probabilistic Sbox DPA attack if $\beta_d(fout_{j,r}) = 1/2$ for any $j = 0, 1, ..., 15$ and any $d = 0, 1, .., 7$.

Lemma 2. In Type I method when $q \geq 2$, the attacker must perform $O(2^{128})$ calculations to analyze K even when using a probabilistic key XORing DPA attack. In Type II method when $q \geq 2$, the attacker must perform $O(2^{128})$ calculations to analyze K even when using a probabilistic key XORing DPA attack if $\beta_d(fk_{j,r}) = 1/2$ for any $j = 0, 1, ..., 15$ and any $d = 0, 1, .., 7$.

(Proof of lemma 1.) From (13), it is clear that an attacker cannot get any information if $\beta_d(fout_{j,r}) = 1/2$, because the size of the spike $\epsilon_{loadS'_{j,r}}$ is 0.

(Proof of lemma 2.) From (15), if $\beta_d(k'_{j,r}) = 1/2$ for some j and d, an attacker cannot analyze the d-th bit of $k'_{j,r}$ because the size of the spikes is 0, and if $\beta_d(k'_{j,r}) = 0$ or 1 for some j and d, the attacker can analyze the d-th bit of $k'_{j,r}$. If $0 < \beta_d(k'_{j,r}) < 1/2$ or $1/2 < \beta_d(k'_{j,r}) < 1$ for some j and d, we assume that the attacker can analyze the d-th bit of $k'_{j,r}$. (This assumption is advantageous for the attacker.) Under this assumption, the attacker can analyze h bits out of a 128-bit $K'_r = k'_{15,r}k'_{14,r}..., k'_{0,r}$ by using a probabilistic key XORing DPA attack, where

$$h = \sum_{j=0}^{15} \sum_{d=0}^{7} h_{j,d}, \ h_{j,d} = \begin{cases} 0 \ (\beta_d(k'_{j,r}) = 1/2) \\ 1 \ (otherwise) \end{cases}$$

The attacker can then analyze K by using a full-search for the possible values of K'_r except for leaked bits, and for the possible values of FK_r for leaked bits by using the relation $K = K'_r \oplus FK_r$ (see figure 4). The calculation for this search is

(possible values of K'_r except for leaked bits)

$\times$ (possible values of FK_r for leaked bits)

$$= \begin{cases} O(2^{128-h}) \times O(2^h) = O(2^{128}) \ \text{(in Type I)} \\ O(2^{128-h}) \times O(2^{h/16}) = O(2^{128-15h/16}) \ \text{(in Type II)} \end{cases} \qquad (16)$$

In the Type I method, the attacker must perform $O(2^{128})$ calculations to analyze K. In the Type II method, the attacker can analyze K with only $O(2^8)$ calculations in the weakest case ($h = 128$), but must perform $O(2^{128})$ calculations in

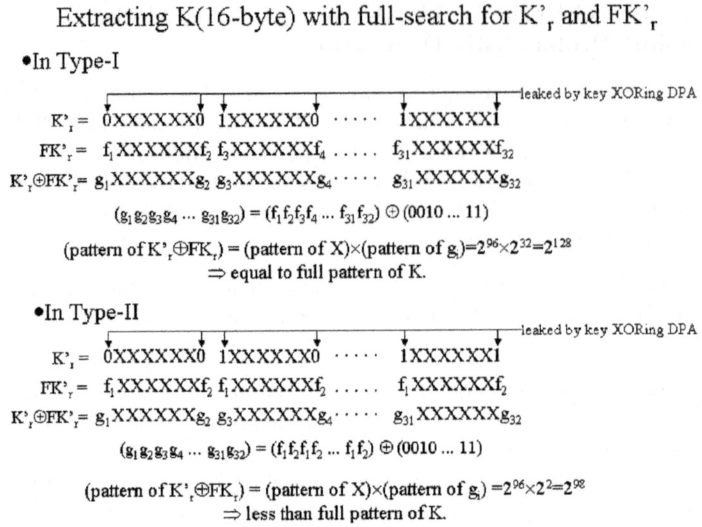

Extracting K(16-byte) with full-search for K'$_r$ and FK'$_r$

•In Type-I

$$(g_1 g_2 g_3 g_4 \cdots g_{31} g_{32}) = (f_1 f_2 f_3 f_4 \dots f_{31} f_{32}) \oplus (0010 \dots 11)$$

(pattern of K'$_r \oplus$ FK'$_r$) = (pattern of X)×(pattern of g$_i$)=$2^{96} \times 2^{32} = 2^{128}$
$\Rightarrow$ equal to full pattern of K.

•In Type-II

$$(g_1 g_2 g_3 g_4 \cdots g_{31} g_{32}) = (f_1 f_2 f_1 f_2 \dots f_1 f_2) \oplus (0010 \dots 11)$$

(pattern of K'$_r \oplus$ FK'$_r$) = (pattern of X)×(pattern of g$_i$) =$2^{96} \times 2^2 = 2^{98}$
$\Rightarrow$ less than full pattern of K.

Fig. 4. Extracting K using a full-pattern search for K'_r and FK'_r

the strongest case $(h = 0)$, which is satisfied by setting $\beta_d(k'_{j,r}) = 1/2$ for any $j = 0, 1, ..., 15$ and $d = 0, 1, ..., 7$. Noting that $k'_{j,r} = k_j \oplus fk_{j,r}$, this condition is equivalent to

$$\beta_d(fk_{j,r}) = 1/2 \text{ (for any } j = 0, 1, .., 15 \text{ and } d = 0, 1, .., 7)$$

If this condition is satisfied, the attacker must perform $O(2^{128})$ calculations to analyze K in Type II method. (end of proof.)

5   Comparison of Fixed- and Random-Value Masking Methods

The fixed- and random-value masking methods are compared for the Rijndael algorithm in table 1.

As shown in table 1, S is the processing time using the straight-forward implementation, M is the size of the Sbox data (M=256 bytes in the Rijndael algorithm), and R is the amount of RAM needed to store the other variables (plaintext data, partial sub-keys, etc.). We assume $R < M$. If q is set to 2 in our method, R bytes of RAM and 512 bytes of ROM are required. If a small memory size is the most important factor in the implementation, it can generally be said that $R < 128$. (An implementation described in [14] showed $R \leq 52$ on Motorola 68HC08.) Therefore, our method can be implemented on low-cost smartcards such as the ST16, which has 128 Bytes of RAM and 6 K Bytes of ROM.

Table 1. Comparison of the running time of a modular multiplication and a modular addition @ 200 MHz.

	NORMAL	RMASK	FMASK
Processing time	S	$\cong 2S^*$	$\cong S$
Size of RAM	R	$\geq R + M$	$\cong R$
Size of ROM	M	M	$\cong 16qM(\text{Type I}), \cong qM(\text{Type II})$

NORMAL...Straight-forward implementation.
RMASK...Random-value masking method.
FMASK...Fixed-value masking method(proposed).
*... Messerges' implementation([12]).

6   Concluding Remarks

We have described a new DPA countermeasure that is an improvement on the random-value masking method in terms of processing speed and required RAM. Our method is thus suitable for implementation on low-cost smartcards. It can protect Rijndael algorithm from DPA attacks and can be extended to other algorithms, such as DES. It is secure against the Sbox DPA attacks and key XORing DPA attacks. We haven't evaluated the security of our method against more powerful DPA attacks, such as DiPA(direct power analysis) by Akkar et al. [1] and high-order DPA attacks by Messerges [13] which are recently introduced. Security evaluation for these DPA attacks is the problem in the future work.

References

1. Mehdi-Larurent Akkar, Regis Bevan, Paul Dischamp and Didier Moyart, "Power Analysis, What is Now Possible...", *Advances in Cryptology − ASIACRYPT 2000*, pp.489-502.
2. Mehdi-Larurent Akkar, Chiristophe Giraud, "An implementation of DES and AES, secure against some attacks", *Cryptographic Hardware and Embedded Systems (CHES 2001)*.
3. Suresh Chari, Charanjit S.Jutla, Josyula R.Rao and Pankaj Rohtagi, "A cautionary note regarding evaluation of AES candidates on smart-cards", *AES round 2*.
4. Suresh Chari, Charanjit S.Julta, Josyula R.Rao and Pankaj Rohtagi, "Towards Sound Apporoaches to Counteract Power-Analysis Attacks", *Advances in Cryptography − CRYPTO'99*, pp.398-412.
5. Christophe Clavier and Marc Joye, "Universal Exponentiation Algorithm − A FirstStep Towards Provable SPA-resistence −", *Cryptographic Hardware and Embedded Systems (CHES 2001)*.
6. Jean-Sébastein Coron, "Resistance against Differential Power Analysis for Elliptic Curve Cryptosystems", *Cryptographic Hardware and Embedded Systems (CHES'99)*, pp.292-302.
7. Jean-Sébastein Coron and Louis Goubin, "On Boolean and Arithmetic Masking against Differential Power Analysis", Cryptographic Hardware and Embedded Systems, *Cryptographic Hardware and Embedded Systems (CHES 2000)*, pp.231-237.

8. Louis Goubin and Jacques Patarin, "DES and Differential Power Analysis. — The Duplication Method —", *Cryptographic Hardware and Embedded Systems (CHES 2001)*.
9. Marc Joye and Christphe Tymen, "Protection against Differential Analysis for Elliptic Curve Cryptography — An Algebraic Apporoach —", *Cryptographic Hardware and Embedded Systems (CHES 2001)*.
10. Paul Kocher, Joshua Jaffe, and Benjamin Jun "Differential Power Analysis", *Advances in Cryptography — CRYPTO'99*, pp.388-397.
11. Thomas S.Messerges, Ezzy A.Dabbish and Robert H.Sloan "Power Analysis Attacks of Modular Exponentiation in Smartcards.", *Cryptographic Hardware and Embedded Systems (CHES'99)*, pp.144-157.
12. Thomas S.Messerges, "Securing the AES Finialists Against Power Analysis Attacks", *Fast Software Encryption (FSE 2000)*, pp.150-164.
13. Thomas S.Messerges, "Using Second-Order Power Analysis to Attack DPA Resistant Software", *Cryptographic Hardware and Embedded Systems (CHES 2000)*, pp.238-251.
14. Rijndael Specification,
    http://csrc.nist.gov/encryption/aes/rijndael/Rijndael.pdf

Author Index

Lecture Notes in Computer Science

For information about Vols. 1–2199
please contact your bookseller or Springer-Verlag